6th International Conference on Accelerated Carbonation for Environmental and Material Engineering (ACEME 2018)

Newcastle, Australia
11 - 14 March 2018

ISBN: 978-1-5108-7281-3

Printed from e-media with permission by:

Curran Associates, Inc.
57 Morehouse Lane
Red Hook, NY 12571

Some format issues inherent in the e-media version may also appear in this print version.

Printed by Curran Associates, Inc. (2018)

For permission requests, please contact AIChE
at the address below.

AIChE
120 Wall Street, FL 23
New York, NY 10005-4020

Phone: (800) 242-4363
Fax: (203) 775-5177

www.aiche.org

Additional copies of this publication are available from:

Curran Associates, Inc.
57 Morehouse Lane
Red Hook, NY 12571 USA
Phone: 845-758-0400
Fax: 845-758-2633
Email: curran@proceedings.com
Web: www.proceedings.com

TABLE OF CONTENTS

LABORATOIRE DE GÉNIE CHIMIQUE
TOULOUSE • UMR 5503

Design of a hybrid leaching process for mineral carbonation of magnesium silicates: learnings and issues raised from combined experimental and geochemical modelling approaches

Carine Julcour[1], Florent Bourgeois[1], Laurent Cassayre[1], Julien Leclaire[2], Solène Touzé[3], Martin Cyr[4], France Bailly[5]

[1]LGC, Toulouse, France ; [2]ICBMS, Villeurbanne, France ; [3]BRGM, Orléans, France ; [4]LMDC, Toulouse, France ; [5]CNRT, Nouméa, France

Université de Toulouse

ACEME2018
6th Accelerated Carbonation for Environmental and Material Engineering Conference, March 11-14, 2018, Newcastle, New South Wales, Australia

I. INTRODUCTION

LGC

CARMEX
French National Research Agency (ANR CO_2 2009-2012)

- Worldwide potential of MC (GIS)
- Examination of MC mechanisms
- Proof of concept of selected MC route
- Environmental assessment (LCA)

New-Caledonia as prime candidate for MC deployment

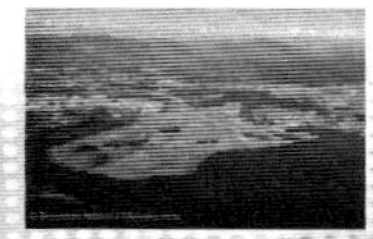

CARBOSCORIES I
National Centre for Technology Research CNRT "Nickel and its environment" (2015-2016)

- Potential of Ni slags for MC hybrid process
- Balance analysis for two metallurgy plants

CO₂EMR
Collaborative work
CO_2 Enhanced Metal Recovery

CARBOSCORIES II
Caledonian Energy Agency (2018-2021)

- Bench-scale continuous pilot
- Valorisation route for MC products

I. INTRODUCTION

LGC

Ex-situ Mineral Carbonation pathways

- Direct Carbonation (single step)
  - Gas-solid
    - *High T, slow kinetics for natural ore*
  - Aqueous
    - No additives: < 50% yield
    - With additives: Up to 85% yield (1h)
    - *Not so simple an approach*
    - *Valorisation routes required for solid mixture (M traces?)*
    - *Net energy penalties: 50-220%**
- Indirect Carbonation (2/multi-steps)
  - Gas-solid
  - Aqueous with additives: 60-80% yield
    - *Flexible approach (pH-swing)*
    - *→ Separate formation of silica & carbonates*
    - *But regeneration of acids/bases required*
    - *→ Higher energy penalties: 50-360%**
- Other approaches (brines, biomineralization ...)
  - *Limited resources*
  - *OR Low process maturity*

Adapted from *Sanna et al. (2014)*

***based on < 2010 literature routes for 154 MWe coal-fired power plant producing 1Mt-CO_2/yr**

3

ACEME 2018, March 11-14, 2018, Newcastle, Australia

I. INTRODUCTION

LGC

Studied MC pathways

- Direct Carbonation (single step)
  - Aqueous
    - No additives: < 50% yield
    - With additives: Up to 85% yield (1h)

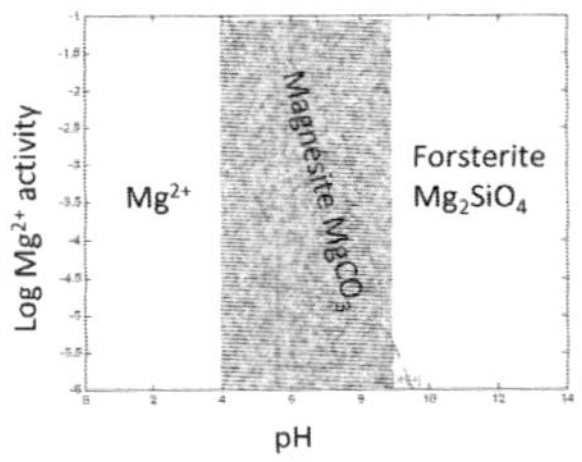

4

ACEME 2018, March 11-14, 2018, Newcastle, Australia

I. INTRODUCTION

Surface leach layer & mechanisms

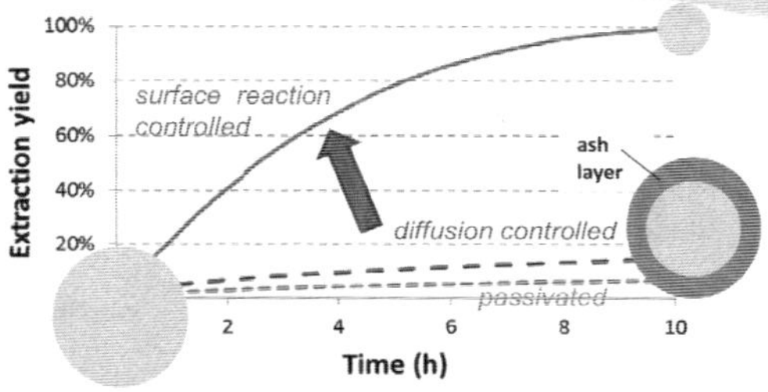

Passivating

→ No carbonate

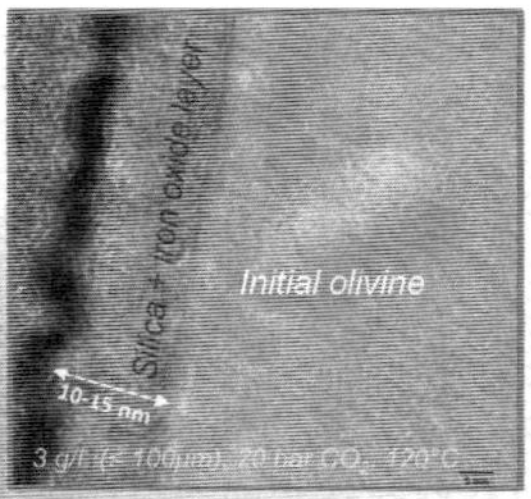

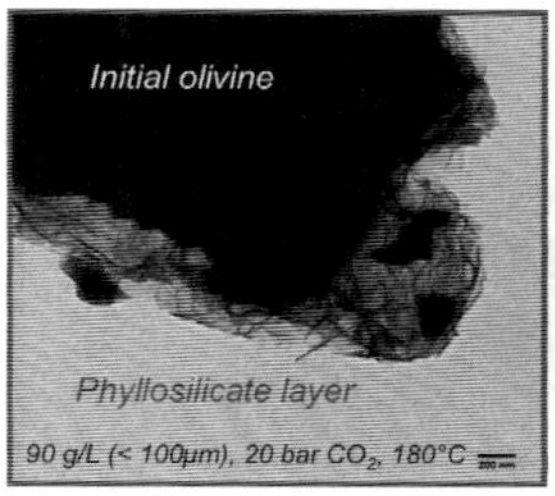

Diffusion controlling

Carbonation yield < 10% (90 g/L, 95 h)

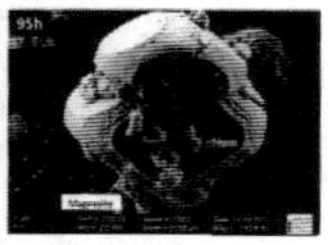

***FIB cross-sections of olivine particle after leaching (TEM)* (Bodenan et al., 2014)**

5

ACEME 2018, March 11-14, 2018, Newcastle, Australia

Studied MC pathways

Direct Carbonation (single step)

Aqueous

No additives	With additives
< 50% yield	Up to 85% yield (1h)

EXPERIMENTAL METHODOLOGY

6

ACEME 2018, March 11-14, 2018, Newcastle, Australia

II. EXPERIMENTAL METHODOLOGY

Ores

Bodenan et al. (2014)

Serpentinisation degree up to 90%

Name	Origin	Dominating phases
Harzburgite (Hz1)	New Caledonia	Serpentine (~ 90%) >> olivine > orthopyroxene
Harzburgite (Hz2)	New Caledonia	Serpentine (~ 90%) >> olivine > orthopyroxene
Wherlite (We)	New Caledonia	Serpentine (~ 50%) > olivine > clinopyroxene
Lherzolite (Lz)	France (Pyrenees)	Serpentine (~ 50%) > olivine > clino/ortho-pyroxene
Olivine	Austria (Magnolithe)	Synthetic olivine (from high T dunite processing)

30.8 < MgO < 47.4%, 39.3 < SiO_2 < 46.3%, 7.3 < Fe_2O_{3tot} < 9.8%, 0.2 < CaO < 5.0%, 0.2 < Al_2O_3 < 4.0%

II. EXPERIMENTAL METHODOLOGY

Slags

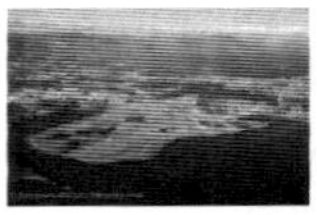

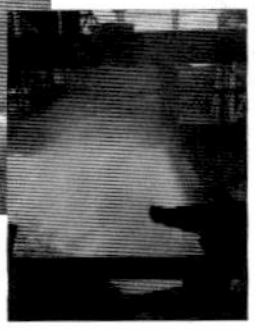

From Ni pyrometallurgy plant (Koniambo)
KNS slowly cooled under ambient conditions, SLN quenched by seawater

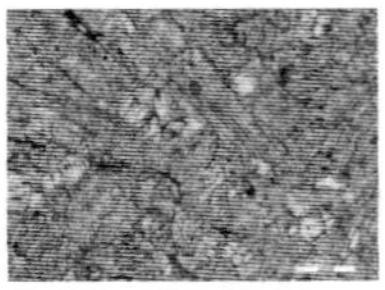

Name	Origin	Dominating phases
KNS	New Caledonia	Pyroxene: proto-enstatite, clinoenstatite, forsterite
SLN	New Caledonia	Vitreous fraction >> olivine

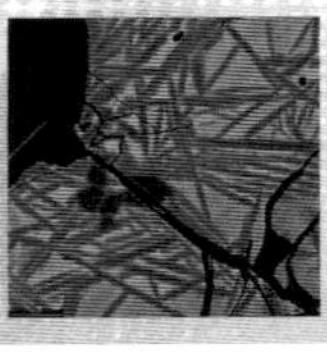

MgO ~ 30%, SiO_2 ~ 52%, Fe_2O_{3tot} ~ 13%, CaO < 0.5%, Al_2O_3 ~ 2%

LGC

Solid preparation

➢ **Grinding** (4h in ball mill) **& sieving** < 100 µm

➢ **Mechanical sampling**

➢ **Initial particle size distribution**

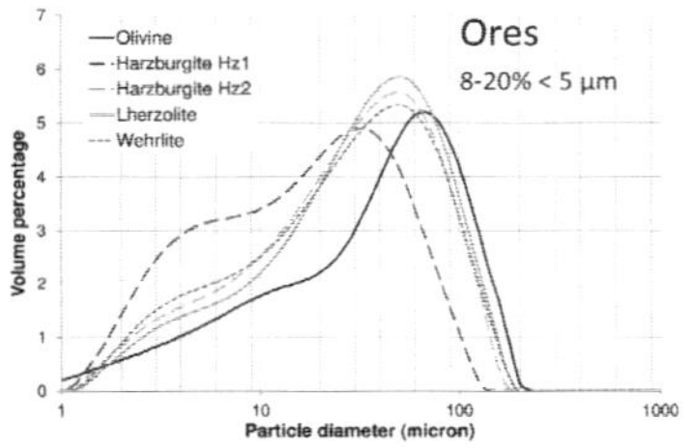

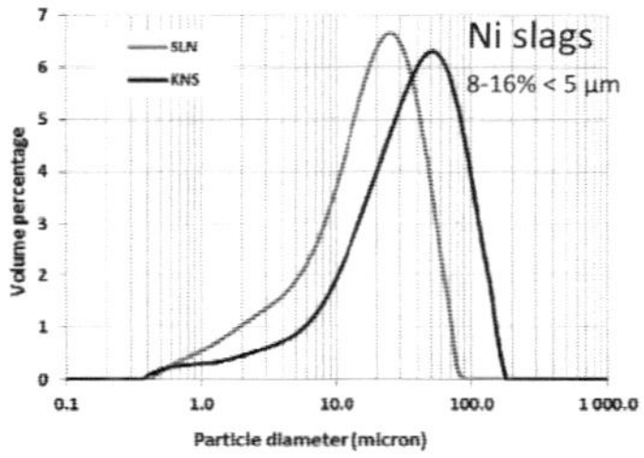

9

II. EXPERIMENTAL METHODOLOGY

LGC

Carbonation set-up & procedure

- 300 mL autoclave reactor
- Controlled atmosphere (P_{CO2}) and T
- Gas auto-dispersing stirrer (800 rpm)
- Continuous recording of CO_2 consumption (24 h)

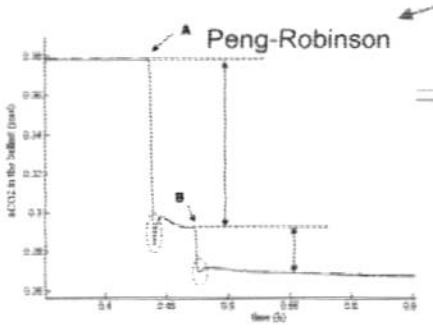

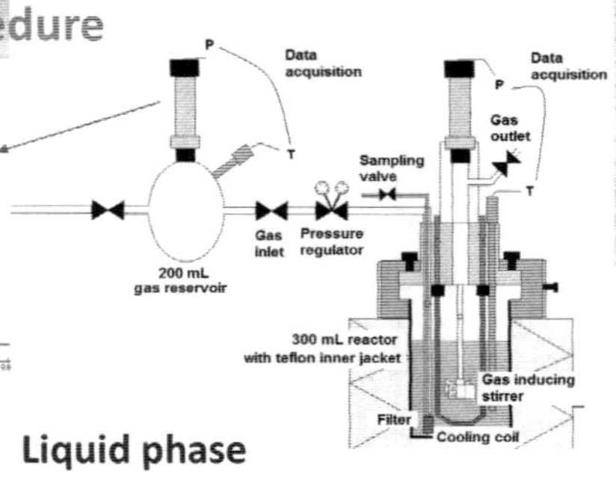

Solid phase

Laser diffraction analysis
Filtration (0.2 µm)

drying (70°C)

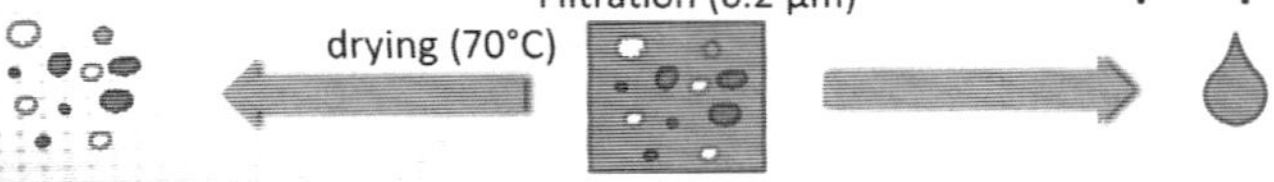

Liquid phase

- Identification of solids formed: SEM(TEM)/EDX and elemental mapping (Mg, Si, O), X-Ray Diffraction, Raman and IR (DRIFT) Spectroscopy, ICP/AES (after acid dissolution)
- Quantification of carbonated products: elemental analysis (C content), ThermoGravimetry Analysis coupled with IR

- Quantification of dissolved elements: ICP/AES (Mg, Si, Fe), inorganic carbon

10

Studied MC pathway

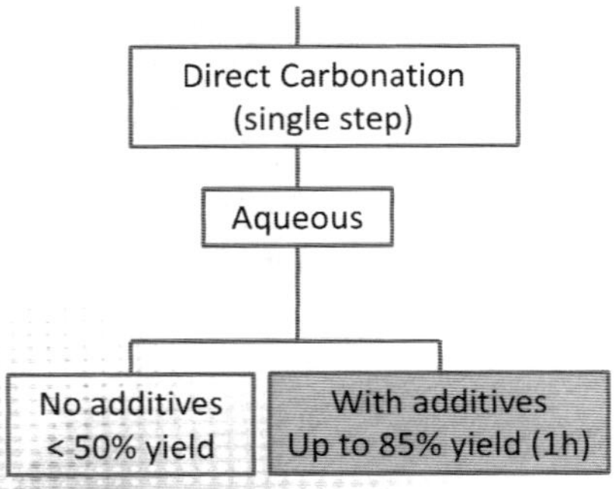

CHELATING AGENT APPROACH

State of the art

➢ **Effect of organic polyacids on dissolution rate**

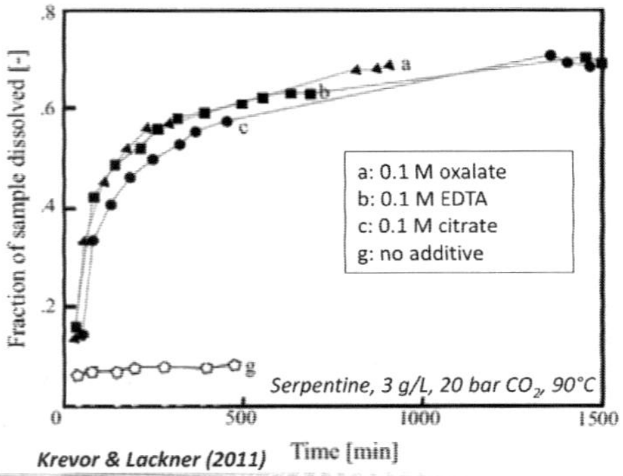

Krevor & Lackner (2011)

➢ **Mechanism**

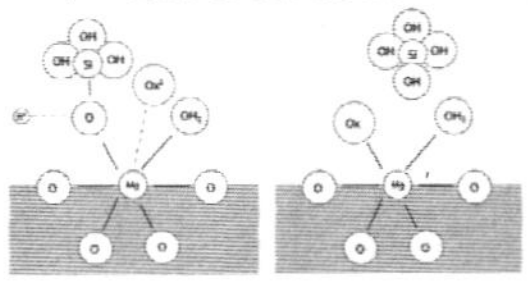

Olsen & Rimstidt (2008)

➢ **Oxalate-enhanced dissolution rate**

$$r = \underbrace{\left(\frac{1+\beta K_x\, a_{ox}}{1+K_x\, a_{ox}} \right)}_{\approx 7 \text{ at } 0.1\text{ M oxalate (olivine, pH 5, 120°C)}} r_{H^+}$$

≈ 7 at 0.1 M oxalate (olivine, pH 5, 120°C)

Prigiobbe & Mazzotti (2011)

III. CHELATING AGENT APPROACH

Oxalate-enhanced dissolution of olivine: leaching yield & Mg speciation

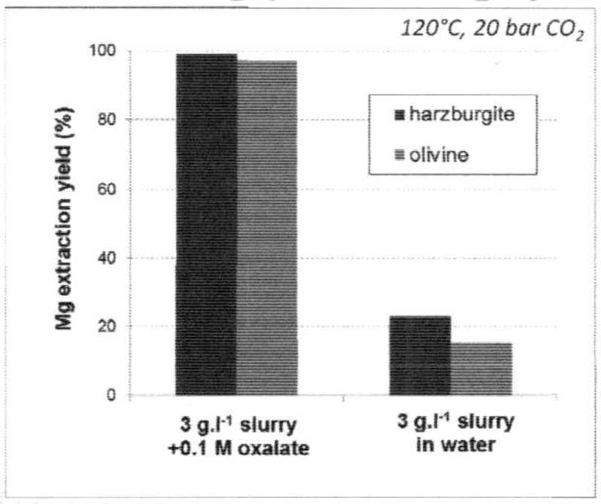

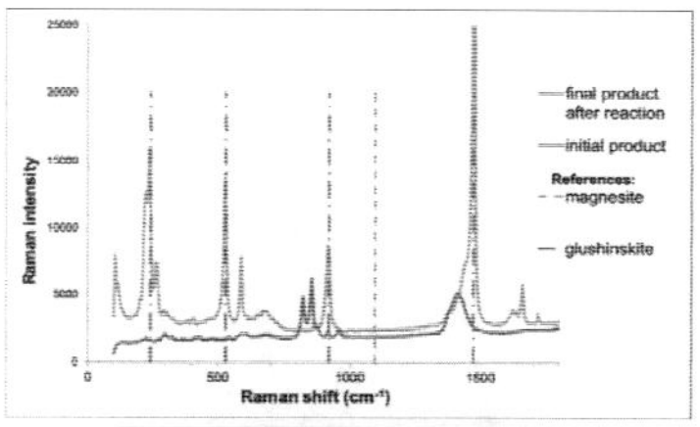

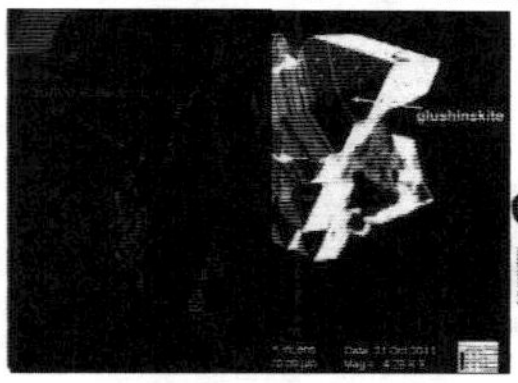

- ➢ Significantly improved extraction yield with oxalate at 3 g/L, but formation of solid by-product: **glushinskite** ($MgC_2O_4.2H_2O$)

- ➢ Any compromise between glushinskite precipitation & Mg carbonation?

13

III. CHELATING AGENT APPROACH

Oxalate-enhanced dissolution of olivine: effect of slurry concentration

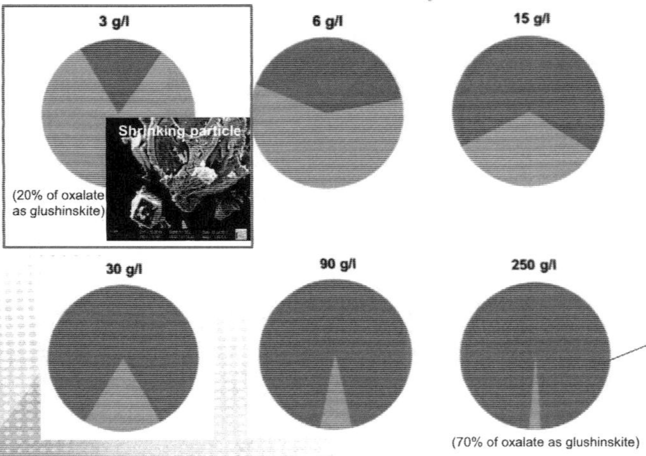

Mg speciation after 24h

120°C, 20 bar CO_2, olivine < 100 µm, speciation after cooling

- Carbonates
- Glushinskite
- Olivine
- Solution

- ➢ Carbonation yield < 1% → not a viable option in these conditions
- ➢ Could geochemical modeling predict such competing effects?

14

LGC

Oxalate-enhanced dissolution of olivine: kinetic modeling

Rate-limiting step: ore dissolution (instant. gas absorption & solid precipitation)

~20 size classes (SSA_{0i}, m_{0i}) generated from Rosin Rammler PSD model

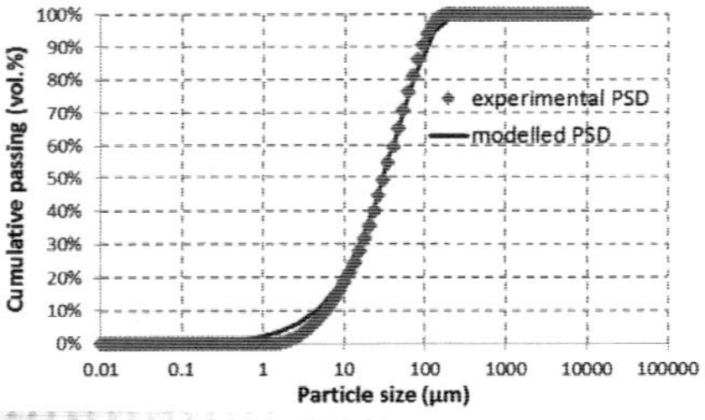

(α & d_{90} parameters fitted to match experimental PSD)

Shrinking particle model

$$\frac{dn_i}{dt} = -r.\,SSA_{0i}.\,m_{0i}\left(\frac{n_i(t)}{n_{i0}}\right)^{2/3}$$

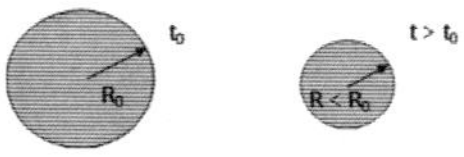

$$r = \left(\frac{1+\beta K_x a_{ox}}{1+K_x a_{ox}}\right) r_{H+} \qquad r_{H+} = k_0.\exp(-E_a/RT).a_{(H+)}^{\,n}\,(1-10^{SI})$$

surface reaction rate from Prigiobbe & Mazzotti (2011)

LGC

Oxalate-enhanced dissolution of olivine: geochemical modeling (with CHESS)

Issues:
- Mg-oxalate species missing in used thermodynamic database (CTDP)
- Glushinskite equilibrium data only available at 25°C

→ dedicated precipitation experiments at 120°C & **database updating**

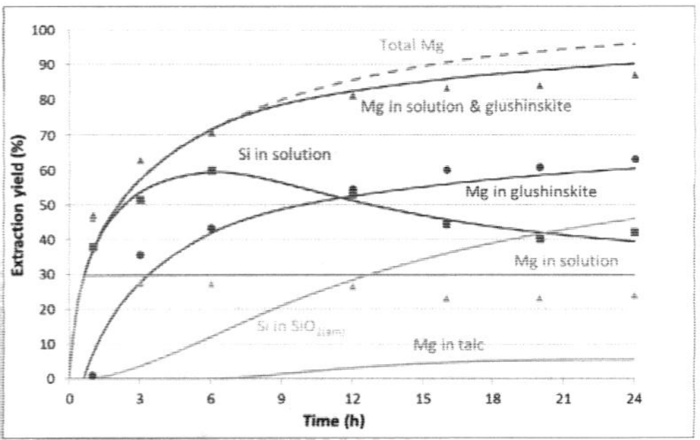

Bonfils et al. (2012)

120°C, 20 bar CO_2, 3 g/L olivine < 100 µm

Simulation with CHESS code *(van der Lee, 2007)* after estimation of precipitation rate constants for silica & talc

➤ Good agreement with experimental data at 3 g/L
→ chemical controlled reaction

LGC

Oxalate-enhanced dissolution of olivine: geochemical modeling (with CHESS)

Issues: • Mg-oxalate species missing in used thermodynamic database (CTDP)
• Glushinskite equilibrium data only available at 25°C
→ dedicated precipitation experiments at 120°C & **database updating**

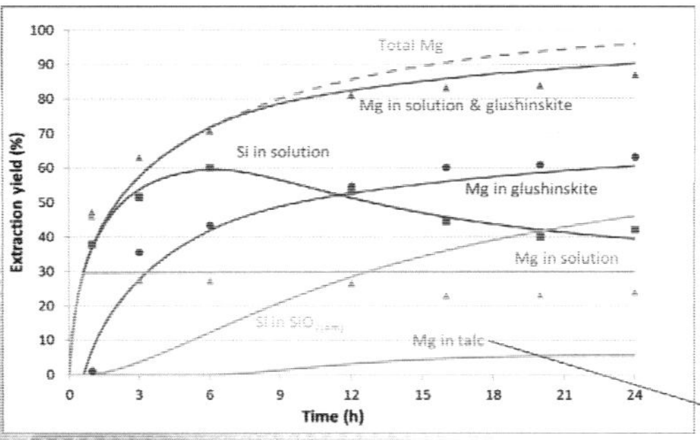

Bonfils et al. (2012)

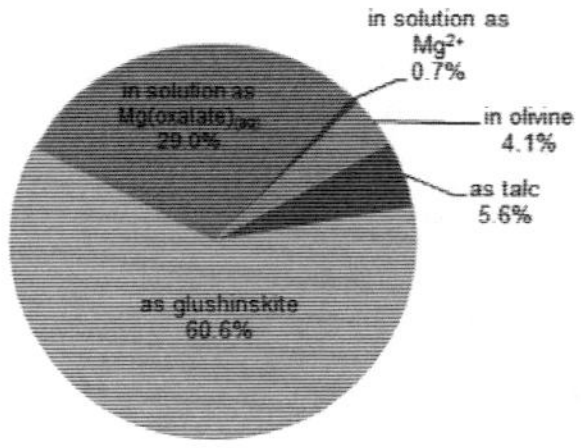

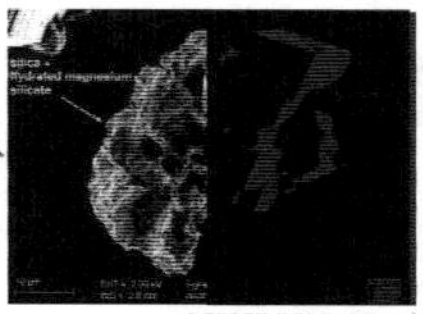

- Good agreement with experimental data at 3 g/L
  → chemical controlled reaction

LGC

Learnings & issues

- Organic polyacid salts (oxalate, citrate, EDTA): too strong Mg binders for MC at moderate P_{CO2}
- Alternative option: silica(te) layer modulation / impeding by ligands

Oxalate

Catechol

Carbonation yield increased from
9.6% after 95 h (water) to
14.5% after 75 h (0.5 M catechol)

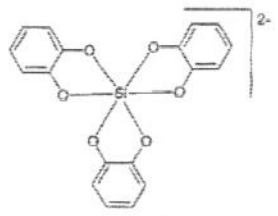

Si-catechol complex
(Barnum, 1970; Russo-Mascioli, 2001)

but might be ore-sensitive (complexation with Fe, Al ...)

- Exhaustive analysis of both liquid & solid products = MANDATORY
- Geochemical modeling as a valuable tool to predict effect of all inputs, but
  * careful selection & analysis of thermodynamic database required
  * going from & to experimental data needed

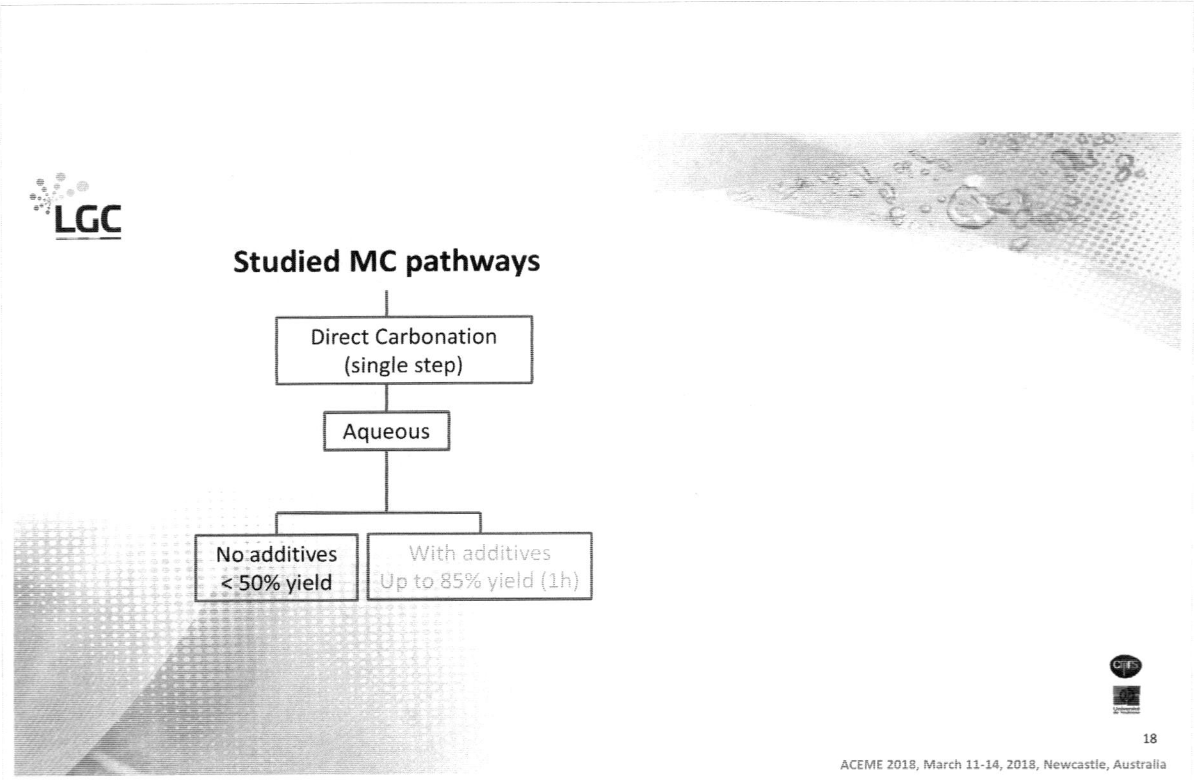
LGC
Studied MC pathways
Direct Carbonation
(single step)
Aqueous
No additives
< 50% yield
With additives
Up to 85% yield (1h)
18
ACEME 2018, March 11-14, 2018, Newcastle, Australia

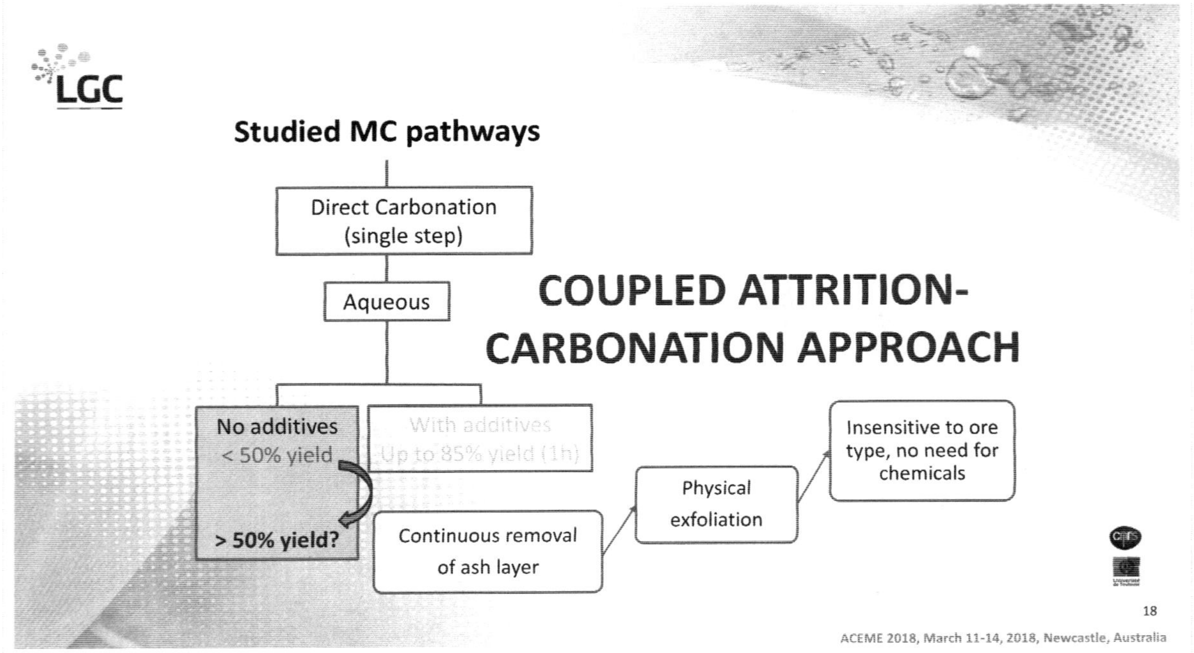
LGC
Studied MC pathways
Direct Carbonation
(single step)
Aqueous
COUPLED ATTRITION-CARBONATION APPROACH
No additives
< 50% yield
> 50% yield?
With additives
Up to 85% yield (1h)
Continuous removal of ash layer
Physical exfoliation
Insensitive to ore type, no need for chemicals
18
ACEME 2018, March 11-14, 2018, Newcastle, Australia

LGC

State of the art

- ***High power ultrasound***

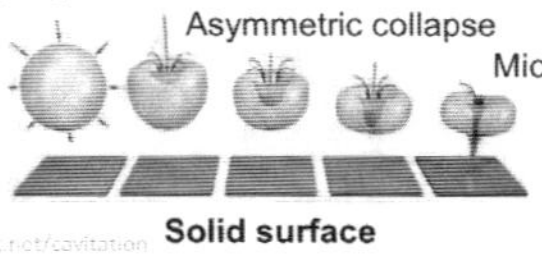

Santos et al., 2011-2013; McKelvy et al., 2004; Park & Fan, 2004

Can be operated at moderate P (< 10 bar), but dampened under high T (> 50 °C)
→ mainly in a sequential process

➢ Mixed results

- ***Fluidized bed with grinding medium***

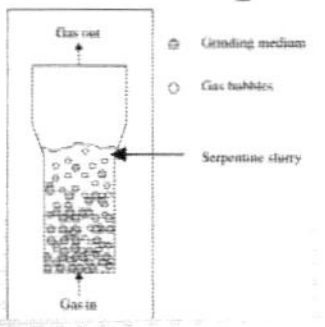

Park & Fan, 2004

Might be difficult to control (density and/or size differences between reactive & inert particles), limited amount of grinding medium (20 wt.%)

➢ Some improvement of ore dissolution

- ***Stirred tank with abrasive particles***

Chizmeshya et al., 2007

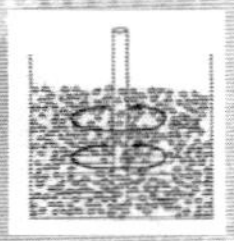

➢ Enhanced carbonation yield (with quartz particles up to 60 wt.%)

LGC

Proposed technical solution: stirred ball mill

IsaMill™ (Netzsch)

Advantages:

✓ proven technology at large scale
✓ slurry conc.: up to 40%
✓ feed PSD: from µm to mm size range
✓ operability under high T & high P
✓ scalability from 4 L to 50 m^3

Issues:

- (long-term) exfoliation efficiency?
  - *significant carbonation yield?*
  - *within a reasonable solid residence time?*
- passivation layer attrition *vs.* breakage of ore particles?

↓

energy efficiency: matching attrition process with passivation process?

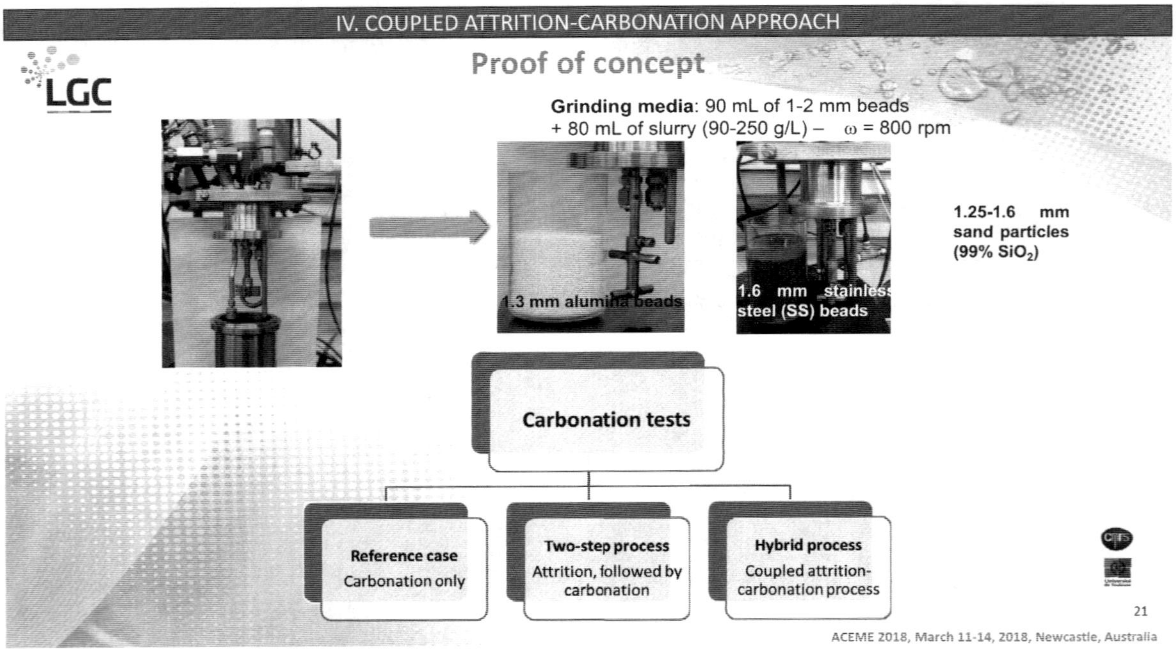
IV. COUPLED ATTRITION-CARBONATION APPROACH
LGC
Proof of concept
Grinding media: 90 mL of 1-2 mm beads
+ 80 mL of slurry (90-250 g/L) – ω = 800 rpm
1.3 mm alumina beads
1.6 mm stainless steel (SS) beads
1.25-1.6 mm sand particles (99% SiO2)
Carbonation tests
Reference case
Carbonation only
Two-step process
Attrition, followed by carbonation
Hybrid process
Coupled attrition-carbonation process
21
ACEME 2018, March 11-14, 2018, Newcastle, Australia

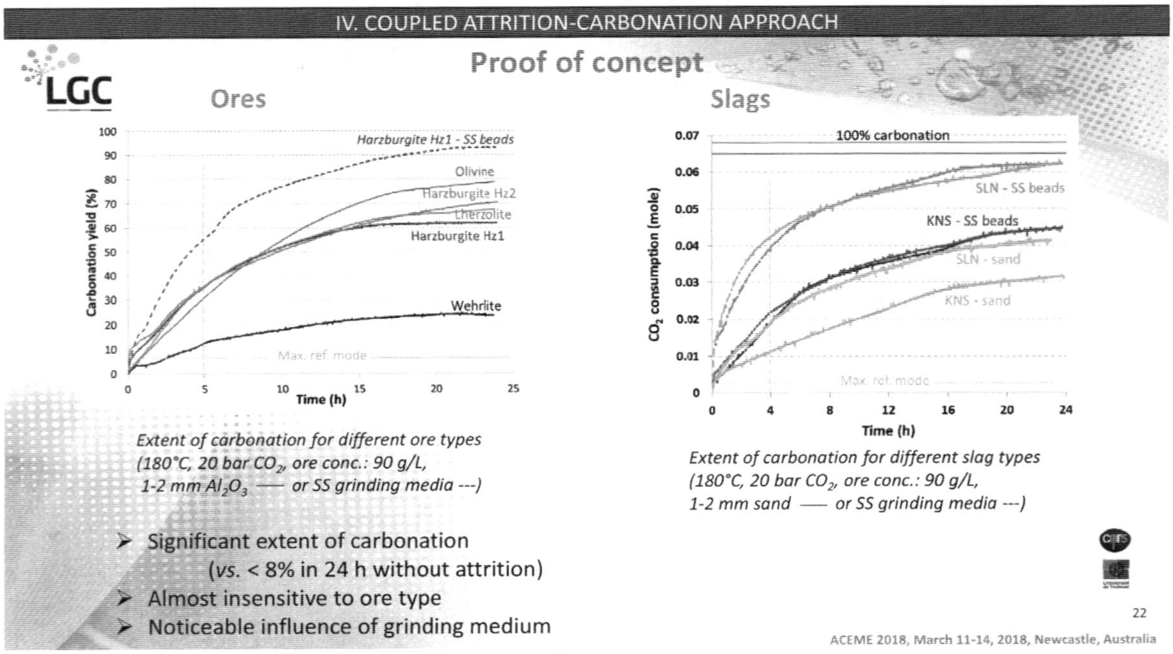
IV. COUPLED ATTRITION-CARBONATION APPROACH
LGC
Proof of concept
Ores
Harzburgite Hz1 - SS beads
Olivine
Harzburgite Hz2
Lherzolite
Harzburgite Hz1
Wehrlite
Max. ref. mode
Carbonation yield (%)
Time (h)
Extent of carbonation for different ore types
(180°C, 20 bar CO2, ore conc.: 90 g/L,
1-2 mm Al2O3 — or SS grinding media ---)
Slags
100% carbonation
SLN - SS beads
KNS - SS beads
SLN - sand
KNS - sand
Max. ref. mode
CO2 consumption (mole)
Time (h)
Extent of carbonation for different slag types
(180°C, 20 bar CO2, ore conc.: 90 g/L,
1-2 mm sand — or SS grinding media ---)
Significant extent of carbonation
(vs. < 8% in 24 h without attrition)
Almost insensitive to ore type
Noticeable influence of grinding medium
22
ACEME 2018, March 11-14, 2018, Newcastle, Australia

IV. COUPLED ATTRITION-CARBONATION APPROACH

LGC

Proof of concept

Size effect only?

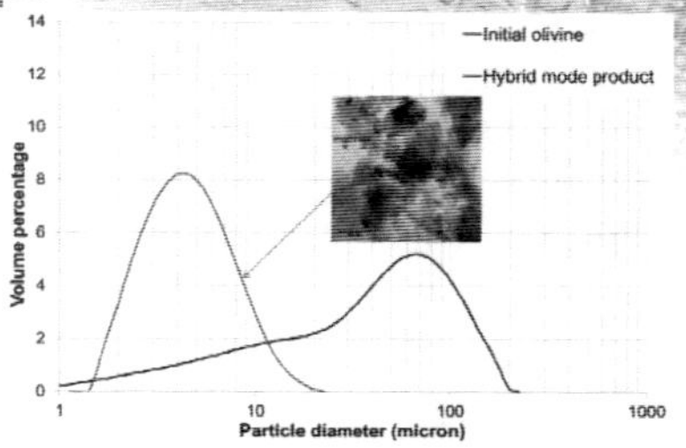

23

IV. COUPLED ATTRITION-CARBONATION APPROACH

LGC

Proof of concept

Size effect only?

- Proof of synergy between attrition and leaching

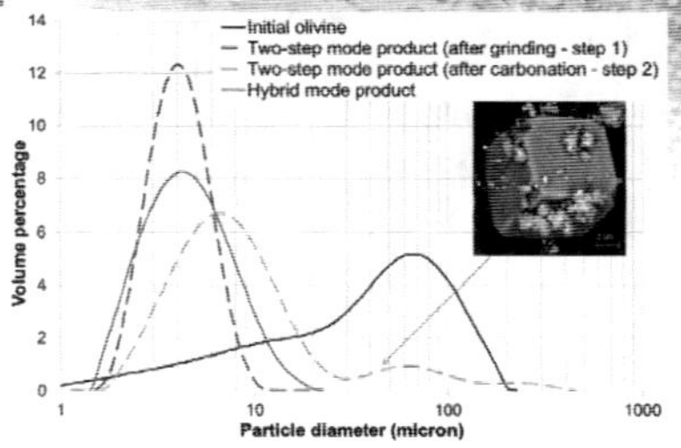

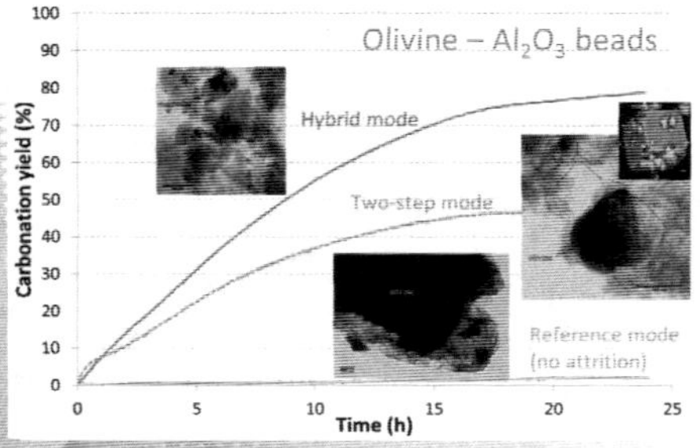

180°C, 20 bar CO_2, ore conc.: 90 g/L

23

Proof of concept

LGC

Size effect only?

➤ Proof of synergy between attrition and leaching

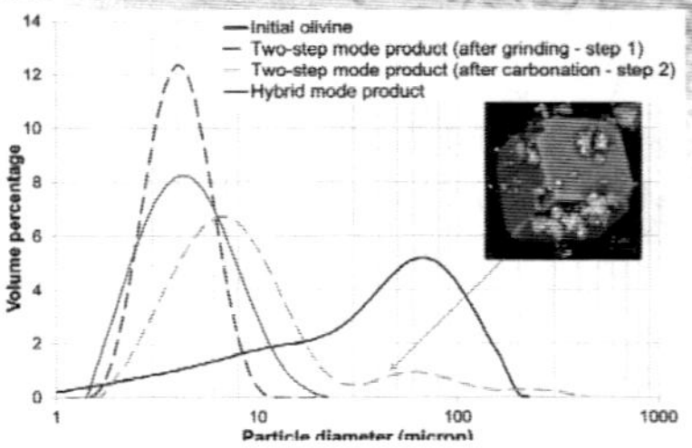

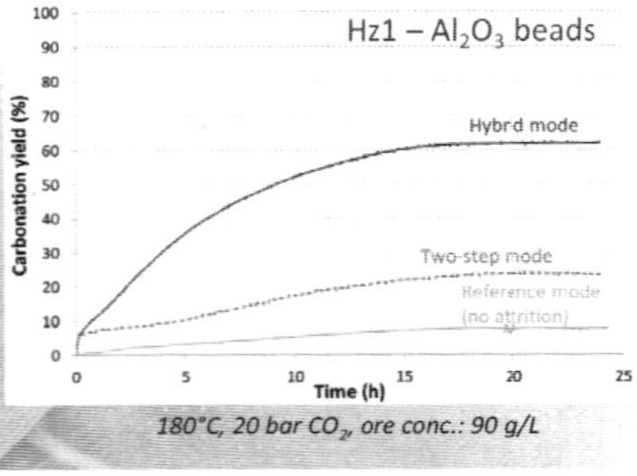

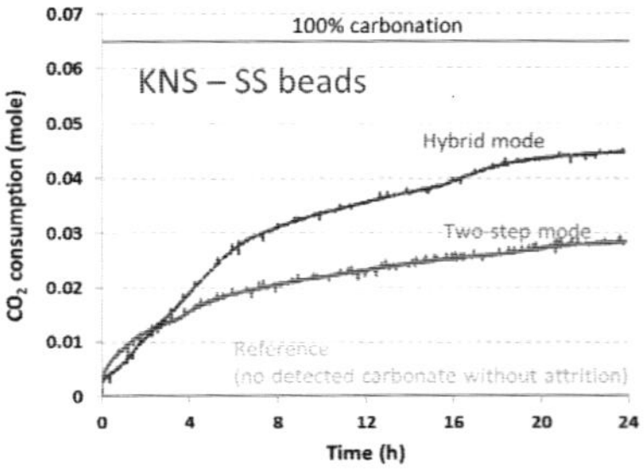

23

Life Cycle Assessment (LCA)

LGC bio Intelligence Service

Functional unit: 1 MWhe with a coal-fired power station

Main hypotheses: - yield of ~ 80% in 24 h at 20 bar CO_2 & 180°C
- CO_2 pipeline transport over a 300 km distance
- no recycling of process solution; no valorization of products

Impacts accounted for:
- CO_2 capture & compression
- crushing & milling of ore from 1 cm down to 100 μm
- mechanical energy expended for attrition
- reactants pre-heating & cooling (after heat integration)

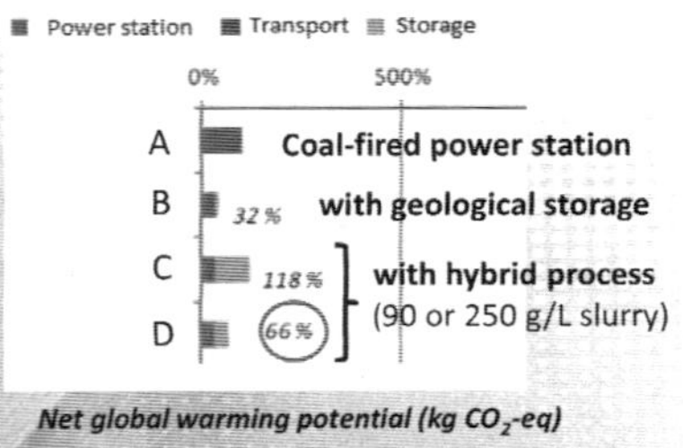

➤ Promising results regarding CO_2 avoided, without any process optimization

➤ Beneficiation of products & water recycling will also improve other LCA criteria (natural resource depletion +110% for case D)

24

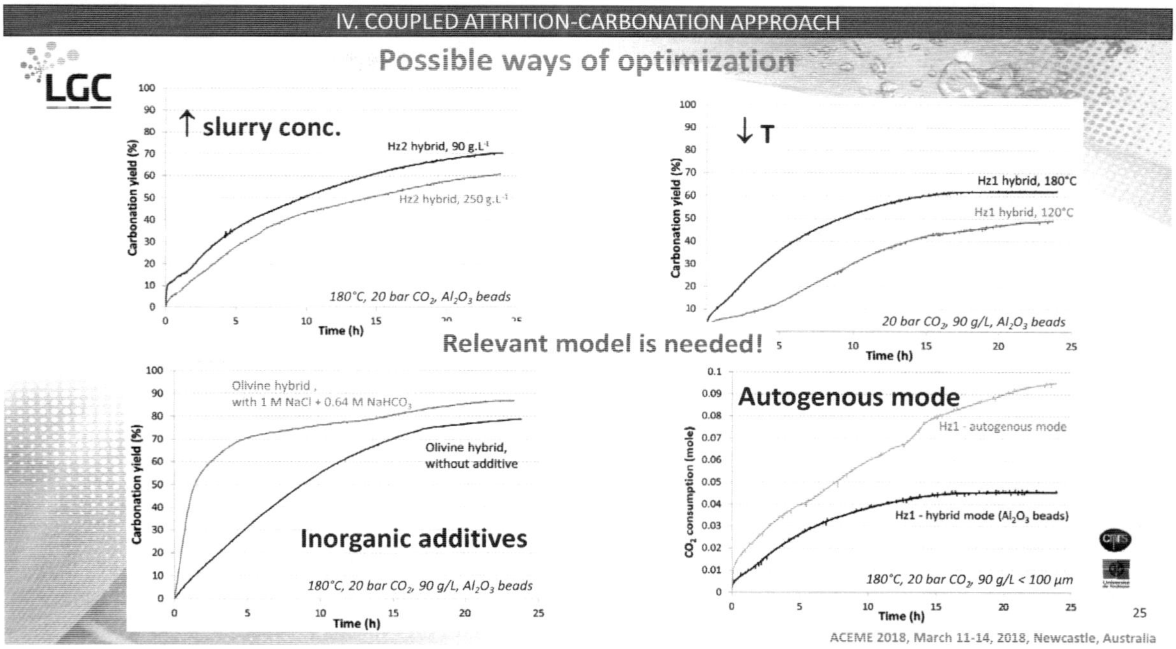
IV. COUPLED ATTRITION-CARBONATION APPROACH
LGC
Possible ways of optimization
↑ slurry conc.
Hz2 hybrid, 90 g.L-1
Hz2 hybrid, 250 g.L-1
180°C, 20 bar CO2, Al2O3 beads
Carbonation yield (%)
Time (h)
↓ T
Hz1 hybrid, 180°C
Hz1 hybrid, 120°C
20 bar CO2, 90 g/L, Al2O3 beads
Relevant model is needed!
Olivine hybrid,
with 1 M NaCl + 0.64 M NaHCO3
Olivine hybrid,
without additive
Inorganic additives
180°C, 20 bar CO2, 90 g/L, Al2O3 beads
Autogenous mode
Hz1 - autogenous mode
Hz1 - hybrid mode (Al2O3 beads)
CO2 consumption (mole)
180°C, 20 bar CO2, 90 g/L < 100 µm
25
ACEME 2018, March 11-14, 2018, Newcastle, Australia

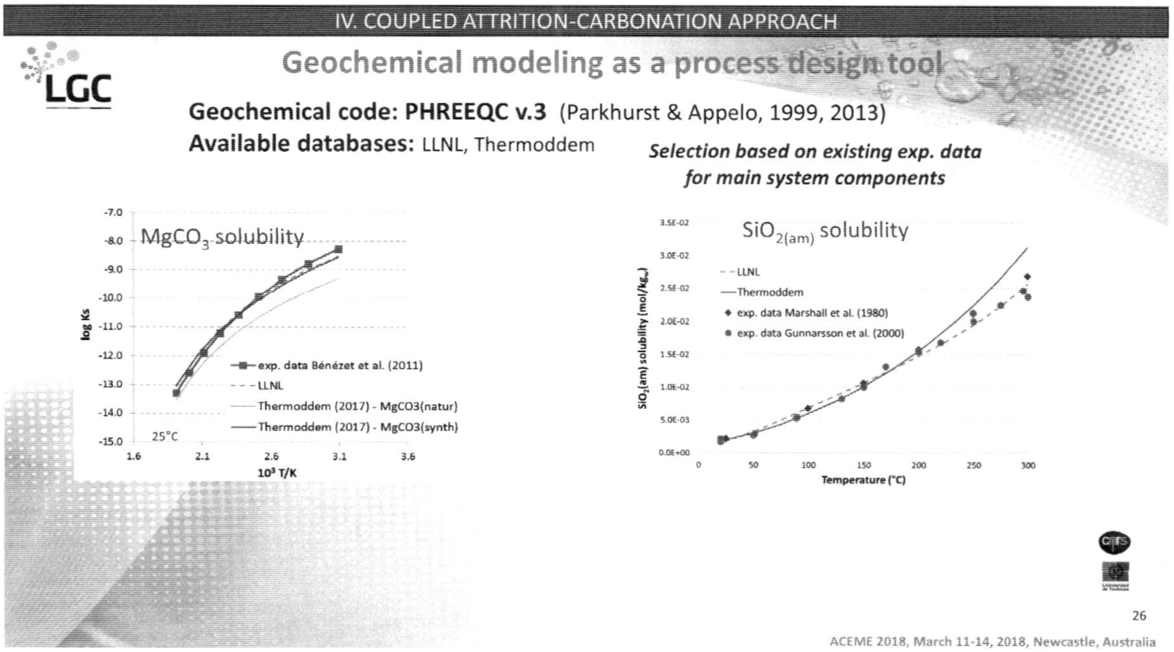
IV. COUPLED ATTRITION-CARBONATION APPROACH
LGC
Geochemical modeling as a process design tool
Geochemical code: PHREEQC v.3 (Parkhurst & Appelo, 1999, 2013)
Available databases: LLNL, Thermoddem
Selection based on existing exp. data
for main system components
MgCO3 solubility
log Ks
exp. data Bénézet et al. (2011)
LLNL
Thermoddem (2017) - MgCO3(natur)
Thermoddem (2017) - MgCO3(synth)
25°C
10³ T/K
SiO2(am) solubility
LLNL
Thermoddem
exp. data Marshall et al. (1980)
exp. data Gunnarsson et al. (2000)
SiO2(am) solubility (mol/kgw)
Temperature (°C)
26
ACEME 2018, March 11-14, 2018, Newcastle, Australia

LGC

Geochemical modeling as a process design tool

Geochemical code: PHREEQC v.3 (Parkhurst & Appelo, 1999, 2013)

Available databases: LLNL, Thermoddem

Selection based on existing exp. data for main system components

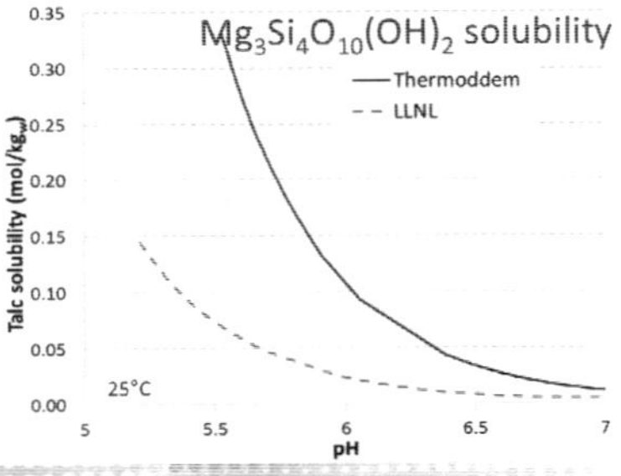

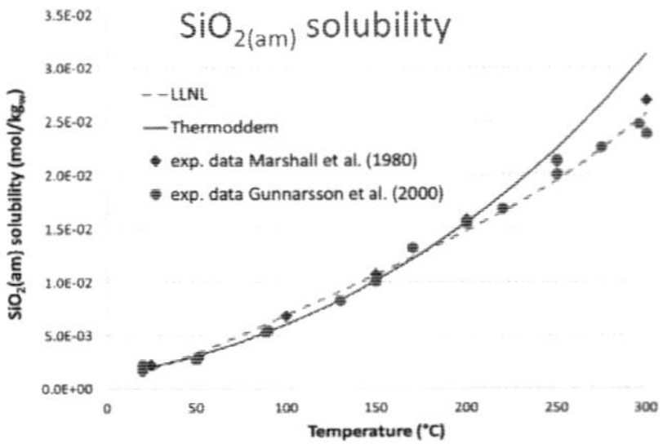

Discrepancies for talc, but no solubility data available (in similar conditions) → assessment on carbonation results

Liquid: extended Debye-Hückel activity coefficient model (low salinity)
Gas: Peng-Robinson equation of state (non-ideal behavior of CO_2 - H_2O mixture at investigated P)
Solid: description of several **solid solutions** (e.g. $(Mg,Fe)CO_3$)

LGC

Case study 1: KNS / "inert" grinding medium

"KNS" (90 g/L) described as an assemblage of MgO, SiO_2 and FeO
All minerals including Mg, Si or Fe are allowed to precipitate, except quartz

Mineral speciation of KNS-H_2O-CO_2

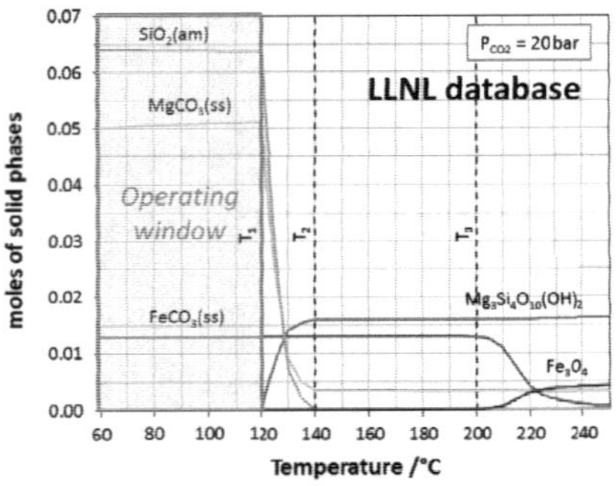

Thermoddem 2017 database

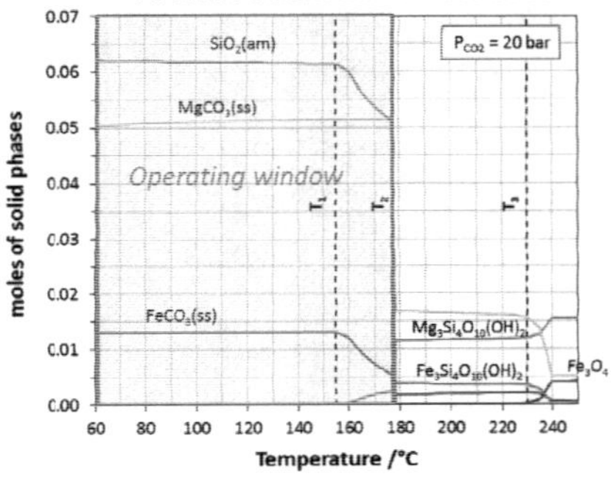

- Threshold T for quantitative carbonation at 20 bar CO_2 depends on the database!
- At T = 180°C and P_{CO2} = 20 bar:
  - ~~LLNL → theoretical max carbonation yield: 25%~~
  - **Thermoddem 2017 → conditions close to the drop in carbonation yield**

IV. COUPLED ATTRITION-CARBONATION APPROACH

LGC

Case study 1: KNS / "inert" grinding medium

"KNS" (90 g/L) described as an assemblage of MgO, SiO_2 and FeO
All minerals including Mg, Si or Fe are allowed to precipitate, except quartz

Mineral speciation of KNS-H_2O-CO_2

Thermoddem 2017

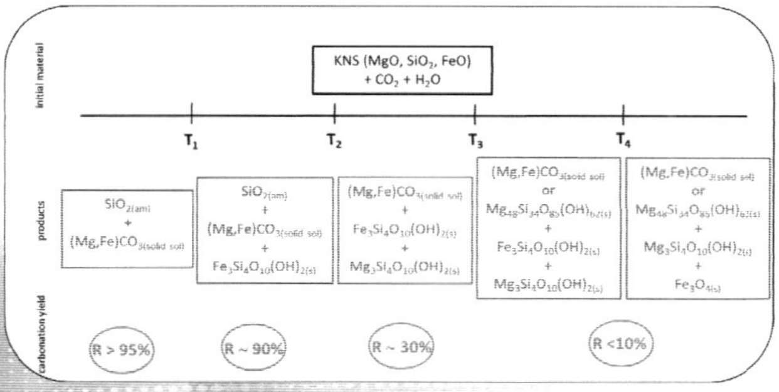

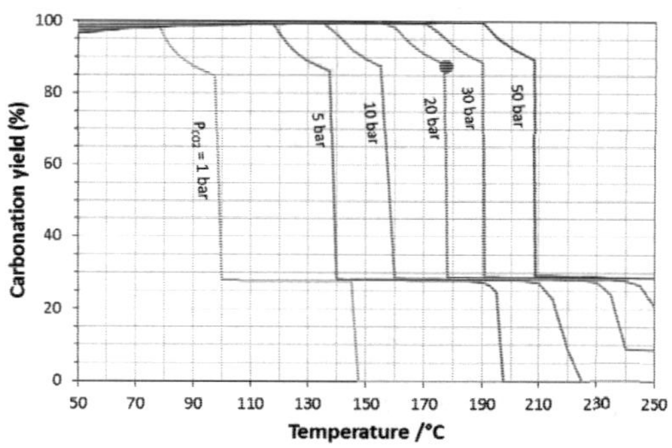

➢ Operating threshold temperature depends on P_{CO2}

28

ACEME 2018, March 11-14, 2018, Newcastle, Australia

IV. COUPLED ATTRITION-CARBONATION APPROACH

LGC

Case study 1: KNS / "inert" grinding medium

"KNS" (90 g/L) described as an assemblage of MgO, SiO_2, FeO, **CaO, Al_2O_3 and MnO**
All minerals including Mg, Si, Fe, **Ca, Al or Mn** allowed to precipitate (except quartz)

Mineral speciation of KNS-H_2O-CO_2

Thermoddem 2017

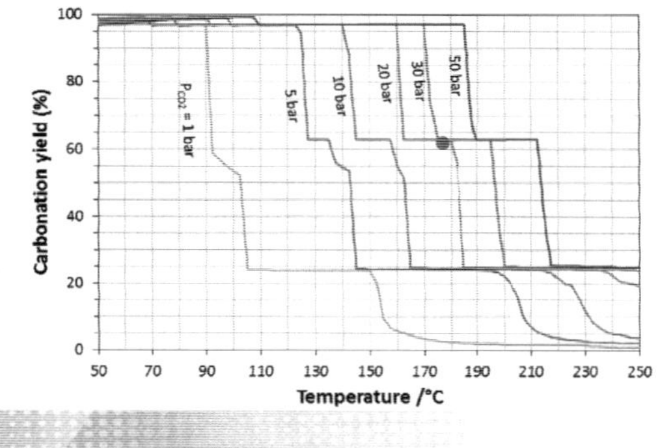

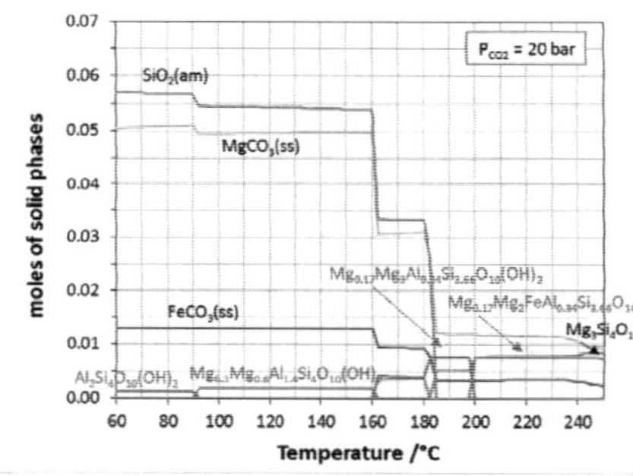

➢ The model predicts a noticeable effect of slag "impurities" (Al_2O_3 ~ 2%) due to the existence of various stable aluminosilicate phases

29

ACEME 2018, March 11-14, 2018, Newcastle, Australia

IV. COUPLED ATTRITION-CARBONATION APPROACH

Case study 2: KNS / effect of grinding medium

Experimental data

– **Carbonation yield ~ 50%** with sand (180°C, 20 bar, 90 g/L KNS)

– **Crystallized phases** = initial mineral phases (enstatite, ferrosilite, augite), quartz (sand) **& mixed carbonates** (dominated by $MgCO_3$ pole)

– **TEM/EDX**

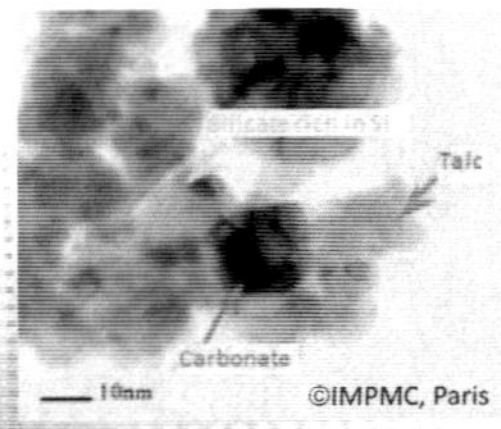

→ **Precipitation of various (amorphous) silicates**

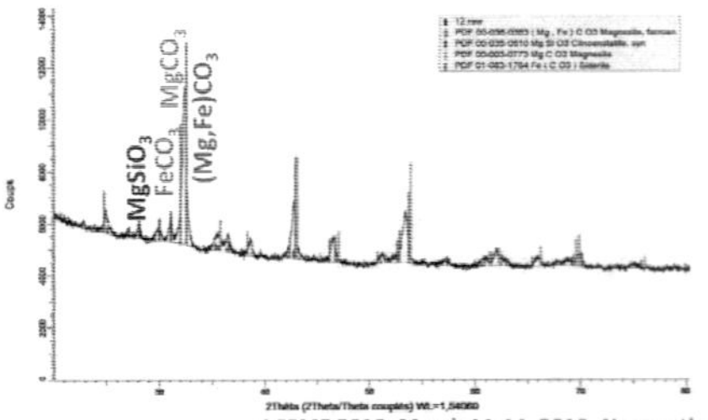

– **Higher yield achieved with stainless steel beads**

... but increase of Fe content in solid product &

IV. COUPLED ATTRITION-CARBONATION APPROACH

Case study 2: KNS / effect of grinding medium

Experimental data

– **Carbonation yield ~ 50%** with sand (180°C, 20 bar, 90 g/L KNS)

– **Crystallized phases** = initial mineral phases (enstatite, ferrosilite, augite), quartz (sand) **& mixed carbonates** (dominated by $MgCO_3$ pole)

– **TEM/EDX**

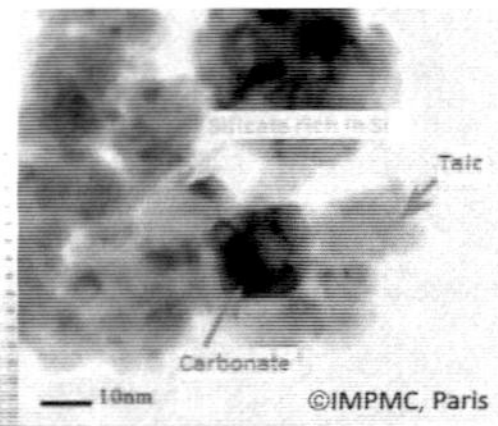

→ **Precipitation of various (amorphous) silicates**

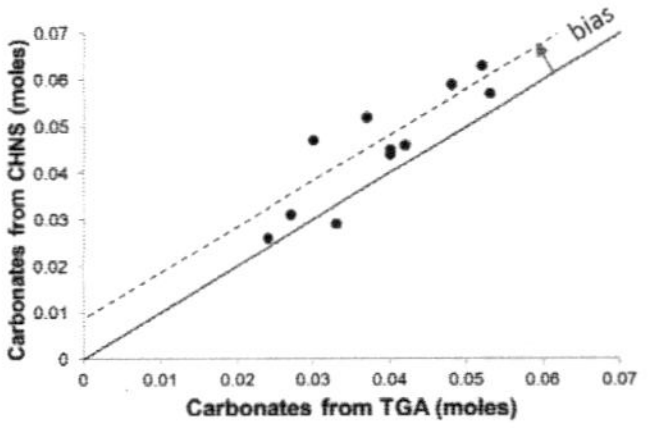

– **Higher yield achieved with stainless steel beads**

... but increase of Fe content in solid product &

a few discrepancies between carbonate amounts calculated from TGA and carbon content

IV. COUPLED ATTRITION-CARBONATION APPROACH

LGC

Case study 2: KNS / effect of grinding medium

"KNS" described as an assemblage of MgO, SiO_2, FeO, CaO, Al_2O_3 and MnO
Steel beads modeled as a $(Fe_{0.87}Cr_{0.13})$ solid solution

Mineral speciation of KNS-H_2O-CO_2

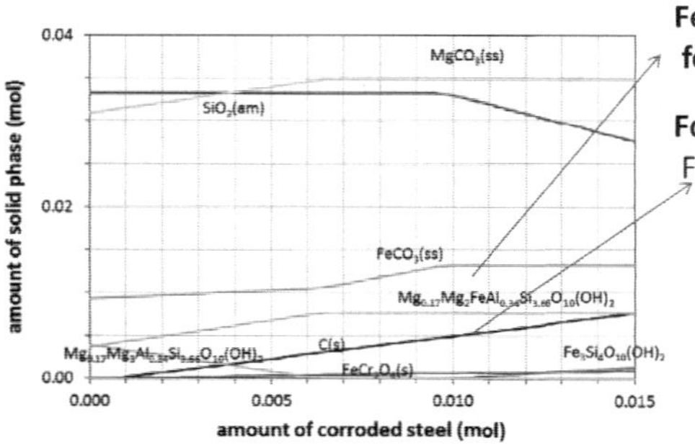

P_{CO2} = 20 bar, T = 175°C, slag conc. = 90 g/L

Fe extracted from beads as $FeCO_3$ & ferro-magnesium-aluminosilicate

Formation of solid carbon:

$Fe_{(steel)} + 3/2\ CO_{2\,(aq)} = FeCO_{3\,(ss)} + 1/2\ C_{(s)}$

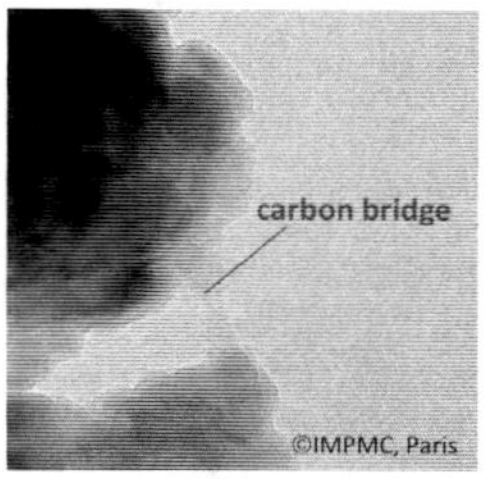

➢ **Stainless steel beads are corroded during attrition-leaching under CO_2**

IV. COUPLED ATTRITION-CARBONATION APPROACH

LGC

Case study 3: Batch simulation for olivine ore

"Synthetic olivine" described as an assemblage of MgO, SiO_2, FeO, CaO, Al_2O_3 and MnO
Initial PSD accounted for, dissolution kinetic parameters from Prigiobbe et al. (2009)
Alumina beads modeled as corundum, kinetic parameters from Palandri & Kharata (2004)

Time-evolution of carbonation yield

"Inert" grinding medium

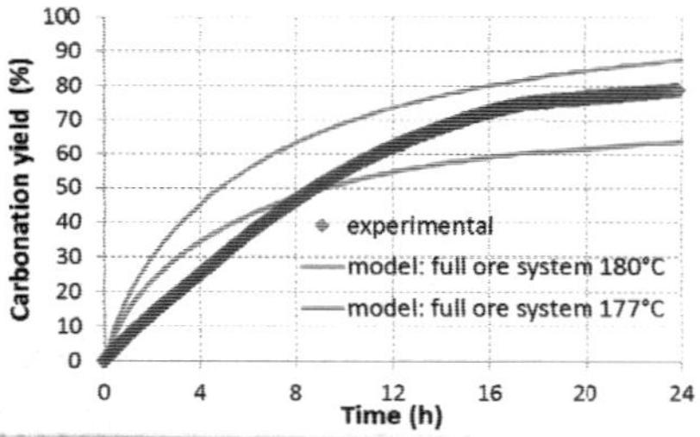

P_{CO2} = 20 bar, T = 180°C, ore conc.= 90 g/L

Effect of alumina grinding beads

Very slow dissolution kinetics
→ Negligible effect on system speciation

➢ Good agreement between experimental data and modeling
➢ **Process dynamics driven by the dissolution rate of fresh ore surface**

IV. COUPLED ATTRITION-CARBONATION APPROACH

LGC

Learnings & issues

- Proof of concept of the attrition-carbonation process with a stirred bead mill; **Synergy between attrition & carbonation**
- **Favorable LCA**; several optimization levers to improve process efficiency, cost & environmental impact
- **Grinding medium to be carefully selected (autogenous mode?)**
- **Geochemical modeling is a powerful tool** for designing the attrition-carbonation process:
  - *equilibrium calculations* → suitable operating window (to be verified experimentally) & material selection for the process equipment (grinding media)
  - *coupling of thermodynamics with chemical kinetics* → process sizing & optimization

SO WHAT'S NEXT ?

- **A continuous scalable demonstrator**
- **Beneficiation of carbonation products**

33

ACEME 2018, March 11-14, 2018, Newcastle, Australia

V. CONCLUSIONS: ONGOING & FUTURE WORK

LGC

CNRT NICKEL & son environnement

Continuous bench-scale process

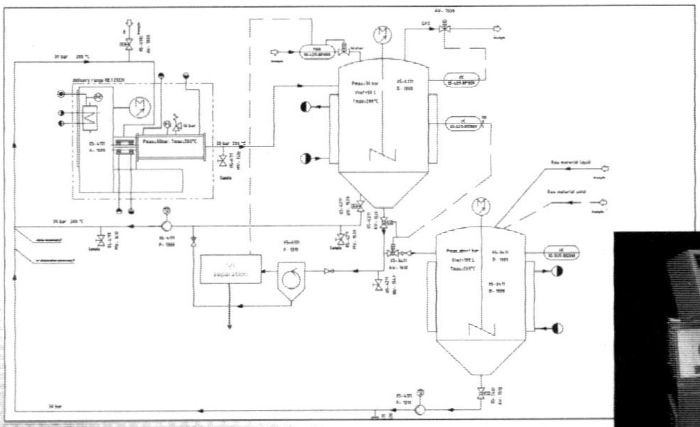

Attrition reactor~4 L operating under T (max 200°C) & P (max 30 bar)

Bench-scale pilot reactor to be built in the coming months

34

ACEME 2018, March 11-14, 2018, Newcastle, Australia

V. CONCLUSIONS: ONGOING & FUTURE WORK

Utilization of MC products as construction or filling materials

- could in theory absorb Gt of CO_2 (approximately 33 billion tons of concrete produced / year)

Hybrid process products:
agglomerates of nano-sized particles of carbonates, silica & silicates

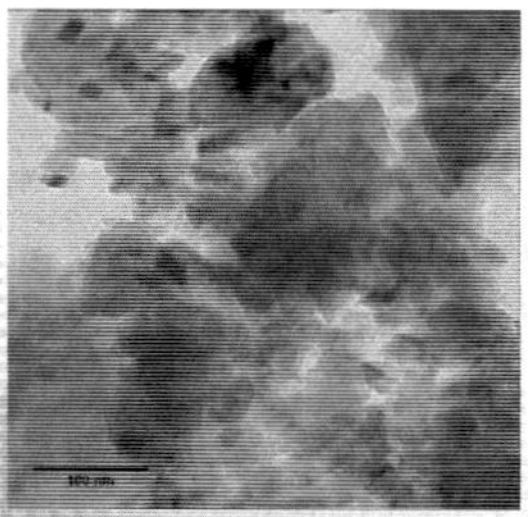

almost "as it is" →

✓ Potentially highly reactive additives

Inert filler or pozzolana
(mainly ultra-fine silica fraction
+ $MgCO_3$)
CO_2 sequestered (& avoided)

after heating →

Hydraulic binder
(mainly Mg rich fraction
+ SiO_2)
CO_2 avoided

Goal: valorization without solid separation & with minimum dewatering

- ✓ Virtuous CO_2 loop
- ✓ Reduction of natural resource depletion (no need for limestone)
- ✓ Reduction of heat demand ($T_{calcination}$ ↓)
- ✓ Local production of construction materials

V. CONCLUSIONS: ONGOING & FUTURE WORK

New concept: CCUS combined with Enhanced Metal Recovery

Innovative chemistry under attrition environment

Attrition-carbonation process

CO_2 →
(Ni) slags →

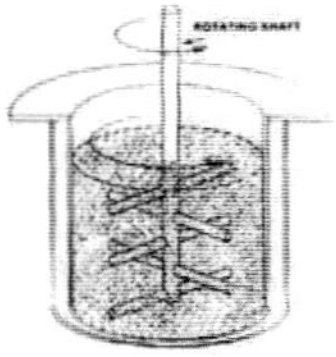

\+

Polyamines

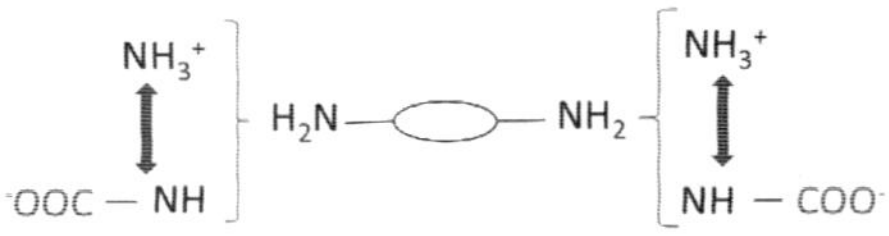

efficient CO_2 capture (dilute flue gases)

\+ catalysis of **CO_2 mineralization**

\+ potential modulation of silica precipitation

\+ selective **metal recovery**

LGC

Overall scheme for CO_2 mineralization

"Historically", CO_2 mineralization was compared to geological storage, in terms of storage cost and avoided CO_2: Limited development ...

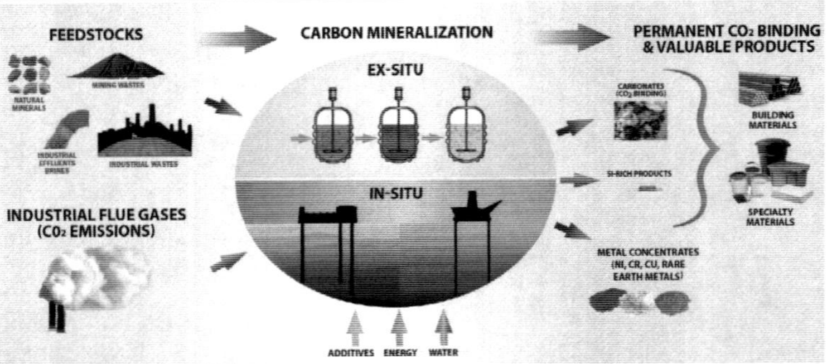

Mission Innovation, CCUS Workshop, Houston, Sept. 2017;* http://mission-innovation.net/our-work/innovation-challenges

In recent years, the mineralization of CO_2 systematically combines CO_2 storage and production of commercial goods, coupling environment and economy:

Booming development!

LGC

Acknowledgements

French National Research Agency,

Total Oil & Gas company,

Center for technological research on Nickel of New Caledonia,

Caledonian Energy Agency

for funding these research works

Our PhD & post-doc students: B. Bonfils, I. Benhamed, J. Diouani

Our technical staff: J.L. Labat, I. Coghe, L. Farhi, C. Rey-Rouch (LGC), A. Moreau (LCC Toulouse), C. Charvillat (CIRIMAT Toulouse) for their support on the reactor pilot and the solid analyses

Our project partners: S. Touzé, T. Augé, F. Bodenan (**BRGM Orléans**)
F. Guyot, F. Ploquin, F. Juillot (**IMPMC Paris**)
C. Petiot, L. Cassowitz, A. Boukary (**BioIS**)
J. Leclaire & J. Septavaux (**ICBMS Villeurbanne**)
for their valuable contribution

LGC

Thank you for your kind attention!

LGC

The actual team

CNRT NICKEL & son environnement

France Bailly, Director

LABORATOIRE DE GÉNIE CHIMIQUE TOULOUSE • UMR 5503

Pr. Florent Bourgeois

Dr. Carine Julcour

Dr. Laurent Cassayre

Institut de Chimie et Biochimie Moléculaires et Supramoléculaires icbms UMR 5246

Pr. Julien Leclaire

brgm Géosciences pour une Terre durable

Dr. Solène Touzé

Lmdc toulouse

Pr. Martin Cyr

41

ACEME 2018, March 11-14, 2018, Newcastle, Australia

ZheJiang University

Kinetic study on accelerated mineral carbonation curing of cement-based materials

Xutao Hu[1], Hao Huang[2], Tao Wang*, Mengxiang Fang[4]
Corresponding author*: Tao Wang - oatgnaw@zju.edu.cn
Presenting author: Xutao Hu
Zhejiang University

BACKGROUND

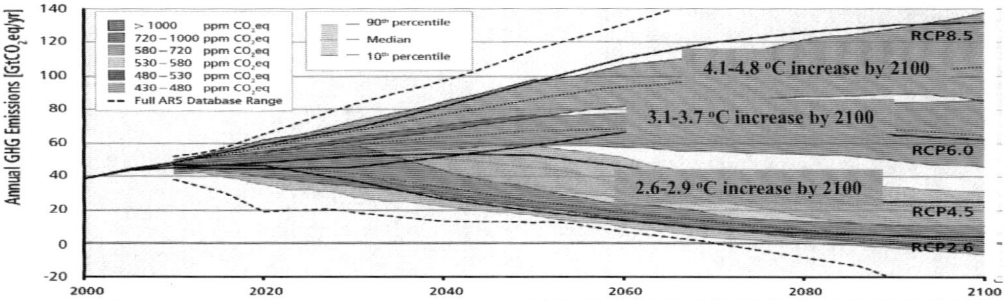

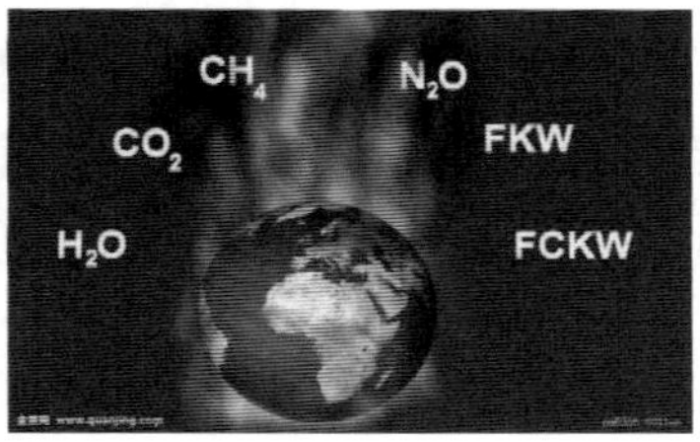

Main greenhouse gas : CO_2, CH_4, N_2O

⟹ Mineral Carbonation Curing

- ◆ Novel curing technology
- ◆ Promising CO_2 utilization route

BACKGROUND

- Increase demand of concrete:
- Rapid development of prefabricated construction : 5% to 35% ,500-1000 million tons precast concrete per year .10 years

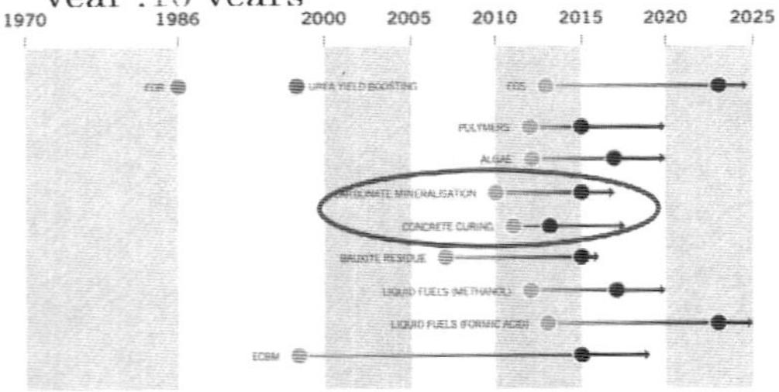

(Ref: G.C. Institute, Accelerating the uptake of CCS: Industrial use of captured carbon dioxide 02.03.2017 (2011))

MC curing of cement-based materials since 1970s

- CO_2 sequestration:30-300 million tons per year globally
- Faster strength gain than natural curing and comparable with steam curing
- Additonal physical properties :
  - thermal insulation
  - fire resistance
  - Acoustical insulation

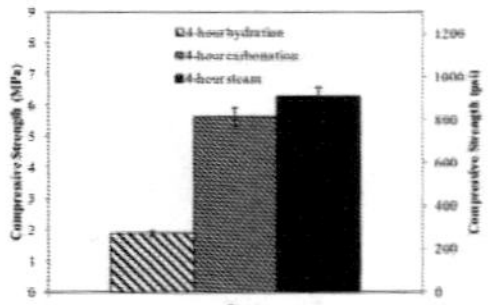

(Ref: Y. Shao, H. El-Hassan, CO2 Utilization in oncrete, International Conference on Sustainable Construction Materials and Technologies, 2013)

EXPERIMENT

Materials and pretreament

Materials:

- P.O.42.5 cement
- Calcium silicate
- Industrial waste :fly ash ,slag,red mug

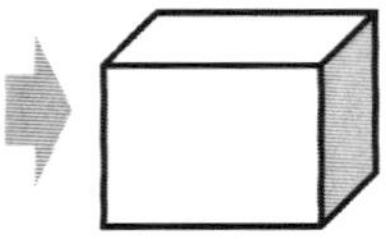

Composition of OPC,CS and fly ash (wt. %)

	CaO	SiO_2	SO_3	Al_2O_3	Fe_2O_3	K_2O	Others
OPC	62.911	21.886	4.649	4.531	4.432	0.934	0.657
CS	42.795	46.641	-	2.062	3.370	-	5.132
Fly ash	12.08	46.86	0.8	25.76	7.54	-	6.96

Controlled pretreatment conditions:

- Temperature
- Humidity
- Pre-curing time

EXPERIMENT

Curing experiment

Different curing conditions:

- Pressure
- Temperature
- Curing time

CO2 uptake calculation formula

$$PV = z \cdot nRT = z \cdot \frac{m}{M_{CO_2}} RT$$

$$m_{CO_2} = \left(\frac{P_1 V_s}{z_1 R T_1} - \frac{P_2 V_s}{z_2 R T_2} - \frac{(P_c - P_w) V_c}{z_c R T_c} \right) \cdot M_{CO_2}$$

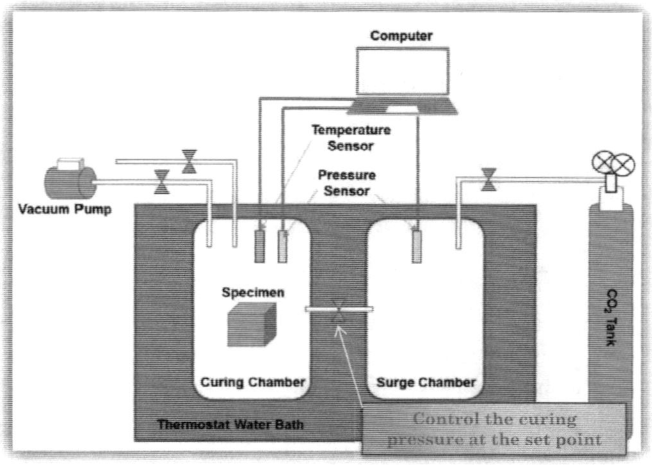

Schematic diagram of the MC curing equipment

EXPERIMENT

characterization analysis

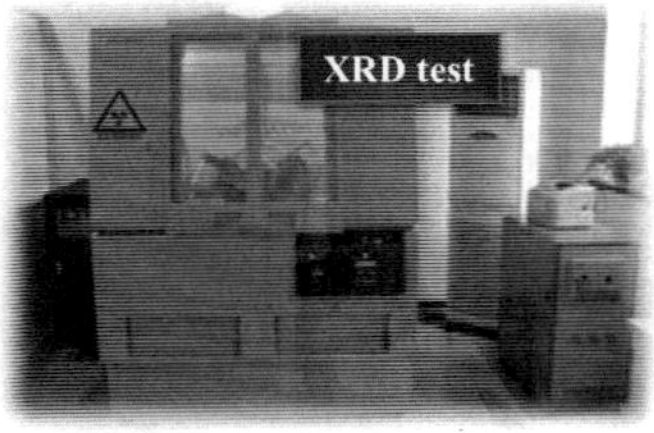

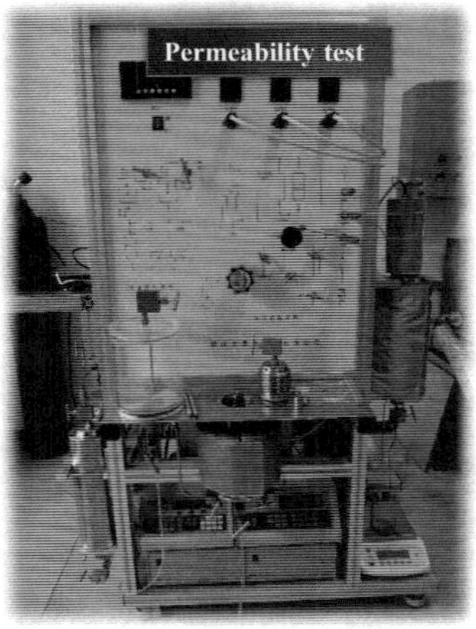

RESULTS AND ANALYSIS

Accelerate MC curing of OPC paste

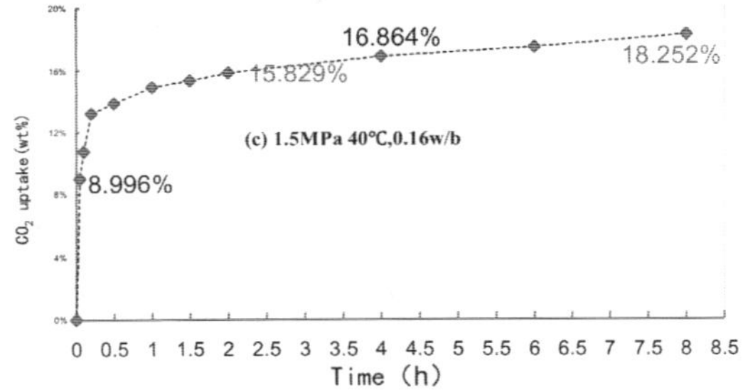

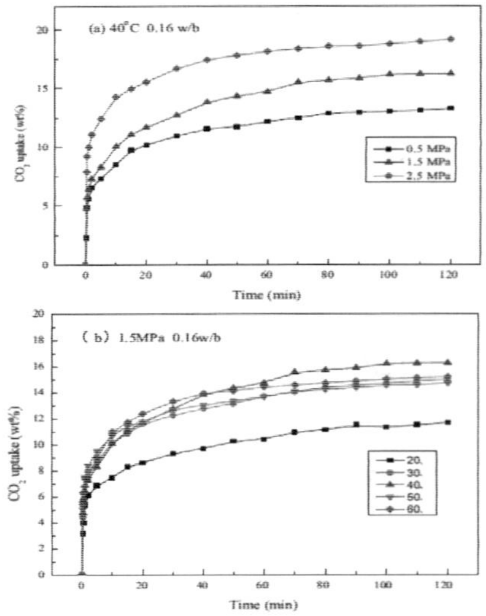

- Modified carbonation time is 2h
- A visible increase of CO2 uptake in first 20min.
- The reaction rate decrease during the remaining time.

RESULTS AND ANALYSIS

Why knetic study?

Relationship between CO2 uptake and compressive strength

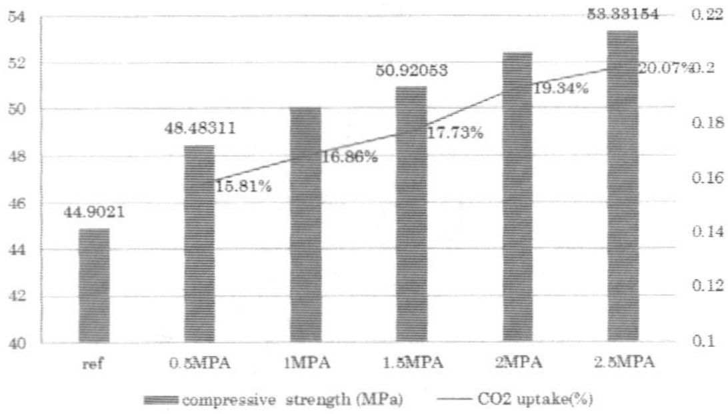

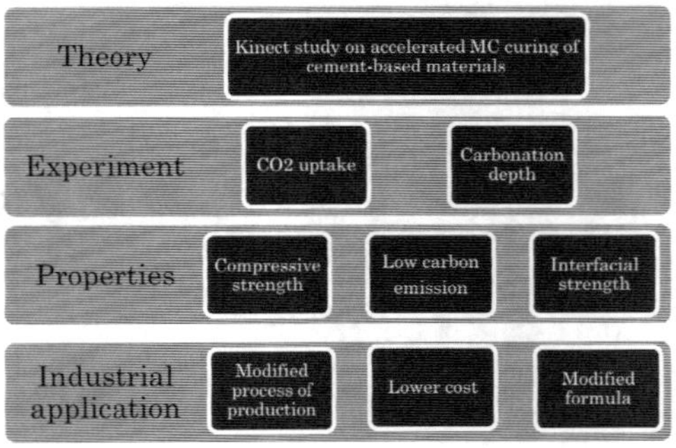

- Higher CO2 uptake and conversion
- Better product properties
- Lower cost and easier process of production

→ determine the rate-controlling step !

RESULTS AND ANALYSIS

Kinetic model of OPC paste

Different gas-solid reaction kinetic models compare with experimental data (2-h curing under 0.5 MPa and 2.5 MPa, 40℃ and 0.16 w/b)

Kinetic equation :

$$[1-(1-\alpha)^{1/3}]^2 = kt$$

α: conversion percent
k: rate constant
t: reaction time

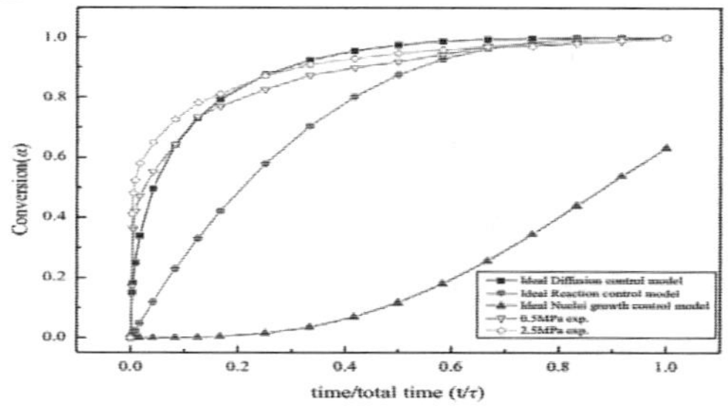

Best matching

Kinetic Function of the Most Common Solid-State Reactions and Mechanism.

Mechanism		Function	Non-dimensional transformation
Based on diffusion	Three-dimensional diffusion	$[1-(1-\alpha)^{1/3}]^2 = kt$	$[1-(1-\alpha)^{1/3}]^2 = \frac{t_d}{\tau_d}$ $\rightarrow \alpha = 1-(1-(\frac{t_d}{\tau_d})^{1/2})^3$
Based on chemical reaction	Contracting sphere	$1-(1-\alpha)^{1/3} = kt$	$1-(1-\alpha)^{1/3} = \frac{t_c}{\tau_c}$ $\rightarrow \alpha = 1-(1-\frac{t_c}{\tau_c})^3$
Based on nuclei growth	Avrami-Erofeev (three-dimensional)	$(-\ln(1-\alpha))^{1/3} = kt$	$(-\ln(1-\alpha))^{1/3} = \frac{t_n}{\tau_n}$ $\rightarrow \alpha = 1-e^{-(\frac{t_n}{\tau_n})^3}$

RESULTS AND ANALYSIS

Kinetic analysis of early stage

$$[1-(1-\alpha)^{1/3}]^2 = kt.$$

$$\ln[1-(1-\alpha)^{1/3}] = (1/2)\ln t + (1/2)\ln k.$$

the slopes of the curves slightly deviated from the given theoretical value (1/2). WHY?

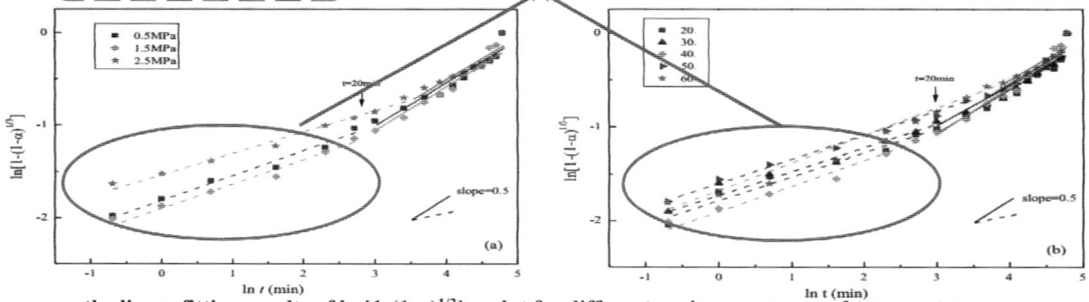

the linear fitting results of ln $[1-(1-\alpha)^{1/3}]$ vs. lnt for different curing pressure and temperature

RESULTS AND ANALYSIS

Kinetic analysis of early stage

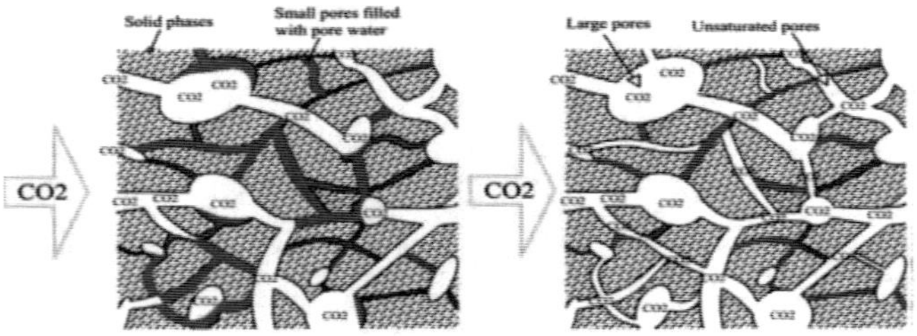

The joint effect of pore diffusion (evaporated H_2O and CO_2) and early temperature rise caused the deviation of fitting curve!

Sharp temperature rise in early stage

outward diffusion of residual pore water

number of hollow microstructure ↑
Gaseous permeability ↑

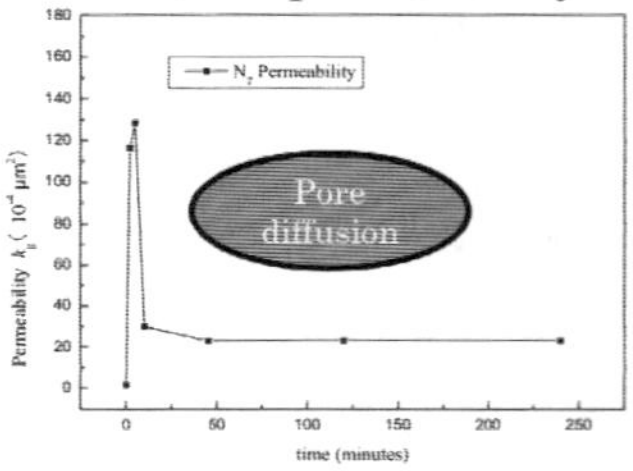

RESULTS AND ANALYSIS

Kinetic analysis of product-layer diffusion stage

The overall fitting line show concave upward

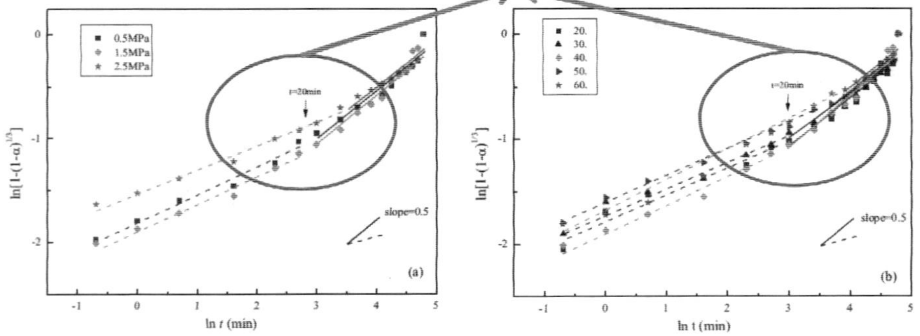

product-layer diffusion model:

$$[1-(1-\alpha)^{1/3}]^n = kt$$

α: Conversion rate
k: Apparent rate constant
t: Time

A progressive product-layer-diffusion limited kinetic !

RESULTS AND ANALYSIS

Kinetic analysis of product-layer diffusion stage

Reaction rate constant k_p for the product-layer diffusion stage

Variables	Pressure	(Almost same)		Temperature	(Positive correlation)			
	0.5MPa	1.5MPa	2.5MPa	20℃	30℃	40℃	50℃	60℃
k_p / 10^{-3} min^{-1}	6.06	6.03	6.09	5.44	5.68	6.03	6.25	6.54
R^2	0.919	0.967	0.796	0.921	0.899	0.967	0.896	0.871

The k_p contains items of:

- ◆ diffusion coefficient → Positive correlation to curing temperature
- ◆ pore structure → Keep relatively stable
- ◆ other physical properties of cement past e.g. particle size, specific surface area . → **More investigation !**

RESULTS AND ANALYSIS

Effect of blended calcium silicate

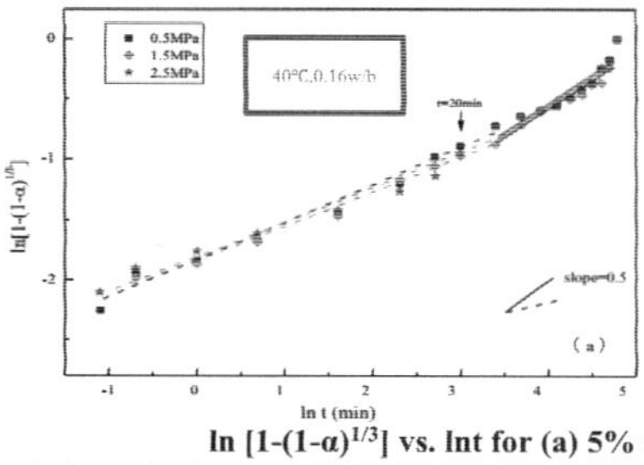

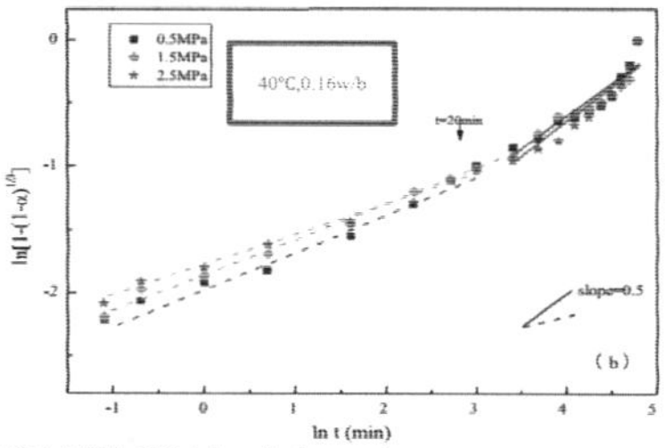

⇨ **Similar to OPC paste**

ln $[1-(1-\alpha)^{1/3}]$ vs. lnt for (a) 5% CS (b) 15% CS blended specimens

k_p for the product-layer diffusion stage with 5% and 15% CS

Variables	5%CS			15%CS		
	0.5MPa	1.5MPa	2.5MPa	0.5MPa	1.5MPa	2.5MPa
k_p / 10^{-3} min^{-1}	5.94	5.85	5.81	5.45	5.58	5.52
R^2	0.880	0.901	0.883	0.913	0.882	0.896

⇨ **Blending rate of CS↑ k_p ↓**

RESULTS AND ANALYSIS

Microstructure properties

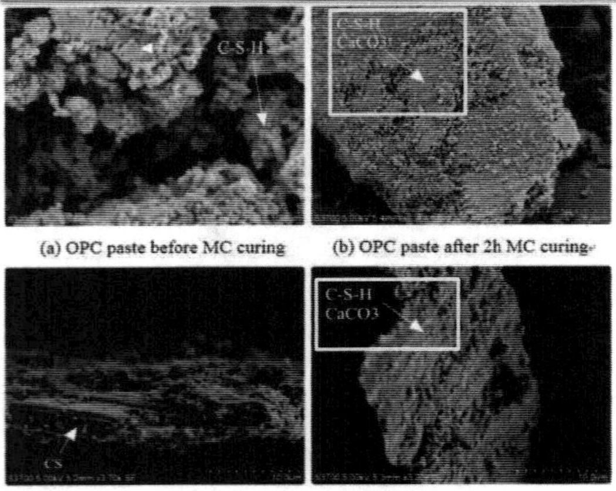

(a) OPC paste before MC curing (b) OPC paste after 2h MC curing

(c) OPC paste with 15% CS before MC curing (d) OPC paste with 15% CS after 2h MC curing

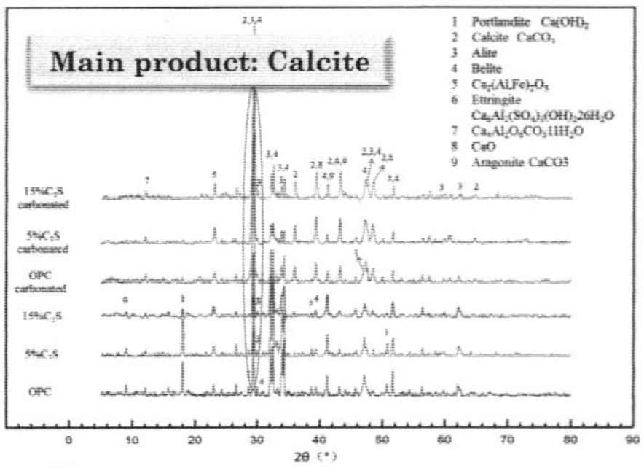

- Denser structure
- Less interfacial transition zone

Matching → Reaction Process

HOW TO ENHANCE THE REACTION ?

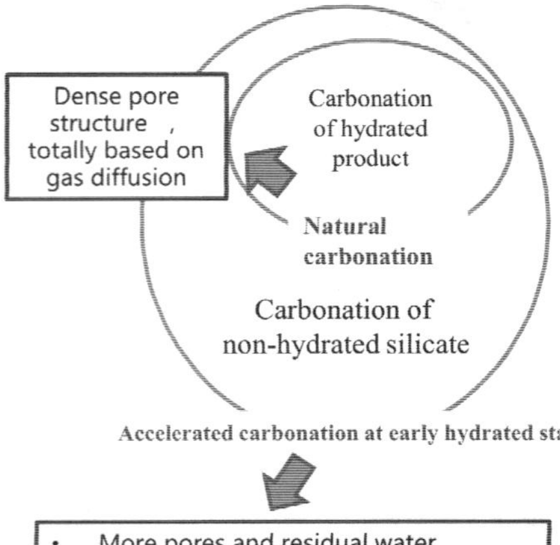

- More pores and residual water
- based on progressive product-diffusion kinetic
- Pore structure affected by reaction dynamically

Enhance the gas diffusion

- **Supplementary Cementitious material**
  - Low Ca/Si ratio silicate: $CaSiO_3$、$Ca_3Si_2O_7$、γ-C_2S
  - $CaCO_3$
  - SiO_2
  - $CaMg(CO_3)_2$
    - Affect the pore structure before MC curing
    - Affect the permeability characteristics
    - Less cement ,less hydrated product ,more pores

- **Aerated concrete**

  - More macro pores
  - Easier gas diffusion during the early stage

DEMONSTRATION PROJECT

50mm×50 mm×50m m cube

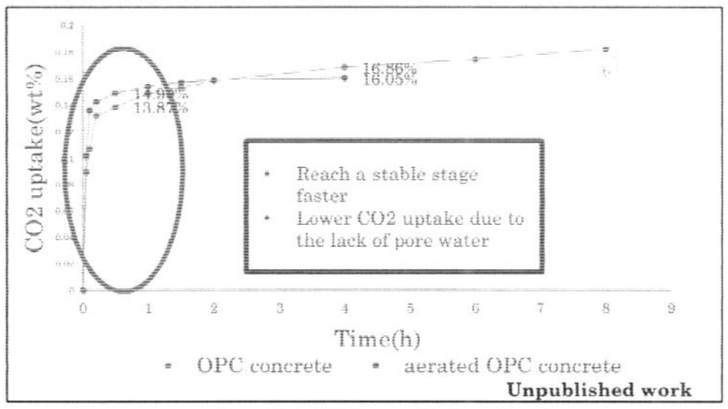

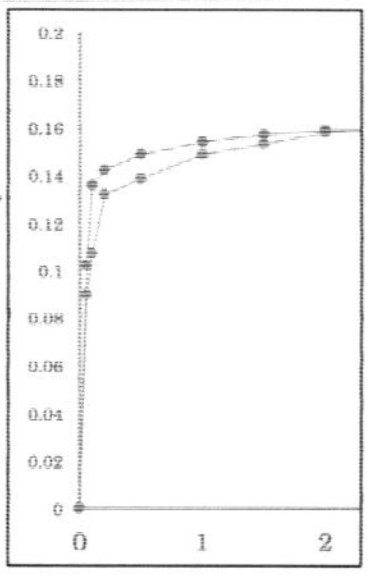

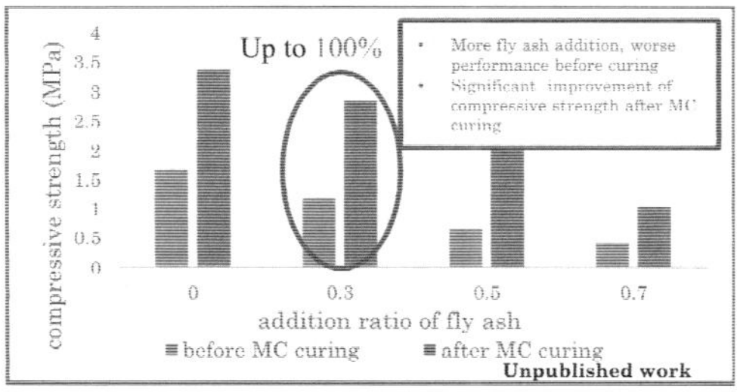

Modified curing condition:

- w/b ratio:0.35-0.4
- Addition ratio of fly ash:no more than 0.5
- Curing time:1-1.5h

DEMONSTRATION PROJECT

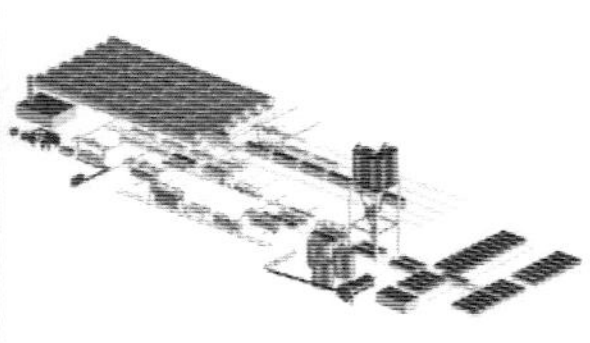

From:
- Laboratory
- small specimen
- OPC paste

To :
- Industry
- Standard product
- Concrete made with industrial waste

(e.g. OPC,FA,slag,red mug)

1000 tons/year LWC MC curing demonstration project with Huanyu company in Henan province

10000tons/year concrete MC curing demonstration project with Aluminum Corporation of China

ACKNOWLEDGEMENT

This study is supported by the National Key Projects for Fundamental Research and Development of China (2016YFB0600904).

THANK YOU FOR LISTENING !

ANY QUESTIONS?

vito

Accelerated carbonation of stainless steel slag compacts at low pCO_2 pressure – microstructure and strength development as a function of temperature and CO_2 content

P. Nielsen, L. Horckmans, R. Snellings, M. Quaghebeur
Newcastle, March 11-14 2018, ACEME

vito

CO_2 UTILIZATION AND MINERAL CARBONATION

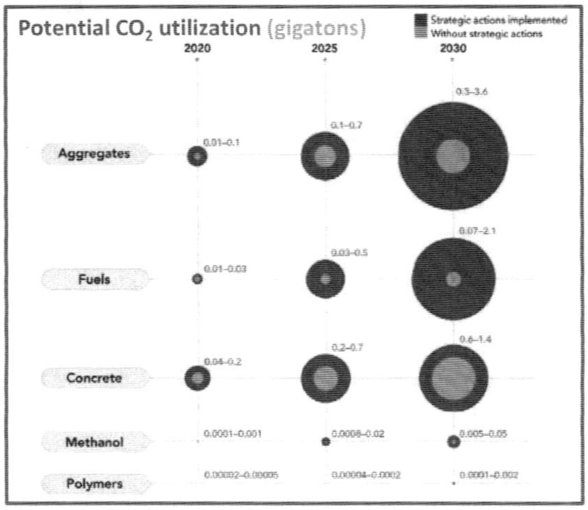

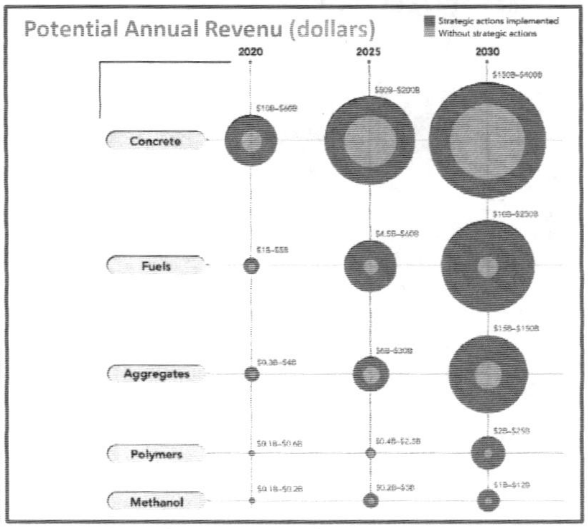

Source: Global Roadmap for Implementing CO_2 Utilization (CO_2 Sciences and The Global CO_2 Initiative)

2

vito

CCU AND BUILDING MATERIALS: SIGNIFICANT PROGRESS

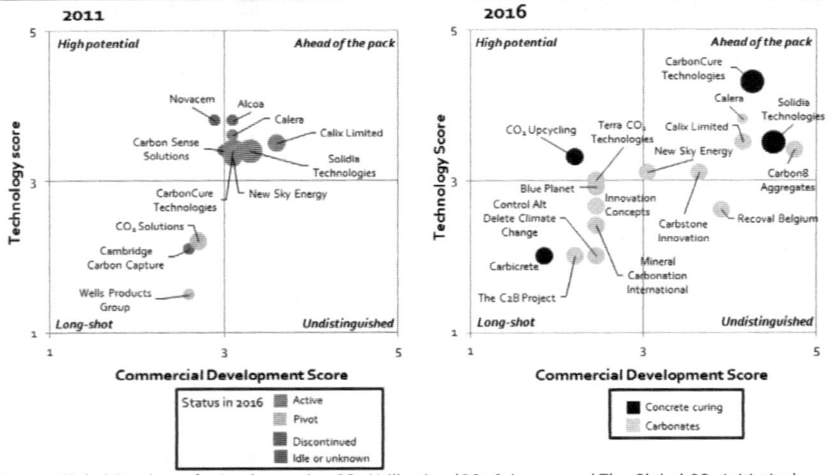

Source: Global Roadmap for Implementing CO_2 Utilization (CO_2 Sciences and The Global CO_2 Initiative)

3

vito

CARBSTONE PROCESS: WASTE-TO-PRODUCT

- Optimal recuperation of metal
- Optimal reuse of fines
- Current restrictions reuse

- Strong materials
- No binders/additives
- Carbonates as cement

4

INTRODUCTION: SCHEMATIC OVERVIEW OF THE CARBONATION PROCESS

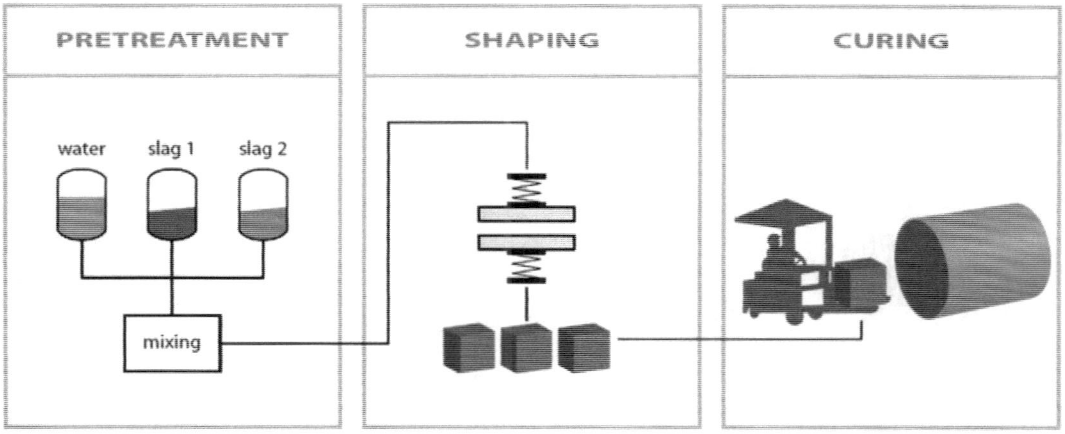

5

COMPACTION STEP AT INDUSTRIAL CARBONATION PLANT: COMPENSATIESTEEN

6

vito

AIM OF THE STUDY: OPTIMIZATION OF STRENGTH VERSUS CO_2-UPTAKE

- In case of carbon sequestration the aim is finding the conditions in which the CO_2-uptake rate and degree is optimized.
- When using CO_2 for the production of construction materials (with a low CO_2 footprint), the compressive strength of the material is generally a key parameter

7

vito

MATERIALS AND METHODS

- Starting material: stainless steel slags (mineral fraction after demetalisation)
  - Mixture of fine grained and sand fraction (laser diffraction/sieving): d10=14 μm, d50=245 μm, d90=1040 μm
  - BET = 3.47 m^2/g
  - Chemical composition (XRF): CaO 41%, MgO 9.8%, Al_2O_3 3.5%, SiO_2: 26%,
  - Mineralogical composition (XRD): Calcio-olivine (γ-Ca_2SiO_4), merwinite ($Ca_3Mg(SiO_4)_2$, bredigite ($Ca_7Mg(SiO_4)_4$, cuspidine ($Ca_4Si_2O_7(F,OH)_2$), gehlenite ($Ca_2Al_2SiO_7$), spinel ($(Mg(Cr, Al)_2O_4$), portlandite ($Ca(OH)_2$), periclase (MgO)
  - Moisture content: 4 to 10wt%

- Compaction pressure: 150 kgf/cm2 (14,7 MPa)

- CO_2-curing in curing chamber or autoclave
  - Carbonation time: 2 to 156 h
  - Temperature: 10, 20 and 60°C
  - Gas feed pressure: 0,05 MPa (0,5 bar)
  - CO_2-concentration: 5, 17, 33, 100% (CO_2-N_2 gas mixture)

14/06/2018

8

COMPRESSIVE STRENGTH AS A FUNCTION OF MOISTURE CONTENT

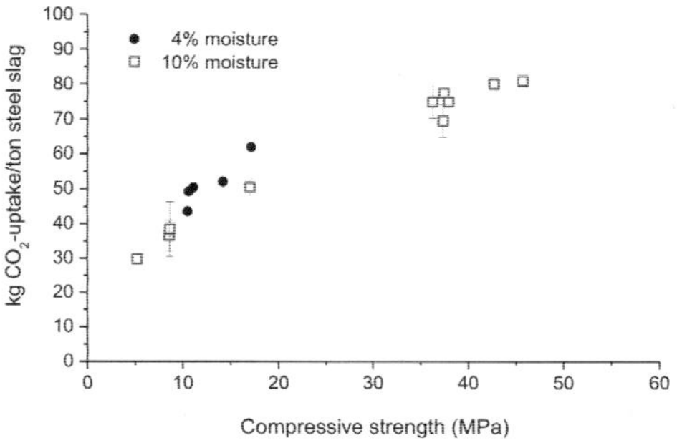

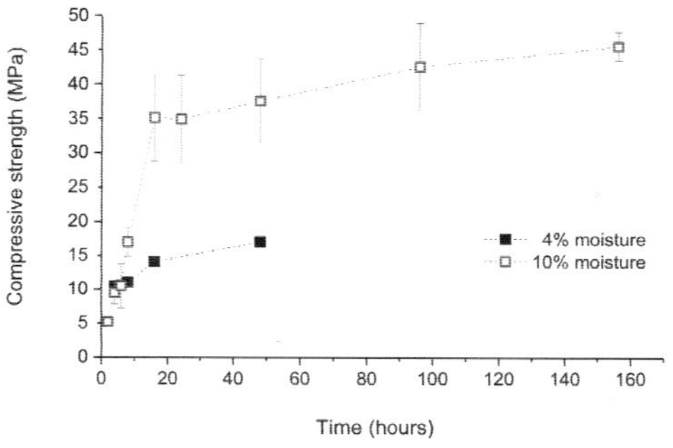

(Carbonation at 17% CO_2 and 20°C)

COMPRESSIVE STRENGTH AS A FUNCTION OF MOISTURE CONTENT

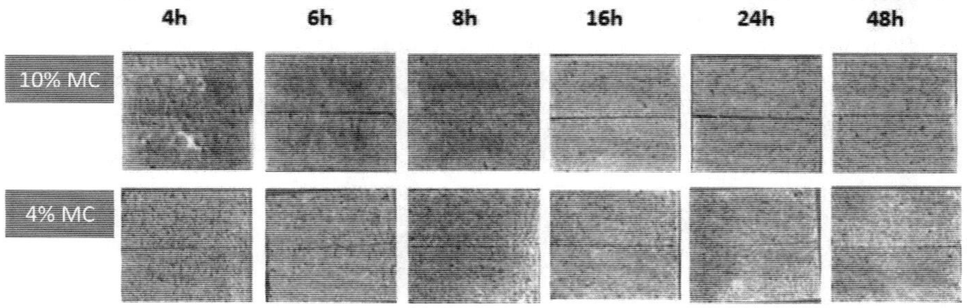

- Evaluation of carbonation front based on phenolphtalein indicator.
- For compacts with high moisture content (10%) a carbonation penetration front was observed after 4, 6 and 8 h. No front for compacts made at low moisture content.

vito

COMPRESSIVE STRENGTH AS A FUNCTION OF MOISTURE CONTENT

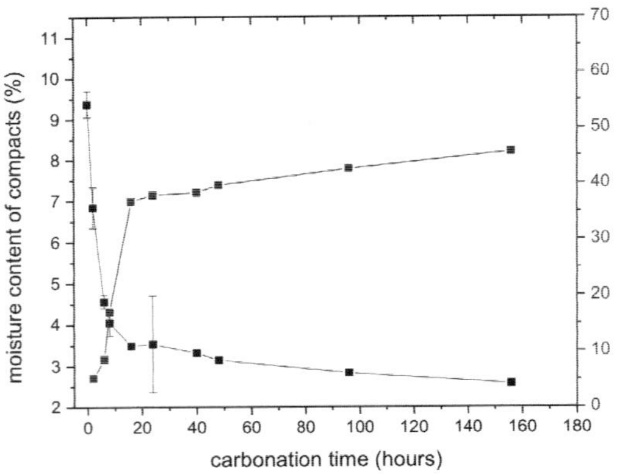

- Mechanisms: decrease in moisture content due to evaporation of water and by the reduction in pore volume, which may push some of the water out of the compacts.
- For compacts made with high moisture content (10%) good correlation between carbonation front and moisture content development
- For compacts made with low moisture content (4%) the loss in moisture content due to evaporation results in too little water to allow carbonation.

14/06/2018
©VITO – Not for distribution

11

vito

COMPRESSIVE STRENGTH AS A FUNCTION OF TEMPERATURE

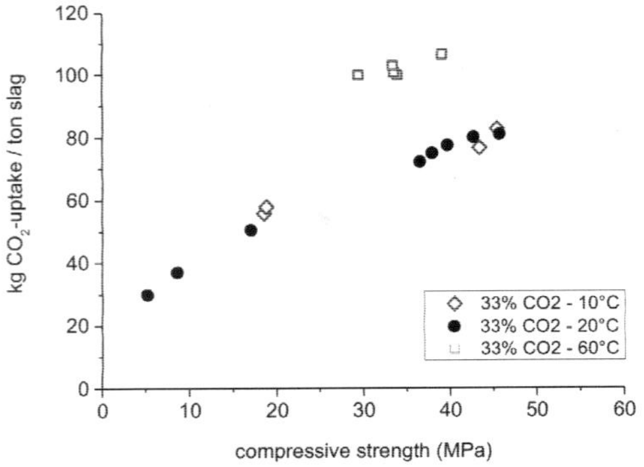

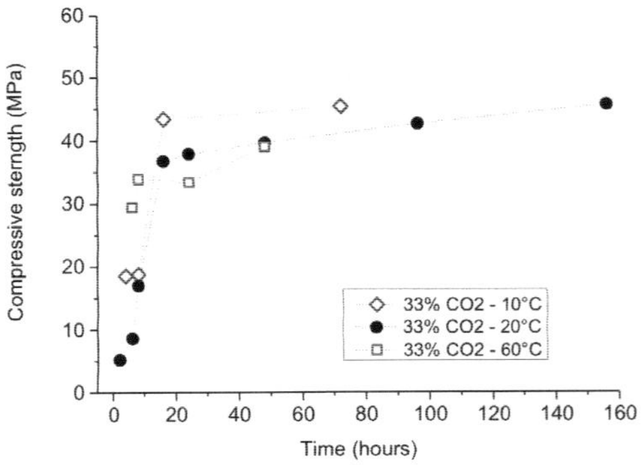

14/06/2018
©VITO – Not for distribution

12

vito

MICROSTRUCTURE: INFLUENCE OF TEMPERATURE

33% CO_2 – 20°C – 16 h

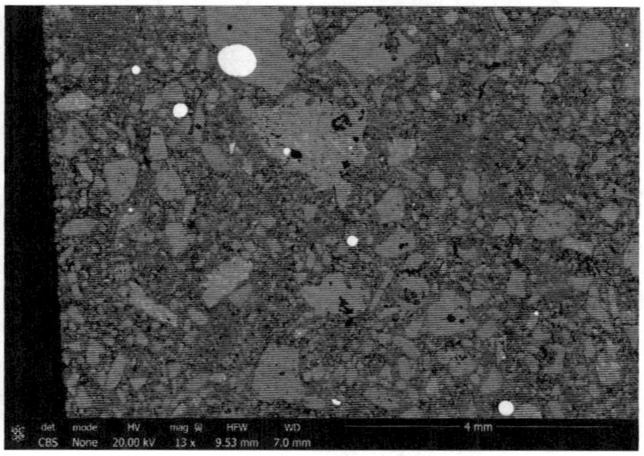

33% CO_2 – 60°C – 16 h

vito

MICROSTRUCTURE: INFLUENCE OF TEMPERATURE

33% CO_2 – 20°C – 16 h

33% CO_2 – 60°C – 16 h

HIGH TEMPERATURE (60-80°C)

High silicate dissolution/leaching rate:
$Ca_2SiO_4 + 2H_2O \rightarrow 2Ca^{2+} + H_2SiO_4^{2-} + 2OH^-$

Low CO_2 adsorption/dissolution rate:
$CO_2 + H_2O \rightarrow H_2CO_3$
$H_2CO_3 + OH^- \rightarrow HCO_3^- + H_2O$
$HCO_3^- + OH^- \rightarrow CO_3^{2-} + H_2O$

Carbonation precipitation
$CO_3^{2-} + Ca^{2+} \rightarrow CaCO_3$
rate determined by CO_2 dissolution
→ precipitation near source of CO_3^{2-}

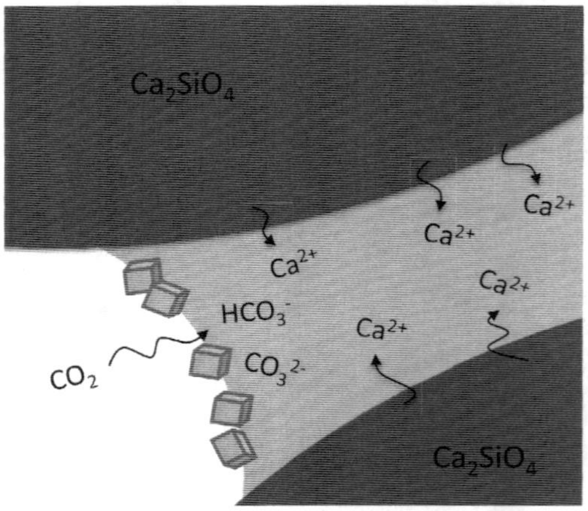

vito

LOW TEMPERATURE (20°C)

Low silicate dissolution/leaching rate:
$Ca_2SiO_4 + 2H_2O \rightarrow 2Ca^{2+} + H_2SiO_4^{2-} + 2OH^-$

High CO_2 adsorption/dissolution rate:
$CO_2 + H_2O \rightarrow H_2CO_3$
$H_2CO_3 + OH^- \rightarrow HCO_3^- + H_2O$
$HCO_3^- + OH^- \rightarrow CO_3^{2-} + H_2O$

Carbonation precipitation
$CO_3^{2-} + Ca^{2+} \rightarrow CaCO_3$
rate determined by Ca^{2+} dissolution
→ precipitation on surface of Ca-silicates

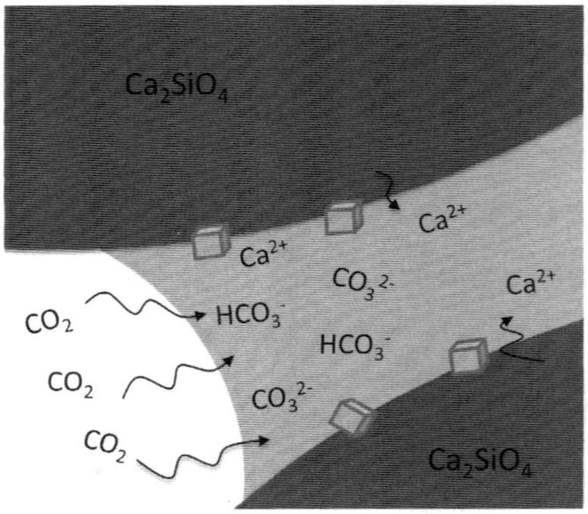

vito

vito

COMPRESSIVE STRENGTH AS A FUNCTION OF CO_2-CONCENTRATION

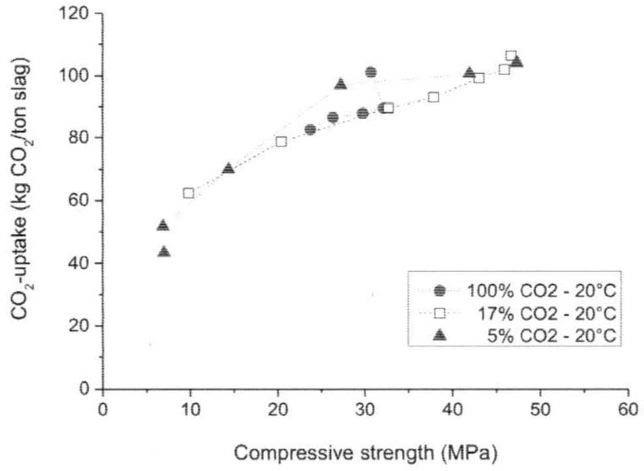

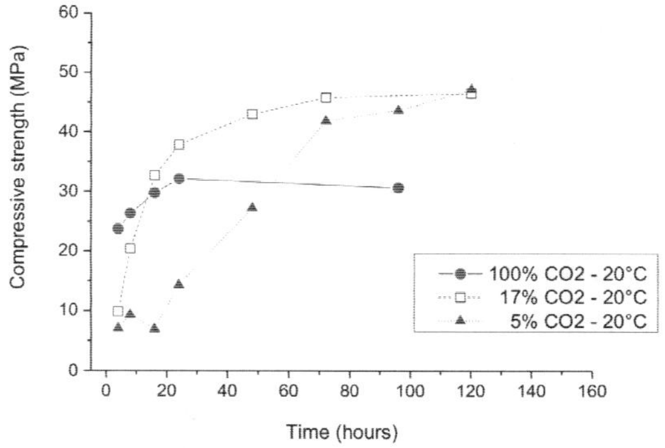

vito

CONCLUSIONS

- In this study both moisture content, temperature and $\%CO_2$ had an impact on the compressive strength development of compacts made from steel slags during carbonation at low pressure.
- CO_2 uptake and strength development do not allows correlate!
- At higher temperature (60°C) microporeus carbonate is precipitated that results in a lower compressive strength for a certain amount of total CO_2-uptake compared to compacts made at lower temperature.
- Microstructure of carbonate precipitation and the rate limiting step during carbonation are important aspects to consider!

Contact: mieke.quaghebeur@vito.be

19

^{29}Si solid state MAS NMR study on the fate of silicon in mineral carbonation processes: A journey from mining to end products

Emad Benhelal[a], M.I.Rashid[a], J. Hook[b], M.Rayson[a,c], G.Brent[c], M.Stockenhuber[a], E.M.Kennedy[a]

[a)] Department of Chemical Engineering, The University of Newcastle
[b)] Nuclear Magnetic Resonance Facility, The University of New South Wales
[c)] Orica Australia Pty Ltd

THE UNIVERSITY OF NEWCASTLE AUSTRALIA

ACEME Conference, Newcastle, Australia March 2018

Mineral Carbonation International

Presentation Overview

- Background
- Aims
- Materials Analysed
- ^{29}Si MAS NMR spectroscopy
- Results and discussion

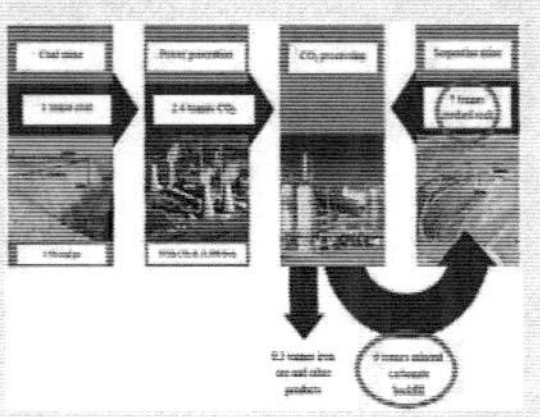

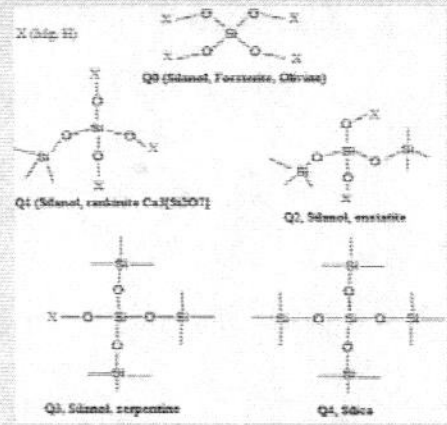

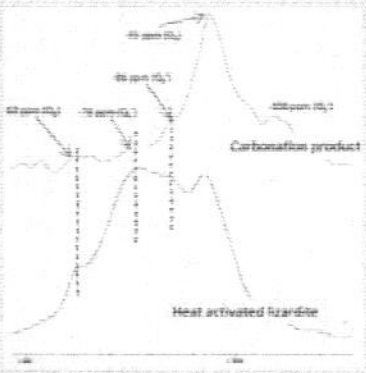

2

Background

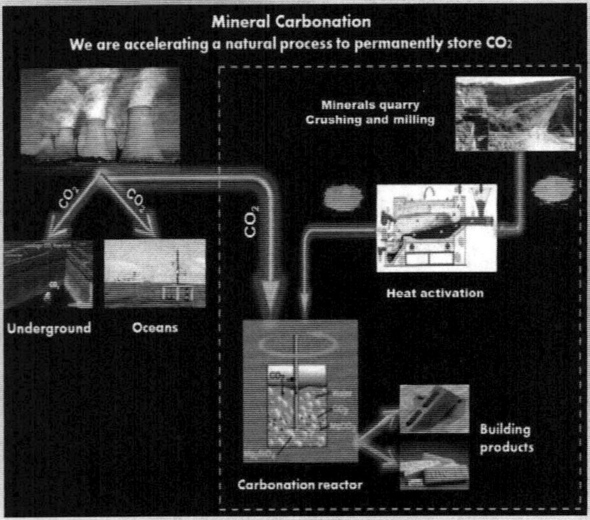

Simple schematic of aqueous mineral carbonation process

The primary challenges MC faces:

1- High capital costs and energy demand;

2- Additional CO_2 emissions; and

3- Accumulation of by-products.

Utilizing the by-products can offset these challenges.

Among by-products, silica enriched residue is more promising which can partially replace cement in concrete.

Understanding the fate of silicon during carbonation processes is essential.

3

Single vs two stage aqueous mineral carbonation

Single stage (ARC process)

- High PCO_2 is required
- Both extraction of Mg and precipitation of $MgCO_3$ occurs in one vessel.
- By-product is a mixture of mainly $MgCO_3$, SiO_2 and unreacted serpentine
- It is difficult to separate.

Two stage

- Much lower PCO_2 is needed.
- Extraction of Mg and precipitation of carbonate occur in two reactors.
- Solid residue of the first stage is mainly amorphous silica and unreacted lizardite.
- Products are relatively pure.

4

Aims

- Due to the importance of **silica** in mineral carbonation process, understanding its fate is very critical.
- ^{29}Si MAS NMR provides information about the **chemical environment of silicon** independent of the particle size and the level of crystallinity.
- The aim is to **characterise** feed and by-products of MC processes using **^{29}Si solid state MAS NMR**.

5

Materials Analysed

(1) Raw lizardite	(2) Heat activated lizardite	(3) Carbonation product	(4) Dissolution residue	(5) ATSER
Crushed lizardite from Great Serpentine Belt, NSW Australia was wet milled in 2 L S.S ball mill and pass through 75 μm sieve.	Dried milled lizardite was heat activated in a S.S rotary kiln at 630 °C, 4 h.	Heat activated lizardite was (15 wt% solid Conc.) carbonated at 150 bar CO_2, 150 °C, 5 h. Product slurry was rinsed with distilled water and dried at 110 °C, 24 h.	Heat activated lizardite was dissolved in carbonic acid at 6.5 bar CO_2, 45 °C, 2 h. Solid residue was dried. A part of solid residue was milled and redissolved. Solid residue from this stage dried and named silica enriched residue (SER).	A part of SER was dissolved in 2 M nitric acid (0.5 wt% solid conc.) for 7 h. Solid residue was dried and named "acid treated silica enriched residue (ATSER)

6

^{29}Si MAS NMR spectroscopy

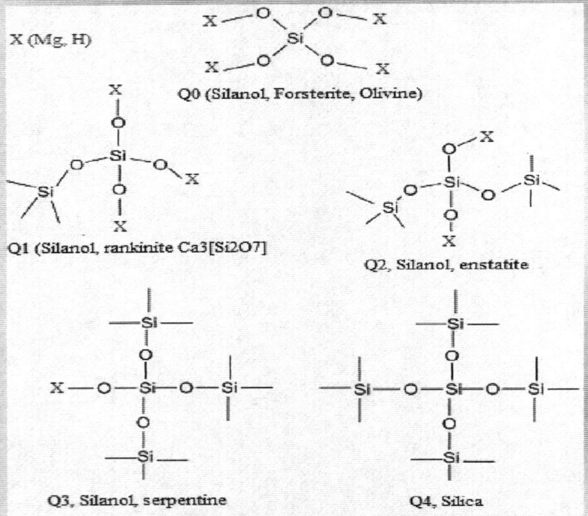

Q_n species in our study

- Technique to investigate the local **chemical environment of silicon** in feed and by-products of MC processes.
- Each degree of polimerisation of silica (Q_n) shows a **peak** at certain ppm **(chemical shift)** with respect to standard (TMS tetramethylsilane).
- *n* refers to the **number of bridging oxygens** that link a tetrahedron to other tetrahedrons

7

Q_n VS Q_n'

Interconnectivity of the $[SiO4]^{4-}$ tetrahedra for **disrupted tetrahedral silica** differs from crystalline materials denoted as *Q_n'*

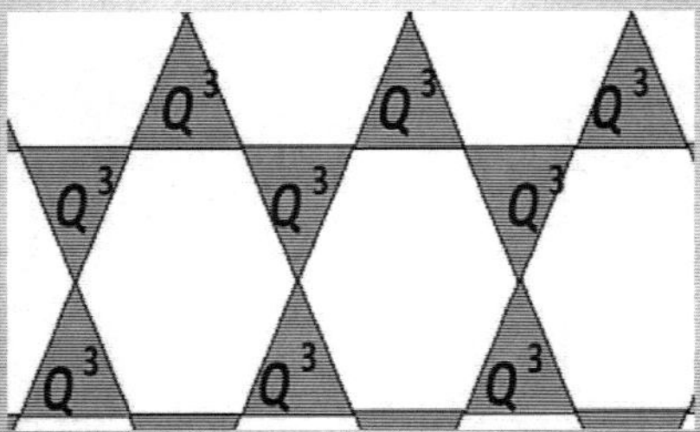

chemical shifts in crystalline minerals are **well established**, but in **intermediate and amorphous phases** are not.

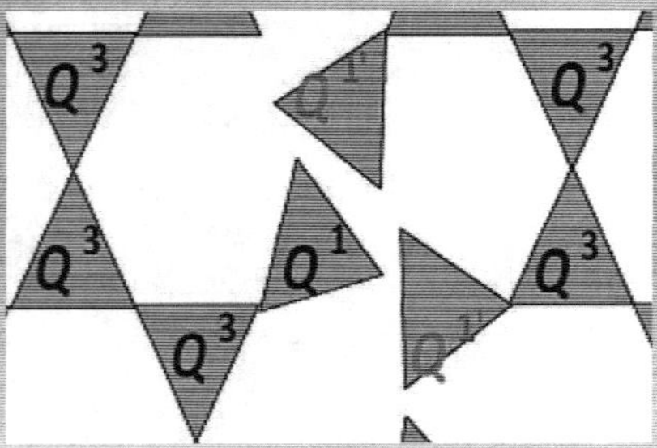

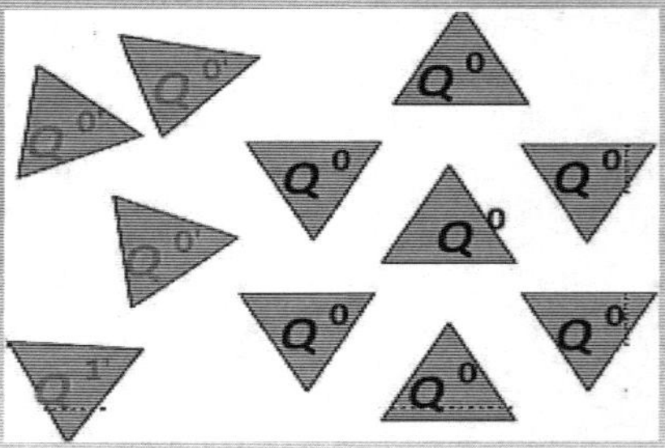

8

Raw lizardite

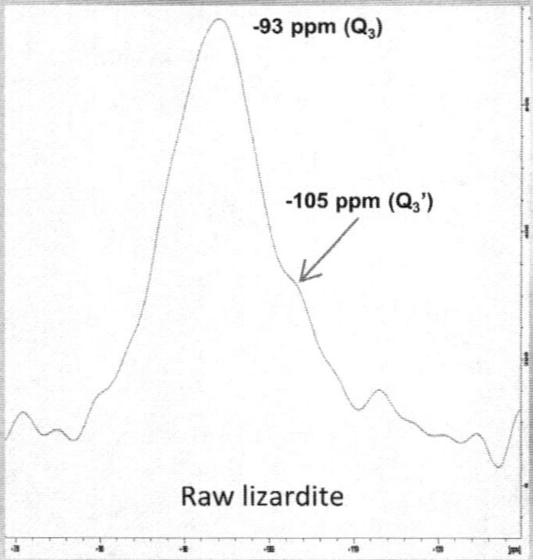

Q_3 accounts for approximately **90%** (relative peak area) of total silica

Attributed to **silica sheets** in serpentinite (Kosuge et al., 1995; Mantovani et al., 2009; Balucan 2013).

What is Q_3'?

It contains 7% of total silica

Indicating the presence of **disrupted silica sheets** possibly due to size reduction by milling to <75 μm (Drief and Nieto, 1999).

9

Heat activated lizardite

Q_0 (6%) attributed to the formation of **forsterite** (MacKenzie and Meinhold, 1994; Balucan, 2013) due to heat activation of raw lizardite.

Q_1' and Q_2' (66 % of total silica) attributed to formation of **intermediate phases**, disrupted silica sheets of raw lizardite (Kirkpatrick, 1988; MacKenzie and Meinhold, 1994; Chizmeshya et al., 2006; Balucan, 2013).

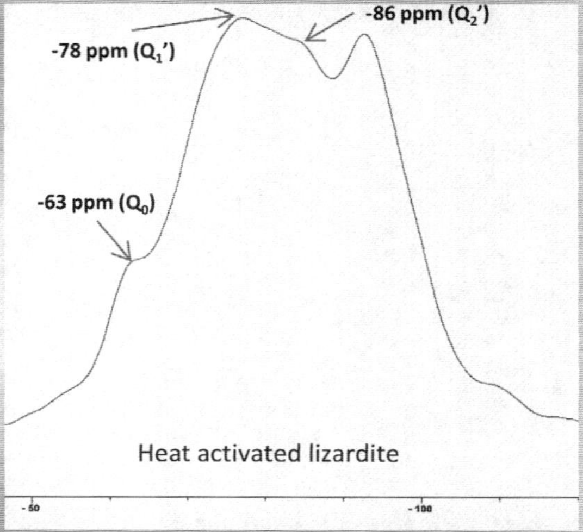

10

Heat activated lizardite

Q_3 (26 % of the silica) attributed to silica sheet structure similar to raw lizardite.

XRD and TGA analysis showed that almost **5 wt% of raw lizardite** remained after heat activation.

Then what is it?

A **talc like structure** or **"α" phase** formed during heat activation contains silica structure of Q_3 similar to silica sheet structure of raw lizardite (Whilton et al., 1998; MacKenzie and Meinhold, 1994; Balucan, 2013).

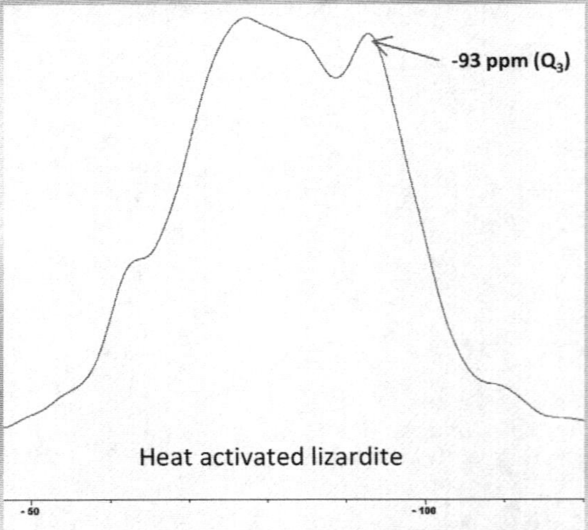

11

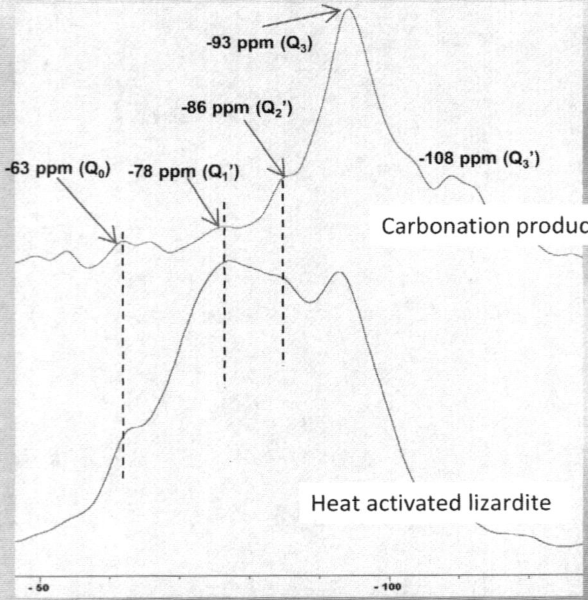

Heat activated lizardite VS carbonation product

Q_0 = **forsterite** remaining after carbonation.

The majority of the intermediate phases are absent. **The intermediate phases are the most reactive components**

Reactivity of Q_1' intermediate phase is **greater** than Q_2' intermediate phase.

Q_3 increased in intensity **(26% to over 50%)**, suggested to be due to **precipitation** of phase/s having Q_3 structure of silica.

Other phases i.e. Q_3' **precipitated** during carbonation.

12

SEM image of carbonated product (Single stage ARC process)

13

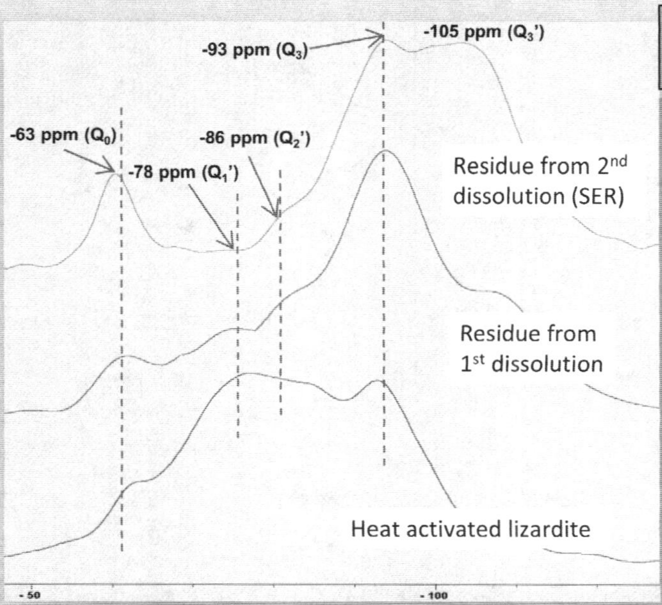

Heat activated lizardite VS dissolution residue

Forsterite (Q_0) did not dissolve in weak acid

Q_1' and Q_2' (intermediates) phases consumed even in weak acid (CO_2+water)

Reactivity of Q_1' intermediate phase is greater than Q_2' intermediate phase.

Q_3 did not dissolve

Q_3' is formed (precipitated) even after 2 h dissolution in weak acid (CO_2+water)

Q_3' increased in quantity after regrinding and re-dissolution

14

NMR showed silica precipitation (Q_3') during dissolution but ICP results show no reduction in Mg and Si concentration.

Then, simultaneous dissolution-precipitation;

Mg extraction does not exceed 75%, even after regrinding; WHY???

Because Q_3 (mainly talc like structure) is the tough guy.

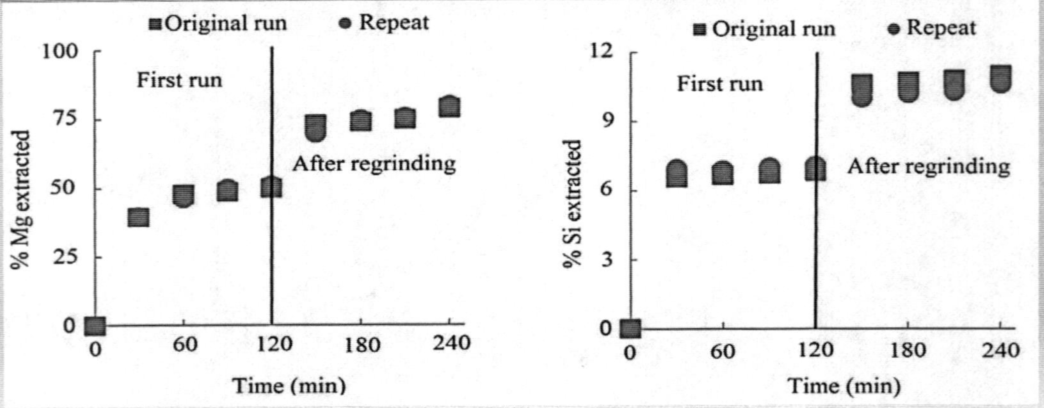

15

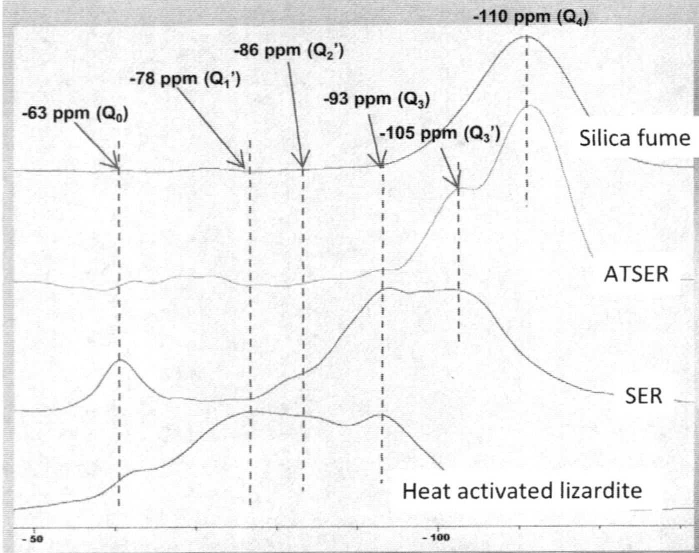

Heat activated lizardite, SER and ATSER

Dissolution in 2 M nitric acid, consumed forsterite and talc like structure (tough guy).

Dominant phases in ATSER are silica (Q_4), 65% and Q_3' 35%

Since ATSER is Mg depleted, Q_3' is silanol groups

ATSER is similar to silica fume

16

Conclusions

- ^{29}Si MAS NMR is a useful technique to study **chemical environment of silicon** in feed and by-products of MC processes;
- The most reactive phases of heat activated lizardite are **intermediate phases Q_1' and Q_2'**;
- Talc like structure is **X-ray amorphous** but it is **not reactive** under tested conditions similar to forsterite;
- The ultimate goal of heat activation should be **producing intermediate phases** and avoiding talc like structure, **possible ???**

17

Conclusions

- In both single and two stage processes, dissolved silicon is precipitated as amorphous **magnesium silicate (Q_3')**;
- **Pure silica** is not produced unless using strong acid;
- **ATSER** is the most similar by-product to silica fume, expecting to have similar **pozzolanic** activity;

18

Acknowledgment

- The University of Newcastle for Postgraduate research scholarships.
- Mineral Carbonation international (MCi).
- EMX (The University of Newcastle) for XRD and SEM analyses.

THE UNIVERSITY OF NEWCASTLE AUSTRALIA

Mineral Carbonation International

19

Dissolution mechanism of heat activated serpentine and the re-precipitation of silica under constant pH conditions

Presenter: Dr Faezeh (Fay) Farhang*[1]
Co-authors: Dr M. Rayson[2], Dr G. Brent[2], Prof. M. Stockenhuber[1], Prof E. Kennedy[1]

1- School of Chemical Engineering, The University of Newcastle, Australia

2-Orica Ltd, Kurri Kurri, NSW 2327, Australia

12 March 2018

Overview:

2

a) Introduction
1. Material Preparation
2. Experimental set-up

b) Results
1. Effect of Solid/Liquid ratio (S/L)
2. Effect of pH
3. Effect of temperature

c) Summary

Material: Material Preparation

3

- **Serpentine (Lizardite)** from the Great Serpentine Belt, NSW, Australia
- Crushed, milled and passed a 75 μm sieve (d_{80} of 45 μm)
- Powdered lizardite was heat activated in a rotary kiln at 630 °C for 3 h

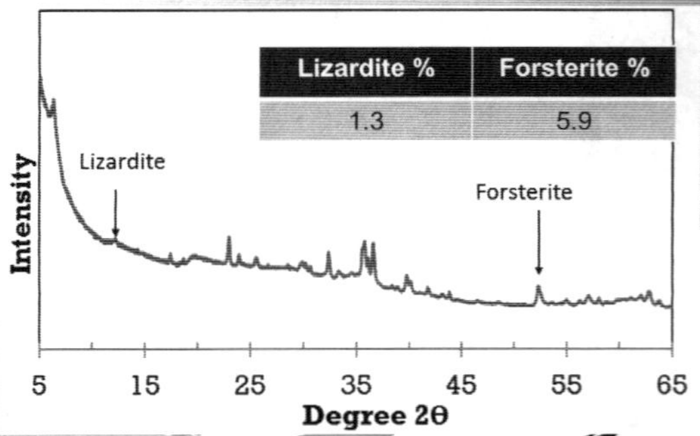

Lizardite %	Forsterite %
1.3	5.9

THE UNIVERSITY OF NEWCASTLE AUSTRALIA

Experimental: Dissolution set-up

4

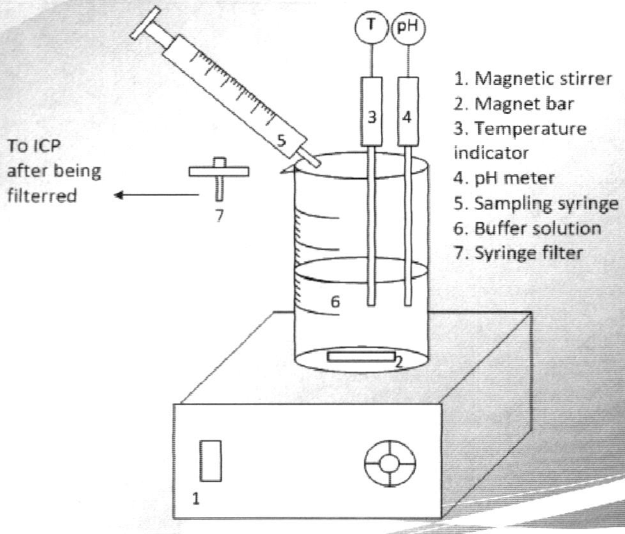

- 200 mL of buffer solution
- 0.4 – 8 g solid (HA lizardite)
- Test duration:7 h
- Sampling (1 mL): over 7 h
- Temp: 25, 45 and 60 °C

THE UNIVERSITY OF NEWCASTLE AUSTRALIA

Experimental: Buffer solutions

5

Buffer	pH	Acid/Base [M]	Conjugate salt [M]
1	1.2	100 mL Na_2SO_4 [0.98]	100 mL $NaHSO_4$ [0.98]
2	3.1	88 mL HCl [0.1]	112 mL $C_8H_5KO_4$ [0.1]
3	4.6	100 mL CH_3COOH [2]	100 mL $C_2H_3NaO_2$ [2]
4	6.1	100 mL CH_3COOH [0.1]	100 mL $C_2H_3NaO_2$ [2]
5	9.8	18 mL NaOH [0.1] +82 mL water	100 mL Borax [0.025]

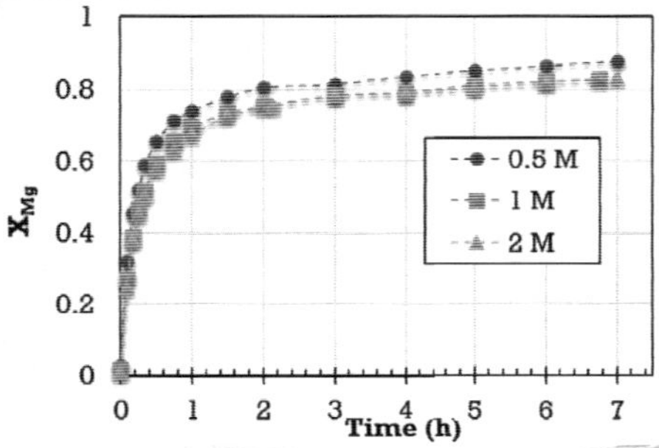

- Acetate does not enhance the dissolution rate of serpentine

Buffer	pH	Acid/Base [M]	Conjugate salt [M]
1	4.6	100 mL CH_3COOH [1]	100 mL $C_2H_3NaO_2$ [1]
2	4.6	100 mL CH_3COOH [2]	100 mL $C_2H_3NaO_2$ [2]
3	4.6	100 mL CH_3COOH [4]	100 mL $C_2H_3NaO_2$ [4]

THE UNIVERSITY OF NEWCASTLE AUSTRALIA

Experimental: Variables studied

6

Variables studied in this work for their effect on the dissolution kinetics of heat activated lizardite:

- Solid to Liquid ratio (S/L)
- pH
- Temperature

THE UNIVERSITY OF NEWCASTLE AUSTRALIA

Results: Effect of solid concentration (S/L)

7

Experimental condition: t= (23±2 °C) & pH 4.6

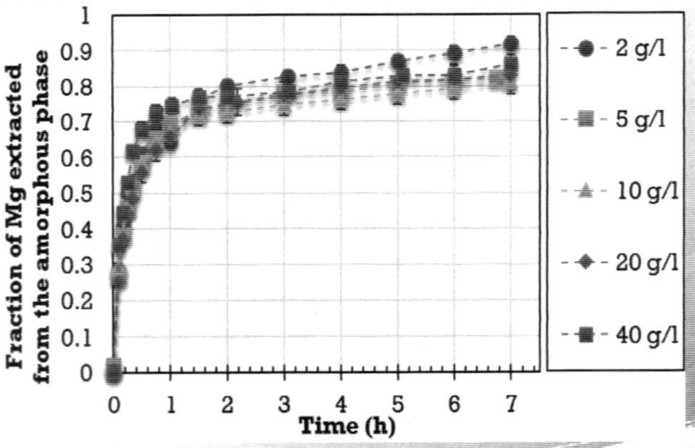

- Solid concentration did not have a significant impact on the extent of Mg extraction.

THE UNIVERSITY OF NEWCASTLE AUSTRALIA

Results: Effect of solid concentration (S/L)

8

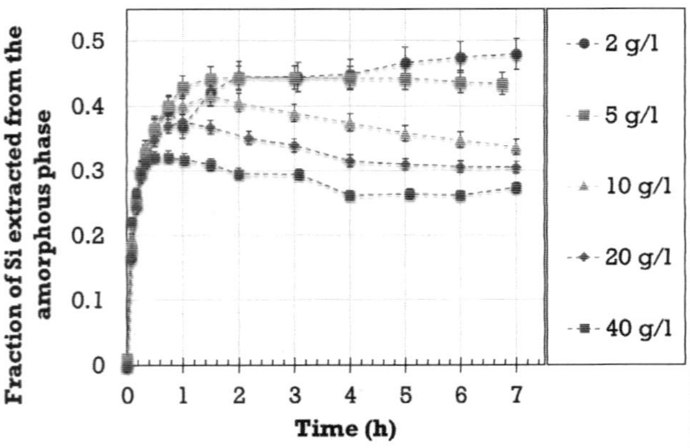

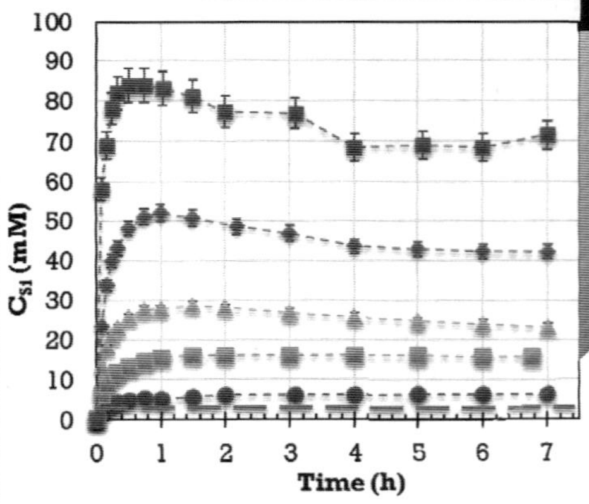

— — — saturation concentration of amorphous silica

- It is hypothesized that the decreasing trend in Si concentrations at higher S/L is due to precipitation of amorphous silica

THE UNIVERSITY OF NEWCASTLE AUSTRALIA

Results: Effect of pH

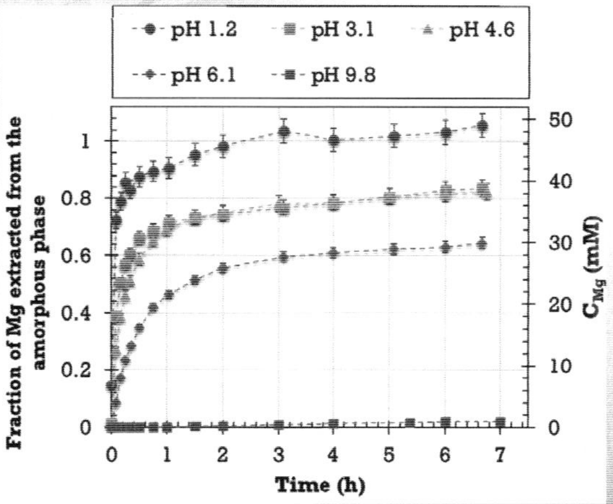

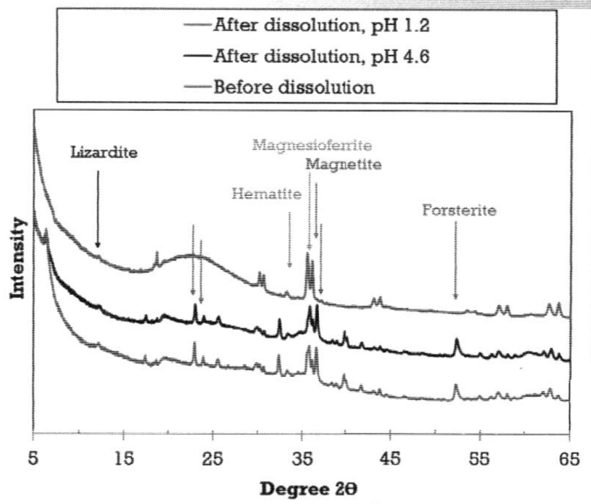

9

- At very low pH (i.e. 1.2) not even all the amorphous phase was dissolved but also the crystalline phases (forsterite and lizardite) were dissolved.

THE UNIVERSITY OF NEWCASTLE AUSTRALIA

Results: Effect of pH

10

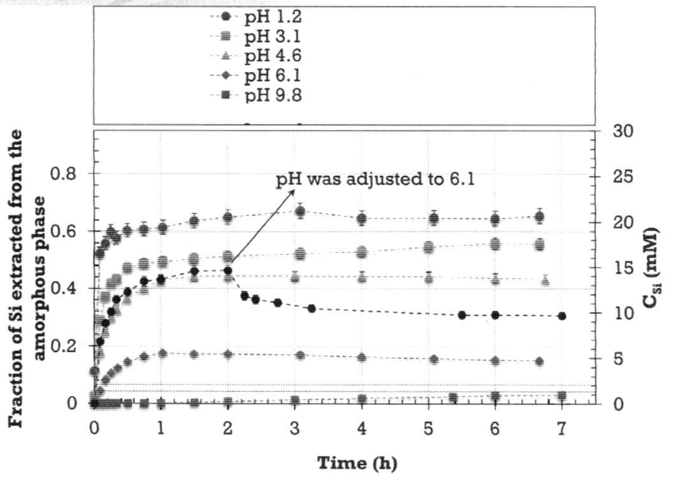

The blue and red dotted lines show the amorphous silica saturation point at pH < 9 and pH 10, respectively.

- By reducing the pH, the rate of silica polymerization is reduced and, therefore, silica precipitation is reduced.
- Drop in Si concentration after the pH was adjusted to 6.1 is another indication that amorphous silica precipitated in the solution.

THE UNIVERSITY OF NEWCASTLE AUSTRALIA

Results: Effect of S/L at pH 6.1

11

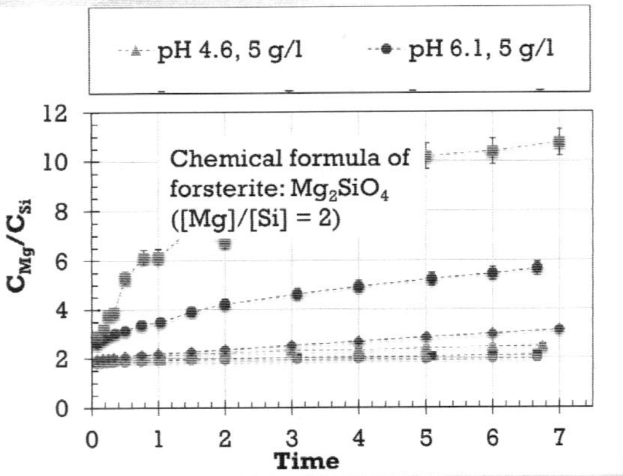

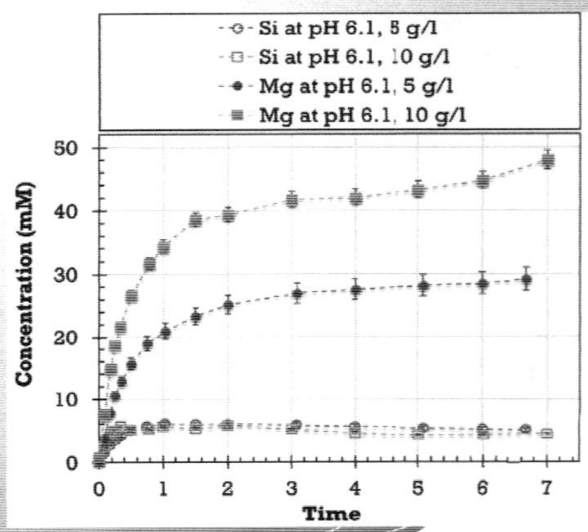

- Despite increasing the S/L ratio, the observed silica concentrations in solution do not increase and the Mg/Si ratio increases dramatically.
- At pH 4.6 the Mg/Si ratio is almost independent of the S/L ratio

THE UNIVERSITY OF NEWCASTLE AUSTRALIA

Results: Effect of Temperature

12

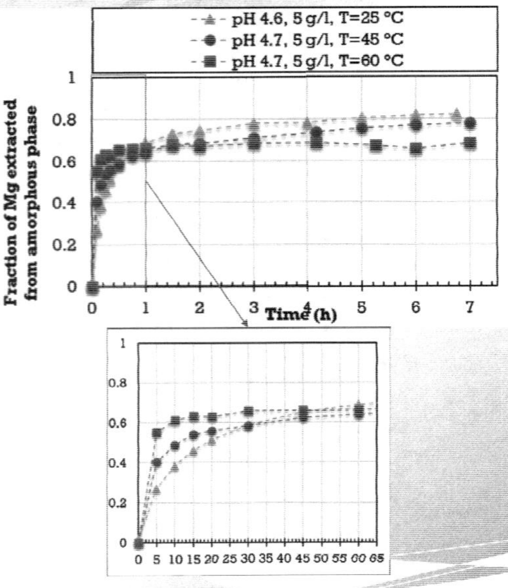

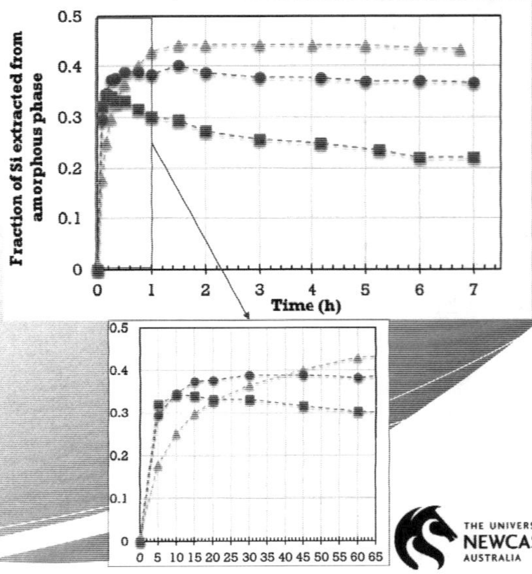

THE UNIVERSITY OF NEWCASTLE AUSTRALIA

Supporting Experiment: Precipitation of silica on the surface of an inert substrate

13

Non-porous alumina spheres

Soft & smooth surface

Alumina spheres added to the dissolution cell (pH 4.6) in the presence of HA serpentine

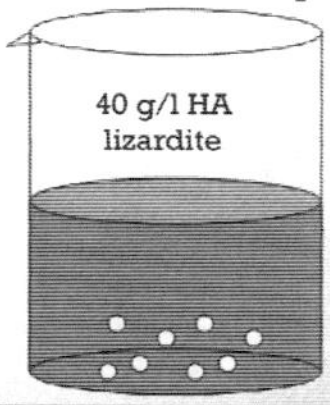

Rough surface indicating deposit on the surface

Alumina spheres added to the buffer solution in the absence of HA serpentine

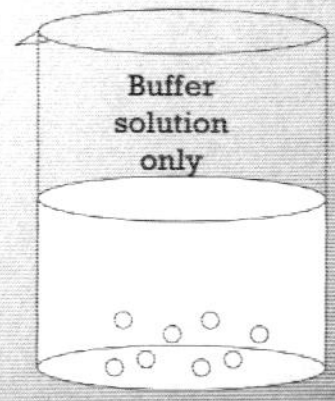

Soft & smooth similar to original

THE UNIVERSITY OF NEWCASTLE AUSTRALIA

Supporting Experiment: Precipitation of silica on the surface of an inert substrate

14

Before

After

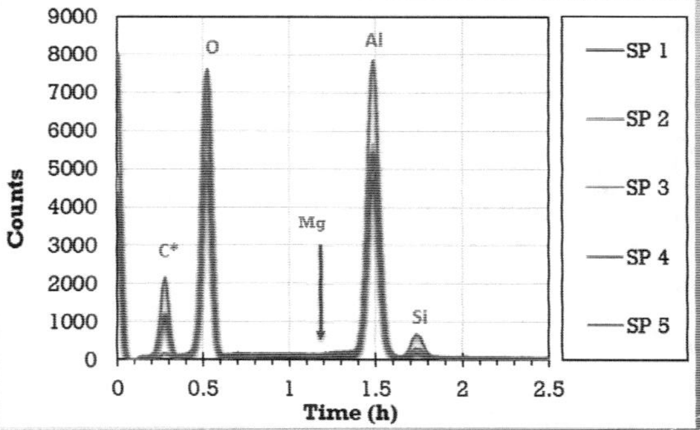

- Only Si was detected at the surface.
- Absence of Mg or Fe at the spots indicates that the deposit is not fragments of dissolved particles and is only silica

THE UNIVERSITY OF NEWCASTLE AUSTRALIA

Supporting Experiment: Identifying polymeric silica in the solution

15

- Molybdosilicate standard method was used to measure the reactive silica
  - ➢ Polymeric silica $-SiO_2-O-SiO_2-O-SiO_2-$
    - Insoluble
    - unreactive
  - ➢ Monomeric silica $(OH)_3Si-O-$
    - Soluble
    - Reactive

Reactive silica (monomeric) + ammonium molybdate → silicomolybdic acid (strongly absorbing yellow complex)

Monomeric silica can be measured by UV-visible using Molybdosilicate method

C_{Si} = 90 ppm

Measured by UV-visible

Total silica = Reactive silica (monomeric) + non-reactive silica (polymeric)

Total silica can be measured by ICP

C_{Si} = 1625 ppm

Measured by ICP

The result of this experiment confirms that in the solution other forms of Si rather than monomeric Si are present.

THE UNIVERSITY OF NEWCASTLE AUSTRALIA

Summary

16

- ➢ S/L did not have a considerable impact on the Mg extraction, however, it had significant effect on the Si concentration.
- ➢ Evidence of precipitation of silica at pH 6.1 and high temperature was found
- ➢ SEM imaging along with EDS analysis provided supporting information on the silica deposits on the surface.
- ➢ The presence of both monomeric and polymeric silica in the filtrate after the dissolution experiment was proved using molybdosilicate standard method.

THE UNIVERSITY OF NEWCASTLE AUSTRALIA

Acknowledgement:

17

- Mineral Carbonation International (MCi) is thanked for funding this work.
- We gratefully acknowledge the facilities, and the scientific and technical assistance, of the Electron Microscope and X-ray unit (EMX) at the University of Newcastle.
- We would like to thank Dr Timothy Oliver for the great discussion on the experimental design, Dr Glenn Bryant for the discussions and data analysis and Ms Jennifer Zobec for XRD analysis and phase identification.

THE UNIVERSITY OF NEWCASTLE AUSTRALIA

18

Thank you for your attention!

THE UNIVERSITY OF NEWCASTLE AUSTRALIA

Results: Effect of S/L at high temperature

19

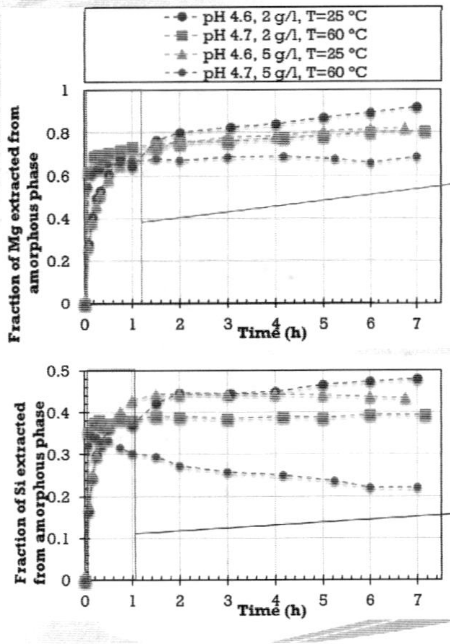

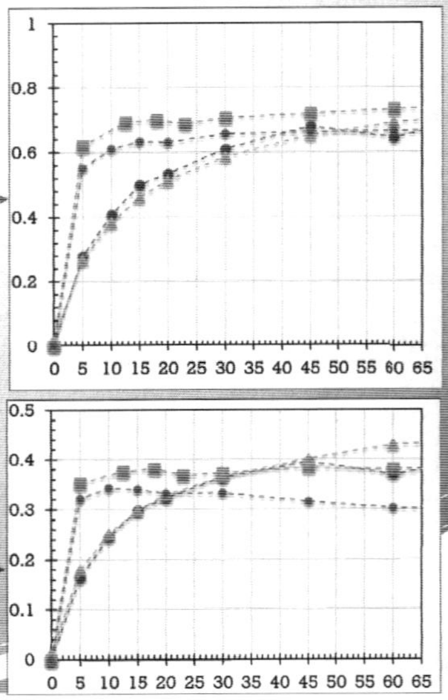

THE UNIVERSITY OF NEWCASTLE AUSTRALIA

Experimental: Dissolution set-up

20

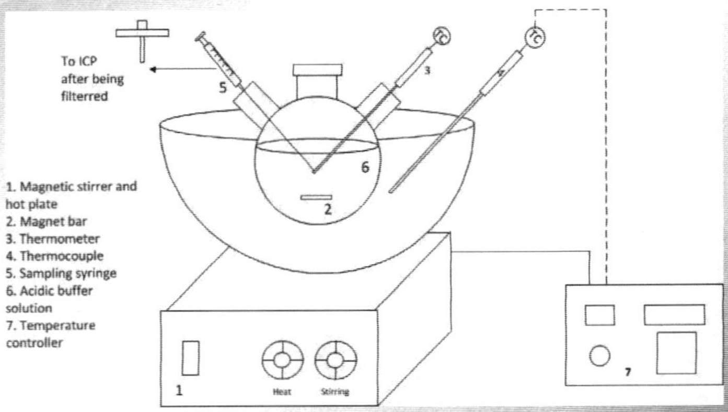

- 200 mL of buffer solution
- 0.4 – 8 g solid (HA lizardite)
- Test duration:7 h
- Sampling (1 mL): over 7 h
- Temp: 25, 45 and 60 °C

THE UNIVERSITY OF NEWCASTLE AUSTRALIA

Kinetic Modelling of CO_2 Degassing from Magnesium Bicarbonate Solutions and Resultant Precipitation

Timothy K. Oliver, Bogdan Z. Dlugogorski, and Eric M. Kennedy

ACEME 2018

11-14 March, 2018, Newcastle

THE UNIVERSITY OF NEWCASTLE AUSTRALIA

Priority Research Centre for Energy

CO_2 degassing?

2

- Controlled removal of CO_2 from solutions
- *Context: Two-stage MC*
- Commonly researched phenomenon (e.g. precipitate struvite through air stripping (Fattah et al., 2010), hardness removal for desalination and water treatment more generally (Cézac et al., 2006; Glade and Al-Rawajfeh, 2008; Mubarak, 1998; Nefzi et al., 2004; Pearson, 1982)
- We explore and model kinetics of degassing combined with precipitation (for MC)
- ***This presentation:*** *Experimental chemical system, Fundamental thermodynamic and kinetic aspects, modelling and experimental results & conclusions*

THE UNIVERSITY OF NEWCASTLE AUSTRALIA

Priority Research Centre for Energy

Chemical system

3

- $MgSO_4$ and $NaHCO_3$ used in preparing Mg bicarbonate solutions (typically 0.1 M, 30 & 50°C)
- Prepared solutions sparged with N_2 (using different bubble generating devices)
- Batch & semibatch experiments (some with carbonic anhydrase (CA))

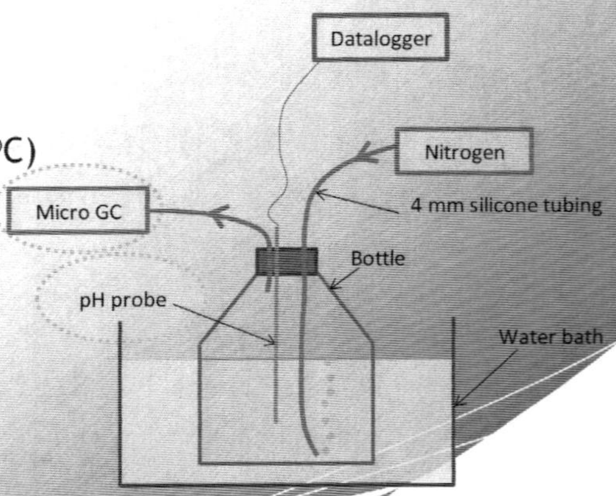

THE UNIVERSITY OF NEWCASTLE AUSTRALIA — Priority Research Centre for Energy

Fundamental thermodynamic aspects

4

- *CO_2 removed from system increases pH of system and increases mineral saturation (=IAP/K)*
- *Quasi-equilibrium enacted*

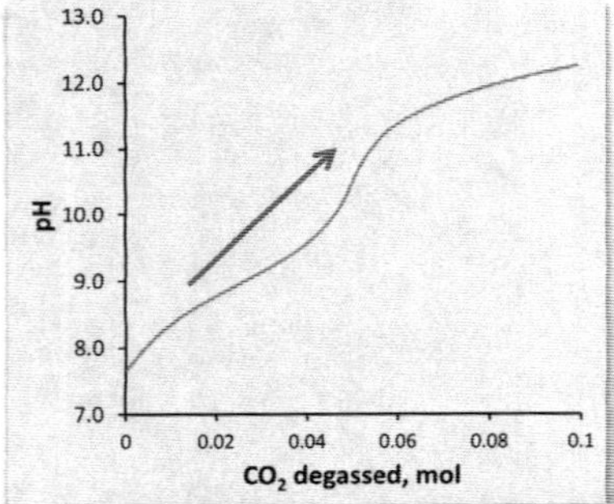

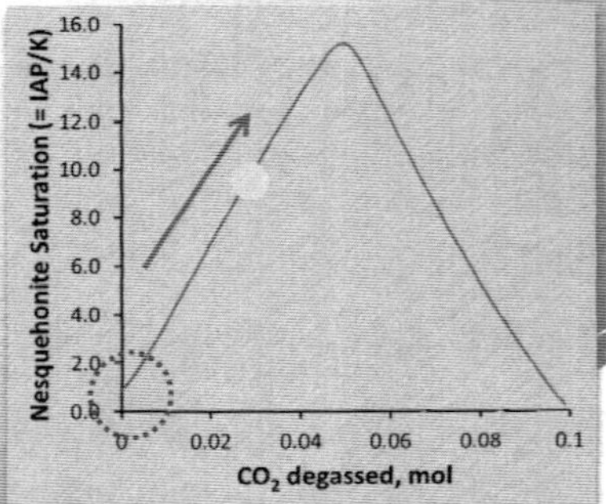

Batch calculations using PHREEQC; 0.1 M $MgSO_4$ and 0.1 M $NaHCO_3$; 30°C

THE UNIVERSITY OF NEWCASTLE AUSTRALIA — Priority Research Centre for Energy

Chemical reactions

5

$$CO_{2(aq)} + H_2O_{(l)} \rightleftharpoons HCO_3^-{}_{(aq)} + H^+{}_{(aq)}$$

$$CO_{2(aq)} + OH^-{}_{(aq)} \rightleftharpoons HCO_3^-{}_{(aq)}$$

$$CO_3^{2-}{}_{(aq)} + H^+{}_{(aq)} \rightleftharpoons HCO_3^-{}_{(aq)}$$

$$HCO_3^-{}_{(aq)} + OH^-{}_{(aq)} \rightleftharpoons CO_3^{2-}{}_{(aq)} + H_2O_{(l)}$$

$$H_2O_{(l)} \rightleftharpoons H^+{}_{(aq)} + OH^-{}_{(aq)}$$

+ Dissociation reactions for $MgOH^+{}_{(aq)}$, $MgCO_{3(aq)}$, $MgHCO_3^+{}_{(aq)}$, $NaHCO_{3(aq)}$, $NaCO_3^-{}_{(aq)}$, $MgSO_{4(aq)}$, $NaSO_4^-{}_{(aq)}$, $HSO_4^-{}_{(aq)}$, $NaOH_{(aq)}$

THE UNIVERSITY OF NEWCASTLE AUSTRALIA

Priority Research Centre for Energy

Precipitation with degassing

6

- Overall reaction applying to system

$$Mg^{2+}{}_{(aq)} + 2OH^-{}_{(aq)} + 2CO_{2\,(g)}$$

$$\uparrow$$

$$Mg^{2+}{}_{(aq)} + 2OH^-{}_{(aq)} + 2CO_{2\,(aq)}$$

$$\uparrow$$

$$Mg^{2+}{}_{(aq)} + 2HCO_3^-{}_{(aq)} + 2H_2O_{(l)} \longrightarrow MgCO_3\cdot 3H_2O_{(s)} + CO_2$$

pH

THE UNIVERSITY OF NEWCASTLE AUSTRALIA

Priority Research Centre for Energy

Kinetic model

7

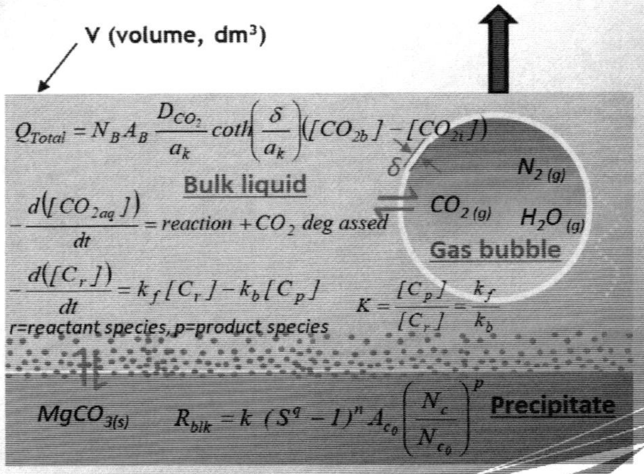

$$Q_{Total} = N_B A_B \frac{D_{CO_2}}{a_k} \coth\left(\frac{\delta}{a_k}\right)([CO_{2b}] - [CO_{2i}])$$

$$-\frac{d([CO_{2aq}])}{dt} = reaction + CO_2\ degassed$$

$$-\frac{d([C_r])}{dt} = k_f[C_r] - k_b[C_p] \qquad K = \frac{[C_p]}{[C_r]} = \frac{k_f}{k_b}$$

$$R_{blk} = k\,(S^q - 1)^n A_{c_0} \left(\frac{N_c}{N_{c_0}}\right)^p$$

Fundamental kinetic aspects

8

- Model formulated by considering the system of chemical reactions acting within the system, G-L-S

1. Modelling degassing as mass transfer

$$Q_{Total} = N_B A_B \frac{D_{CO_2}}{a_k} \coth\left(\frac{\delta}{a_k}\right)([CO_{2b}] - [CO_{2i}])$$

2. Modelling aqueous system

$$-\frac{d(C_r)}{dt} = k_f[C_r] - k_b[C_p] \qquad K = \frac{[C_p]}{[C_r]} = \frac{k_f}{k_b}$$

$$-\frac{d(CO_{2\,(aq)})}{dt} = reaction + CO_2\ degassed$$

Fundamental kinetic aspects

9

3. Modelling precipitation

$$R_{blk} = k\,(S^{q} - 1)^{n} A_{c_0} \left(\frac{N_c}{N_{c_0}}\right)^{p} \qquad S = \frac{IAP}{K} = MINERAL\ SATURATION$$

$$Mg^{2+}_{(aq)} + CO_3^{2-}{}_{(aq)} + 3H_2O_{(l)} \rightleftharpoons MgCO_3 \cdot 3H_2O_{(s)}$$

Dominant reaction under reactor conditions

- Together parts 1 & 2 (from previous slide) & 3 (from this slide) give a series of coupled ordinary differential equations (ODEs)
- Each ODE = rate change in species concentration with time
- Solved using Polymath

THE UNIVERSITY OF NEWCASTLE AUSTRALIA — Priority Research Centre for Energy

Model Validation

10

- Perturbation analysis -
  - Stepped reduction in CO_2
  - Residual state (i.e. speciation & pH)
  - Compared to PHREEQC
- Precipitation reaction -
  - Stepped reduction in CO_2
  - Solution equilibrated with nesquehonite (zero initial moles)
  - Residual state (i.e. speciation & pH & moles nesquehonite)
  - Two cases: Threshold S = 1 or 5
  - Compared to PHREEQC

THE UNIVERSITY OF NEWCASTLE AUSTRALIA — Priority Research Centre for Energy

Model Calibration

11

- Fitted Parameters (to Batch experimental data) -
  1. Factor to adjust mass transfer
  2. Time to initiation of precipitation = induction (directly taken from experiments)
  3. Threshold mineral saturation for precipitation
  4. Precipitation rate constant and Geometric growth exponent (p)
- Rate constants for aqueous reactions (from published data) and equilibrium constants (PHREEQC) and activity coefficients (PHREEQC) were fixed for a given T
- Selected values for all fitted model parameters <u>unchanged</u> for the full term of each simulated batch experiment

THE UNIVERSITY OF NEWCASTLE AUSTRALIA

Priority Research Centre for Energy

Batch Experimental & Modelling Results

12

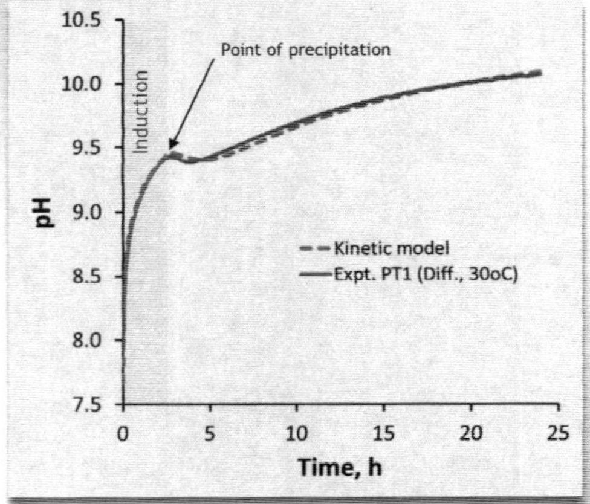

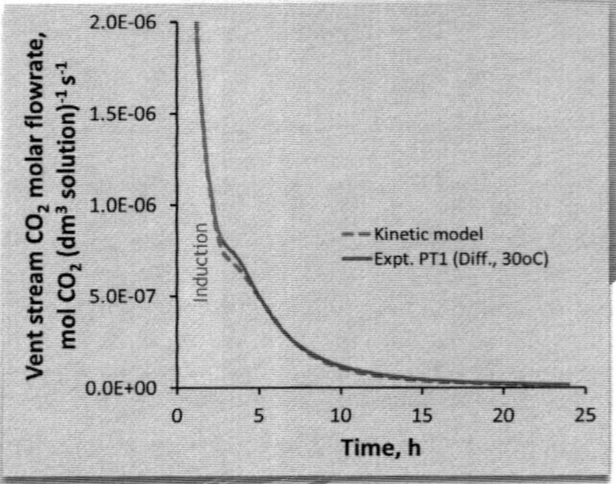

THE UNIVERSITY OF NEWCASTLE AUSTRALIA

Priority Research Centre for Energy

Batch Experimental & Modelling Results

13

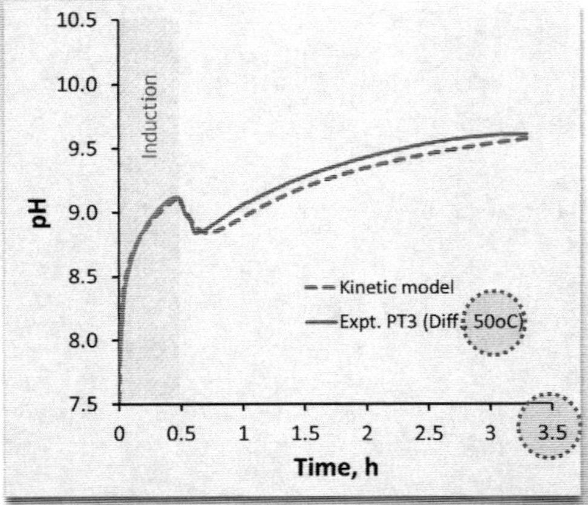

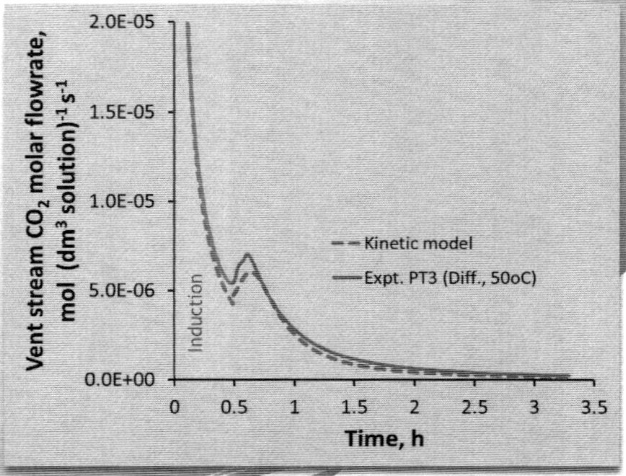

THE UNIVERSITY OF NEWCASTLE AUSTRALIA — Priority Research Centre for Energy

Effect of rate of CO_2 degassing (modelled)

14

- Model used to test effect of threshold level of saturation and intrinsic rate constant on precipitation, also effect of rate of aeration as shown here

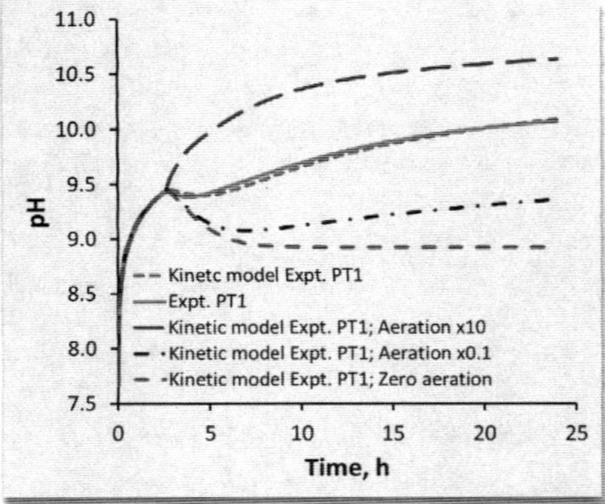

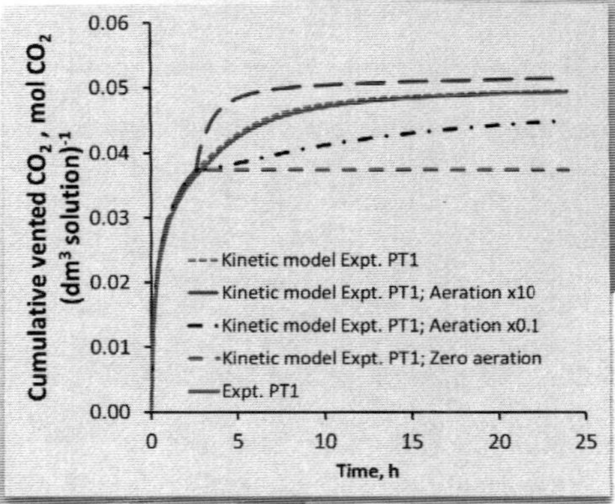

THE UNIVERSITY OF NEWCASTLE AUSTRALIA — Priority Research Centre for Energy

Effect of rate of CO_2 degassing (modelled)

15

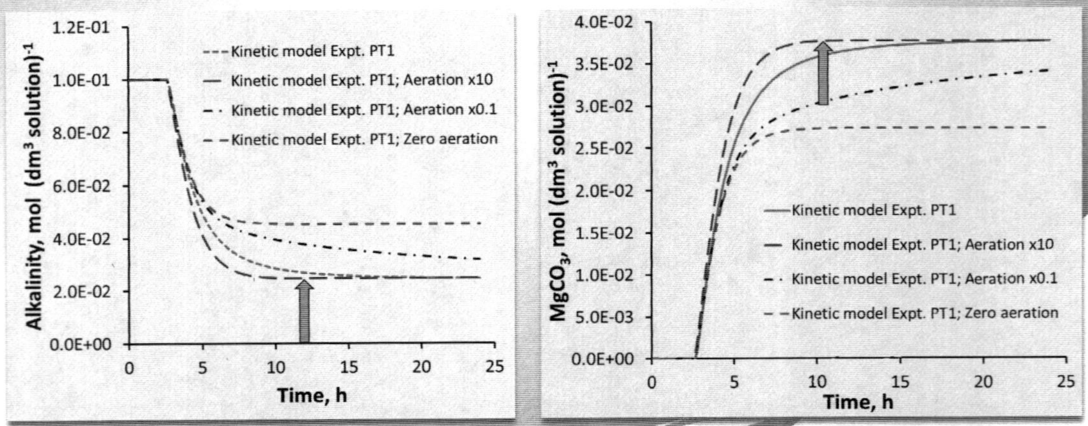

THE UNIVERSITY OF NEWCASTLE AUSTRALIA
Priority Research Centre for Energy

Effect of Carbonic Anhydrase (CA)

16

- Geometric growth exponent reflects homogeneous nucleation
- Experiments at differing T ► precipitation diffusion-controlled

$$R_{blk} = k\ (S^q - 1)^n A_{c_0} \left(\frac{N_c}{N_{c_0}} \right)^p$$

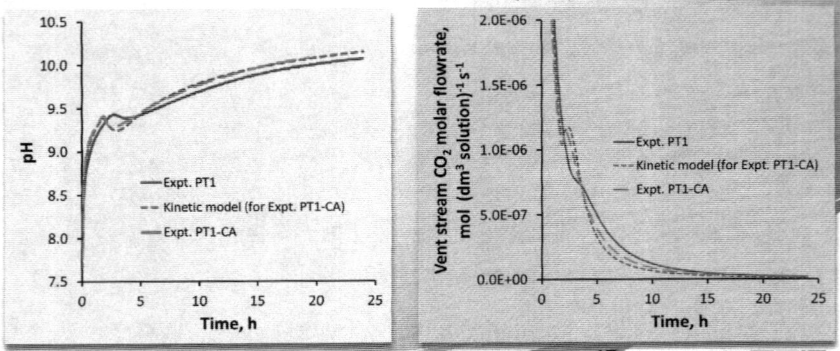

THE UNIVERSITY OF NEWCASTLE AUSTRALIA
Priority Research Centre for Energy

Semibatch experiments, & modelling semibatch & continuous configurations

17

Semibatch experiments & modelling:

- Continuous feed of solution to reactor (at a fixed rate after the induction period) & N_2 is continuously bubbled through the solution

Modelling continuous configuration:

- Constant V; Extractive molar flowrates based on reactor composition

THE UNIVERSITY OF NEWCASTLE AUSTRALIA

Priority Research Centre for Energy

Kinetic model - semibatch & continuous

18

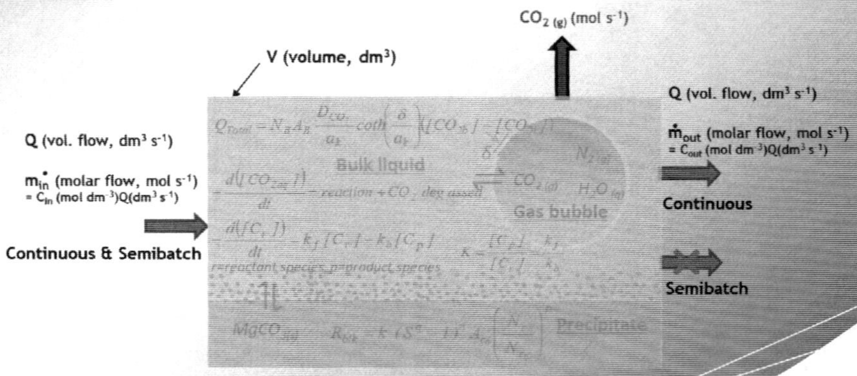

THE UNIVERSITY OF NEWCASTLE AUSTRALIA

Priority Research Centre for Energy

Kinetic model - semibatch & continuous

19

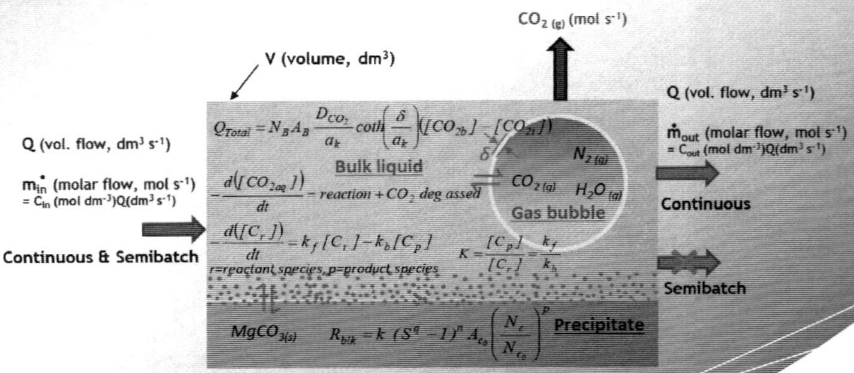

THE UNIVERSITY OF NEWCASTLE AUSTRALIA — Priority Research Centre for Energy

Semibatch experiment & model

20

- Example semibatch experiment & model results

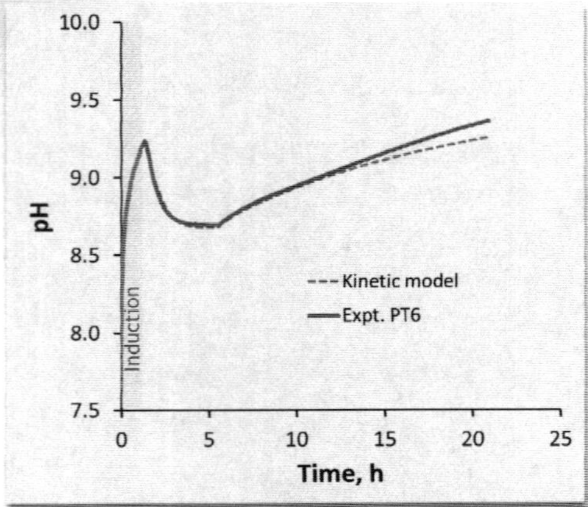

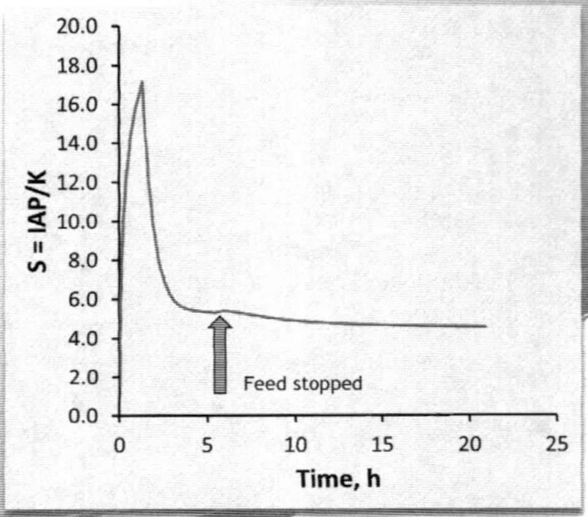

THE UNIVERSITY OF NEWCASTLE AUSTRALIA — Priority Research Centre for Energy

Semibatch experiment & model

21

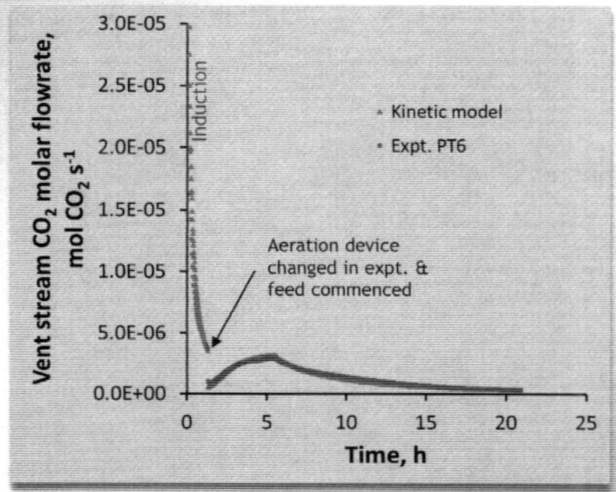

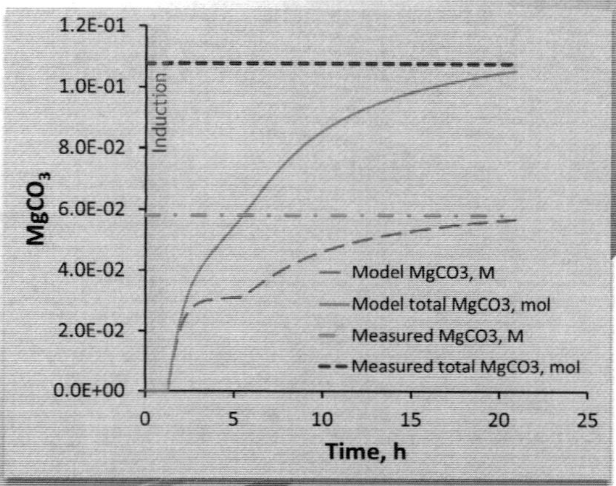

THE UNIVERSITY OF NEWCASTLE AUSTRALIA — Priority Research Centre for Energy

Continuous model

22

- Results for different feed flows and rates of aeration
- Rate aeration ⬆ pH ⬆ Steady-state rate of production ⬆⬇ Energy ⬆
- Physicochemical constraints may be limiting

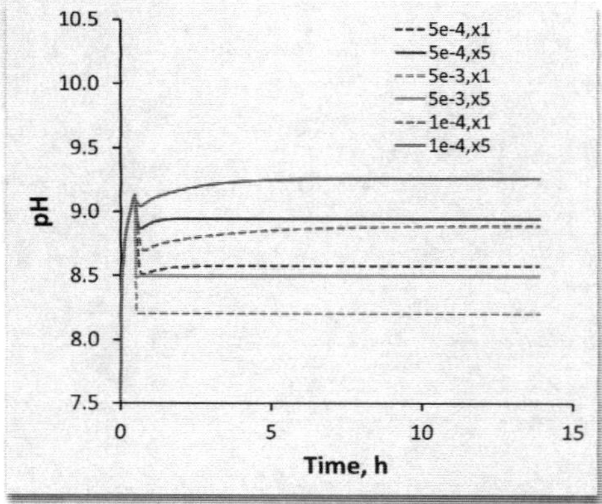

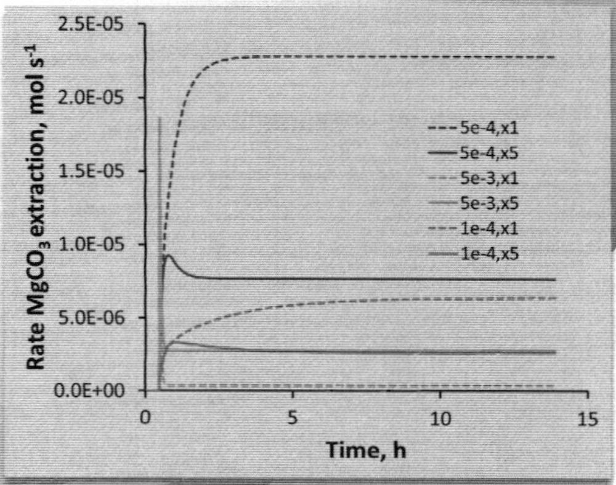

THE UNIVERSITY OF NEWCASTLE AUSTRALIA — Priority Research Centre for Energy

Conclusions

23

- CO_2 degassing reactors are appropriate to two-stage MC
- Mass transfer through the gas-liquid interface represents the kinetic control to degassing
- Specific findings about the precipitation reaction
- Process (degassing + precipitation) can be kinetically modelled and extended to different reactor configurations
- Modelling showed that rate of degassing is controlling to steady-state rate of precipitation in continuous reactor mode

THE UNIVERSITY OF NEWCASTLE AUSTRALIA

Priority Research Centre for Energy

24

Thank you, any questions?

THE UNIVERSITY OF NEWCASTLE AUSTRALIA

Priority Research Centre for Energy

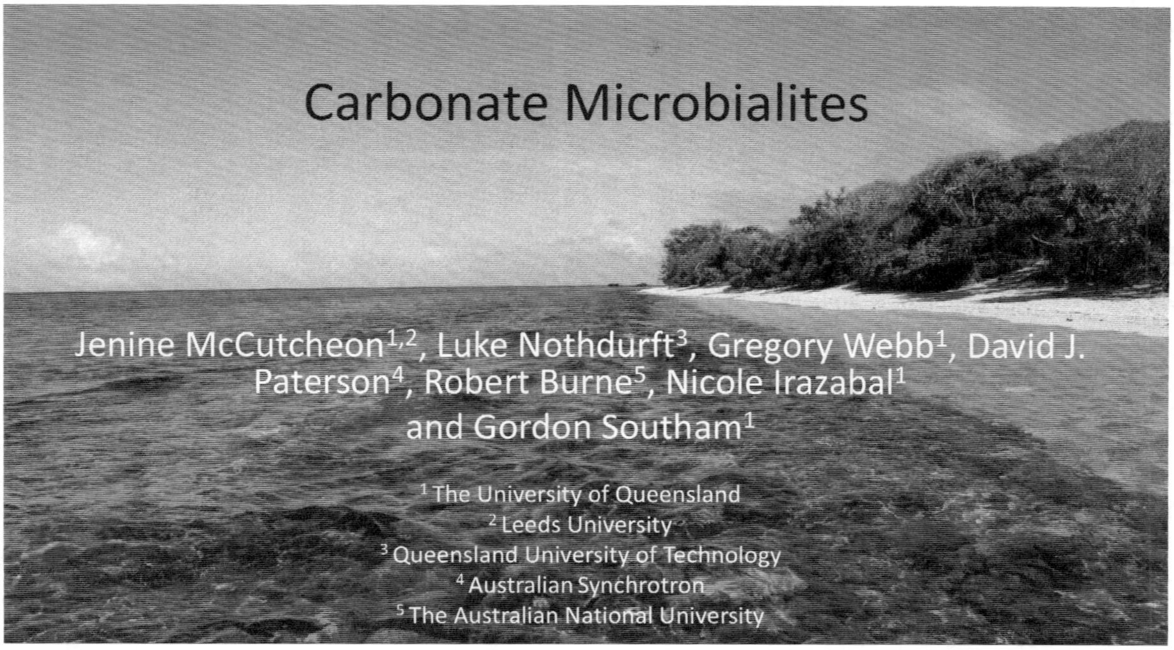
Carbonate Microbialites
Jenine McCutcheon[1,2], Luke Nothdurft[3], Gregory Webb[1], David J. Paterson[4], Robert Burne[5], Nicole Irazabal[1] and Gordon Southam[1]
[1] The University of Queensland
[2] Leeds University
[3] Queensland University of Technology
[4] Australian Synchrotron
[5] The Australian National University

Hydromagnesite
$Mg_5(CO_3)_4(OH)_2{\cdot}4H_2O$
Atlin, BC, Canada

2.5 µm

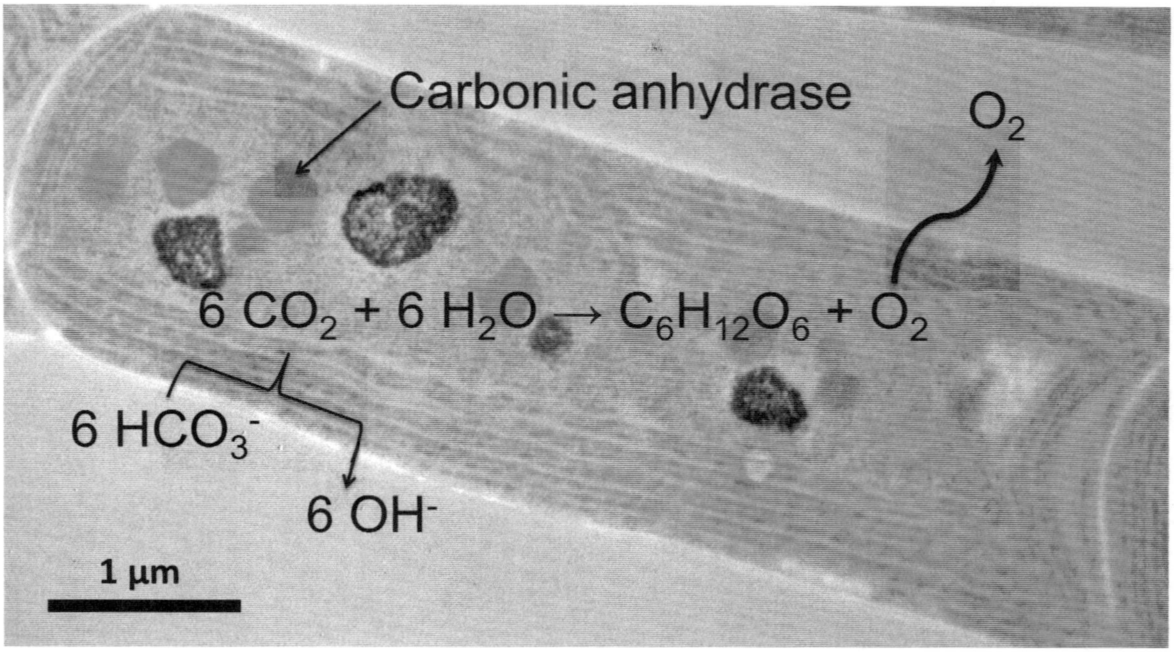
Carbonic anhydrase
O_2
6 CO_2 + 6 H_2O → $C_6H_{12}O_6$ + O_2
6 HCO_3^-
6 OH^-
1 µm

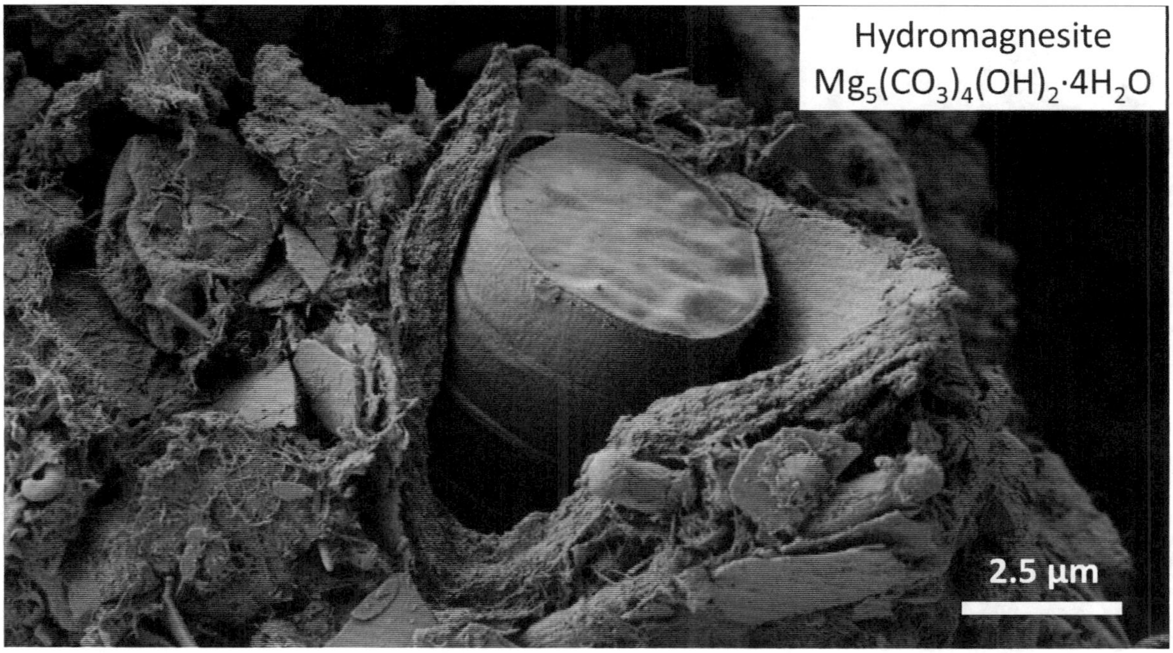
Hydromagnesite
$Mg_5(CO_3)_4(OH)_2 \cdot 4H_2O$
2.5 µm

Dypingite
Mg5(CO3)4(OH)2·5H2O
1 µm

Woodsreef Mine, near Barraba, NSW

1 cm

Pyroaurite + Fe oxides
($Mg_6\ Fe^{3+}{}_2(OH)_{16}[CO_3]\cdot 4H_2O$)

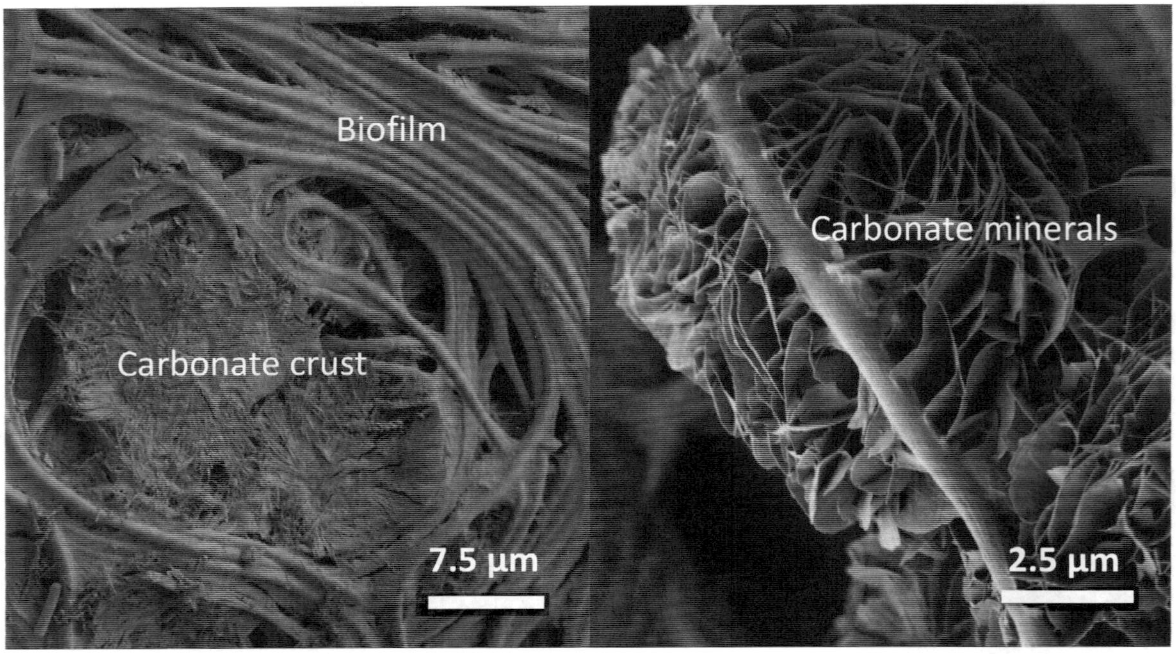
Biofilm
Carbonate crust
7.5 µm
Carbonate minerals
2.5 µm

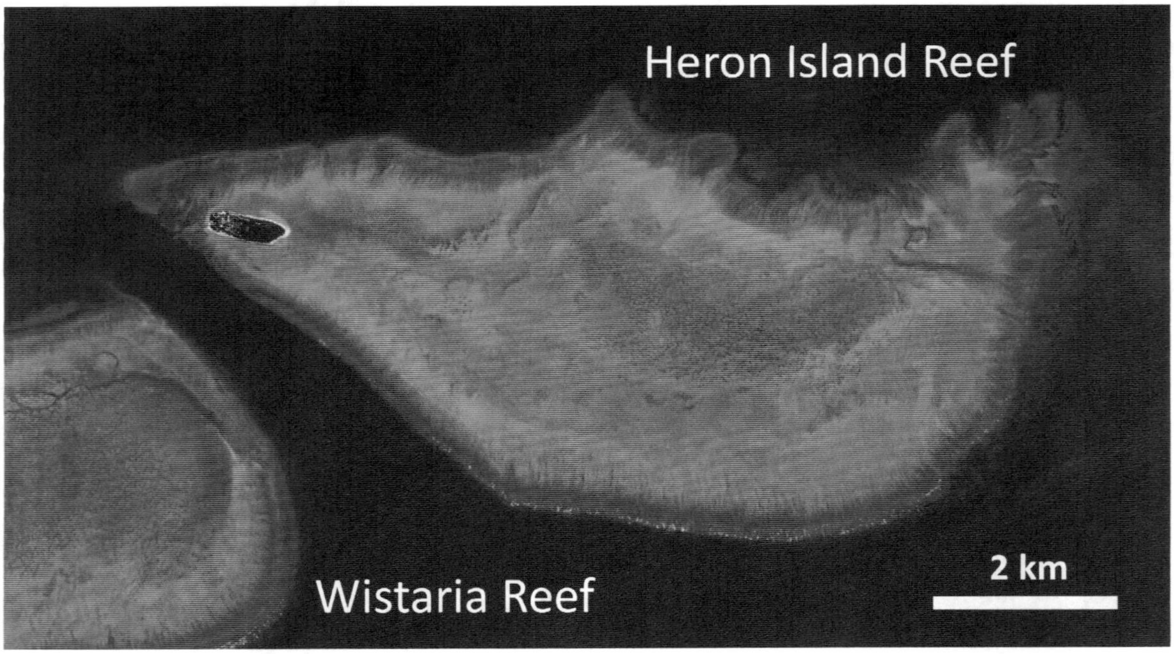
Heron Island Reef
Wistaria Reef
2 km

Heron Island Field Station,
The University of Queensland

1 mm

4 µm

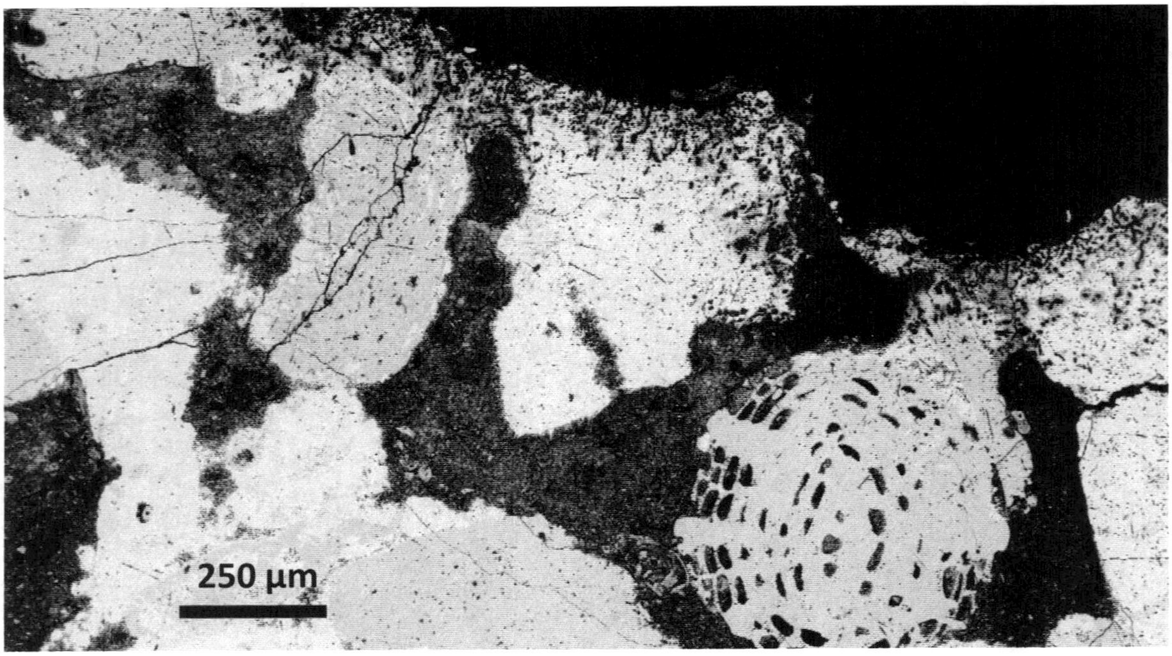
250 µm

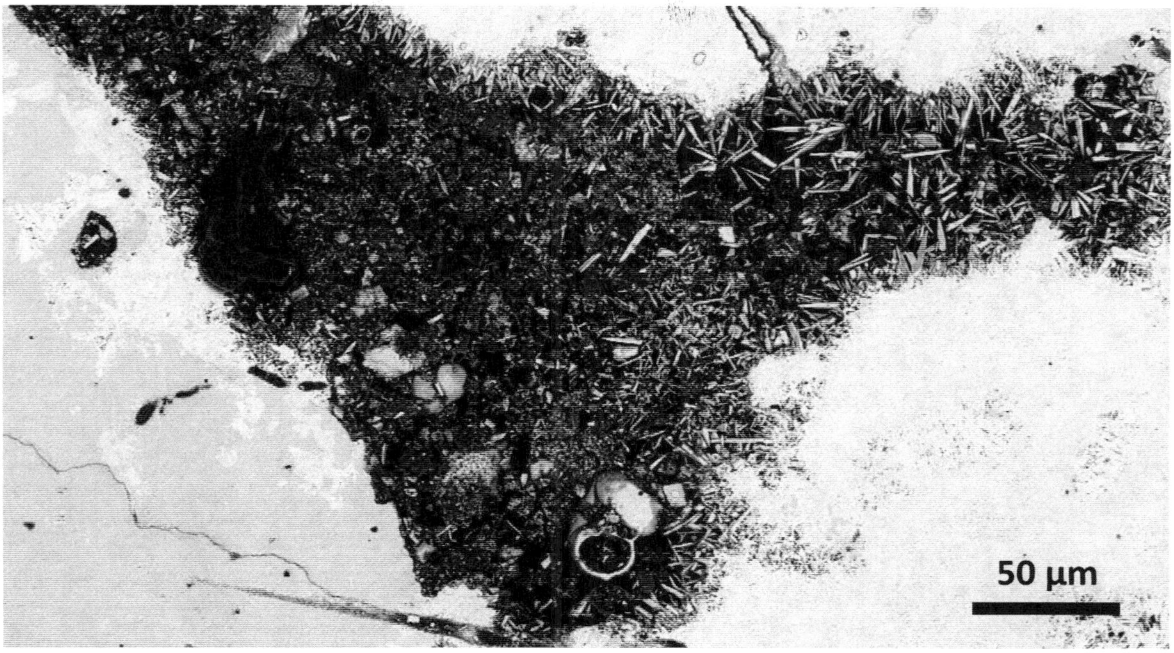
50 µm

20 µm

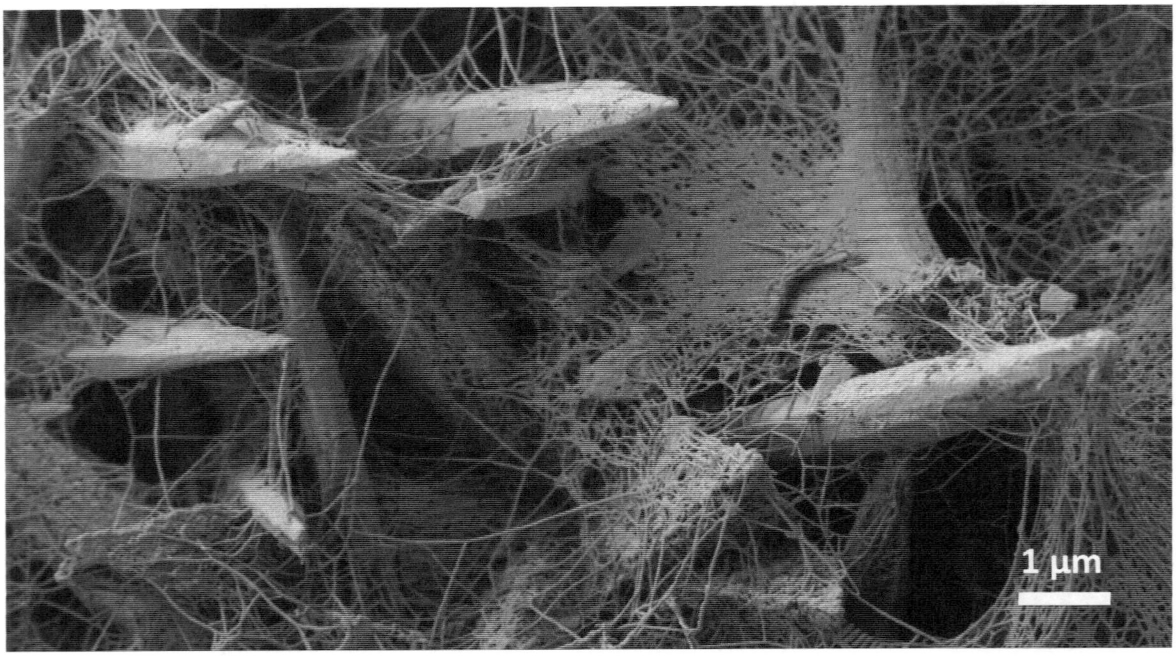
1 µm

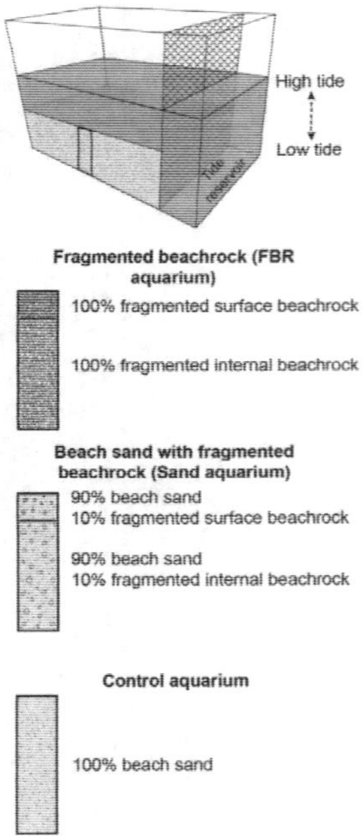
High tide
Low tide
Fragmented beachrock (FBR aquarium)
100% fragmented surface beachrock
100% fragmented internal beachrock
Beach sand with fragmented beachrock (Sand aquarium)
90% beach sand
10% fragmented surface beachrock
90% beach sand
10% fragmented internal beachrock
Control aquarium
100% beach sand

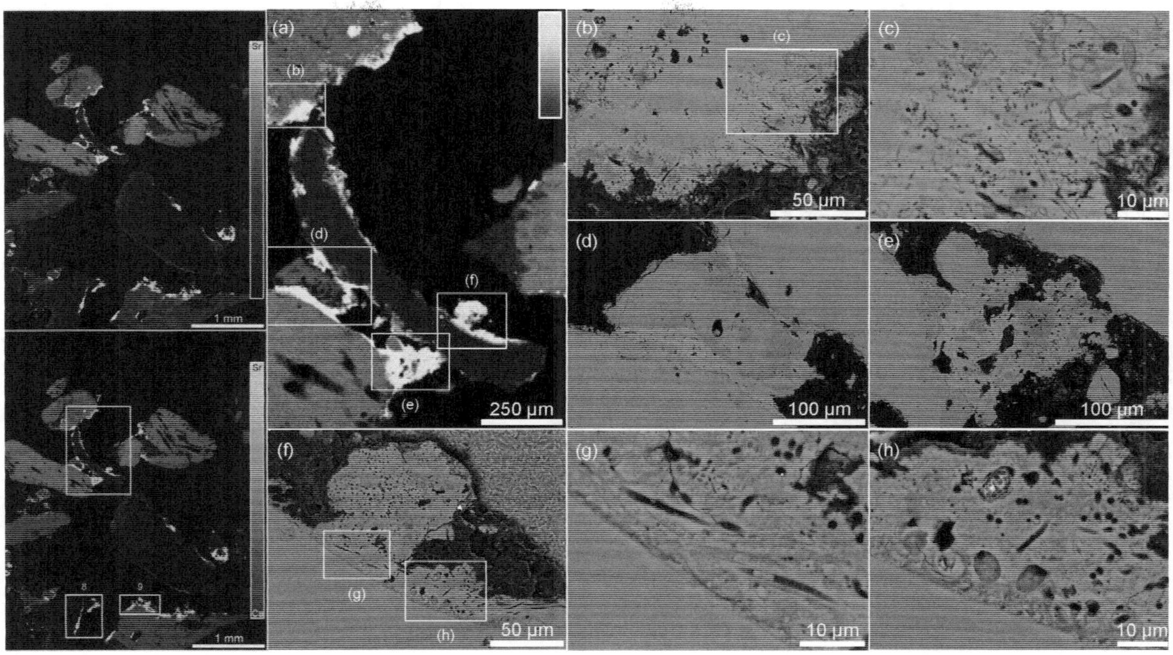
Sr
1 mm
Sr
Ca
1 mm
(a)
(b)
(d)
(f)
(e)
250 µm
(f)
(g)
(h)
50 µm
(b)
(c)
50 µm
(c)
10 µm
(d)
100 µm
(e)
100 µm
(g)
10 µm
(h)
10 µm

2 mm

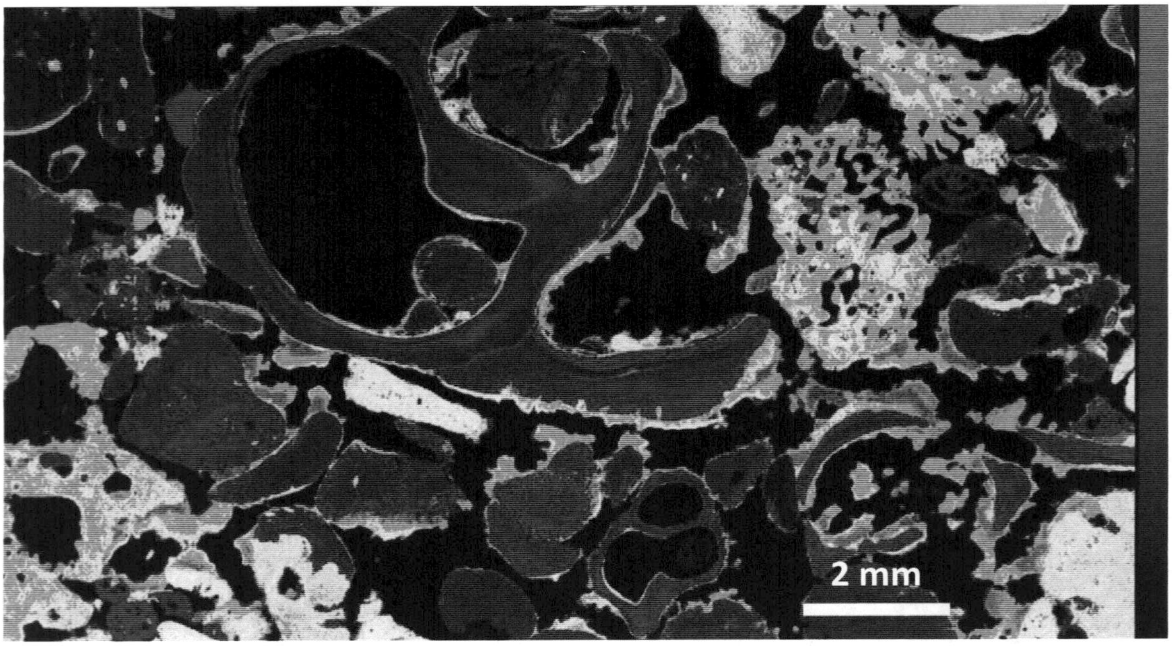
2 mm

Shark Bay, Australia
Wikimedia Commons

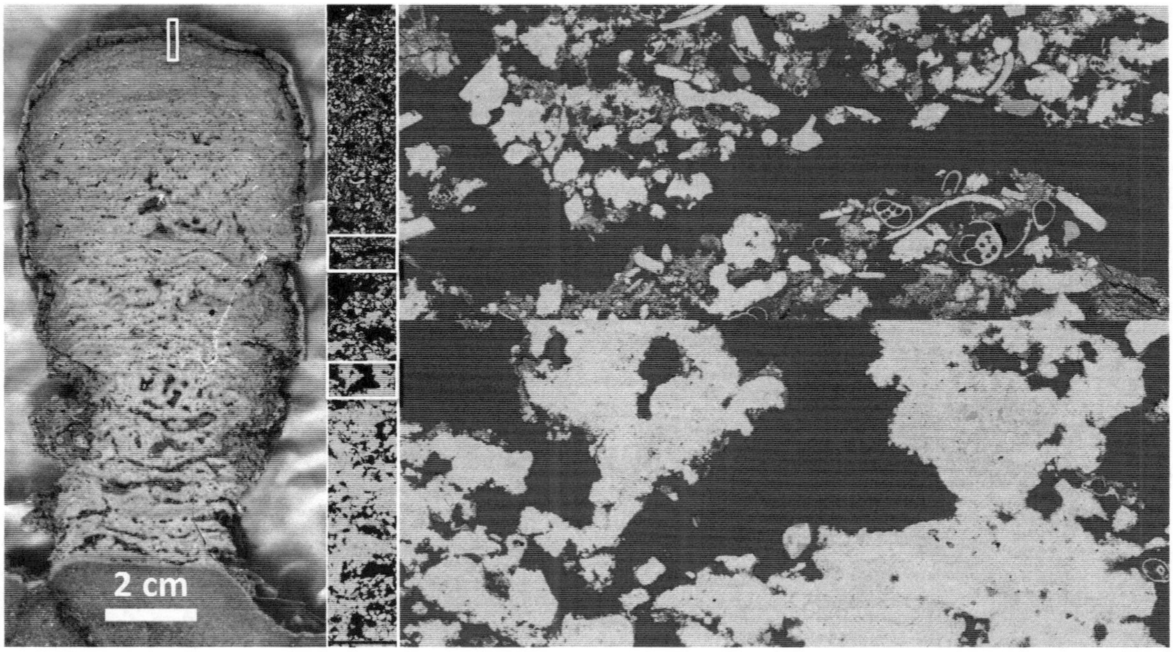
2 cm

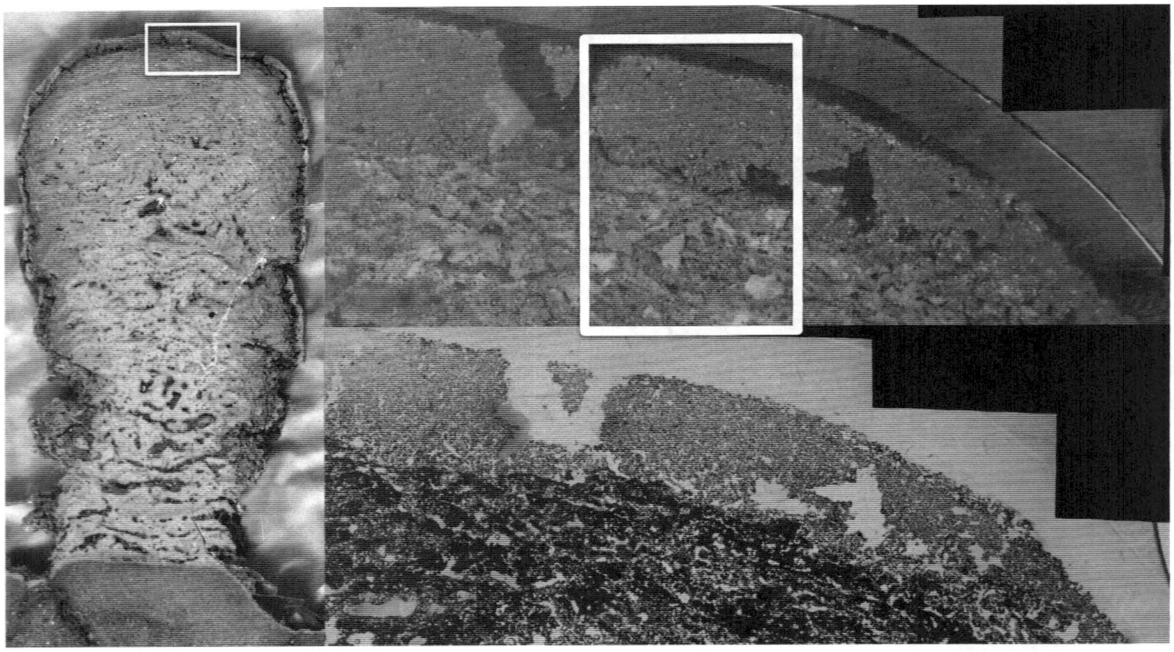

Sr maps

Sr maps

Bitter Springs Formation, Australia

2.5 µm

Visit Heron Island.

Accelerated Carbonation and Granulation for Treatment of Mine Tailings Using MgO/GGBS Binder

Authors:

Kim, T. Y., Pusan National University
Hwang, I., Pusan National University
Ahn, J. Y., Pusan National University
Choi, S. J., Pusan National University

About 4.92 million tons of abandoned mine tailings exist in Korea. Accelerated carbonation and granulation treatment were simultaneously applied to mine tailings using MgO (magnesium oxide) and GGBS (grounded granule blast furnace slag) mixed binder, in order to remediate the mine tailings and reuse them as materials for construction and mine back filling. The MgO/GGBS binder has shown high compressive strength under the accelerated carbonation condition and great stabilization capacity for heavy metals when it was applied to contaminated mine tailings in the previous study. To obtain the optimal granule size and the high granule strength, operational factors (rotational speed and time) and binder dosages were tested as the following: water/solid ratio (0.16-0.21), binder dosage (10-30% to the mine tailings) and MgO and GGBS weight ratio (1:1, 1:4, 3:7). The 50% MgO and 50% GGBS ($MgO_{0.5}GGBS_{0.5}$) at 30% binder dosage was the best conditions, which yielded the greatest granule strength. The $MgO_{0.5}GGBS_{0.5}$ granule was cured under 0%, 20%, 50% and 100% CO_2 environments for 60 days to evaluate its physical and chemical characteristics. The granule strength was greater in accelerated CO_2 than in 0% CO_2 and the strength dramatically increased as the curing time passed until 28th day. The increase of the granule strength was faster in the higher CO_2 conditions on the early stage of curing days. However, the strength decreased on 60th day except the granule cured in 20% CO_2. The highest granule strength was obtained at 20% CO_2 with a value of at 0.47 MPa on 28th days of curing. The maximum amount of CO_2 uptake was 0.025 kg CO_2/ kg binder for the 100% CO_2 cured granule on 28th day. The porosity of 20% CO_2 cured $MgO_{0.5}GGBS_{0.5}$ granule was lower (0.19) than the 100% CO_2 cured (0.26), showing the moderate formation of hydromagnesite in the structures. The $MgO_{0.5}GGBS_{0.5}$ treated granule also showed a greater acid-buffering capacity at around pH 9 than the granule treated with ordinary Portland cement (OPC) treated. This implies the heavy metal leaching around pH 9 is possibly lower for $MgO_{0.5}GGBS_{0.5}$ granules than for OPC granules. The XRD results showed the decreases in peaks for pyrite and arsenopyrite in $MgO_{0.5}GGBS_{0.5}$ treated granules at 20% CO_2. This implies that those minerals could be transformed into the different minerals binding with $MgO_{0.5}GGBS_{0.5}$.

AN ACID RECOVERY AND MINERAL CARBONATION PROCESS FOR THE NICKEL INDUSTRY

LAUREN ZAPPALA, REYDICK BALUCAN, JAMES VAUGHAN AND KAREN STEEL
THE UNIVERSITY OF QUEENSLAND, AUSTRALIA

THE UNIVERSITY OF QUEENSLAND AUSTRALIA

1

NICKEL LATERITES

Future of nickel

~1 wt.% Ni low grade ore

70% global Ni reserves are laterites

12.5 Mt of reserves globally

- Located in tropical climates e.g. Philippines, New Caledonia, Australia

Growth in utilisation due to advances in technology and increasing demand

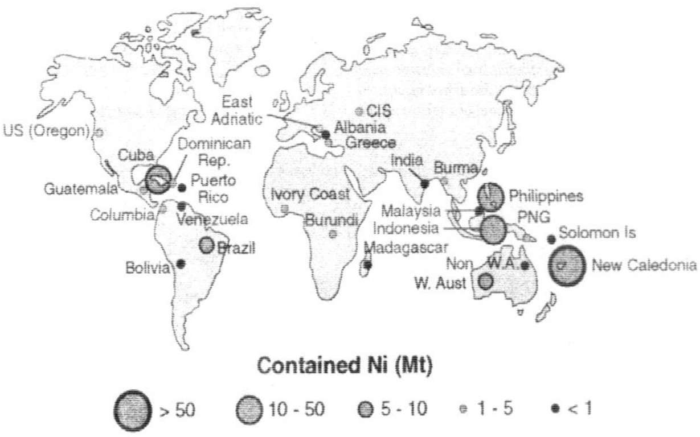

Global distribution of nickel laterite reserves (Brand et al., 1998)

THE UNIVERSITY OF QUEENSLAND AUSTRALIA

2

NICKEL LATERITES CONT.

Contain magnesium silicates

- Present in saprolite layer
- Refining produces ~140 t Mg-rich waste/ t Ni metal

Small/medium scale emitter

- 22 t CO_2/ t Ni metal

Use of acid and base

- Acid cost ~ $4 000/ t Ni metal (assuming 40 t H_2SO_4/ t Ni metal, $100/ t H_2SO_4)
- Neutralization produces ~8 t CO_2/ t Ni metal

Data from Environmental Protection Agency (1996)

THE UNIVERSITY OF QUEENSLAND AUSTRALIA

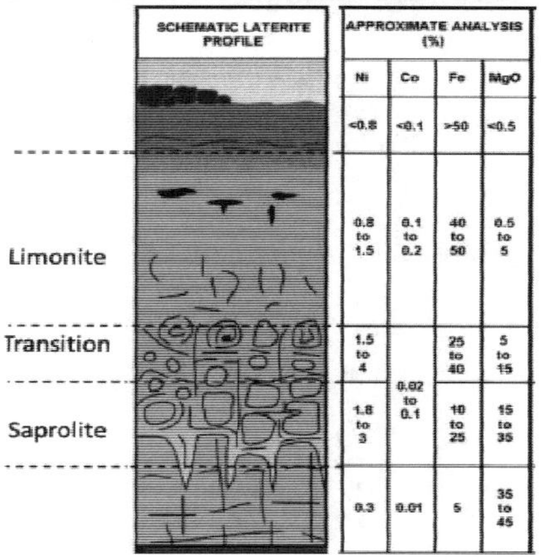

Nickel laterite profile (adapted from Brand et al., 1998)

3

MINERAL CARBONATION PH SWING PROCESSES

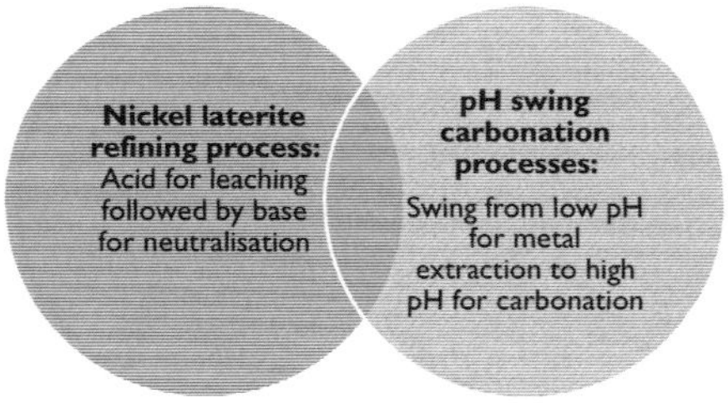

- Good combinations of extraction and carbonation efficiencies without the need for extreme operating conditions because conditions are favourable for each stage of the reaction.

Challenge of pH swing processes: To make processes more economical acid and base can be recycled, but **few energy efficient regeneration processes have been developed.**

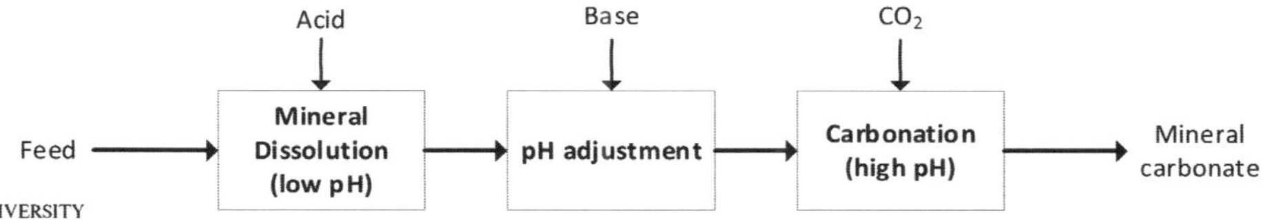

THE UNIVERSITY OF QUEENSLAND AUSTRALIA

4

AMINES

Tertiary amines

- Not conventionally used
- Low capture efficiency conventionally due to equilibrium limitations since CO_2 is not removed
- Reaction of metal ions with carbonate ions in mineral carbonation process removes CO_2
- Low energy input for regeneration
- Triethylamine (Et_3N) found promising in previous work (Balucan and Steel, 2015)

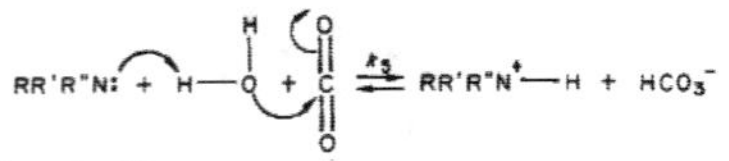

OUT OF SCOPE

Primary and secondary amines

- Used in conventional carbon capture
- High capture efficiency due to formation of a strong carbamate bond which shifts equilibrium so more CO_2 is consumed
- High energy input for regeneration, produced using fossil fuels

$$RR'NH + CO_2 \rightleftharpoons RR'\overset{+}{N}HCO_2^-$$
$$RR'\overset{+}{N}HCO_2^- + R'NH \rightleftharpoons RR'NCO_2^- + RR'\overset{+}{N}H_2$$

R = alkyl
R' = alkyl or H.

Et_3N = triethylamine

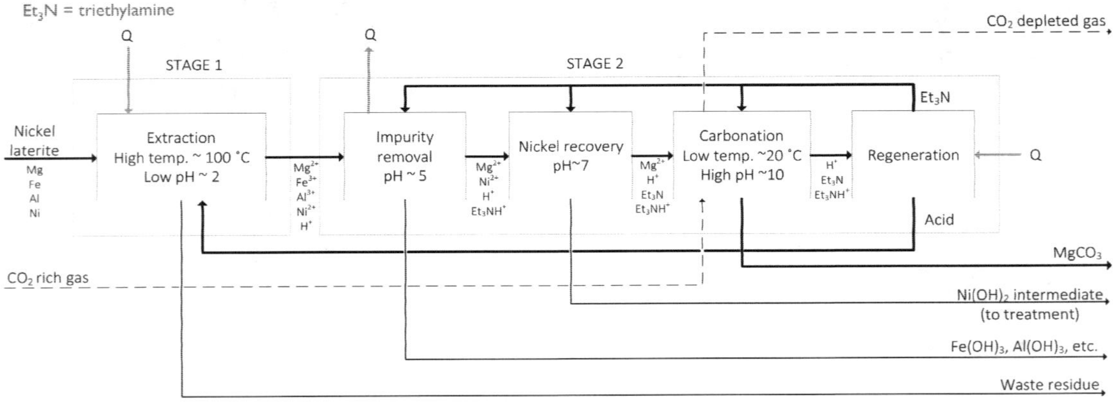

Extraction:

$$Mg_3Si_2O_5(OH)_{4(s)} + 6H^+_{(aq)} + heat \rightarrow 3Mg^{2+}_{(aq)} + H_2O_{(l)} + SiO_{2(s)}$$

Impurity removal:

$$Et_3N_{(org)} + H^+ \rightarrow R_3NH^+_{(aq)}$$
$$Et_3N_{(aq)} + H_2O_{(l)} \rightarrow Et_3NH^+_{(aq)} + OH^-_{(aq)}$$
$$X^{2+}_{(aq)} + OH^-_{(aq)} \rightarrow X(OH)_{2(s)}$$

Carbonation:

$$CO_{2(g)} \leftrightarrow CO_{2(aq)}$$
$$CO_{2(aq)} + OH^-_{(aq)} \leftrightarrow HCO^-_{3\,(aq)}$$
$$HCO^-_{3\,(aq)} \leftrightarrow CO^{2-}_{3\,(aq)} + H^+$$
$$Mg^{2+}_{(aq)} + CO^{2-}_{3\,(aq)} \leftrightarrow MgCO_{3(s)} + heat$$

Regeneration:

$$Et_3NH^+_{(aq)} + heat \rightarrow Et_3N_{(aq)} + H^+_{(aq)}$$
$$Et_3N_{(aq)} \rightarrow Et_3N_{(org)}$$
$$Et_3N_{(org)} \rightarrow Et_3N_{(g)}$$

RESULTS

THE UNIVERSITY
OF QUEENSLAND
AUSTRALIA

7

PH SWING: ET_3N AND H_2SO_4

Titration of 40 mL 0.5 M H_2SO_4 with Et_3N at 25 °C

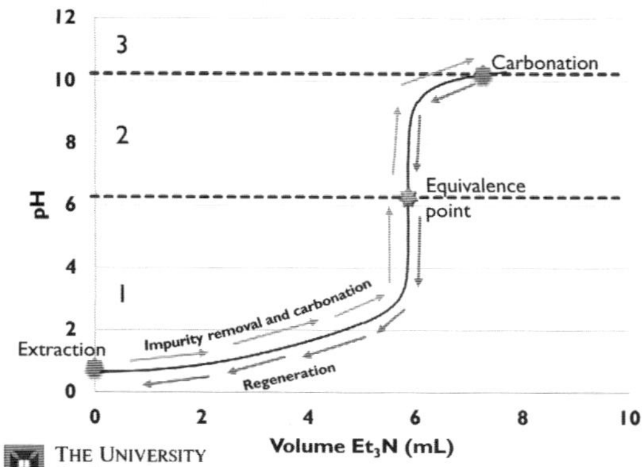

1. Acid neutralisation, formation of acid-protonated amine

$$2Et_3N_{(aq)} + H_2SO_{4(aq)} \leftrightarrow (Et_3NH)_2SO_{4(aq)}$$

Equivalence point

$$pH{\sim}6, V_{Et_3N}{\sim}5.85\ mL\ (42\ mmol)$$

2. Alkalisation, formation of water-protonated amine

$$Et_3N_{(aq)} + H_2O_{(l)} \leftrightarrow Et_3NH^+_{(aq)} + OH^-_{(aq)}$$

3. Formation of immiscible amine layer

$$Et3N_{(org)}$$

THE UNIVERSITY
OF QUEENSLAND
AUSTRALIA

8

PH SWING VS. MASS REMOVED

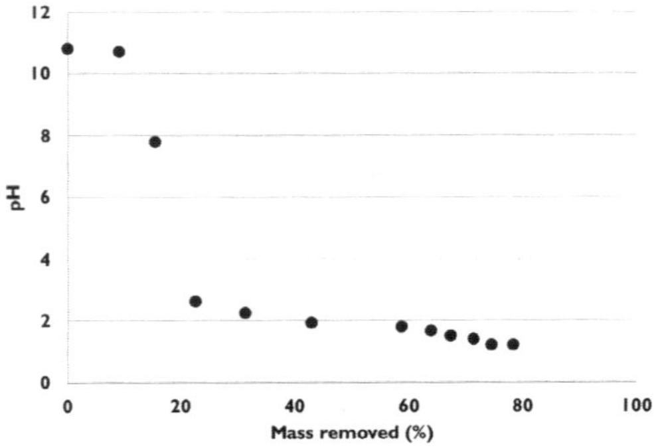

- Batch distillation of 50 mL mixture with Et3N:H+ of 2:1 (V_{Et3N} ~ 10.9 mL, $V_{0.5\ M\ H2SO4}$ ~ 39.1 mL)

Key results:

1. pH swing from 10.7 to 2 at 25 ˚C with 40% decrease in mass
2. Swing from pH 10.7 to 2.5 at 25 ˚C with 20% decrease in mass

The magnitude of pH swings have generally not been quantified in the literature for other processes, but most processes use multiple stages to regenerate whereas we are achieving a large swing in a single unit.

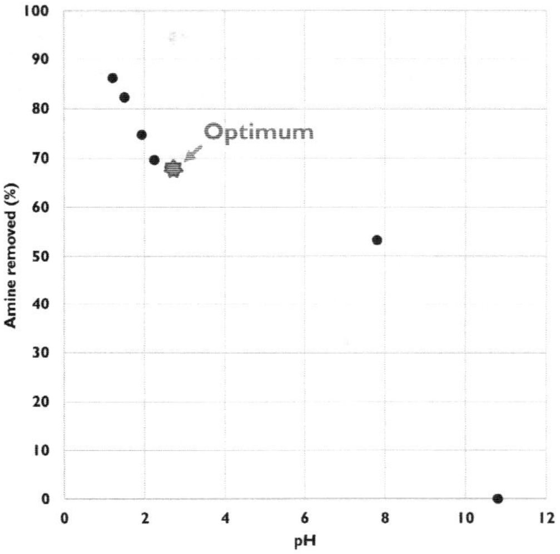

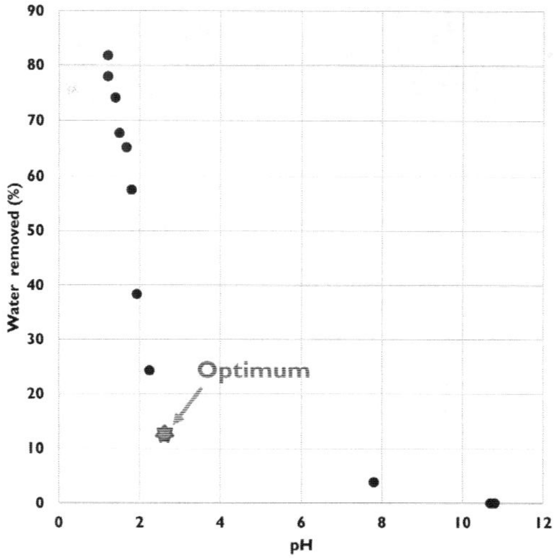

● pH ● $[H^+]_{free}$

THE UNIVERSITY OF QUEENSLAND
AUSTRALIA

11

EQUILIBRIUM MODELLING

Purpose: To compare the measured sulphur with theoretical sulphur.

Assumptions:

- The amount of bound sulphur is proportional to the amount of bound amine calculated using Equations 1 and 2
- Equation 3 goes to completion

Eqtn. 1: $Et_3N + H^+ \leftrightarrow Et_3NH^+$, pKa = 10.8 at 25 °C

Eqtn. 2: $[Et_3N] + [Et_3NH^+] = C_{tot}/6$

Eqtn. 3: $H_2SO_4 \leftrightarrow HSO_4^- + H^+$, Ka = very large

Eqtn. 4: $HSO_4^- \leftrightarrow SO_4^{2-} + H^+$, Ka = 10^{-2}

Eqtn. 5: $[HSO_4^-]_{free} + [SO_4^{2-}]_{free} = 3[H^+]_{free}$

Eqtn. 6: $[S]_{bound} = \alpha[Et_3NH^+]$

Eqtn. 7: $[S]_{tot} = [S]_{bound} + [HSO_4^-]_{free} + [SO_4^{2-}]_{free}$

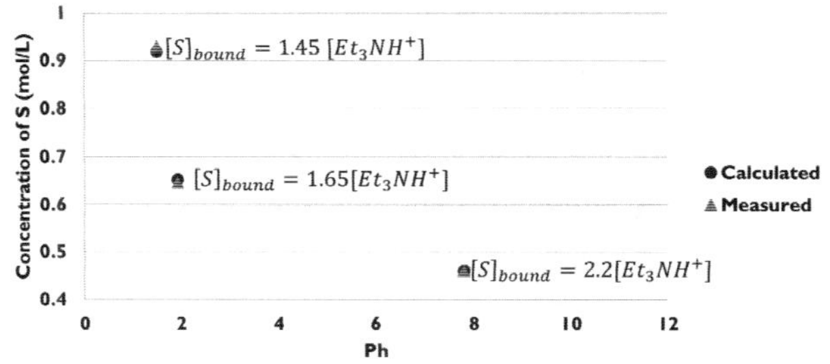

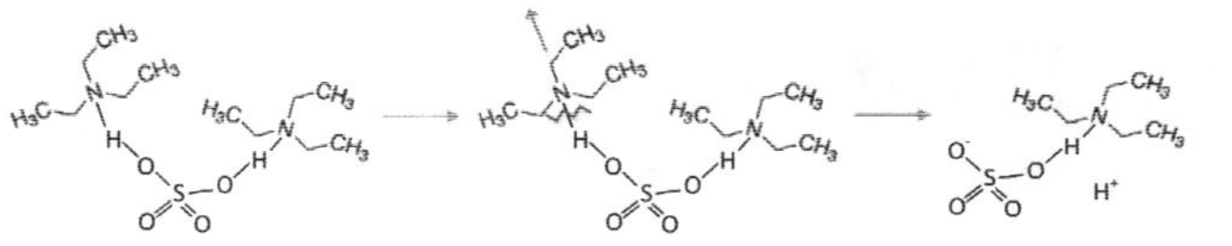

$[S]_{bound} = 2[Et_3NH^+]$ $[S]_{bound} = 1 - 2[Et_3NH^+]$ $[S]_{bound} = [Et_3NH^+]$

THE UNIVERSITY OF QUEENSLAND
AUSTRALIA

12

CONCLUSION

1. The optimal pH of the acid we regenerate is about 2.5 and this occurs with a 20% decrease in mass.
   a) 70% removal of amine
   b) 13% removal of water
2. When amine and acid are mixed $(Et_3NH)_2SO_4$ is formed. During distilling one molecule of Et_3N is freed, leaving behind Et_3N-HSO_4^- which is a stable salt and weak acid.
3. Next studies will involve reactive distillation where extraction and regeneration are combined so that the reaction of the mineral with H^+ drives more H^+ into the aqueous phase, hence releasing more bound amine.
4. Repeat studies using other tertiary amines to compare recovery and leaching performance.

THE UNIVERSITY OF QUEENSLAND AUSTRALIA

13

THANK YOU FOR YOUR ATTENTION

QUESTIONS?

THE UNIVERSITY OF QUEENSLAND AUSTRALIA

14

REFERENCES

Brand, N.W., Butt, C. R. M., & Elias, M. (1998). Nickel laterites: Classification and features. *AGSO Journal of Australian Geology and Geophysics, 17*(4), 81-88.

Environmental Protection Authority (1996). *Nickel/Cobalt ore mining and processing operations, Murrin Murrin, 60km east of Leonora*. Retrieved from http://www.epa.wa.gov.au/sites/default/files/EPA_Report/B816.pdf

THE UNIVERSITY OF QUEENSLAND AUSTRALIA

Carbon Capture, Utilization and Storage (CCUS) via Innovative Mineralization Pathways

A.-H. Alissa Park, Guanhe Rim, MArk Rayson, Geoff Brent, Donghyun Lee
Department of Earth and Environmental Engineering
Department of Chemical Engineering
Lenfest Center for Sustainable Energy
Columbia University in the City of New York

ACEME, Newcastle Australia
March 12, 2018

COLUMBIA | ENGINEERING
The Fu Foundation School of Engineering and Applied Science

Climate Change

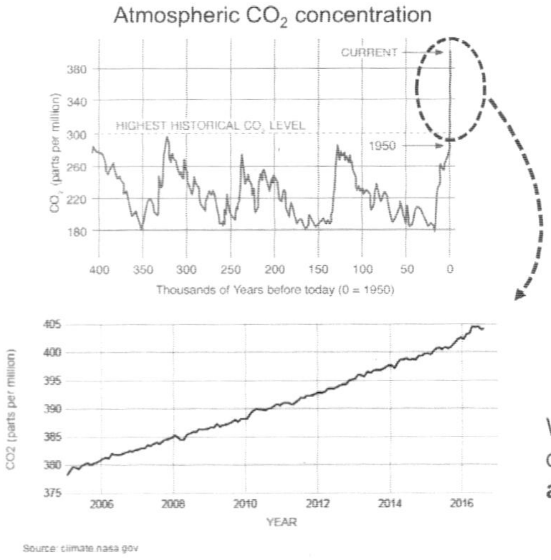

National Research Council of The National Academies

We need "… substantial multi-century climate change commitment created by **past**, **present and future** emissions of CO_2." (IPCC, 2013)

CO_2 Utilization Schemes: Working fluids

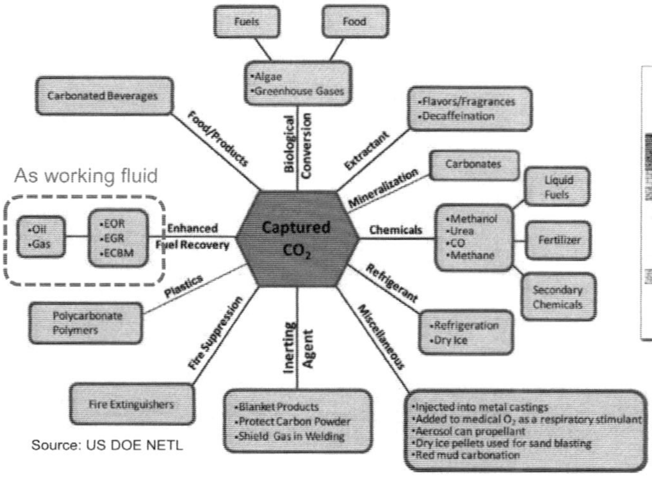

Source: US DOE NETL

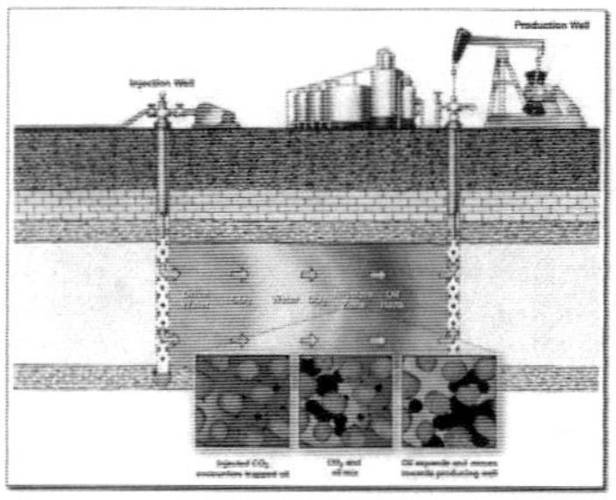

~80% of CO_2 is from natural resources

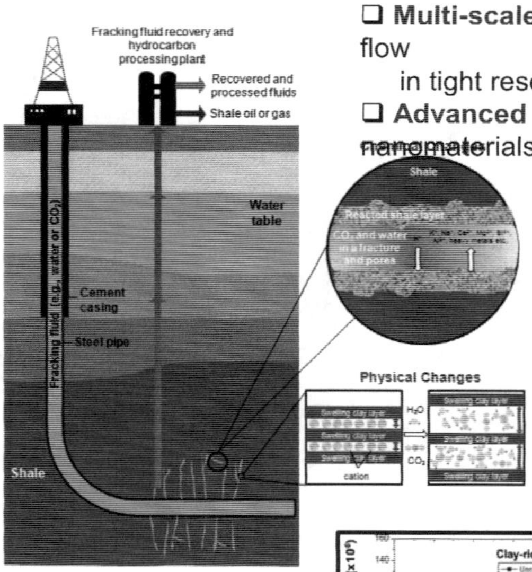

❑ **Multi-scale chemo-physical understanding** of CO_2 flow
in tight reservoirs

❑ **Advanced materials discovery** (e.g., foams, nanomaterials)

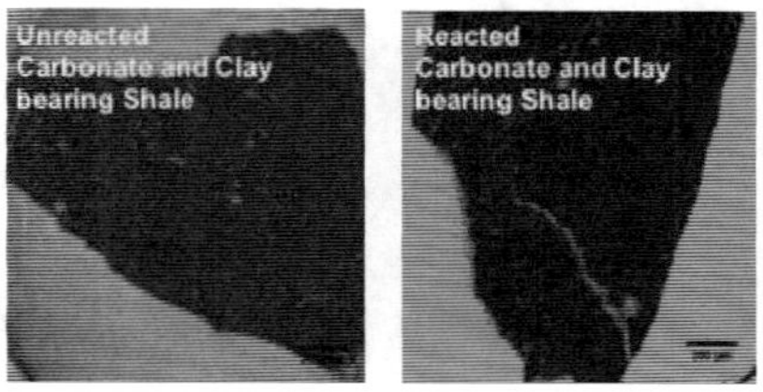

Shale-water-CO_2 interactions

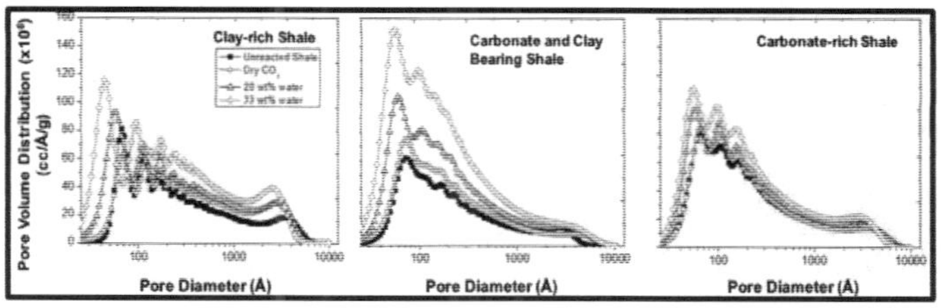

CO_2 Utilization Schemes

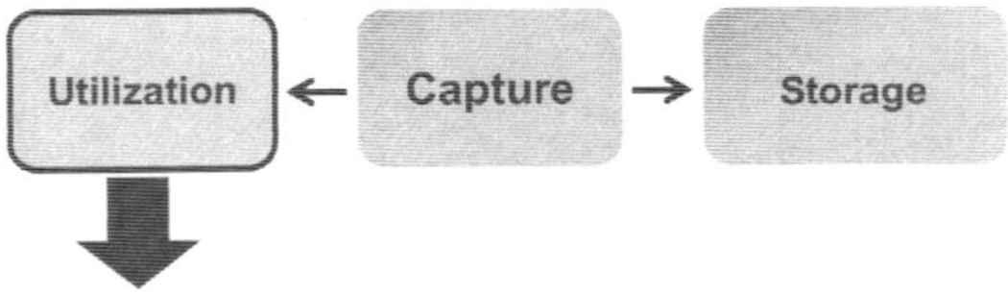

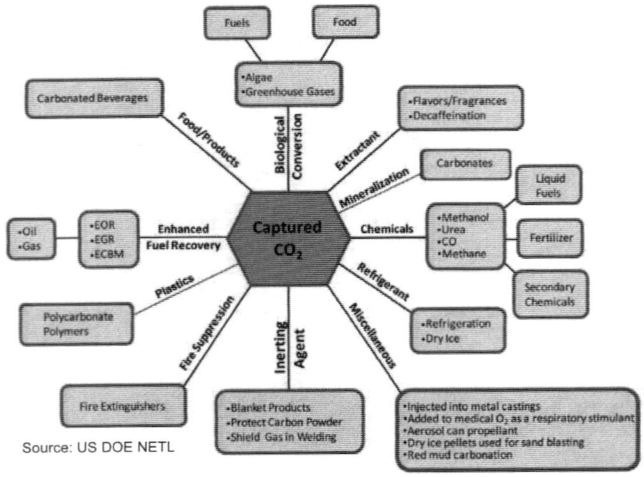

Source: US DOE NETL

Potential Annual Revenue (dollars)

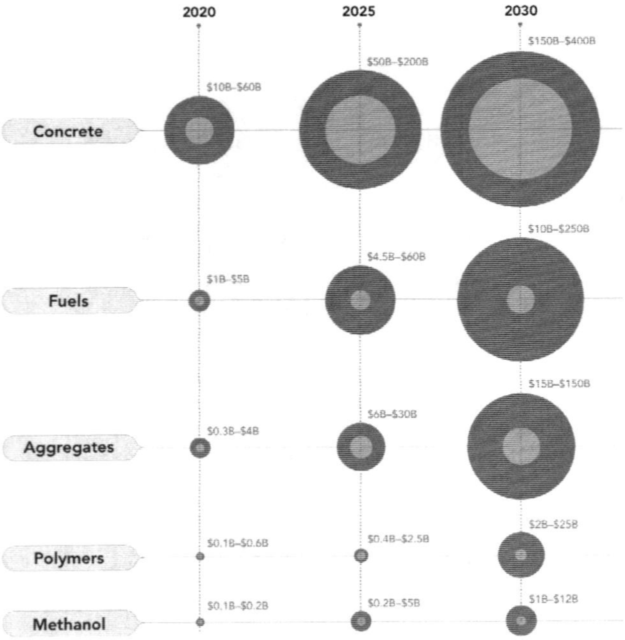

What should we make using CO_2?

(CO_2 Sciences & The Global CO_2 Initiative, Nov. 2016)

Potential Reduction in CO_2 Emissions (tons)

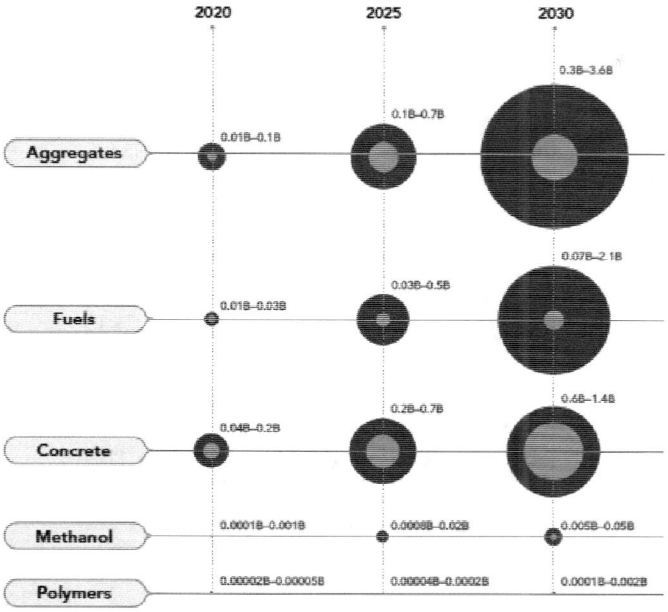

What should we make using CO_2?

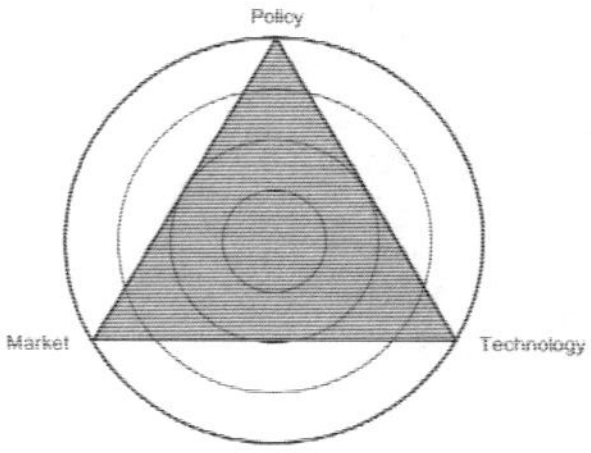

(CO_2 Sciences & The Global CO_2 Initiative, Nov. 2016)

CO_2 Utilization Schemes: Mineral carbonation

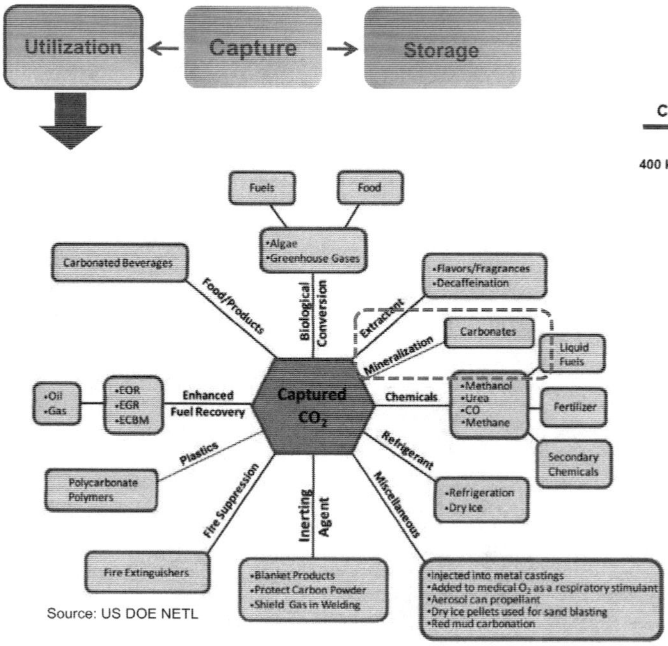

Source: US DOE NETL

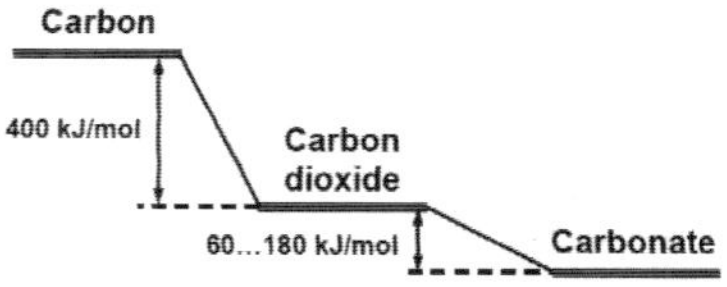

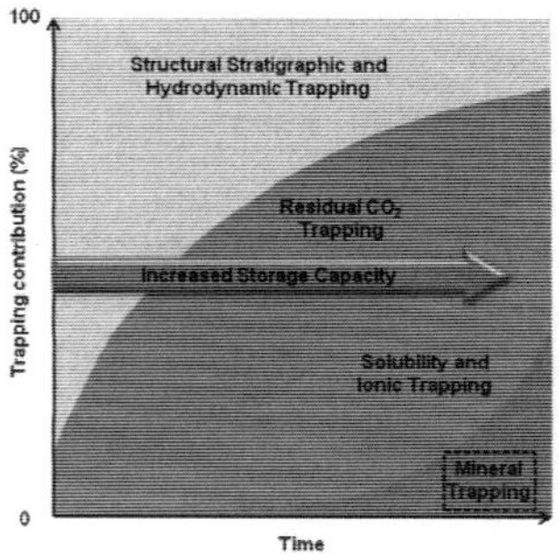

CarbFix Project in Iceland

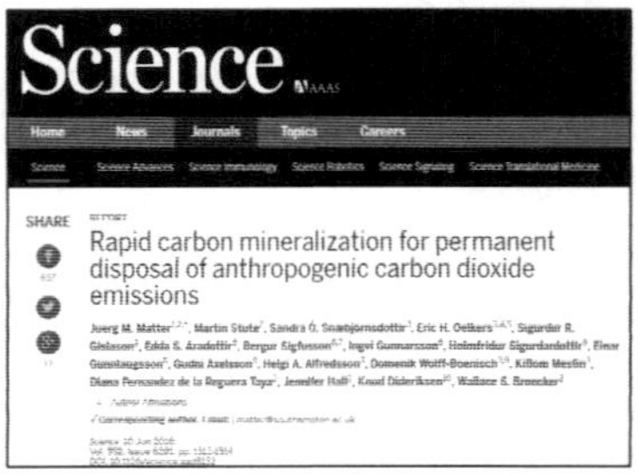

The experiment injected 220 tonnes of carbon dioxide several hundred meters underground (J. MATTER)

Core from CarbFix site

https://www.or.is/en/projects/carbfix/news/carbfix-paper-science

Availability of Minerals

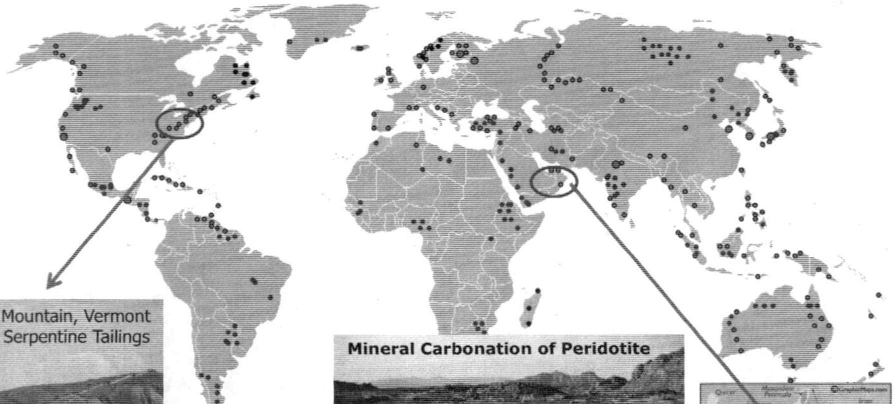

- Basalt
- Labradorite
- Magnesium-based Ultramafic Rocks (Serpentine, Olivine)

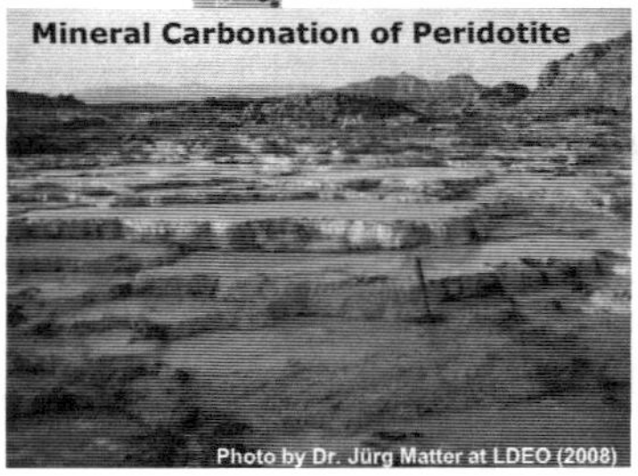

Effect of Mineralogy on CO_2 Storage

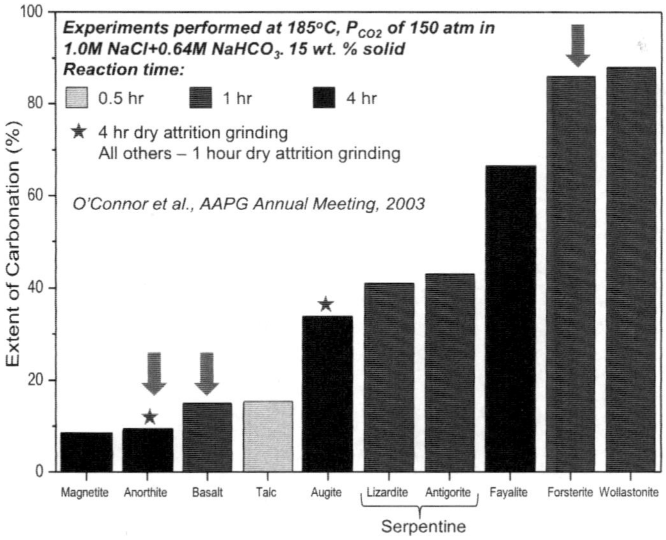

Abundance of less reactive minerals vs. limited availability of highly reactive minerals

Carbonation efficiency defines whether mineral is utilized for ex-situ or in-situ storage

Ex-situ CO_2 Storage	In-Situ CO_2 Storage
Shorter time scales (~hours)	Longer time scales (~years)
Limited spatial scale	Larger spatial scale with utilization of earth as a reactor (~hundreds of miles)
Relatively homogenous mineralogy	Heterogeneous mineralogy
More flexible tuning in reaction conditions	Not limited by reactor size; Use of geothermal gradient
Possible production of value-added products	Multiple CO_2 trapping mechanisms
No monitoring required	Relatively economical at this time

Incongruent Dissolution of Silicate Minerals

Serpentine, $Mg_3Si_2O_5(OH)_4$

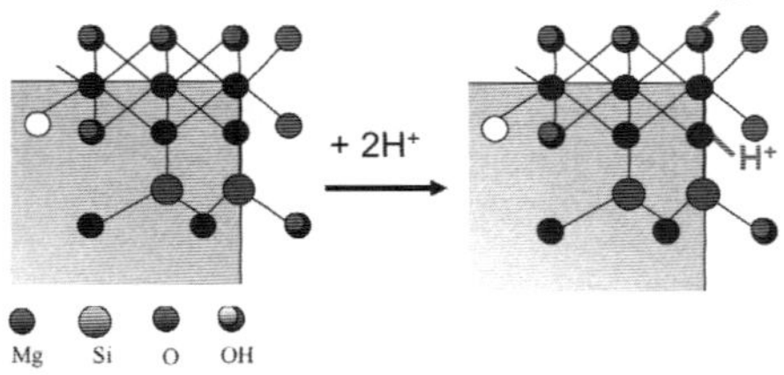

Challenges:

- Slow dissolution kinetics of silicate minerals in nature
- Diffusion-limiting ash layer formation

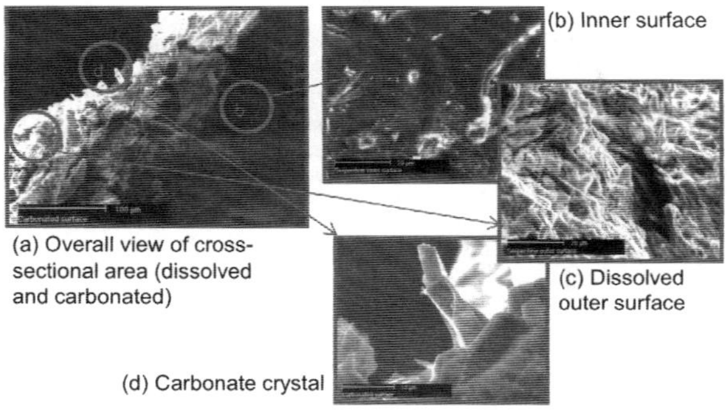

(a) Overall view of cross-sectional area (dissolved and carbonated)

(b) Inner surface

(c) Dissolved outer surface

(d) Carbonate crystal

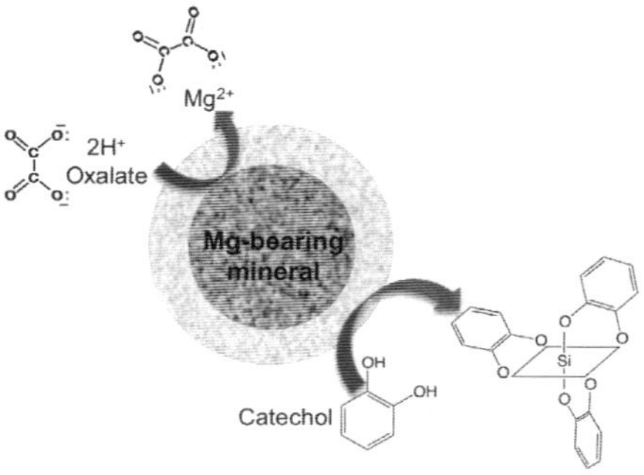

Mg-targeting Ligands and their activities

- **Ligand properties in the solution**

Chelating agent	Structure	Mg-Complex	Log(β)	pKa
Acetate	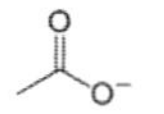	MgL	3.4	1.25 4.14
Citrate	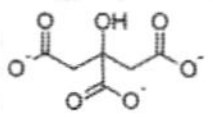	MgL^- MgHL MgH_2L^+	4.7 9.2 1.1	3.13 4.76 6.40
Oxalate	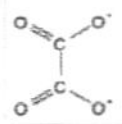	MgL^+	1.26	4.76
Glutamate	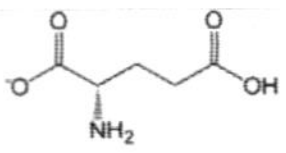	MgL	2.8	2.1 4.07 9.47

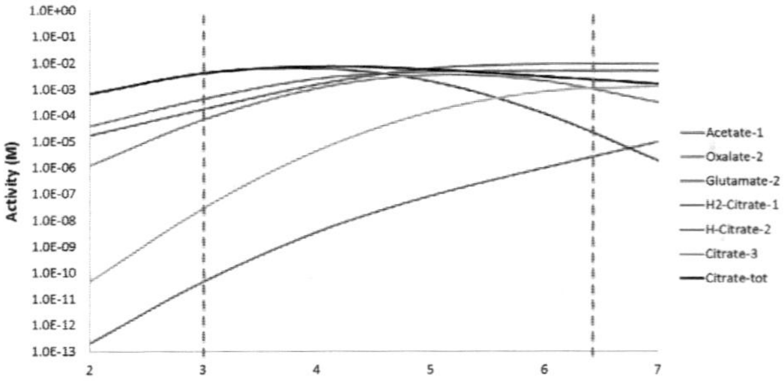

Cit>Oxa>Ace>Glu **Ace>Oxa>Cit>Glu**

- ➢ Citrate has a larger number of binding sites, relatively higher pKa values and larger stability constants to form complex

- ➢ Citrate the highest total ligand activity at pH 3
- ➢ Total ligand activity of citrate doesn't change significantly with pH
- ➢ The activity of glutamate change significantly with pH

13

Catalytic Si Removal – Sustain Reaction Rates

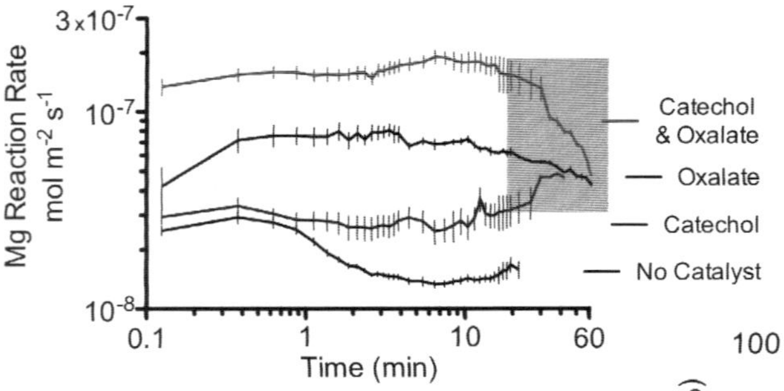

Moderate Process Conditions:
90°C, pH 3.0 ,10^{-2} M Catalyst

Highest literature conversion for this material[1]: 9%

Mg recovery of greater than 85% in < 1 hour

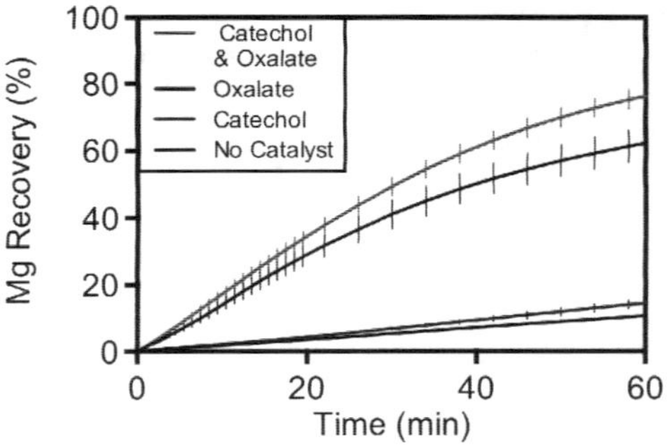

1. Gerdemann et al., *Environ Sci Technol* **41**, 2587–2593 (2007).

Significant Effect on Microstructures

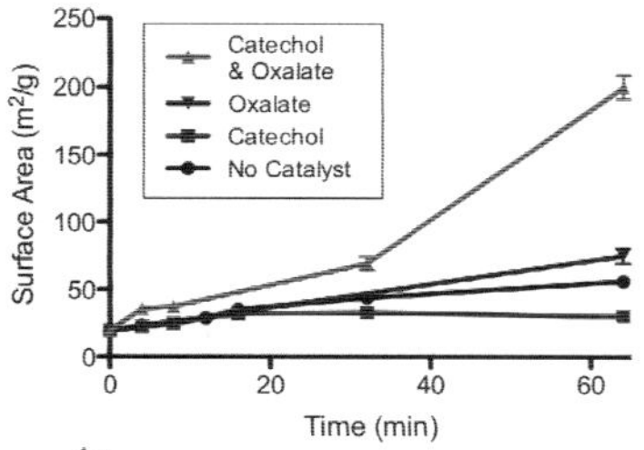

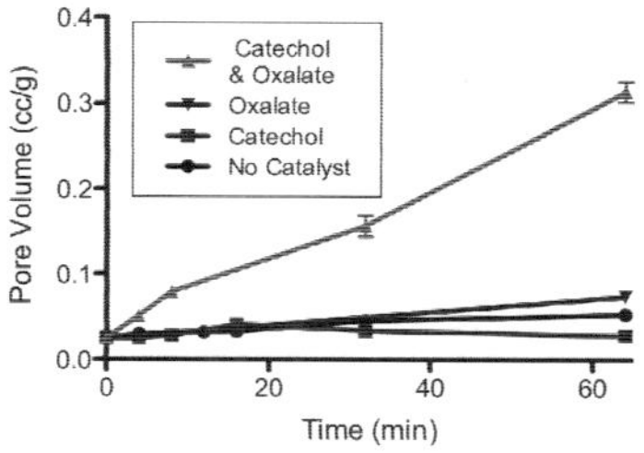

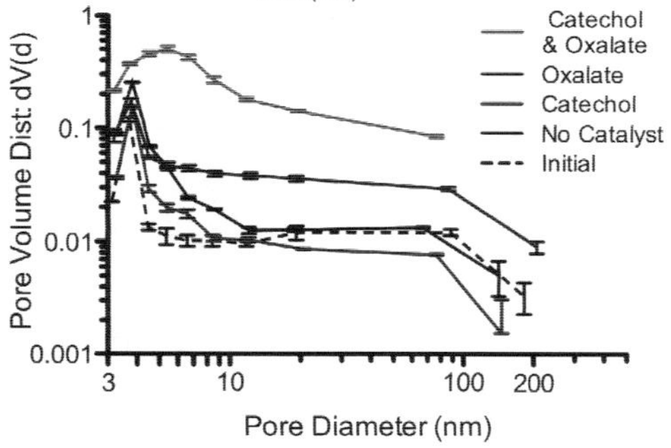

Enhanced reactivity related to Pore Growth

→ Not removal of passivation layer

Kinetics Modeling of Incongruent Silicate Material Dissolution

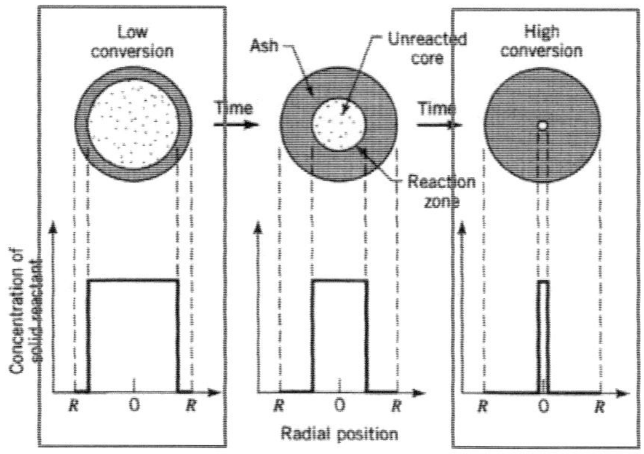

$$\frac{1}{SA(t)} \cdot \frac{dN_{Mg}}{dt} = k \cdot [H^+]^n$$

Chemical Reaction Control Stage of HT Serpentine

$$\frac{1}{SA(t)} \cdot \frac{dN_{Mg}}{dt} = D_e \cdot \frac{dC_{H^+}}{dr} \approx D_e \cdot \frac{[H^+]_0}{h}$$

Mass Transfer Control Stage of HT Serpentine

Levenspiel, Chemical Reaction Engineering, 2nd Edition. (1972.)

Kinetics Modeling of Incongruent Silicate Material Dissolution

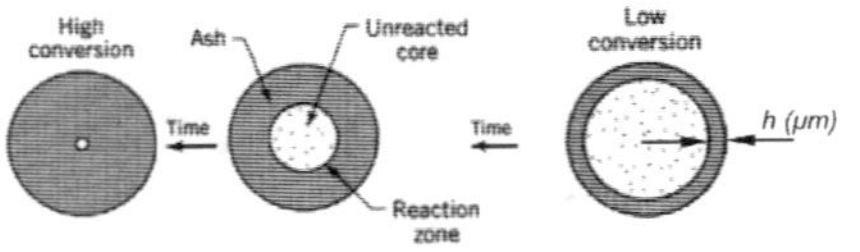

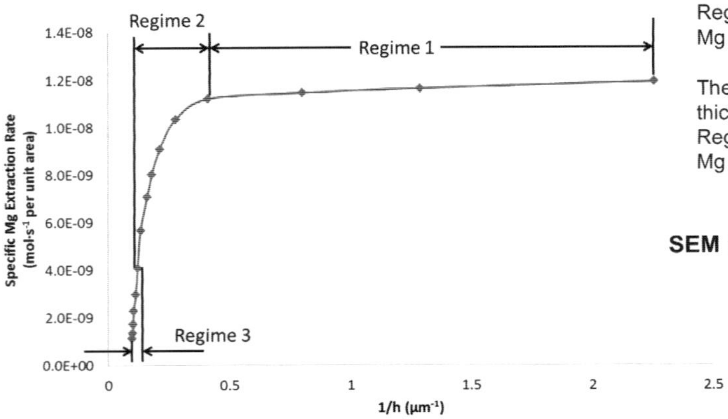

Specific Mg Extraction Rate vs. 1/h

The transition point of ash layer thickness between Regime 1 and Regime 2: ~2 μm. Corresponding Mg extraction is about 16%.

The transition point of ash layer thickness between Regime 2 and Regime 3: ~7 μm. Corresponding Mg extraction is about 44%.

SEM Result: 46% Mg extraction

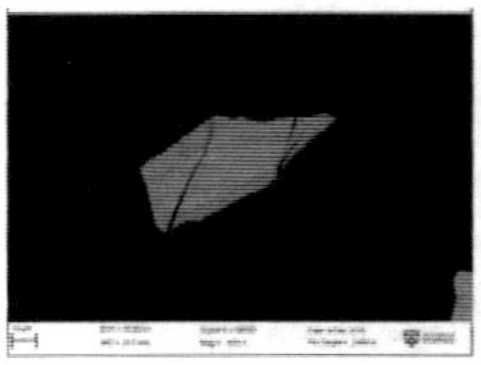

J. Mann and B. Haynes. University of Sydney. (2016)

Activation Energy E_a

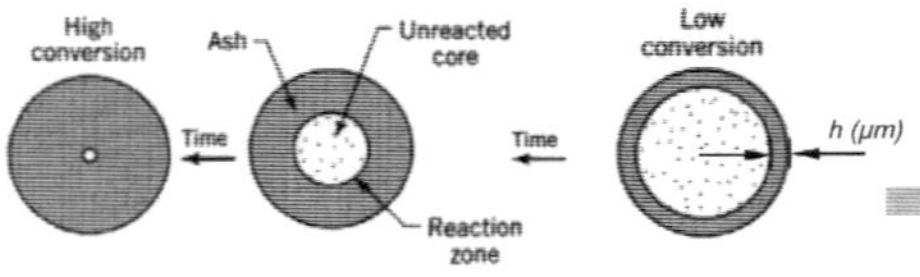

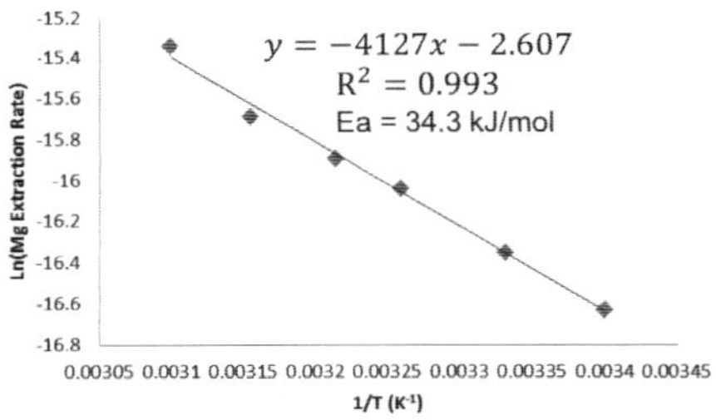

ln(Mg extraction rate) and 1/T

Specific Mg Extraction Rate vs. 1/h

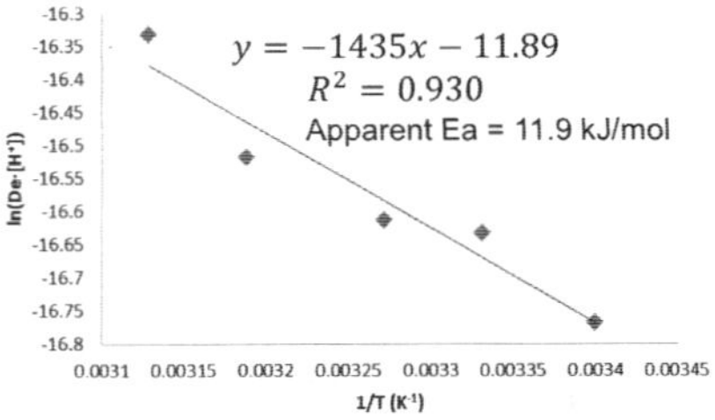

ln(D_e·[H^+]) vs. 1/T

Rate Equation of the chemical reaction control regime:

$$\frac{1}{SA(t)} \cdot \frac{dN_{Mg}}{dt} = A \cdot e^{-\frac{Ea}{RT}} \cdot [H^+]^n$$

$$ln\left(\frac{dN_{Mg}}{dt}\right) = \ln(SA(t)) + \ln(A) + n \cdot \ln[H^+] - \frac{Ea}{R} \cdot \frac{1}{T}$$

Rate Equation of the mass transfer control regime:

$$\frac{1}{SA(t)} \cdot \frac{dN_{Mg}}{dt} = D_e \cdot \frac{d[H^+]}{dr} \approx D_e \cdot \frac{[H^+]_0}{h} = A_{De} \cdot e^{-\frac{Ea}{RT}} \cdot \frac{[H^+]_0}{h}$$

$$ln(D_e \cdot [H^+]_0) = ln(A_{De}) + ln([H^+]_0) - \frac{Ea}{R} \cdot \frac{1}{T}$$

Heat-treated Serpentine Mineral

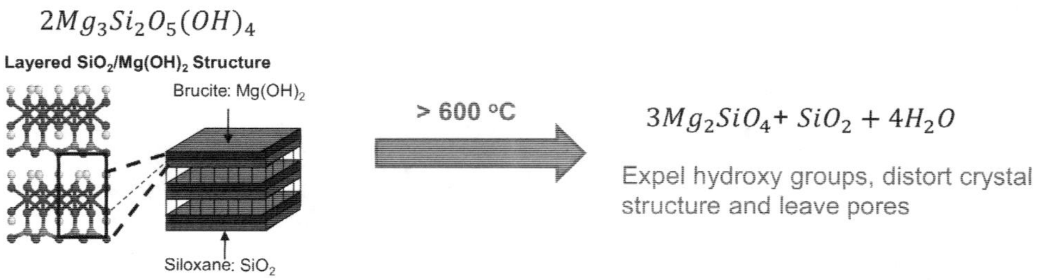

XRD

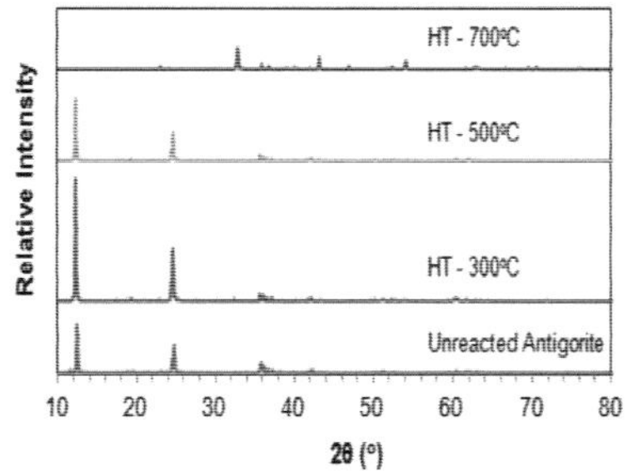

Pore volume distribution change (Synchrotron data from ANL)

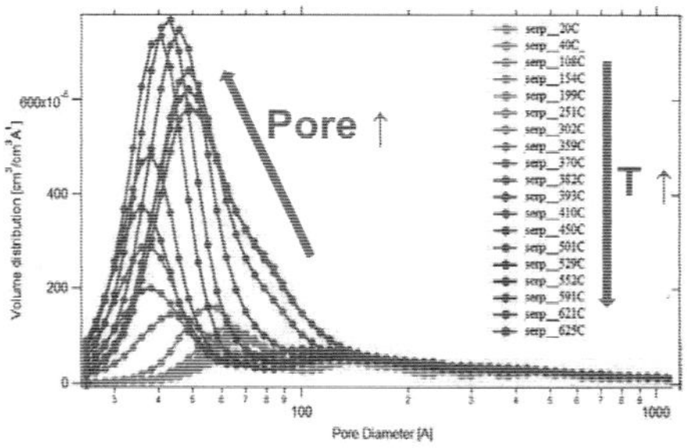

pH Swing by Acid and Base

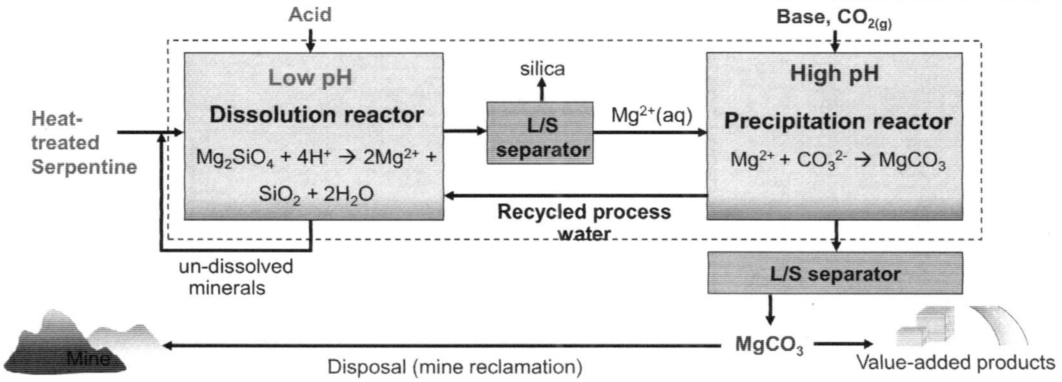

pH swing by base

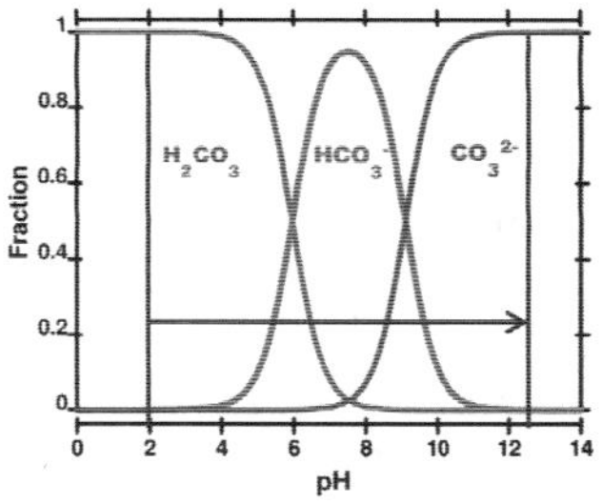

The need for acid and base has limited technical and economic feasibility of carbon mineralization

Two-step Carbonation of HT serpentine via Pco₂ swing

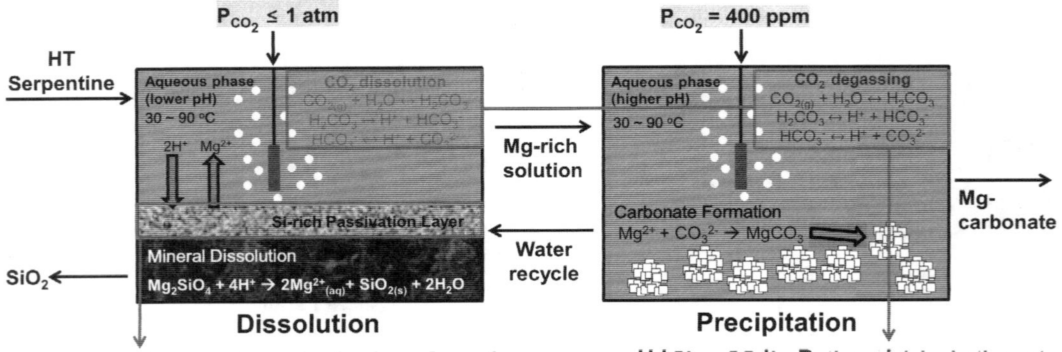

Diffusion-limiting Si-rich passivation layer formation

pH increase by Pco₂ swing

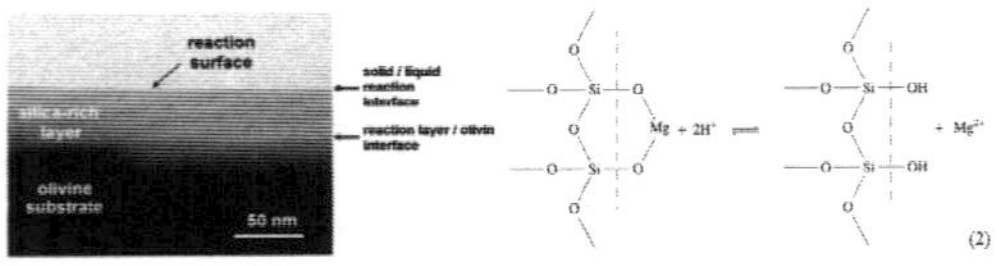

Mg-O bond < Si-O bond

$$> \mathrm{Si-OH} + \mathrm{HO-Si} < \; \rightarrow \; > \mathrm{Si-O-Si} < \; + H_2O$$

silanol group oxide bonds

Ref. Bearat et al., 2006

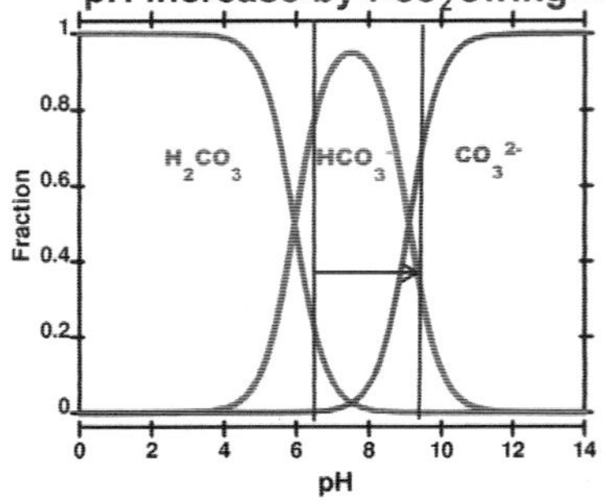

Formation of Passivation Layer

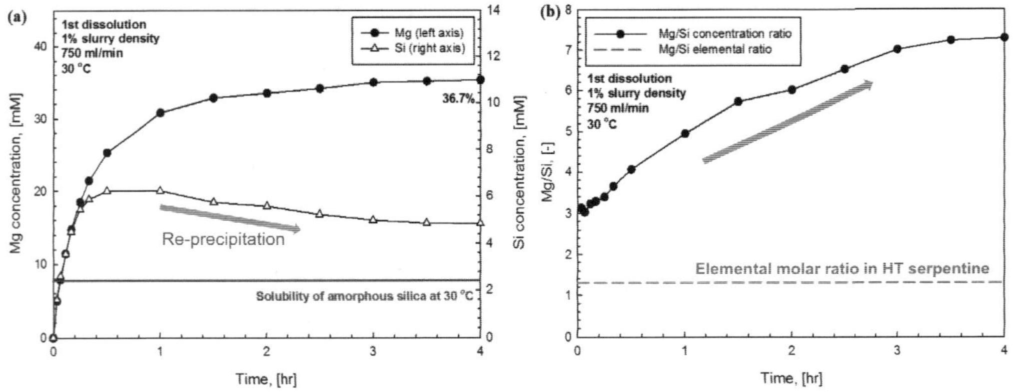

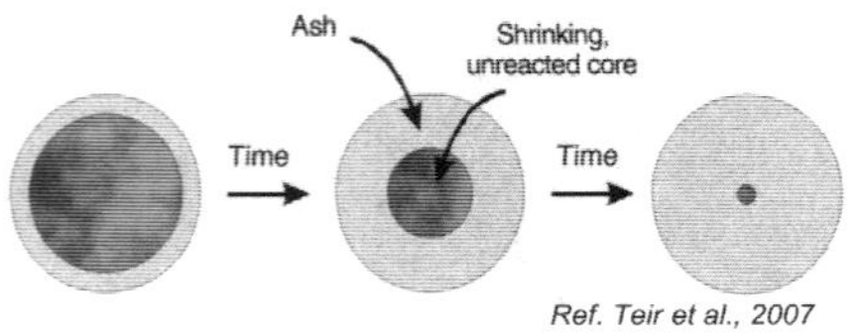

Ref. Teir et al., 2007

Mg-O bond < Si-O bond

⬇

Non-stoichiometric dissolution between Mg^{2+} and $Si_{(aq)}$

⬇

Formation of Si-rich layer

Effect of Ligand on Mg Leaching and Carbonate Precipitation

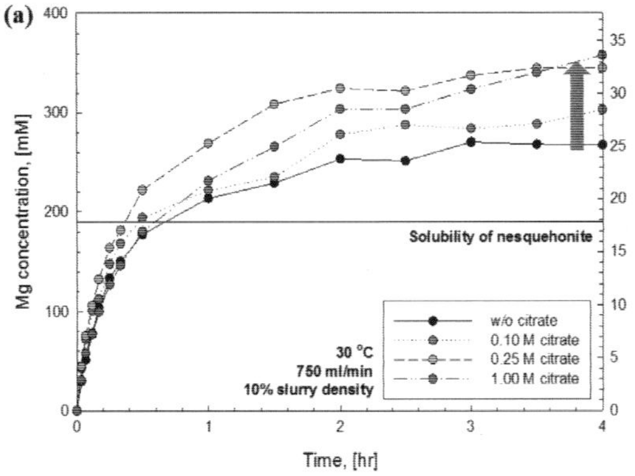

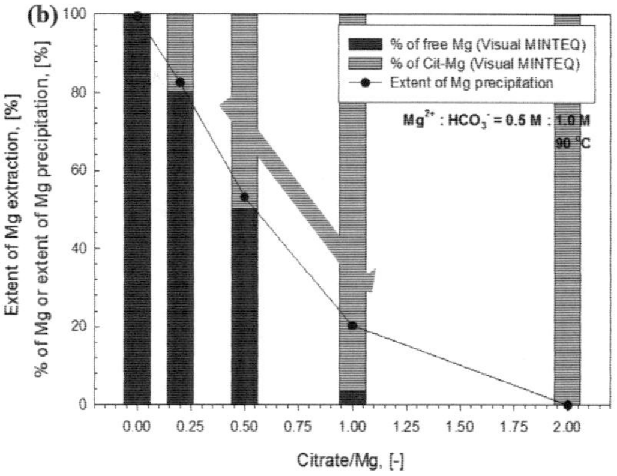

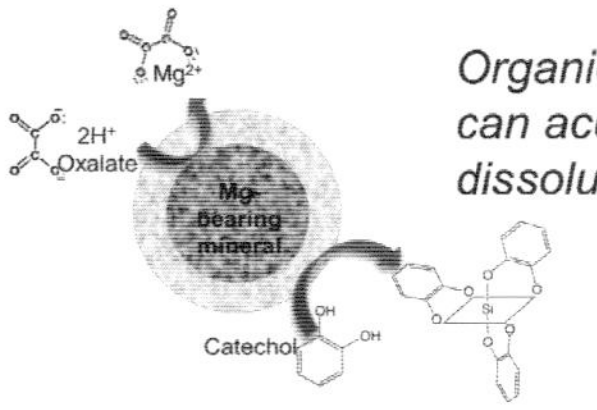

Organic ligands can accelerate dissolution

Formation of Mg-citrate complex

Negative effect on Mg-carbonate precipitation in P_{CO_2} swing method

Minimize use of ligand
Further study of ligand effect is required

Internal Grinding Test

- GB only

- GB with HTS

Internal Grinding Test

- **1st leaching**

- pH profile

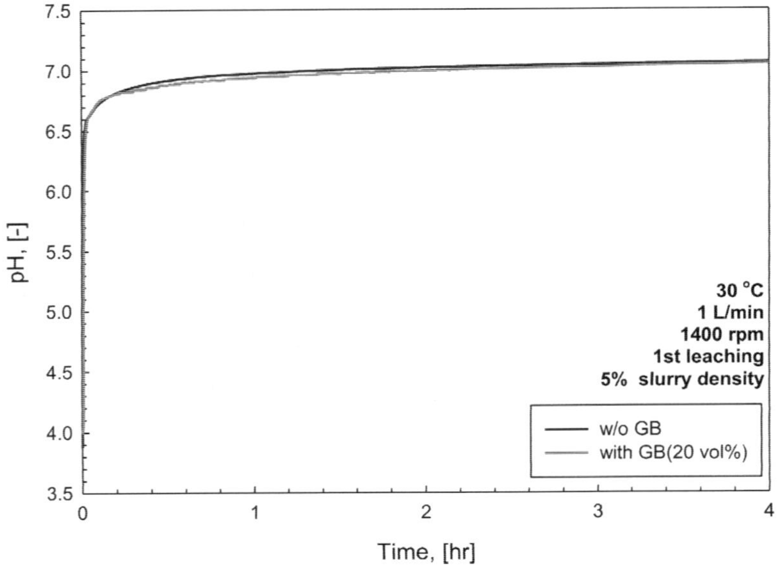

Internal Grinding Test

- **1st leaching**

- Mg concentration profile

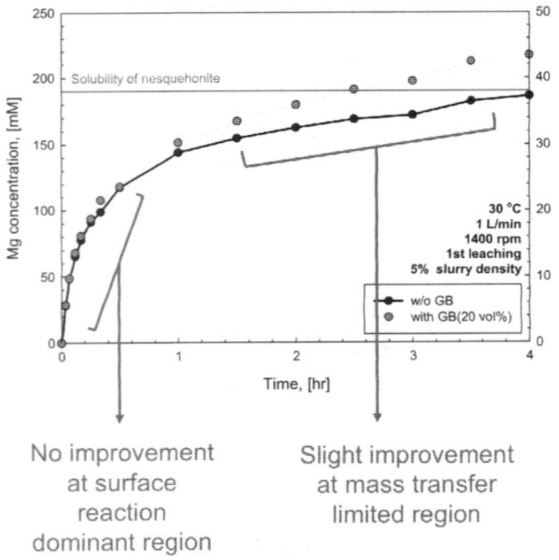

- Si concentration profile

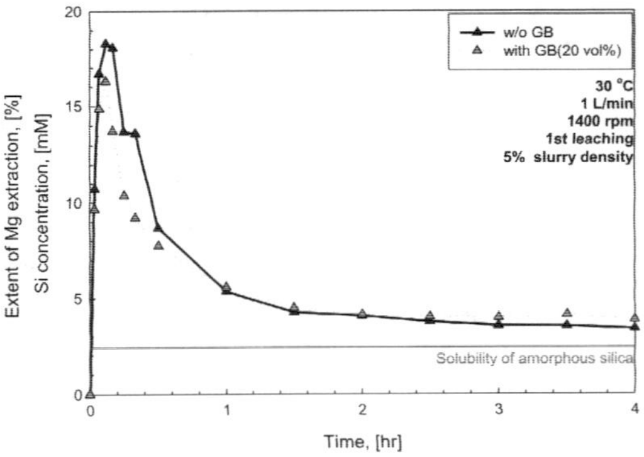

Internal Grinding Test

- **2nd leaching**

- pH profile

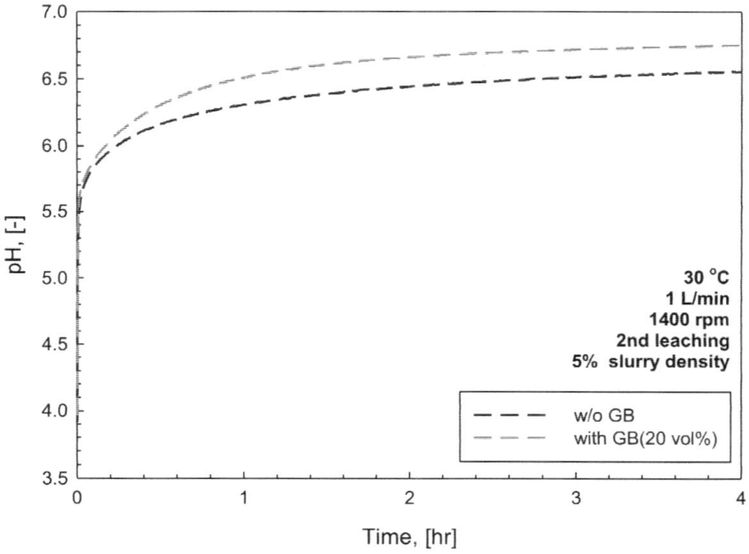

Internal Grinding Test

- **2nd leaching**

- Mg concentration profile

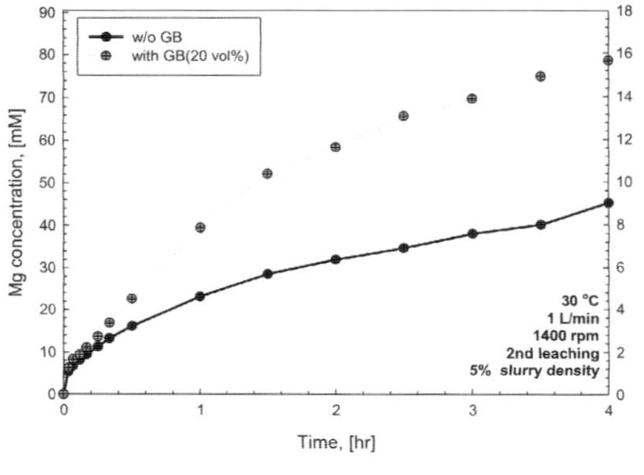

- Si concentration profile

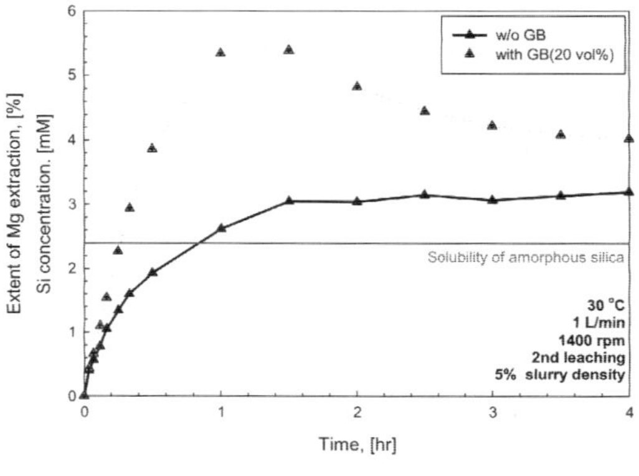

Effect of Physical Activation on Mg leaching

※ After 1st leaching, HT serpentine residue was dried and ground for 1 min

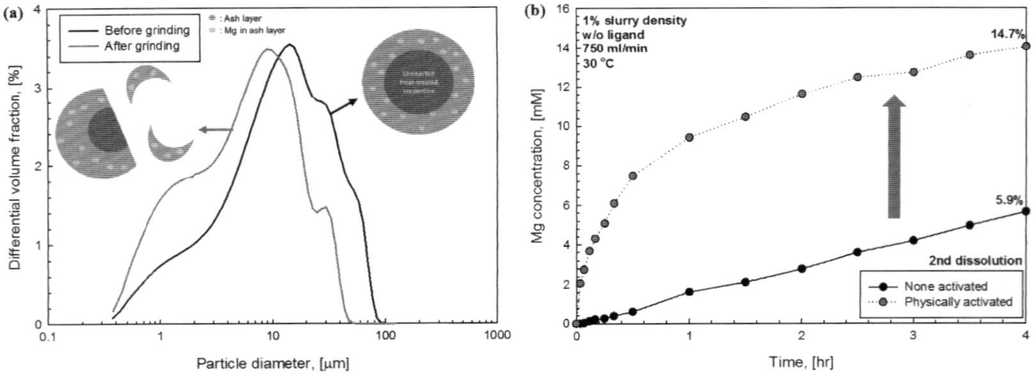

Activated vs **Non-activated**: Improve Mg leaching behavior by ash layer removal

Combined Effect of Ligand and Physical Activation on Mg Leaching

※ After 1st leaching, HT serpentine residue was dried and ground for 1 min

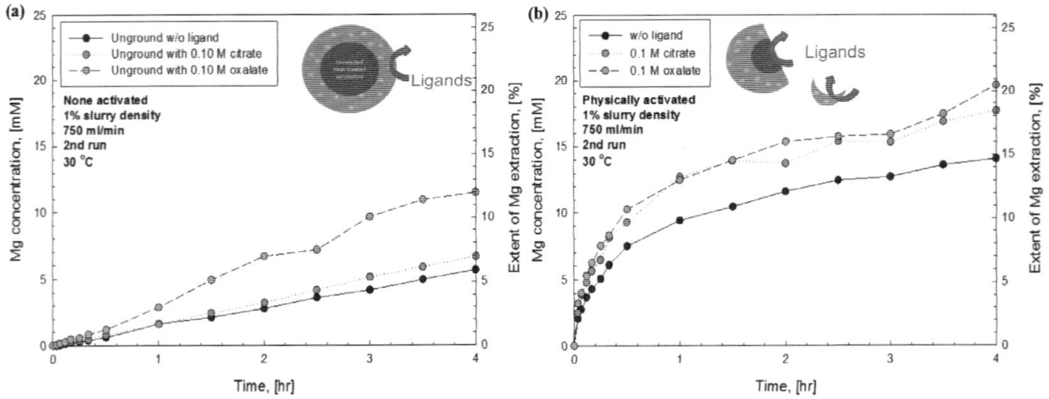

Non-activated: Limited initial Mg leaching kinetic even with ligand

Activated + Ligand: Improved initial kinetic and leaching efficiency

Two-Step Leaching

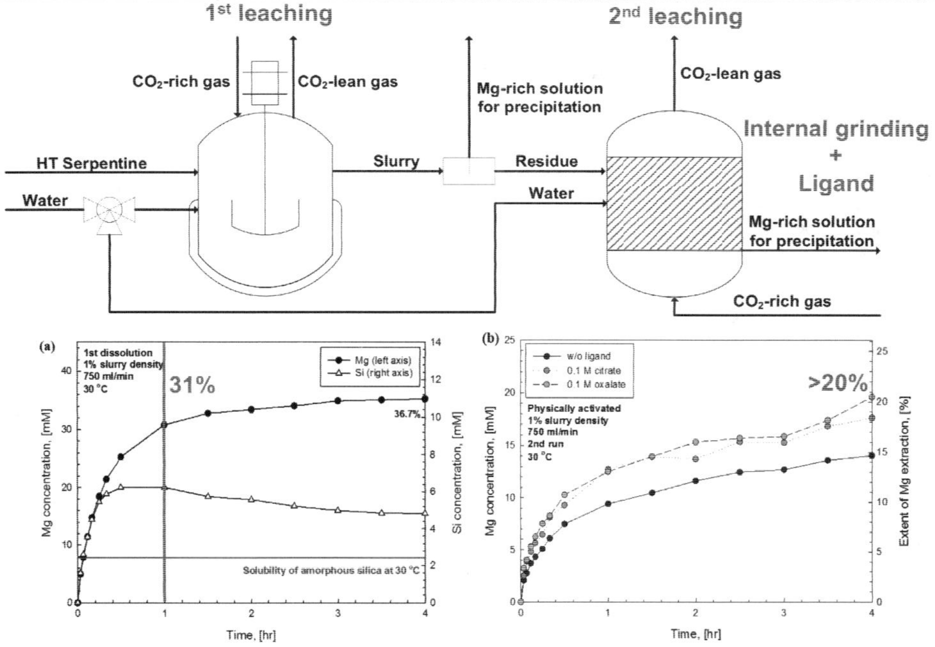

Two-Step Leaching

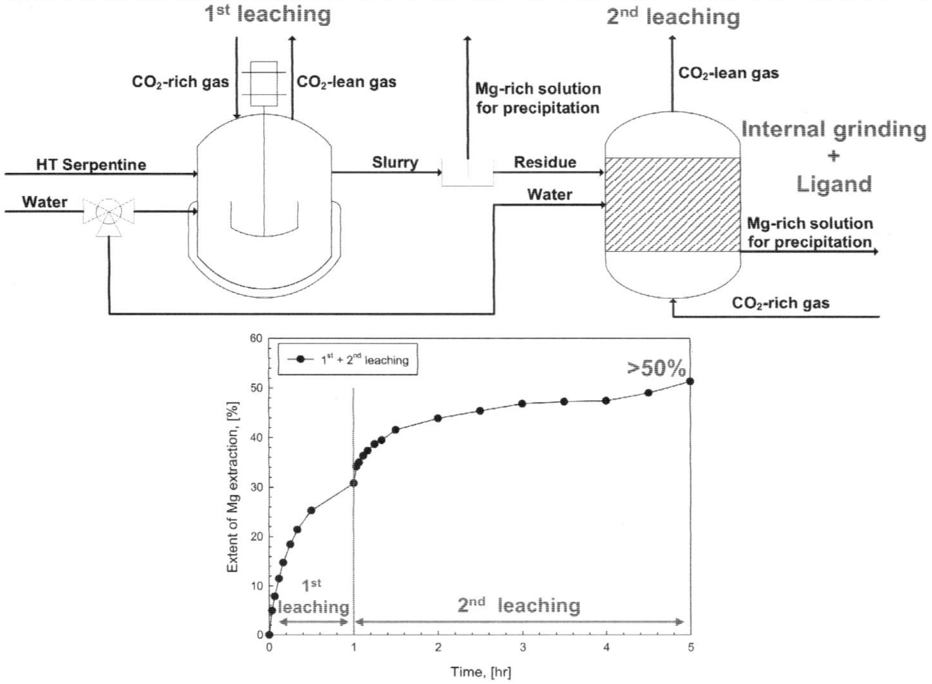

Metastability in $MgO-CO_2-H_2O$ System

- Magnesite is the most stable and least soluble carbonate under most conditions (including P_{CO2})
- Despite this, magnesite is seldom the main product reported in literature:
  - Brucite: $Mg(OH)_2$
  - Lansfordite: $MgCO_3 \cdot 5H_2O$
  - Nesquehonite: $MgCO_3 \cdot 3H_2O$
  - Hydromagnesite: $Mg_5(CO_3)_4(OH)_2 \cdot 4H_2O$
  - Magnesite: $MgCO_3$
- Driven by reaction kinetics, given enough time, magnesite should form

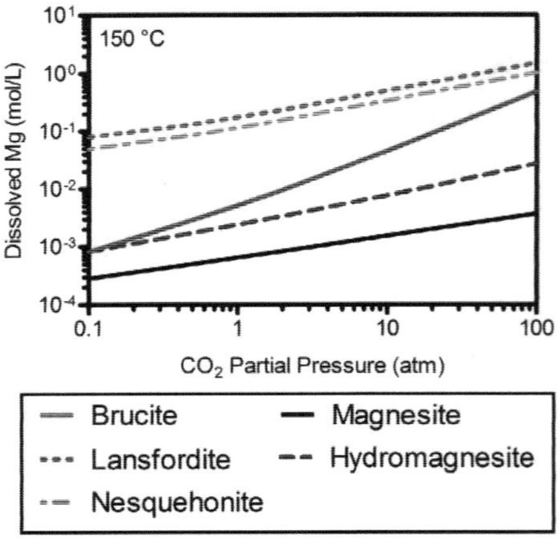

Swanson et al., Directed Precipitation of Hydrated and Anhydrous Magnesium Carbonates for Carbon Storage, **2014**, *In Preparation.*

Carbonation Behaviors of Minerals

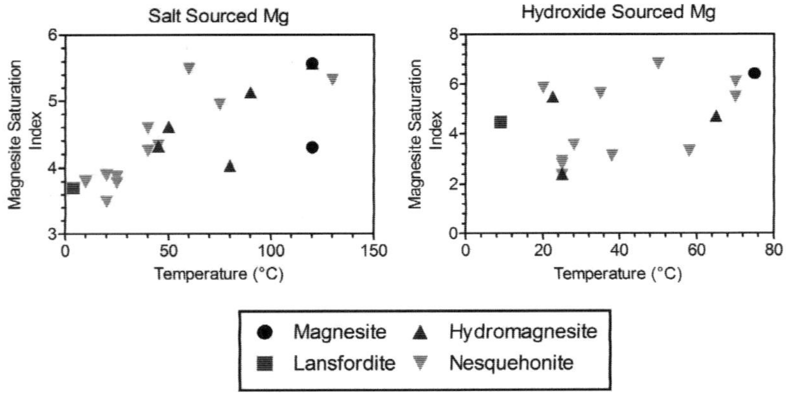

- Maximum saturation index (Ω) achievable under the experimental conditions reported in 23+ publications
- Significant variety in carbonate formation over similar reaction conditions

$$\Omega = \log\left(\frac{IAP}{K_{SP}}\right)$$

Metastable Carbonate Phases at lower T

At 30 °C Make Nesquehonite | **At 150 °C Make Hydromagnesite** | **At 200 °C Make Magnesite**

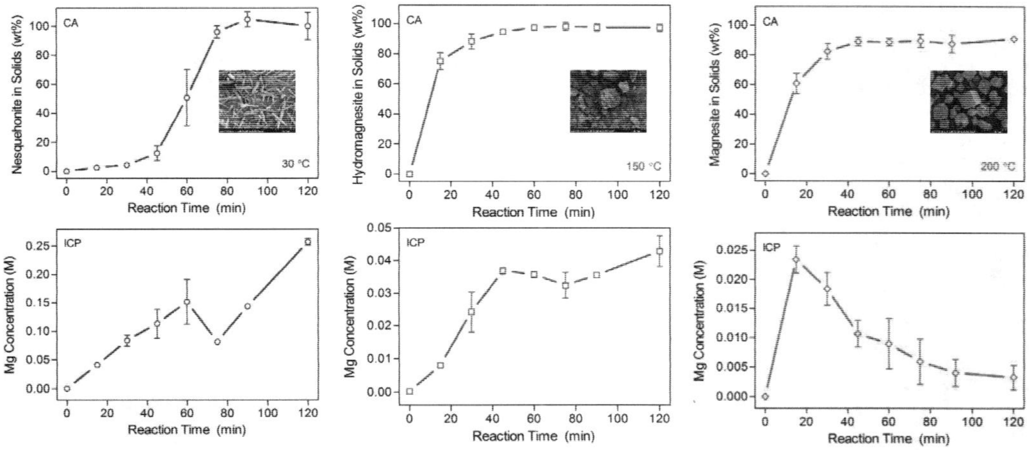

Effect of Temp on Slurry Phase $Mg(OH)_2$ Carbonation

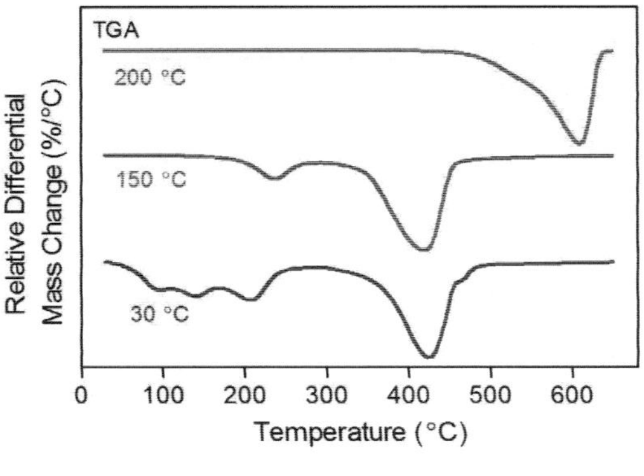

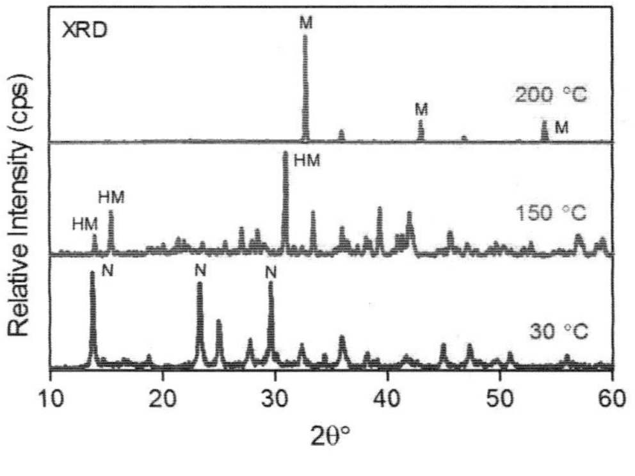

→ Biggest factor influencing phase is Temperature

→ Easily distinguishable by TGA and XRD analyses

Magnesite:	$MgCO_3$
Hydro**m**agnesite:	$Mg_5(CO_3)_4(OH)_2 \cdot 4H_2O$
Nesquehonite:	$MgCO_3 \cdot 3H_2O$

Fricker et al., Investigation of the Different Carbonate Phases and their Formation Kinetics during $Mg(OH)_2$ Slurry Carbonation, *In Review.*

Controlled Precipitation of Mg-carbonates

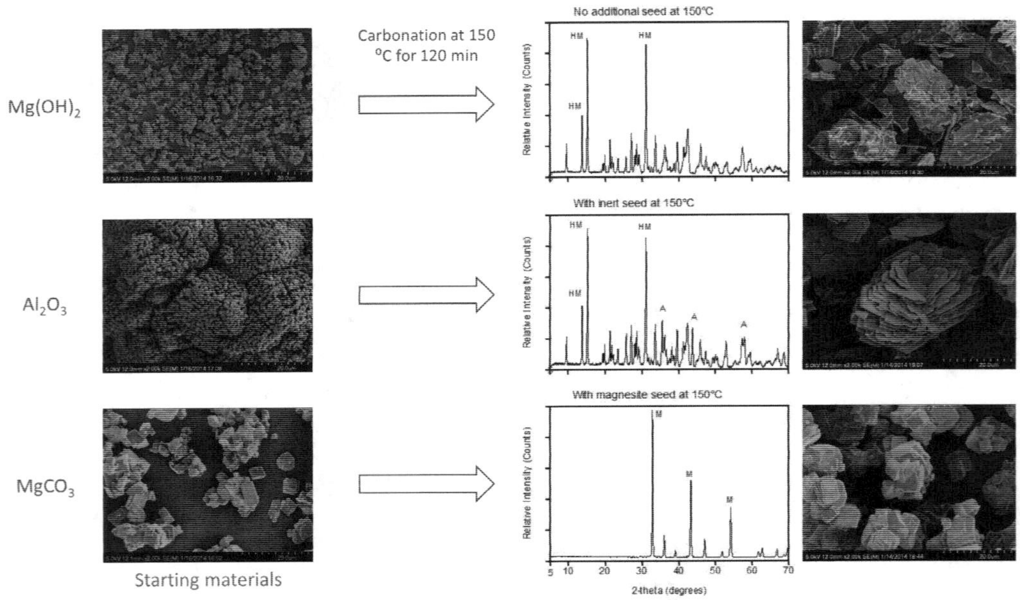

Combined CCUS

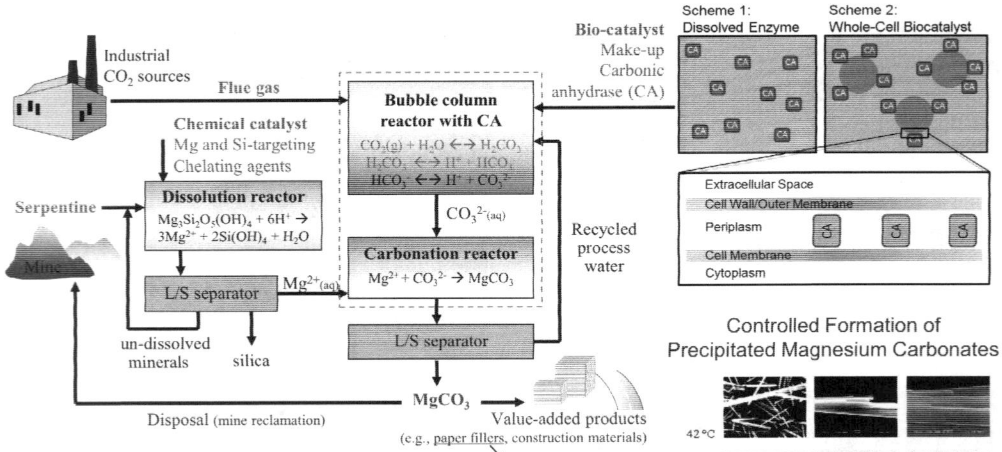

- ❑ No need for the solvent regeneration and CO_2 compression, straightforward MVA
- ❑ Alternative CO_2 utilization option with improved economic feasibility

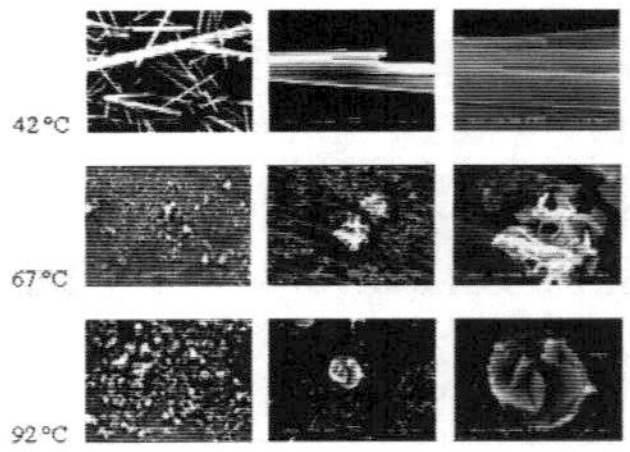

Various Feedstocks for Carbon Mineralization

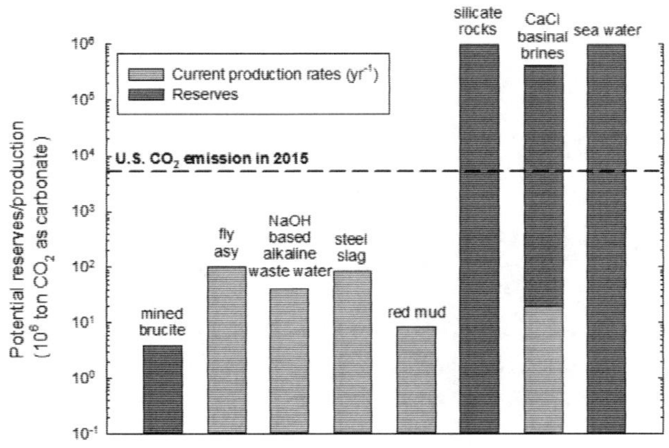

Carbonation of industrial wastes results in reclassification of these materials as non-hazardous hence <u>safe for long-term carbon storage</u> or for <u>sustainable utilization</u>

Availability
minerals >> industrial wastes

Transportation
industrial wastes << minerals

Pre-processing requirements
(e.g., mining, crushing etc.,)
industrial wastes < minerals

Crystallinity
industrial wastes < minerals

⬇

Reactivity
industrial wastes > minerals

Economic feasibility and Net environmental impact
industrial wastes > minerals

Generation and Utilization of Slag

- **Chemistry of Iron Ore Reduction & Slag Formation**

Preheated air and coke	$C + O_2 \rightarrow 2CO$
540 ~ 650 °C	$2CO \rightarrow C + CO_2$
400 ~ 600 °C	$3Fe_2O_3 + CO \rightarrow 2Fe_3O_4 + CO_2$
600 ~ 800 °C	$Fe_3O_4 + CO \rightarrow 3FeO + CO_2$
800 ~ 1100 °C	$FeO + CO \rightarrow Fe + CO_2$
900 °C	$CaCO_3 \rightarrow CaO + CO_2$
Onset of slag formation at ~ 1100 °C	
	$CaO + SiO_2 \rightarrow CaSiO_3$
	$CaO + S + C \rightarrow CaS + CO$
1200-1800 °C	$FeO + C \rightarrow Fe + CO$
	$CO_2 + C \rightarrow 2CO$
	$C + O_2 \rightarrow CO_2$

- **Slag Utilization in Europe and China**

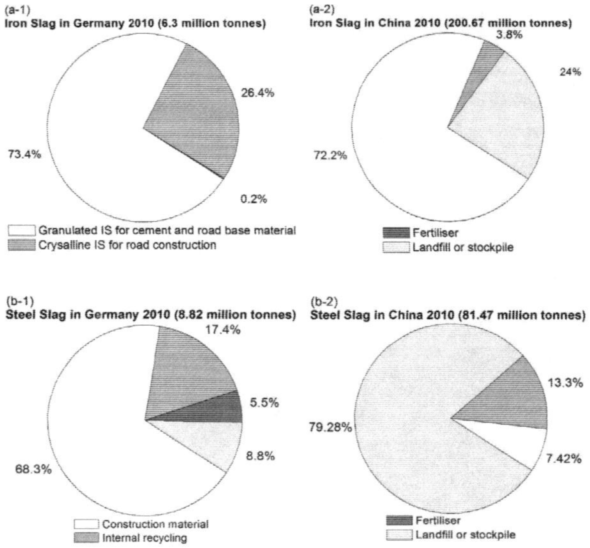

Slags – Big financial & real estate burden for steel industries

"In 2015, the company had a net loss of $200 million for 10MT steel business."
"$20 million was spent on 5 MT slag management."
"An area of 0.3 km^2 will be filled up by June 2016 if slag can not be sold (in 4 years). We need to lease more land."
- Strategy manager at a major China-based steel company

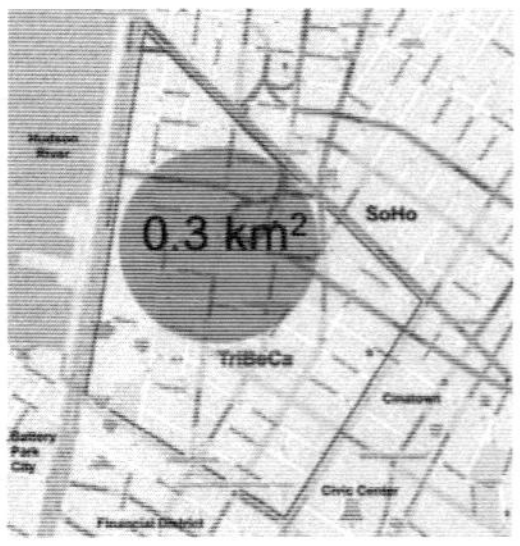

- **Main chemical compositions of procured Basic Oxygen Furnace (BOF) steel slags from Baotou Steel**

Elemental concentration (wt%)					
CaO	SiO_2	MgO	Al_2O_3	MnO	FeO
40.3	13.1	9.03	1.26	4.39	29

Sustainable Construction Materials

(Gunning et al al. 2010)

(Carbstone-innovation)

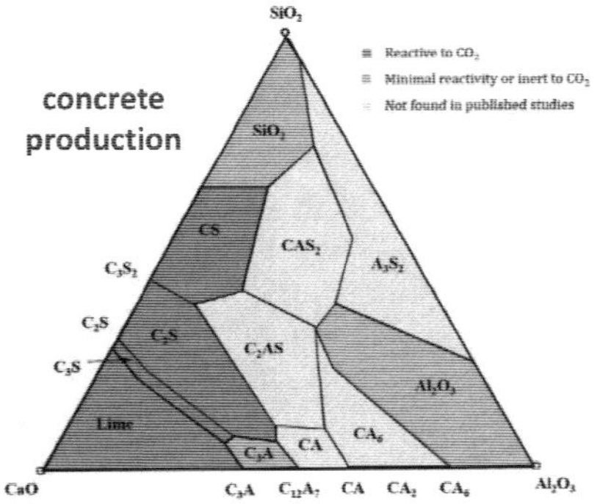

(Zhang et al. 2017)

Carbonated Slags as Sustainable Construction Materials

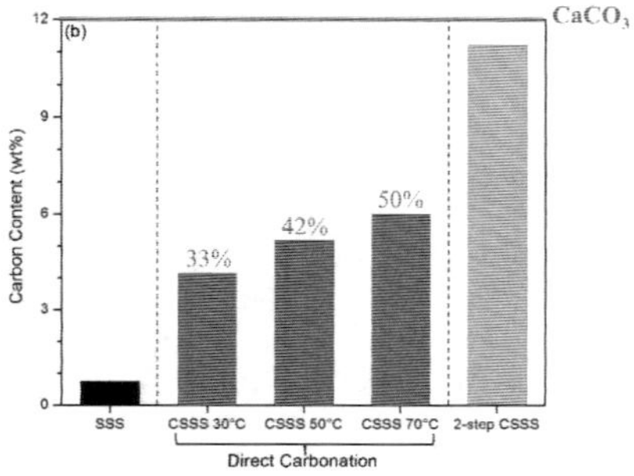

Mechanical Strength of Mixed Cement Mortar (Collaboration with Prof. Yin)

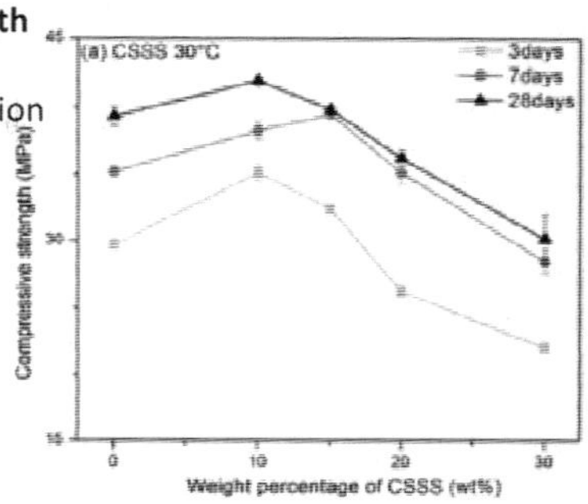

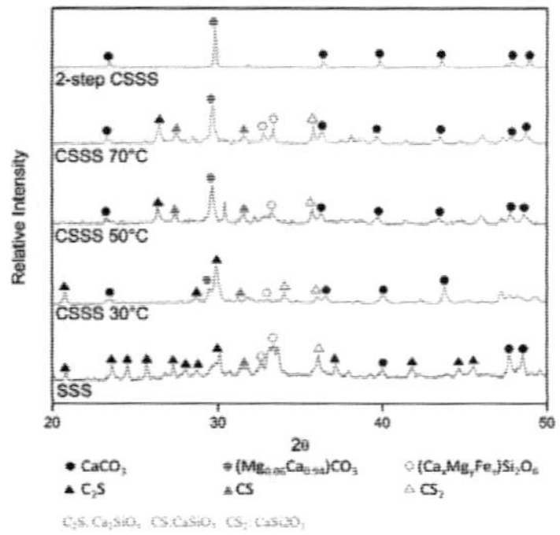

Calorimetric Test During Hydration (Collaboration with Prof. Kawashima)

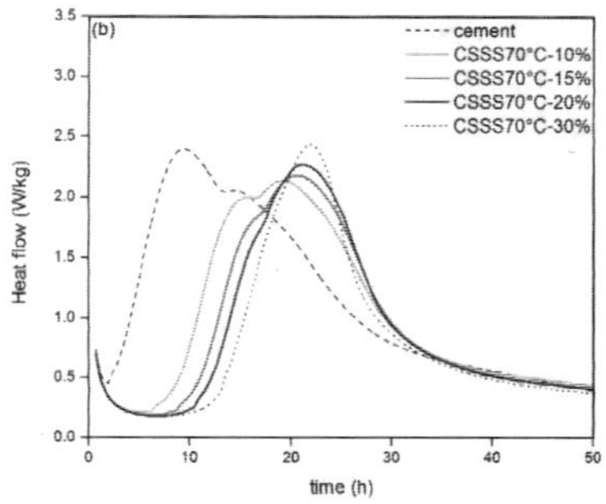

Summary and Outlook

- We need to consider all the CO_2 sources and multifaceted solutions for CCUS
- Important factors for utilization: capacity and economic feasibility, sustainable reactants and energy sources, permanence
- We can consider new pathways to existing chemicals or even new pathways to new chemicals.
- Tailored properties of products – need for new certifications and standards

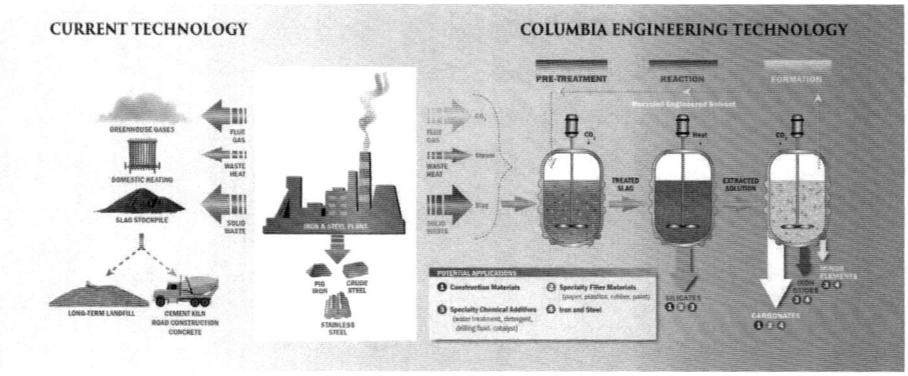

Acknowledgement

Mineral Carbonation International

National Science Foundation
WHERE DISCOVERIES BEGIN

U.S. DEPARTMENT OF ENERGY

arpa·e

ORICA

Sandia National Laboratories

REACTION ENGINEERING INTERNATIONAL

posco

COLUMBIA | ENGINEERING
The Fu Foundation School of Engineering and Applied Science

Lenfest Center for Sustainable Energy
EARTH INSTITUTE | COLUMBIA UNIVERSITY

NYSTAR
Empire State Development's Division of Science, Technology & Innovation

nyserda
Energy. Innovation. Solutions.

Water-Gas Shift Reaction with In-situ Carbon Capture

	ΔHrxn
$Mg(OH)_2 \longrightarrow MgO + H_2O$	~ 80 kJ/mol
$MgO + CO_2 \longrightarrow MgCO_3$	~ -100 kJ/mol
$H_2O + CO \longleftrightarrow CO_2 + H_2$	~ -40 kJ/mol

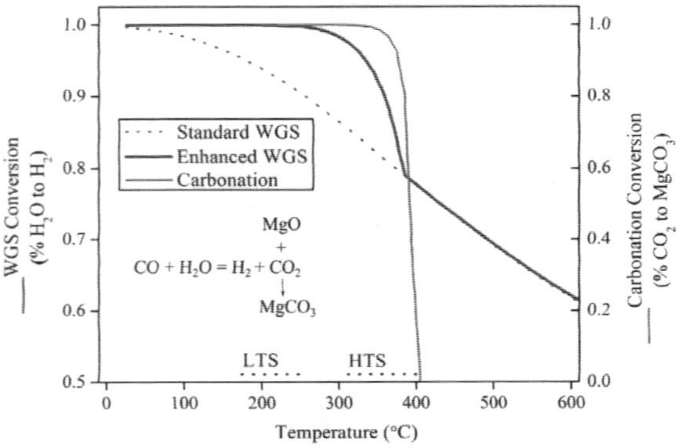

- ✓ Shifts equilibrium to products
- ✓ Provides Water
- ✓ Economic prospects

Mechanism: Effect of Water on Carbonation

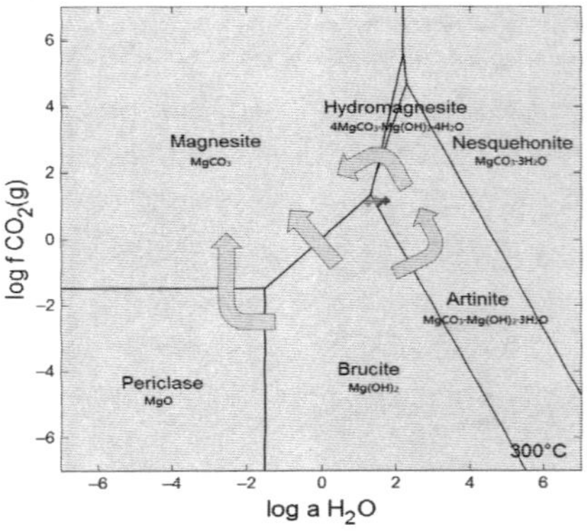

The Geochemist's Workbench (Release 7.0.2)

Bio-Energy with Carbon Capture and Storage (BECCS)

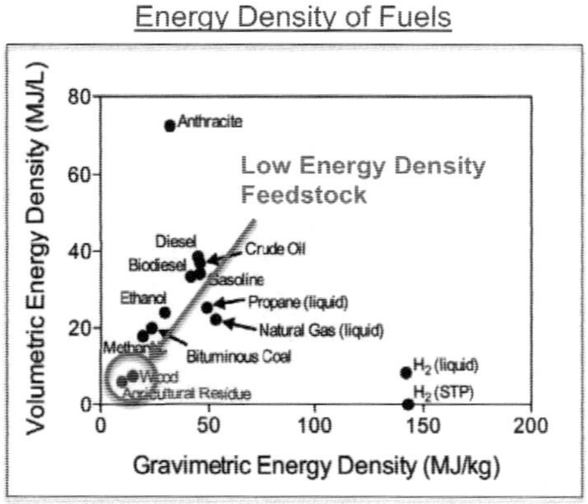

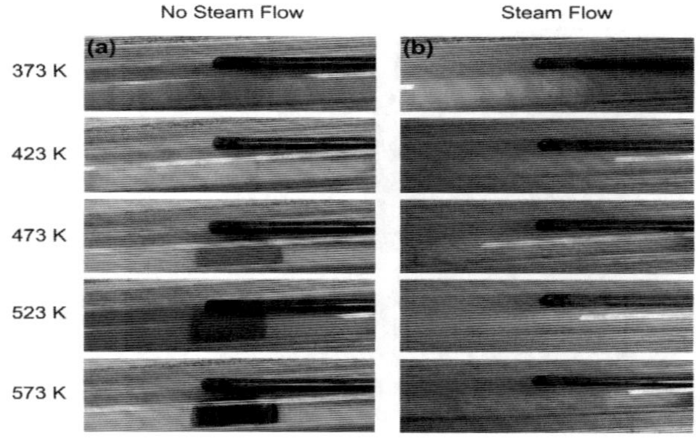

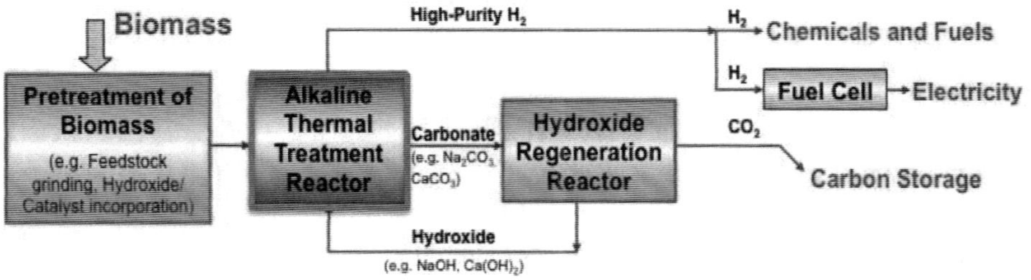

Alkaline Thermal Treatment: $C_xH_yO_z + M_a(OH)_b + H_2O \rightarrow MCO_3 + \mathbf{H_2}$

MONASH University

The Utilisation of Victorian Brown Coal Fly Ash in Mineral Carbonation

Dr. Tahereh (Tara) Hosseini*
Assoc. Professor Lian Zhang

*Email: Tahereh.Hosseinidashtikhani@monash.edu

ACEME Conference

March 2018

GROUP OF EIGHT AUSTRALIA

Clean Solid Fuel Laboratory (CSFL)

Agenda

- Introduction
- Past studies on Mineral carbonation
- Advantage and disadvantage
- New Process developed
- Ongoing Works

Brown coal as a source of energy

MONASH University

- Latrobe Valley mines (Yallourn, Hazelwood and Loy Yang) account for approximately 98% of Australia's brown coal production.
- In 2015 Australia produced approximately 65.4 million tonnes of brown coal (8.1% of global production).
- At current rates the Victorian reserves are estimated to power the stations for a further 500 years.

3

Lignite Fly Ash Utilisation in Victoria

MONASH University

Approx 1.3 million tonnes generated per year;
1.07 million tonnes produced in 2016.

Little is being used, whereas most of the fly ash is pumped with water together into ash dam.

Currently Latrobe Valley's ash ponds contain 25 million tonnes of fly ash

Apart from the expensive maintenance of ash dam, the **environmental pollutions** such as leaching of toxic/trace metals into water is also of concern.

It is becoming more strict for the disposal of fly ash in a disposal site in Australia and the rest of the world.

Pumping of ash slurry

Construction of ash dam

4

Black Coal Fly Ash Utilisation

(Australian Asia Association, 2016)

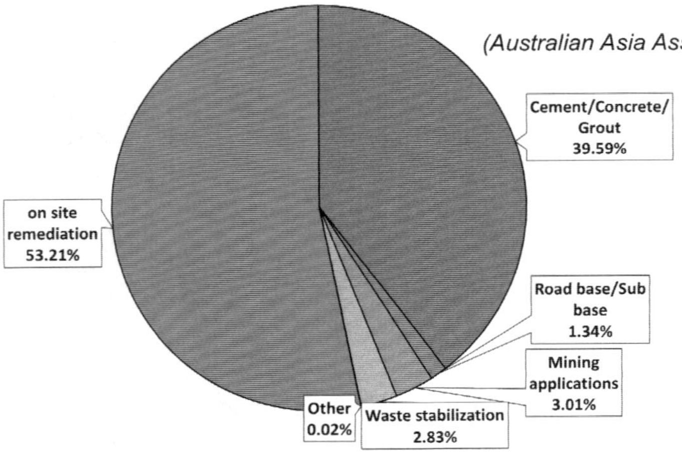

- ✓ 44% of the total fly ash produced in 2016 were utilized;
- ✓ This chart only applies to high-rank coal fly ash which is rich in silica;
- ✓ Silica in fly ash reacts with lime and moisture/water (pozzolanic reaction) to form hard concrete.
- ✓ Lignite fly ash, however, is lean in silica and rich in free MgO, thus cannot be used as additive to cement at all.

5

Comparison of Fly Ash Compositions

Typical Black Coal Fly Ash

	Wt%
SiO_2	41.3
Al_2O_3	30.4
CaO	5.7
MgO	2.4
Fe_2O_3	2.4
Na_2O	0.4
K_2O	0.1
SO_3	15.3
P_2O_5	0.2

Victorian brown coal fly ash

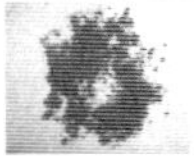

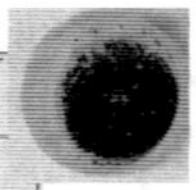

	Hazelwood 1	Hazelwood 2	Yallourn
SiO_2	5.36	2.11	11.3
Al_2O_3	2.6	2.14	6.98
CaO	30	27.4	8.49
MgO	25.8	23.5	23
Fe_2O_3	11.2	9.66	43.9
Na_2O	6.58	8.91	1.01
K_2O	0.48	0.46	0.32
SO_3	15.1	22.8	2.63
P_2O_5	0.04	0	0.07

Abundant alkali (Na, K), alkaline earth metals (Ca, Mg) and iron (Fe)

6

Hazelwood and Yallourn Fly Ash

MONASH University

Chemical Structures

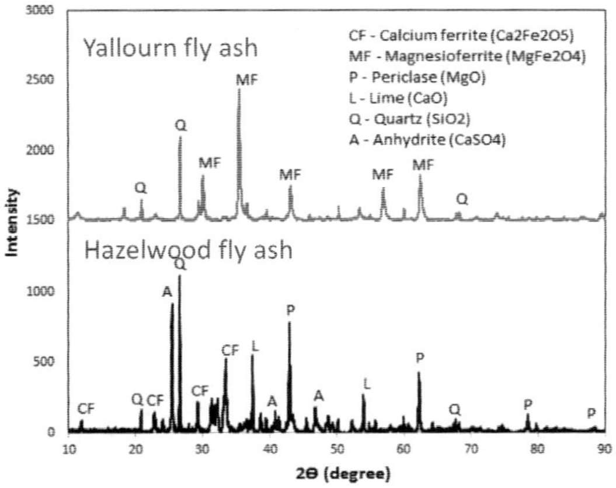

Yallourn fly ash: Rich in spinel formed by the reaction between MgO and Fe_2O_3 in the coal-fired boiler, which is very tough to break down.

Hazelwood fly ash: Rich in free oxides (MgO);

7

The products from Fly Ash Utilisation

MONASH University

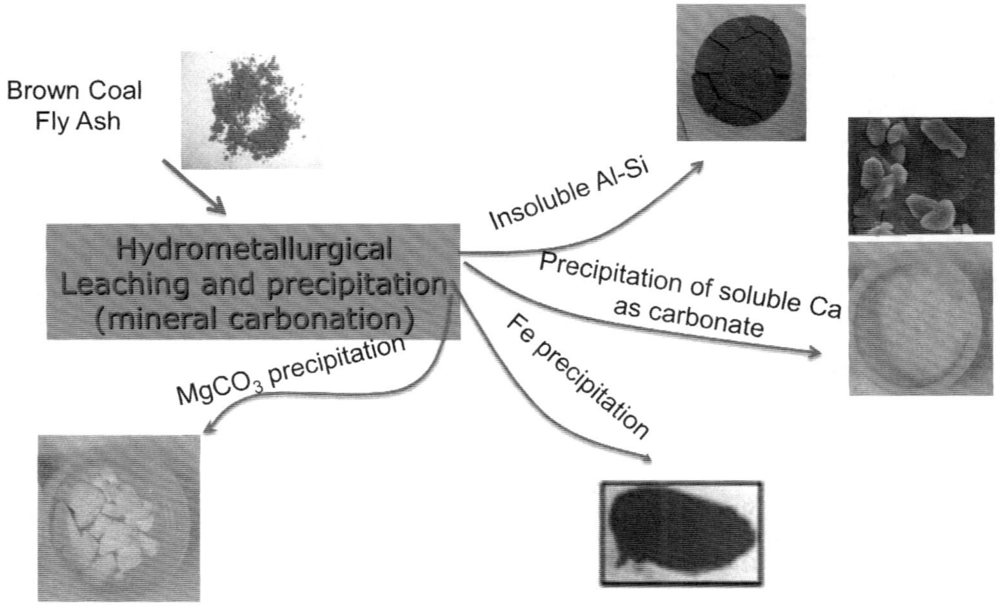

8

High-Value Use of Individual Metals

MONASH University

High-purity $CaCO_3$/$CaSO_4$ (formed via mineral carbonation) : useful in pulping and cement industry; sorbent for desulfurisation and ash slagging flux additive in power plant.

High-Purity $MgCO_3$ and MgO: manufacturing of MgO – board; precursor for magnesium metal – the lightest but strong metal to be used in Al alloy;

$Fe(OH)_3$: precursor for spongy iron metal generation, and good catalyst precursor in a variety of energy processes.

Aluminasilicate-rich residue: additive into cement or precursor for zeolite; or to make geo-polymer.

9

Mineral Carbonation Schematic process

MONASH University

Leaching Reaction : $Mg/CaO + 2NH_4Cl \rightarrow Mg/CaCl_2 + H_2O + 2NH_3(g)$

Carbonation Reaction :
$Mg/CaCl_2+CO_2+2NH_3+H_2O \rightarrow Mg/CaCO_3(s)+2NH_4Cl$

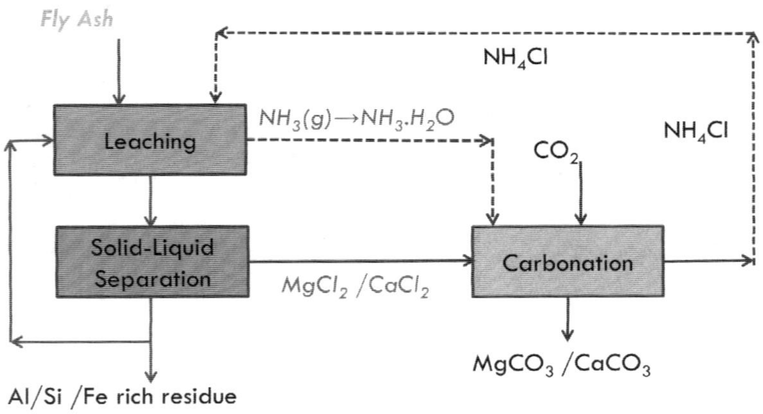

1

Advantage and disadvantage

MONASH University

Advantage:

- The leaching reagent is regenerated simultaneously and recycled back to the process
- The process doesn't release any waste back to the environment
- It doesn't need any extra steps to remove impurities
- The final product is high-purity magnesium carbonate together with calcium carbonate

Drawbacks:

- Only works well for Hazelwood fly ash that is rich in free oxides;
- The use of weak leaching reagent requires min 80°C for the leaching;
- Low yields for the leaching process (i.e. 33% for Mg extraction and 36% for Ca extraction)

11

A Process for high-purity MgO Production

MONASH University

- ✓ Room temperature leaching with a higher Mg extraction yield;
- ✓ Suitable for BOTH Hazelwood and Yallourn fly ashes;
- ✓ High-Purity (≥97%) Epsom salt (MgSO4), and MgO
- ✓ No Cl, No NH3, less steps; and the new leaching reagent can also be regenerated and reused;
- ✓ residue can be used as additive into cement;

Provisional Patent lodged in June 2017

12

New Process Flow-sheet

MONASH University

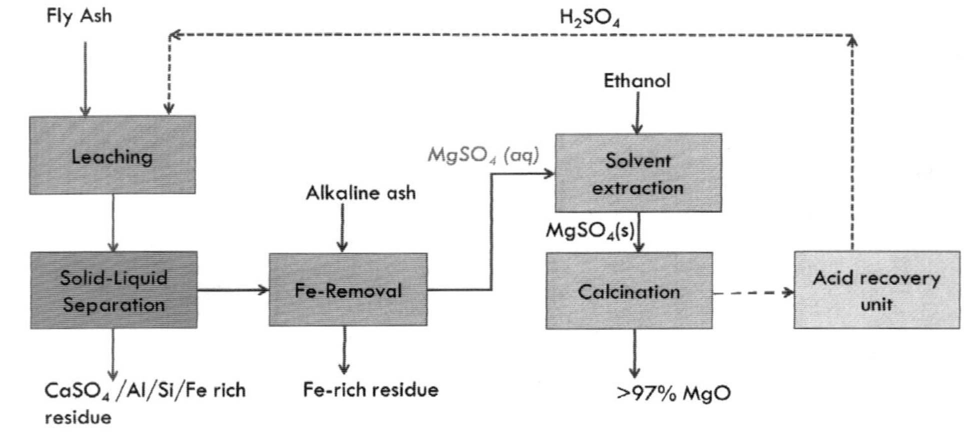

13

Leachate and final product composition

MONASH University

Leachate Compositions

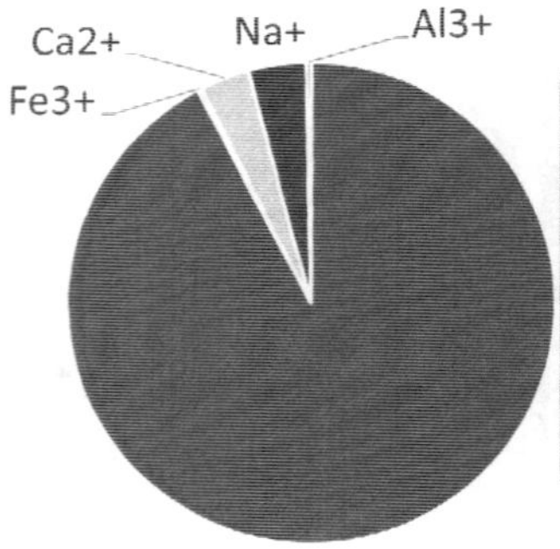

Primary Precipitate

	Wt%
MgO	51.1
SO_3	45.9
Al_2O_3	0.75
SiO_2	0.64
P_2O_5	0.16
K_2O	0.4
CaO	0.49
TiO_2	0.05
Fe_2O_3	0.01

- ✓ Chemical formula: sanderite and kieserite - $MgSO_4 \cdot 4H_2O$;
- ✓ Purity: ≥97 wt%;
- ✓ Submicron size.

~60% MgO extraction for this process VS 33% for ammonia single-stage leaching
It was improved to about ~80% using two stage leaching

14

Ongoing works

MONASH University

Detailed techno-economic analysis for the profit versus cost, payback time

- ✓ Different scenarios;
- ✓ Different markets;
- ✓ Detailed sizing;
- ✓ Detailed cost/profit.

Tweak the precipitation conditions for the making of MgO nanowires (nanosheets)

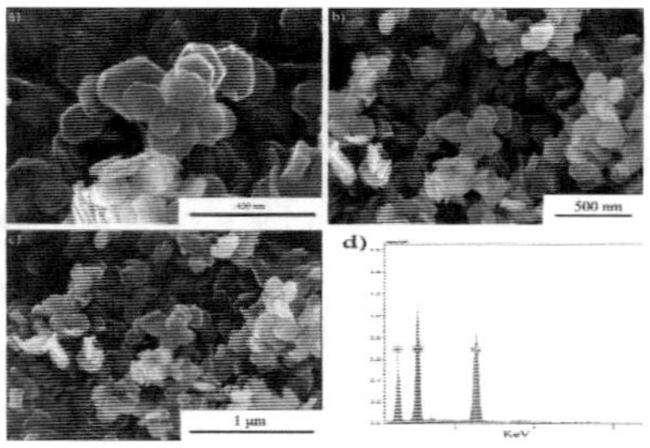

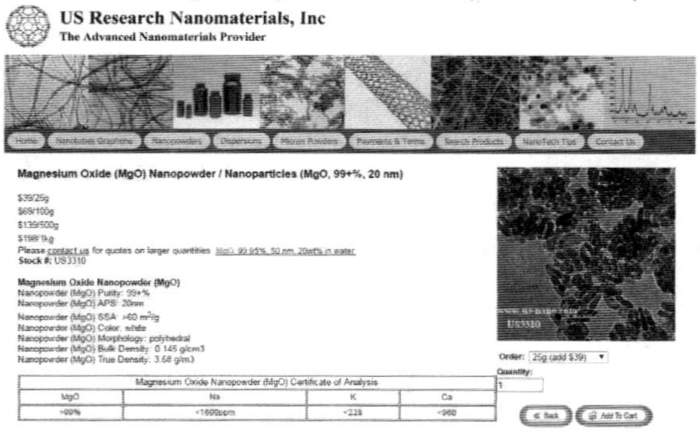

15

MONASH University

Thank you for your attention

Contact details:

Dr Tahereh Hosseini Tahereh.Hosseinidashtikhani@monash.edu

Assoc. Professor Lian Zhang Lian.zhang@monash.edu

Mineral Carbonation Curing of Wollastonite-Blends Paste: Feasibility of Low Lime Calcium Silicate

Authors:

Guo, R., Zhejiang university
Huang, H., Zhejiang university
Wang, T., Zhejiang university
Fang, M., Zhejiang university

The mineral carbonation (MC) curing of cement-based materials is a promising CO_2 utilization route, which could fix CO_2 as stable products, and meantime develop building materials. This work aims to investigate the feasibility of applying non-hydraulic materials (low lime calcium silicates for instance) as substitute binder in blended paste to reduce traditional cement usage. Herein the wollastonite-Portland cement blends were employed for MC curing. The impact of mineral blending proportions (5-25%) and mineral particle sizes (10-150Î¼m) on the carbonation reaction rate and final CO_2 uptake were systematically discussed, accompanied with the optimization of curing conditions. Results indicated that the blended minerals show â€œdiluted effectâ€ under low curing pressure and intensify the MC reaction. Decreasing the particle size of wollastonite mineral could also promote the CO_2 fixation. Tests of mercury intrusion porosimetry, scanning electron microscope and X-ray diffraction were conducted to further reveal the compositional and structural changes. Finally, the carbon footprint assessments were carried out to provide a more comprehensive insight on CO_2 emission reduction of MC curing technology, especially in terms of the composition and consumption of cement-based material. The total carbon footprint includes the CO_2 emission during cement production and minerals processing, as well as the CO_2 uptake from MC curing. It is shown that both the optimized CO_2 uptake and reduced consumption of Portland cement contribute to the reduction of carbon footprint, proving the emission reduction benefits of MC curing technology.

Murdoch
UNIVERSITY
Perth | Singapore | Dubai

Mg isotope signatures for tracing of natural carbonation reactions

[a]H.C. Oskierski, [b,c]V. Mavromatis, [d]A. Beinlich, [a]M. Altarawneh, [a]B.Z. Dlugogorski

[a]School of Engineering and Information Technology, Murdoch University, 90 South Street, Murdoch WA, 6150, Australia
[b]Institute of Applied Geosciences, Graz University of Technology, Rechbauerstraße 12, 8010 Graz, Austria
[c]Géosciences Environnement Toulouse (GET), CNRS, UMR 5563, Observatoire Midi-Pyrénées, 14 Av. E. Belin, 31400 Toulouse, France
[d]Institute for Geoscience Research (TIGeR), Curtin University, 6845 Perth, Australia

Murdoch

Presentation outline

Introduction

- Aims and relevance
- Study sites
- Mg-isotopes

Weathering

- Mg-isotopes in saprolite/soil
- Chemical changes
- Mineralogical changes

Carbonation

- Mg-isotopes in magnesite
- Temperature
- Depositional environment

Presentation title – Arial Bold 8

Murdoch

Aims and relevance

Aims

- Test utility of Mg-isotope signatures to trace natural carbonation of ultramafic rocks
- Constrain factors controlling Mg-isotope fractionation

Relevance

- Magnesite deposits are natural, large scale analogues for mineral carbonation
- Understanding helps development of low energy processes for CO_2 storage
- Mg-isotopes used in models of global Mg cycle
- Mg-cycle is closely associated with carbon cycle and the regulation of climate by weathering

Mg isotope signatures for tracing of natural carbonation reactions

Murdoch

Introduction
Study sites overview

New South Wales (Great Serpentinite Belt)
- Abras serpentine quarry (ASQ)
- Attunga magnesite deposit (AMD)
- Woodsreef asbestos mine (WAM)
- Piedmont magnesite deposit (PMD)
- Hanging Rock (HR)
- Trevena Mine (TM)

Queensland
- Kunwarara magnesite deposit (KMD)

Western Australia
- Ravensthorpe 8 pit (RAV8)
- Bandalup magnesite deposit (BMD)
- Yerrilla (YE), Yundamindera (YU)

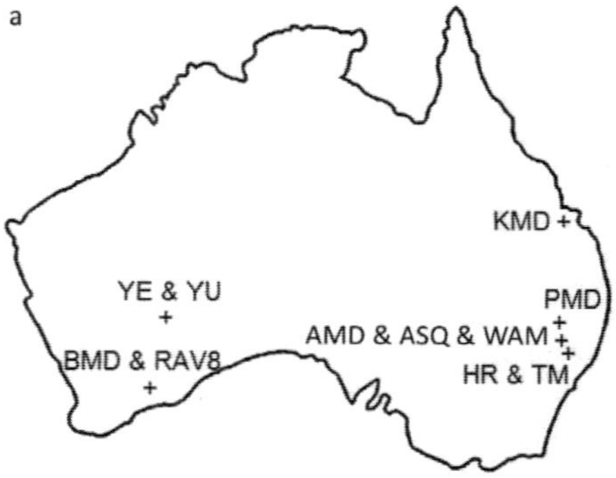

Introduction
Study sites overview

New South Wales
- Abras serpentine quarry (ASQ)
- Attunga magnesite deposit (AMD)
- Woodsreef asbestos mine (WAM)
- Piedmont magnesite deposit (PMD)
- Hanging Rock (HR)
- Trevena Mine (TM)

Queensland
- Kunwarara magnesite deposit (KMD)

Western Australia
- Ravensthorpe 8 pit (RAV8)
- Bandalup magnesite deposit (BMD)
- Yerrilla (YE), Yundamindera (YU)

Serpentinite and weathered serpentinite

Relatively fresh serpentinite at ASQ

Weathered serpentinite (saprolite) above serpentinite at AMD

Introduction
Study sites overview

New South Wales
- Abras serpentine quarry (ASQ)
- Attunga magnesite deposit (AMD)
- Woodsreef asbestos mine (WAM)
- Piedmont magnesite deposit (PMD)
- Hanging Rock (HR)
- Trevena Mine (TM)

Queensland
- Kunwarara magnesite deposit (KMD)

Western Australia
- Ravensthorpe 8 pit (RAV8)
- Bandalup magnesite deposit (BMD)
- Yerrilla (YE), Yundamindera (YU)

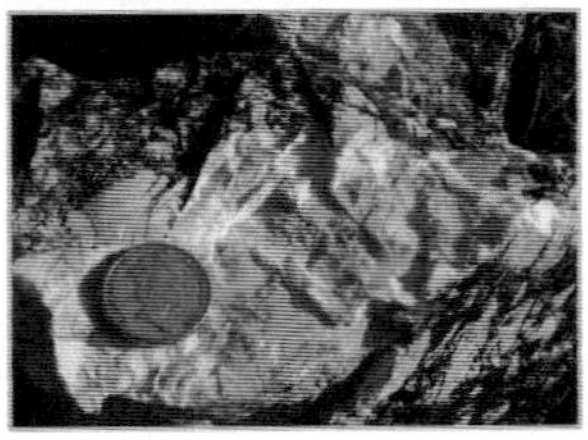

Ultramafic-hosted, hydrothermal magnesite and dolomite

Magnesite and dolomite at PMD and silica-carbonate rock at HR And TM
- Larger grainsize
- Straight, sub-parallel veins
- Fe-rich
- Low $\delta^{18}O$

➢ Hypogene fluids
➢ Formed at temperatures > 200 °C

Source: Ashley, 1995, 1997; Oskierski et al., 2013a

Murdoch

Mg isotope signatures for tracing of natural carbonation reactions

Introduction
Study sites overview

New South Wales
- Abras serpentine quarry (ASQ)
- Attunga magnesite deposit (AMD)
- Woodsreef asbestos mine (WAM)
- Piedmont magnesite deposit (PMD)
- Hanging Rock (HR)
- Trevena Mine (TM)

Queensland
- Kunwarara magnesite deposit (KMD)

Western Australia
- Ravensthorpe 8 pit (RAV8)
- Bandalup magnesite deposit (BMD)
- Yerrilla (YE), Yundamindera (YU)

Ultramafic-hosted magnesite

Magnesite at AMD
- Cryptocrystalline
- Cauliflower nodules
- Irregular veins
- Low Fe-content
- High $\delta^{18}O$

➢ Supergene fluids (meteoric water)
➢ Formed at low temperatures < 50 °C

Source: Oskierski et al., 2013a,b

Murdoch

Mg isotope signatures for tracing of natural carbonation reactions

Introduction
Study sites overview

New South Wales
- Abras serpentine quarry (ASQ)
- Attunga magnesite deposit (AMD)
- Woodsreef asbestos mine (WAM)
- Piedmont magnesite deposit (PMD)
- Hanging Rock (HR)
- Trevena Mine (TM)

Queensland
- Kunwarara magnesite deposit (KMD)
    - 500 Mt $MgCO_3$ = 260 Mt CO_2 stored

Western Australia
- Ravensthorpe 8 pit (RAV8)
- Bandalup magnesite deposit (BMD)
- Yerrilla (YE), Yundamindera (YU)

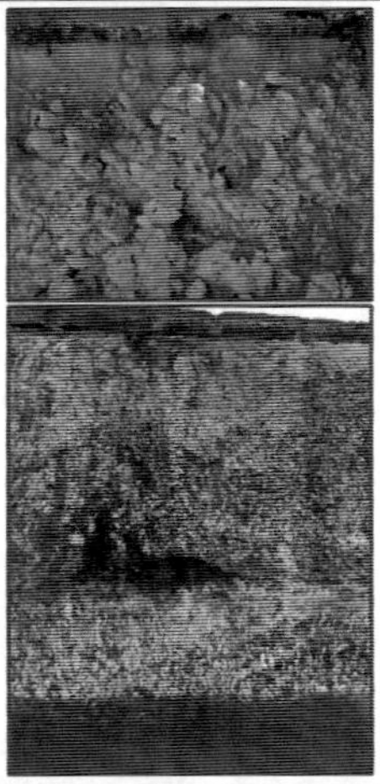

Sediment-hosted magnesite

Magnesite at BMD and KMD
- Cryptocrystalline, bone-magnesite
- Sedimentary beds up to 6 m (BMD) and up to 26 m thick (KMD)
- No deformation
- High $\delta^{18}O$

➢ Supergene weathering fluids
➢ Formed at temperatures < 50 °C

Source: Wilcock, 1998; Photo below: Scott Szulc

Introduction
Mg-isotopes

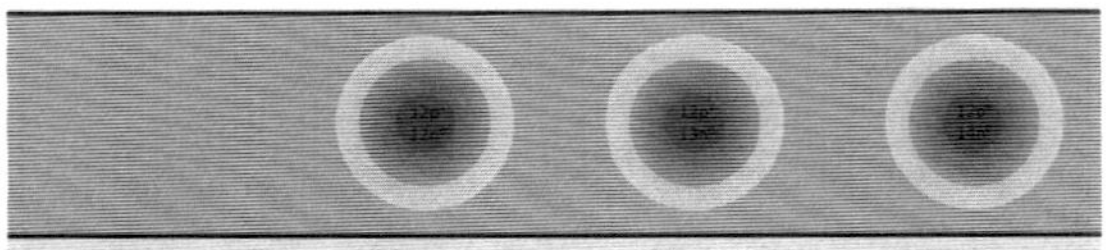

Atomic symbol	^{24}Mg	^{25}Mg	^{26}Mg
Number of protons	12	12	12
Number of electrons	12	12	12
Number of neutrons	12	13	14
Abundance	78.992 %	10.003 %	11.005 %

Definitions:

$\delta^{X}Mg = [(^{X}Mg/^{24}Mg)_{sample}/^{X}Mg/^{24}Mg_{standard}-1]*1000$

With X = either mass 25 or 26

- Standard is DSM-3 (Dead sea magnesium)
- Predominant use of $\delta^{26}Mg$ as isotope fractionation is mass dependent

Mg-isotopes
Overview

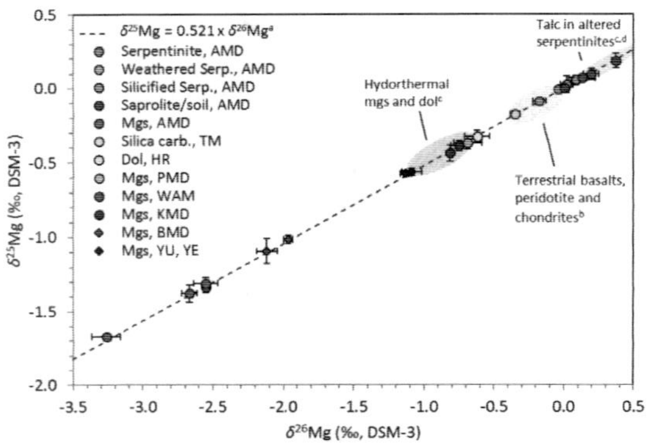

- Ultramafic rocks/serpentinites have relatively homogeneous Mg-isotope signature
- Bulk silicate Earth has δ^{26}Mg = -0.23 ‰[b]

Sources: [a]Young et al., 2002; [b]Bourdon et al., 2010; [c]Beinlich et al., 2014; [d]Beinlich et al., 2018

Mg isotope signatures for tracing of natural carbonation reactions

Murdoch University

Mg-isotopes
Overview

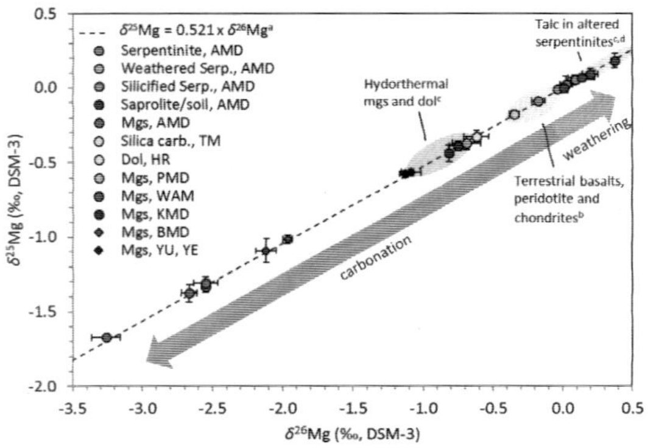

- Ultramafic rocks/serpentinites have relatively homogeneous Mg-isotope signature
- Bulk silicate Earth has δ^{26}Mg = -0.23 ± 0.20 ‰[e]
- Secondary silicates such as talc tend to have higher δ^{26}Mg (preferential incorporation of heavy ^{26}Mg)
- Carbonates have lower δ^{26}Mg (preferential incorporation of light ^{24}Mg)

Sources: [a]Young et al., 2002; [b]Bourdon et al., 2010; [c]Beinlich et al., 2014; [d]Beinlich et al., 2018

Mg isotope signatures for tracing of natural carbonation reactions

Murdoch University

Weathering of ultramafic rocks: Mg-isotopes
Attunga magnesite deposit

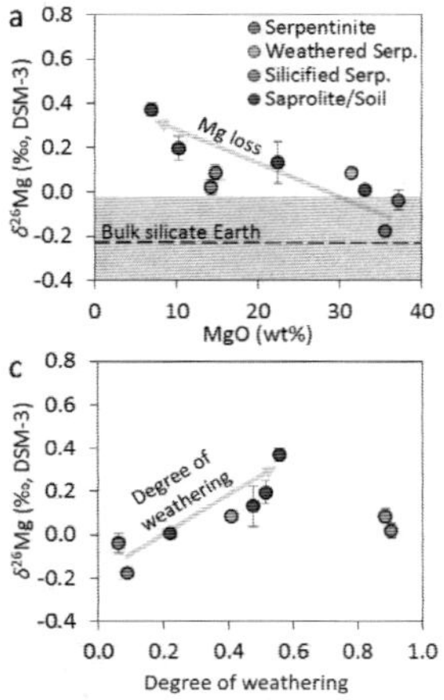

- Degree of weathering increases from serpentinite, over weathered serpentinite to saprolite and soil
- Mg loss from weathering profile is associated with increase in δ^{26}Mg
- δ^{26}Mg increases coherently with degree of weathering

Weathering of ultramafic rocks: Mg-isotopes
Attunga magnesite deposit

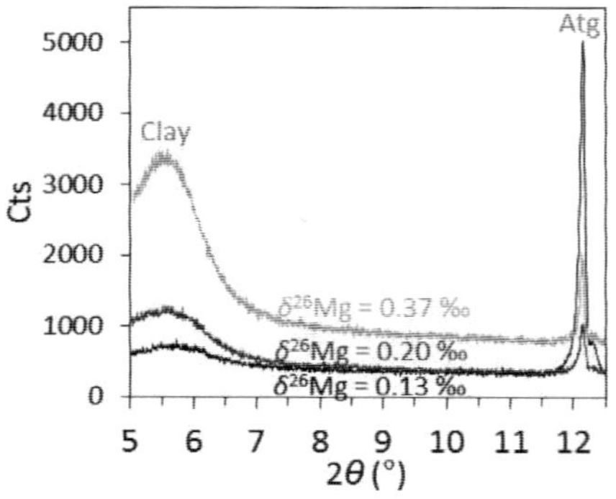

- X-ray diffraction shows both antigorite and clay mineral are present soil/saprolite samples
- δ^{26}Mg increases with clay mineral content

➢ Clay minerals preferentially retain heavy ^{26}Mg

➢ Conversion of serpentine into clay controls Mg-isotope fractionation during continental weathering of ultramafic rocks

Carbonation of ultramafic rocks: Mg-isotopes
Overview

- Magnesite samples span wide range of Mg-isotope signatures (-3.26 ‰ < δ^{26}Mg < -0.62 ‰)
- Ultramafic-hosted, hydrothermal magnesite and dolomite with δ^{26}Mg of -0.62 ‰ to -0.69 ‰
- Ultramafic-hosted magnesite (AMD, WAM) has lowest δ^{26}Mg < -2.55 ‰; lowest δ^{26}Mg observed up to date
- Sediment-hosted magnesite (KMD, BMD, YU, YE) with δ^{26}Mg between -2.12 ‰ and -0.75 ‰

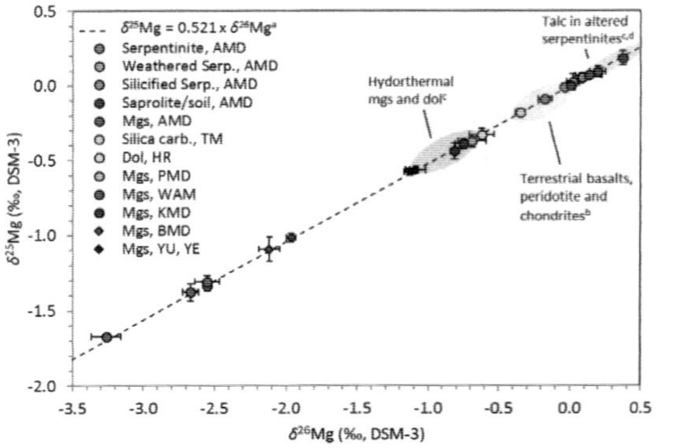

Sources: [a]Young et al., 2002; [b]Bourdon et al., 2010; [c]Beinlich et al., 2014; [d]Beinlich et al., 2018;

Murdoch

Carbonation of ultramafic rocks: Mg-isotopes
Overview

- Magnesite samples span wide range of Mg-isotope signatures (-3.26 ‰ < δ^{26}Mg < -0.62 ‰)
- Ultramafic-hosted, hydrothermal magnesite and dolomite with δ^{26}Mg of -0.62 ‰ to -0.69 ‰
- Ultramafic-hosted magnesite (AMD, WAM) has lowest δ^{26}Mg < -2.55 ‰; lowest δ^{26}Mg observed up to date
- Sediment-hosted magnesite (KMD, BMD, YU, YE) with δ^{26}Mg between -2.12 ‰ and -0.75 ‰

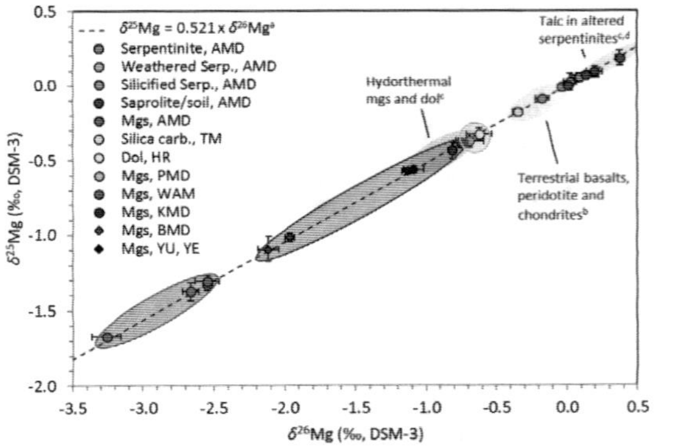

Sources: [a]Young et al., 2002; [b]Bourdon et al., 2010; [c]Beinlich et al., 2014; [d]Beinlich et al., 2018;

Murdoch

Carbonation of ultramafic rocks: Mg-isotopes
Boundary conditions

Do ultramafic-hosted magnesite samples provide boundary conditions for Mg-isotope fractionation?

Low temperature ultramafic-hosted magnesite

- Lowest $\delta^{26}Mg$ (and $\delta^{13}C$) observed for magnesite
- Source rock = host rock
- Ultramafic-rock buffered system
- Minimal overprinting (serpentinite at AMD uplifted)

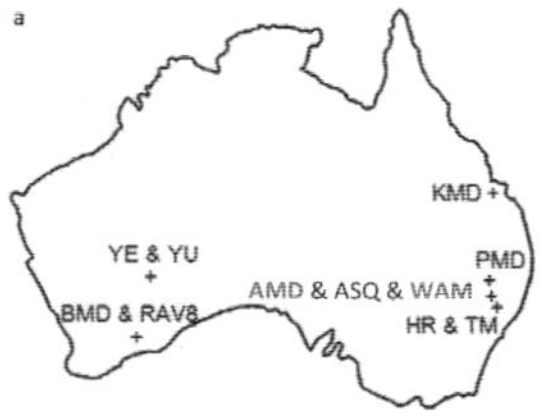

Murdoch

Carbonation of ultramafic rocks: Mg-isotopes
Boundary conditions

Do ultramafic-hosted magnesite samples provide boundary conditions for Mg-isotope fractionation?

Low temperature ultramafic-hosted magnesite

- Lowest $\delta^{26}Mg$ (and $\delta^{13}C$) observed for magnesite
- Source rock = host rock
- Ultramafic-rock buffered system
- Minimal overprinting (serpentinite at AMD uplifted)

Low and high temperature ultramafic-hosted magnesite

- Same host-rock = isotopically homogeneous Mg source
- Different formation temperatures
  - AMD, WAM around 30 °C
  - PMD above 200 °C

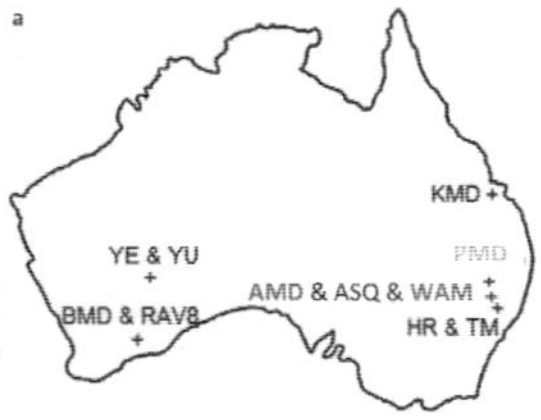

Murdoch

Carbonation of ultramafic rocks: Mg-isotopes

Temperature dependence

Temperature dependence of Mg-isotope fractionation expressed as $\Delta^{26}Mg$ (= $\delta^{26}Mg_{fluid}$ - $\delta^{26}Mg_{mineral}$)

Linear through ultramafic-hosted samples

- Defines temperature dependence of the Mg-isotope fractionation during precipitation of magnesite
- Coincides with experimentally determined values for Mg-isotope fractionation
- Temperature is a primary control of Mg-isotope fractionation during magnesite formation

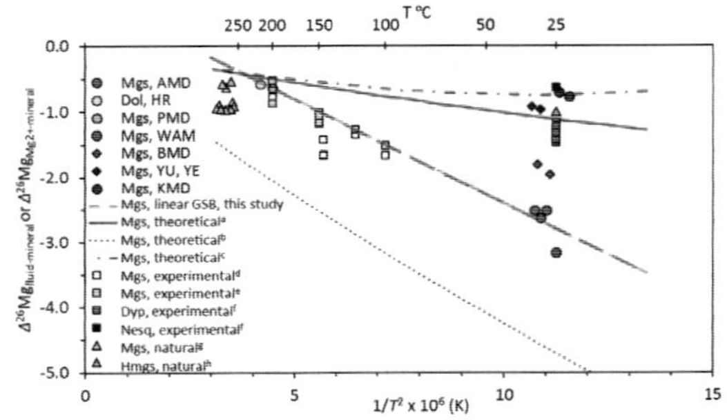

Sources: [a]Rustad et al., 2010; [b]Schauble, 2011; [c]Pinilla et al., 2015; [d]Schott et al., 2016; [e]Pearce et al., 2012; [f]Mavromatis et al., 2012; [g]Beinlich et al., 2014 ; [h]Shirokova et al., 2013.

Carbonation of ultramafic rocks: Mg-isotopes

Temperature dependence

Superimposed variation in $\delta^{26}Mg$ in sediment hosted magnesite reflects

- Fluid evolution during transport (?)
- Host rock (?)
- Lower pH and supersaturation, changes in speciation (?)
- Hydromagnesite pre-cursor (?)
- Diagenetic overprinting (?)

- Formation conditions/depositional environment (!)

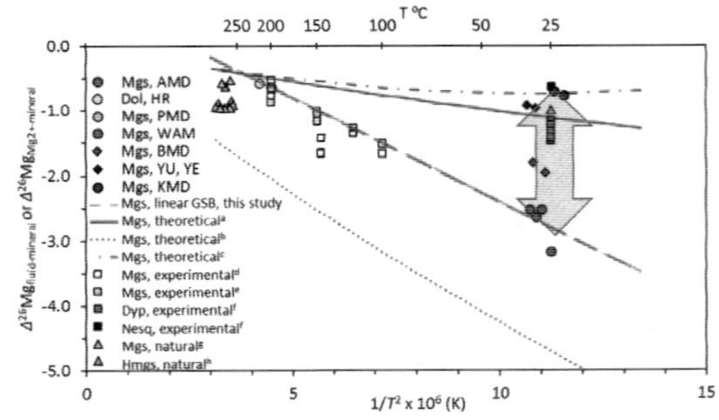

Sources: [a]Rustad et al., 2010; [b]Schauble, 2011; [c]Pinilla et al., 2015; [d]Schott et al., 2016; [e]Pearce et al., 2012; [f]Mavromatis et al., 2012; [g]Beinlich et al., 2014 ; [h]Shirokova et al., 2013.

Conclusions

- Mg-isotopes are well suited to trace weathering and carbonation of ultramafic rocks
- Mg-isotope fractionation during weathering is controlled by mineralogy of secondary phases
- Mg-isotope fractionation during carbonation is controlled by temperature and modulated by depositional environment

Conclusions

Thank you very much for your attention !

Mg isotope signatures for tracing of natural carbonation reactions
Murdoch

References

Ashley, P.M., 1997. Silica-carbonate alteration zones and gold mineralisation in the Great Serpentinite Belt, New England Orogen, New South Wales. In: PM, A., PG, F. (Eds.),Tectonics and Metallogenesis of the New England Orogen. Geol. Soc. Aust. Spec. Pub. 19, pp. 212–225.

Ashley P. M. (1995) The Piedmont hydrothermal magnesite deposit, Great Serpentinite Belt, northern NSW: geochemical and stable isotopic constraints. Research Report 1993–94, Centre for Isotope Studies North Ryde, NSW, Australia. pp. 12–18.

Beinlich , A., Mavromatis, V., Austrheim H., Oelkers, E.H., 2014. Inter-mineral Mg isotope fractionation during hydrothermal ultramafic rock alteration—Implications for the global Mg-cycle. Earth Planet. Sci. Lett. 392, 166–176.

Beinlich A., Austrheim, H., Mavromatis, V., Grguric, B., Putnis, C.V., Putnis, A. (2018) Peridotite weathering is the missing ingredient of Earth's continental crust composition. Nature Comm. 9, 634.

Bourdon, B., Tipper, E. T., Fitoussi, C., Stracke, A., 2010. Chondritic Mg isotope composition of the Earth. Geochim. Cosmochim. Acta 74, 5069–5083.

Mavromatis, V., Pearce, C. R., Shirokova, L. S., Bundeleva, I. A., Pokrovsky, O. S., Benezeth, P., Oelkers, E. H., 2012. Magnesium isotope fractionation during hydrous magnesium carbonate precipitation with and without cyanobacteria. Geochim. Cosmochim. Acta 76, 161–174.

Oskierski, H. C., Dlugogorski, B. Z., Jacobsen, G., 2013a. Sequestration of atmospheric CO_2 in chrysotile mine tailings of the Woodsreef Asbestos Mine, Australia: Quantitative mineralogy, isotopic fingerprinting and carbonation rates. Chem. Geol. 358, 156-169.

Oskierski, H. C., Dlugogorski, B. Z., Jacobsen, G., 2013b. Sequestration of atmospheric CO_2 in a weathering-derived, serpentinite-hosted magnesite deposit: ^{14}C tracing of carbon sources and age constraints for a refined genetic model. Geochim. Cosmochim. Acta 122, 226-246.

Oskierski, H.C., Bailey, J. G., Kennedy, E. M., Jacobsen, G., Ashley, P. M., Dlugogorski, B. Z., 2013c. Formation of weathering-derived magnesite deposits in the New England Orogen, New South Wales, Australia: Implications from mineralogy, geochemistry and genesis of the Attunga Magnesite Deposit. Mineralium Deposita 48, 525-541.

Pearce, C. R., Saldi, G. D., Schott, J., Oelkers, E. H., 2012. Isotopic fractionation during congruent dissolution, precipitation and at equilibrium: Evidence from Mg isotopes. Geochim. Cosmochim. Acta 92, 170–183.

Pinilla, C., Blanchard, M., Balan, E., Suresh, K., Natarajan, S.K., Vuilleumier, R., Mauri, F.,
2015. Equilibrium magnesium isotope fractionation between aqueous Mg^{2+} and carbonate minerals: insights from path integral molecular dynamics. Geochim. Cosmochim. Acta 163, 126–139.

Rustad, J. R., Casey, W. H., Yin, Q. Z., Bylaska, E. J., Felmy, A. R., Bogatko, S. A., Jackson, V. E., Dixon, D. A., 2010. Isotopic fractionation of Mg^{2+}(aq), Ca^{2+}(aq), and Fe^{2+}(aq) with carbonate minerals. Geochim. Cosmochim. Acta 74, 6301–6323.

Schauble, E. A., 2011. First-principles estimates of equilibrium magnesium isotope fractionation in silicate, oxide, carbonate and hexaaquamagnesium(2+) crystals. Geochim. Cosmochim. Acta 75, 844–869.

Schott, J., Mavromatis, V., Fujii, T., Pearce, C. R., Oelkers, E. H., 2016. The control of carbonate mineral Mg isotope composition by aqueous speciation: Theoretical and experimental modelling. Chem. Geol. 445, 120-134.

Shirokova, L. S., Mavromatis, V., Bundeleva, I. A., Pokrovsky, O. S., Benezeth, P., Gerard, E., Pearce, C. R., Oelkers, E. H., 2013. Using Mg isotopes to trace cyanobacterially mediated magnesium carbonate precipitation in alkaline lakes. Aquat. Geochem. 19, 1–24.

Wilcock, S., 1998. Sediment-hosted magnesite deposits. AGSO J. Aust. Geol. Geophys. 17–4, 247–251

Young, E. D., Ash, R. D., Galy, A., Belshaw, N. S., 2002. Mg isotope heterogeneity in the Allende meteorite measured by UV laser ablation-MC-ICPMS and comparisons with O isotopes. Geochim. Cosmochim. Acta 66, 683–698.

Murdoch

UNIVERSITY

Perth | Singapore | Dubai

Greener Pervious Concrete incorporating surface modified Activated Carbon

Heba Hamad
University of Cambridge

ACEME 2018 March 11-14

Contents

- Introduction
- Objectives
- Project Key Elements
- Activated Carbon Surface Modification
- Activated Carbon in Pervious Concrete
- Preliminary Results
- Expectations

Introduction

Objectives

The aim of this study is to maximise the sustainability potential of pervious concrete by introducing modified activated carbon as a partial replacement of cement and aggregate for the purpose of CO_2 sequestration and concrete protection.

Objectives

- Investigate the efficiency of introducing modified activated carbon to the pervious concrete.
- Examine the change of the carbonation rate after activated carbon incorporation.
- Study the impacts of AC pervious concrete on the properties of concrete.

Key Elements

Activated Carbon

Pervious Concrete

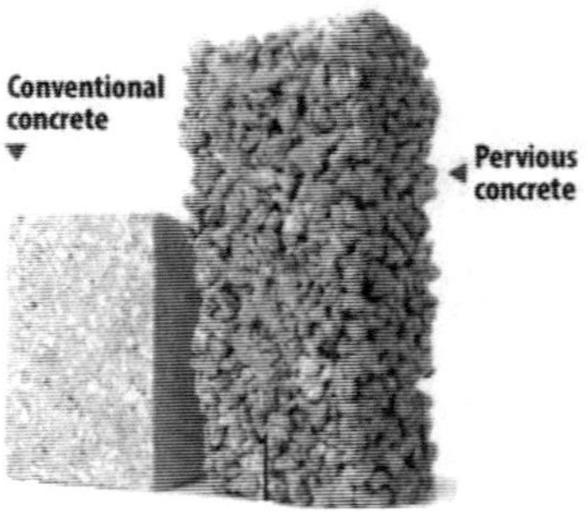

Key Elements

Activated Carbon

- High surface area (> 2000 m2/g).
- Prepared from a wide range of materials.
- Applications in purifying soil, air solutions and chemical products.
- Sorption capacity >800 mg CO2/g AC

Key Elements

- High Porosity > 15%.
- High drainage rate
- Interconnected network of voids.

Pervious Concrete

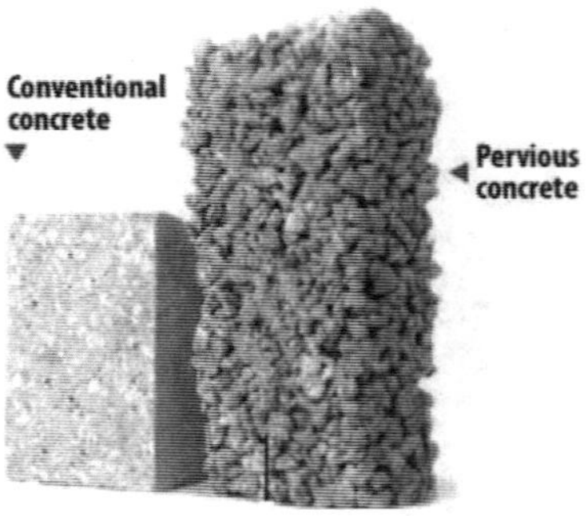

Activated Carbon Modification

Heat Treatment | Chemical Treatment | Microwave Treatment | Plasma Treatment | Ozone Treatment

Activated Carbon Modification

Heat Treatment | **Chemical Treatment** | Microwave Treatment | Plasma Treatment | Ozone Treatment

Different Approaches → **Best modified technique**

Activated Carbon Modification

CO2 sequestration by Activated Carbon

CO2 Incubator

TGA

CO2 Sensor

Activated Carbon Modification

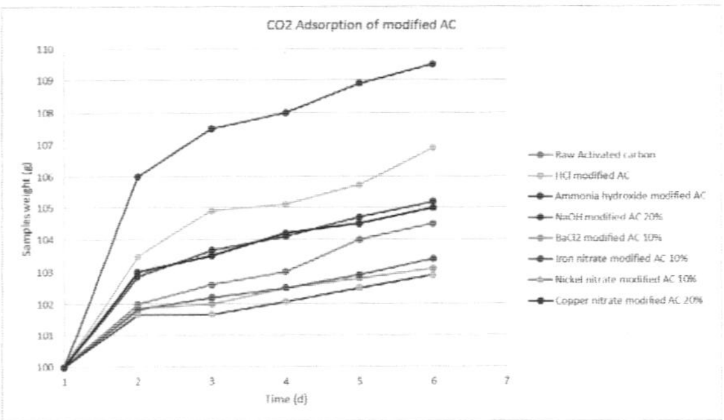

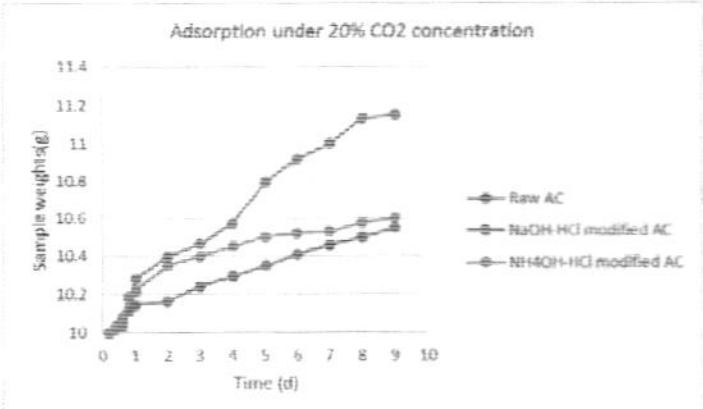

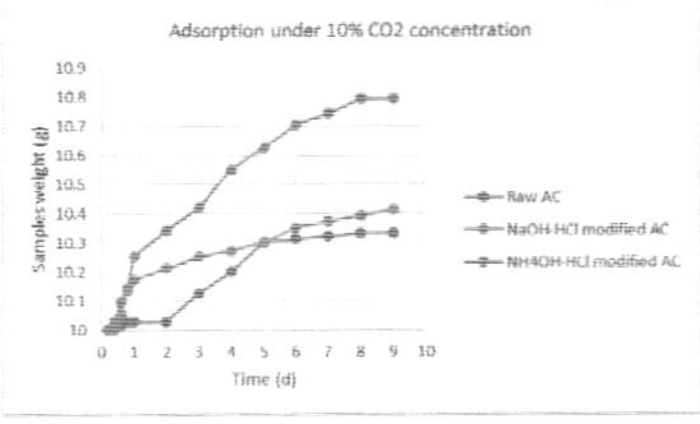

Activated Carbon Modification

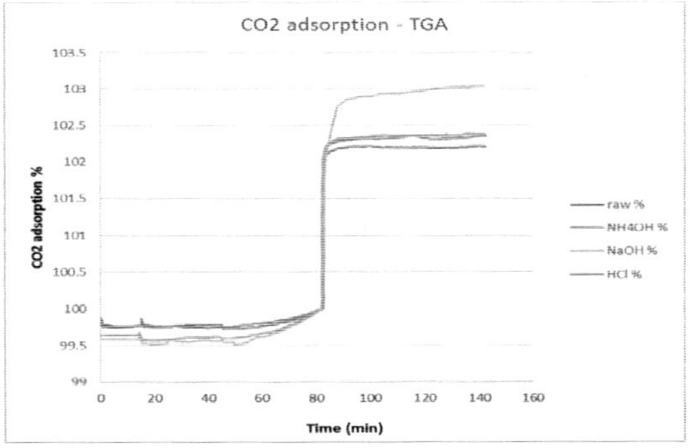

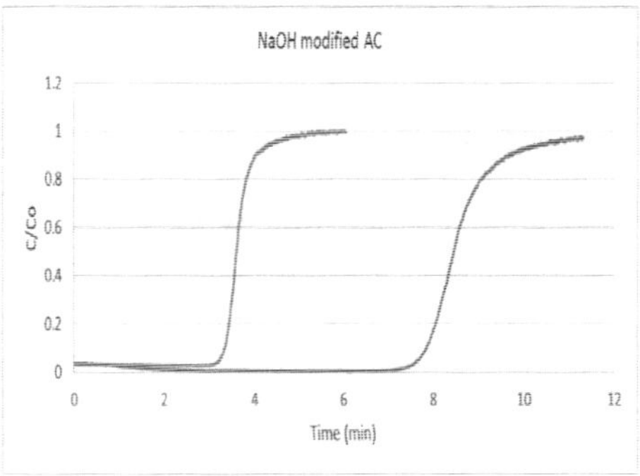

Activated Carbon in Pervious Concrete

- Certain amounts of **modified AC (powder/ granular)** will replace specific percentage of **(cement/aggregate)**.
- The amount of **CO_2** sequestered in **concrete** will be measured.
- The impact of **AC** incorporation in **concrete** will be investigated.

CO2 sequestration by Pervious Concrete

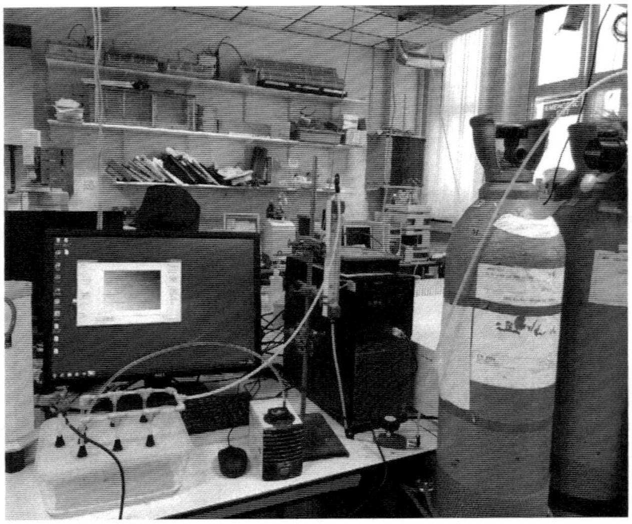

Preliminary Results

Ultimate Compressive Strength

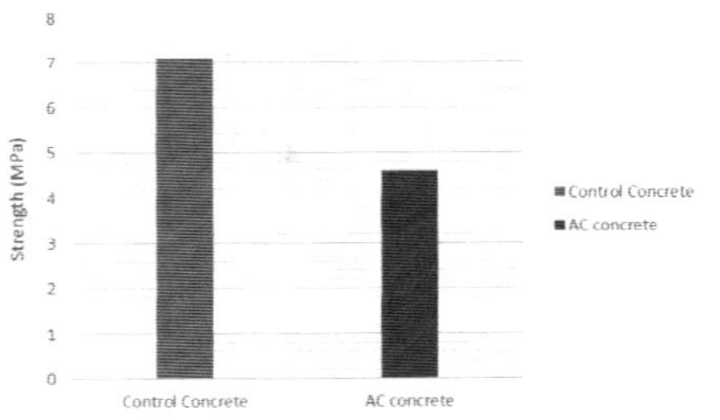

Permeability (water filtration)

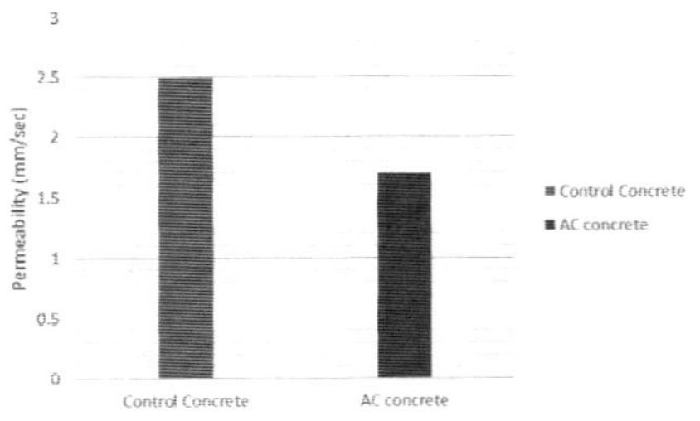

Expectations

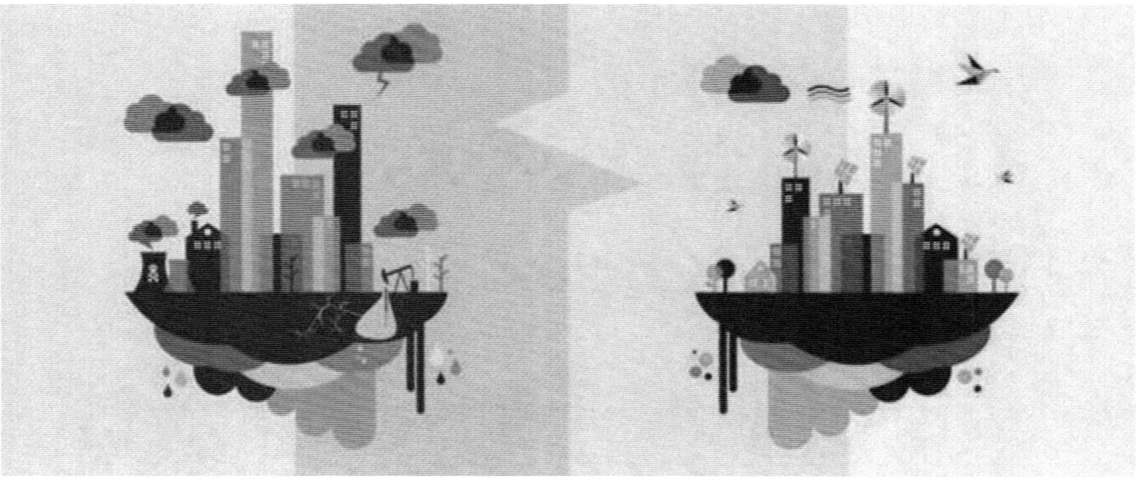

*** Formatting Issue - Best Available Paper/Graphic ***

Dissolution of Heat Activated Lizardite for Direct Aqueous Carbonation at elevated pressures and at low temperatures

Supervisors: Profs. Kennedy and Stockenhuber

Advisors: Drs. Geoff Brent and Mark Rayson

Ammar Abu Fara 12th March 2018

1

Context

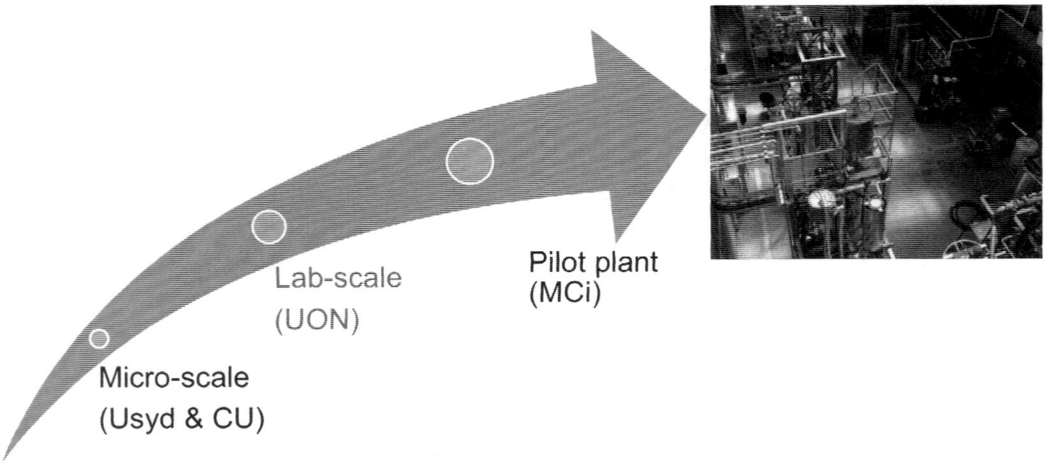

2

Aim of this work

- ❑ Exploring ways to enhance the Mg extraction under minimum precipitation conditions (high CO_2 solubility)
- ❑ To understand the mechanism of the dissolution process.
- ❑ To provide an insight into the effect of solid ratio and particle size on the dissolution process.
- ❑ To understand the dissolution limiting step.

Direct aqueous mineral carbonation principles

Direct aqueous carbonation: Single-stage process (so called "ARC process") at elevated P and T

Dissolution of CO_2:

$CO_{2\,(g)} + H_2O_{\,(l)} \rightleftharpoons CO_{2(aq)} + H_2O \rightleftharpoons HCO^{3-}{}_{(aq)} + H^+{}_{(aq)} \rightleftharpoons CO_3{}^{2-}{}_{(aq)} + 2H^+{}_{(aq)}$

Dissolution of mineral:

$Mg_2SiO_{4\,(s)} + 4H^+{}_{(aq)} \rightarrow 2Mg^{2+}{}_{(aq)} + SiO_{2\,(s)} + 2H_2O_{\,(l)}$

Precipitation of carbonate:

$Mg^{2+}{}_{(aq)} + CO_3{}^{2-}{}_{(aq)\,)} \rightarrow MgCO_{3\,(s)}$

Reactions of dissolution and precipitation occur simultaneously and in the same reactor

Feed Characterisation

Lizardite
$Mg_3Si_2O_5(OH)_4$

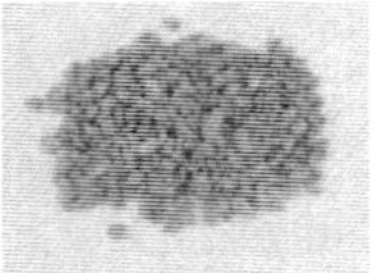

Phases (Powder XRD):
Lizardite 95%
Clinochlore 4%
Magnetite 1%

Source: Great Serpentinite Belt, NSW, Australia

5

Particle Size Distribution (PSD)

Malvern Mastersizer 2000

Method
- Wet grinding of the crushed raw lizardite (< 1mm)
  - 4 hours in the ball mill
  - 50% solid ratio
- Wet sieving
- Cleaning protocol to remove fine particles

Sieve bin (of raw lizardite), µm	d_{10}	d_{50}	d_{90}
45-75	47.2	73.2	112.2
20-45	27.5	43.0	66.8
Sub 20	2.1	8.2	19.2

Based on d_{90} (larger than the bin size); the particles are not spherical

6

Lizardite Heat Treatment

- To remove chemically bound water to promote carbonation (to disrupt the long range order and create reactive amorphous phase)
- Lizardite was heat activated in an electrically heated rotary kiln at 630 °C for 4 hours. $2Mg_3Si_2O_5(OH)_{4\,(s)} \rightarrow 3Mg_2SiO_{4\,(s)} + SiO_{2\,(s)} + 4H_2O_{\,(g)}$ (Lizardite) (Forsterite)

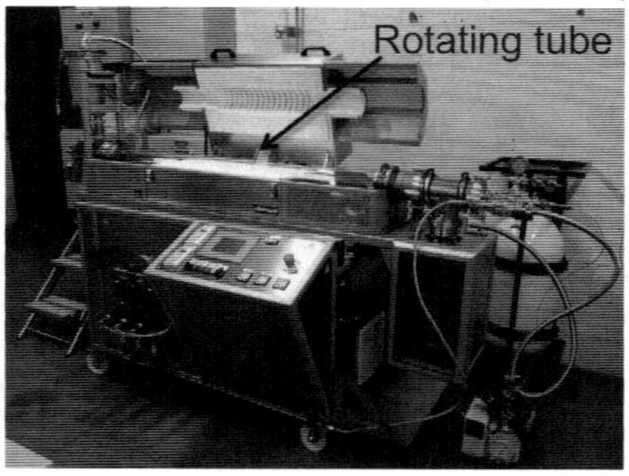

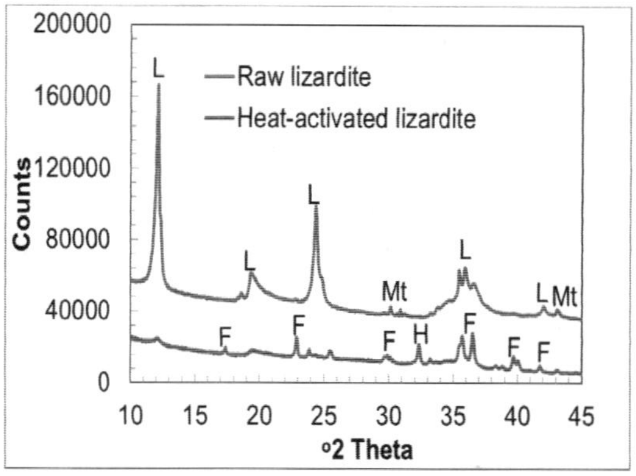

▶ 7

Comparison of all size fractions after the heat activation process

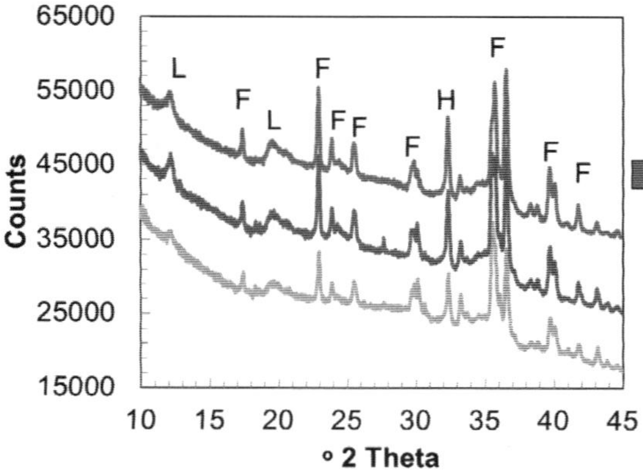

- Chemical compositions are very similar for all fractions
- Reactivity changes can be attributed to the particle size effect

QXRD

Reference Intensity Ratio (RIR) method: Addition of internal standard (20% Si) to quantify the phases

Particle size(µm)	Crystalline phases %					Total crystalline %	Amorphous phase %
	Cl	L	F	Mt	H		
45-75	3.68	7.19	14.1	0.26	0.61	25.9	74.1
20-45	8.13	5.83	12.8	0.32	0.58	27.7	72.3
< 20	6.82	0.94	8.67	0.24	0.58	17.2	82.8

▶ 8

Materials and Methods

PARR reactor

Sampler

Heat Jacket

Operating conditions:

- ❑ Temperature : 40°C
- ❑ Total pressure: 100 bar
- ❑ Reaction time: 90 min (sampling t_0, t_{15}, t_{30}, t_{60}, and t_{90} min)
- ❑ Slurry concentration: 1, 2 & 5% solid ratio
- ❑ Additive concentration ($NaHCO_3$): 0 M

9

Novel sampling system

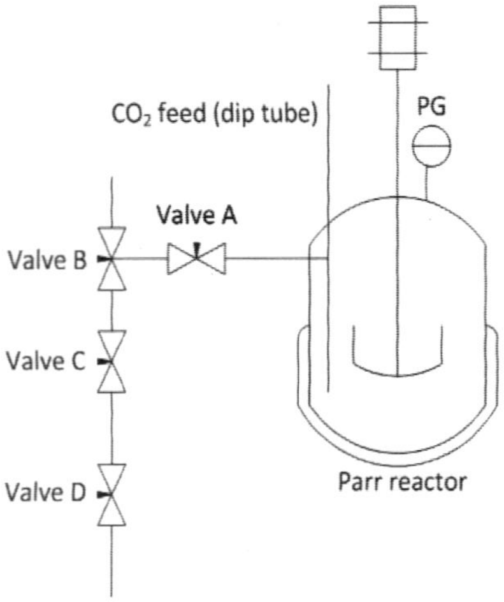

- ➢ A novel sampling system was designed to collect slurry samples from the reactor during the course of each experiment
- ➢ to investigate and study the mechanism and the kinetics of the reaction.

10

Analysis Methodology of the Reaction System

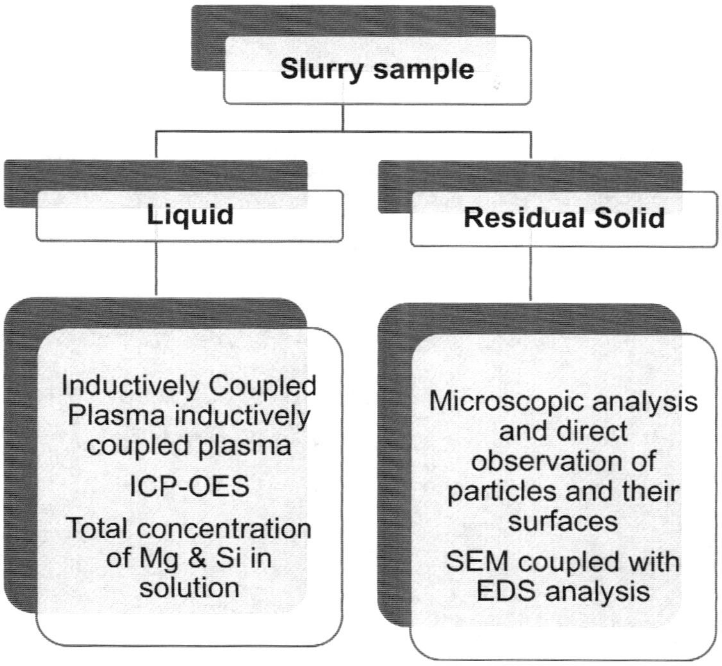

Liquid Analysis: ICP-OES Total concentration of Mg & Si in solution

Effect of solid ratio on Mg extraction

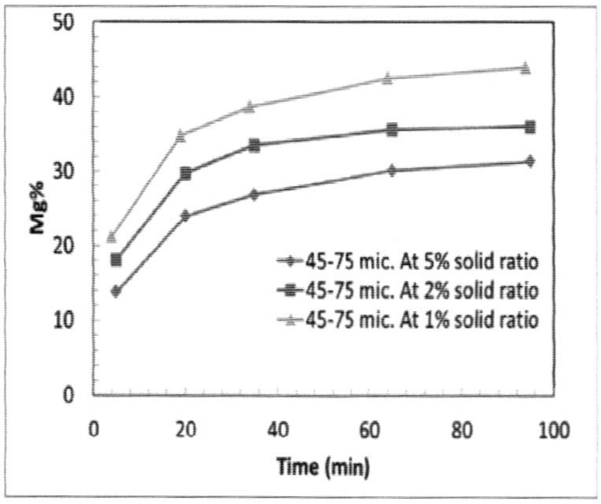

- Rapid Mg extraction initially then the dissolution rate decreased dramatically.
- The solid concentration had a significant effect on Mg extraction.
- Mg extraction from the low solid ratio (1 wt%) is higher compared with the high solid ratio.
- This is possibly due to the lower pH (less alkalinity of lower solid ratio)

Effect of solid ratio on Si extraction

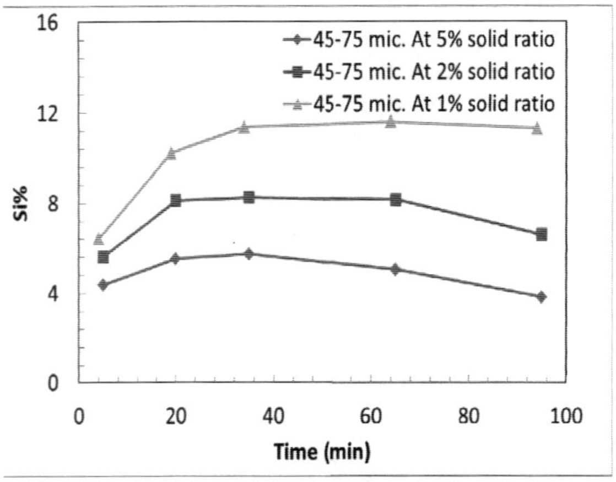

- All samples are super saturated with respect to amorphous Si
- A rapid rate of Si extraction was observed initially.
- The extent of dissolved Si decreased with increasing solid ratio
- Presence of Si as a colloidal as well as aqueous Si.
- The concentration of Si reached a maximum value and started to decline (precipitation of Si).
- A faster rate of precipitation at the higher silica concentrations (hypothesis)
- Big with size exceeding 0.22 µm will be filtered out.
- Precipitation could be the reason of Si passivating layer formation.
- Si measured is aqueous Si smaller than 0.22 µm.

13

Effect of particle size on Mg extraction

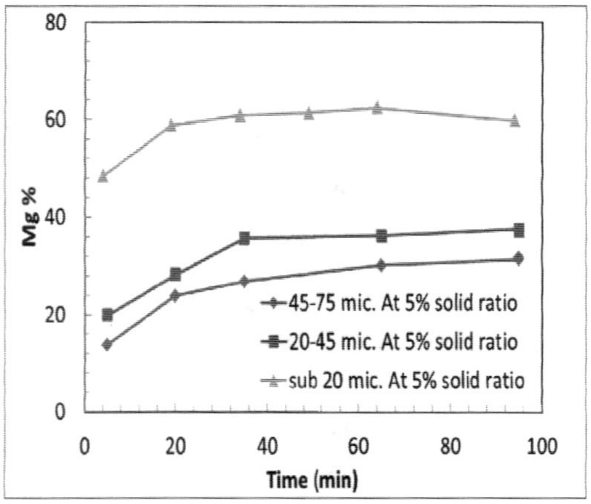

- The highest Mg% extracted from the fines due to the enlarged surface area
- A rapid Mg extraction initially, due to high extent of CO_2 dissolution, which leads to a lower pH.
- Formation of a Si passivating layer could be the barrier for further Mg extraction

14

Effect of particle size on Si extraction

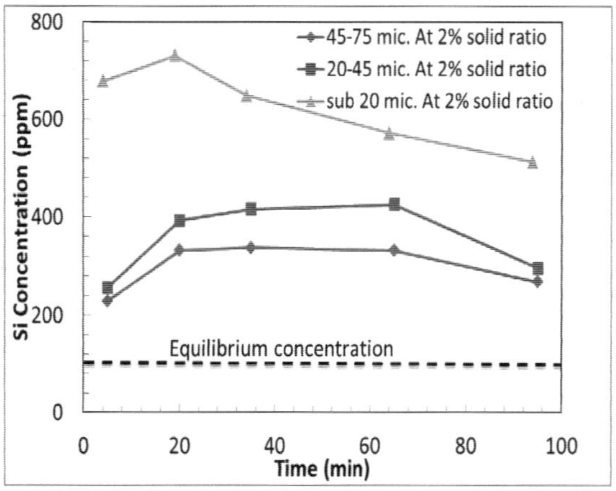

- A rapid rate of Si extraction was observed initially.
- The concentration of Si reached a maximum value and started to decline (precipitation of Si).
- The fine particles have the highest Si concentration (enlarged surface area).
- Precipitation could be the reason of Si passivating layer formation, (same as solid ratio effect)

15

Microscopic analysis and direct observation for particles and their surfaces (SEM-EDS)

- Transverse polished section of a thin blade cut off the particle surface
- SEM-EDS observation of the dissolved samples after 90 min

45-75 µm

Elemental Mapping with EDX

SEM image

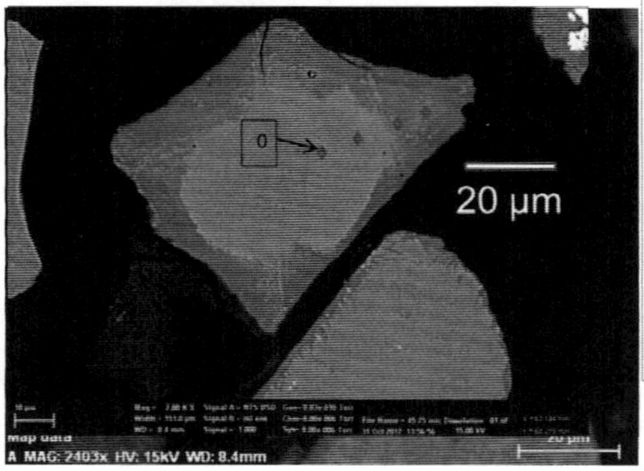

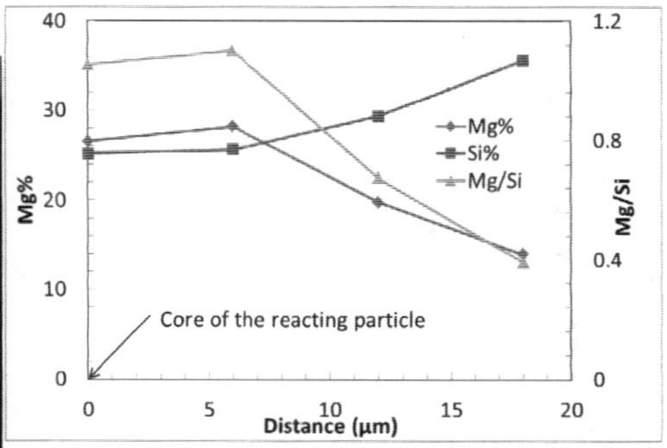

16

Continue microscopic analysis

20-45 µm

SEM image

Elemental Mapping with EDX

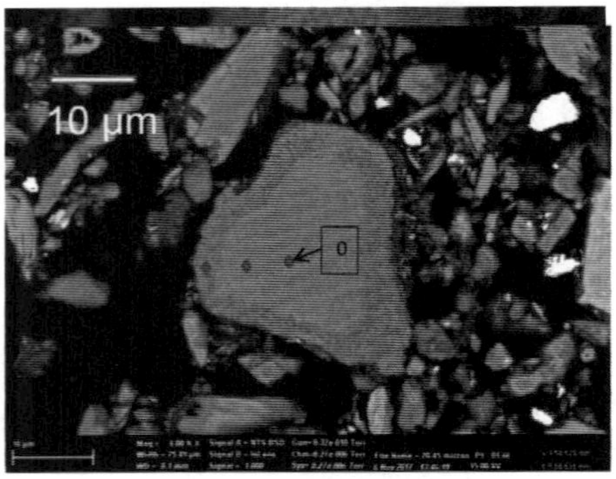

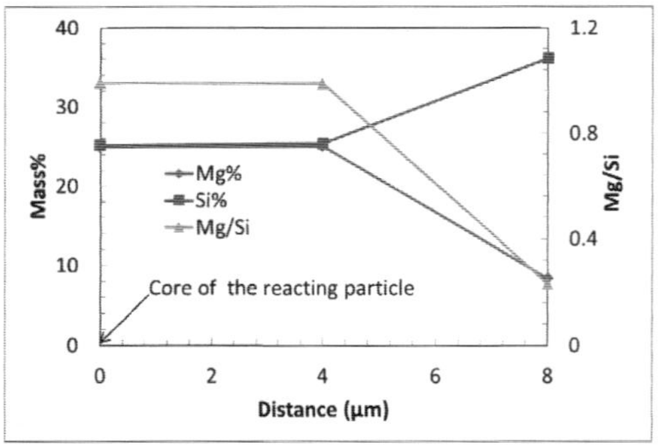

▶ 17

Continue microscopic analysis

Sub 20 µm

Elemental Mapping with EDX

SEM image

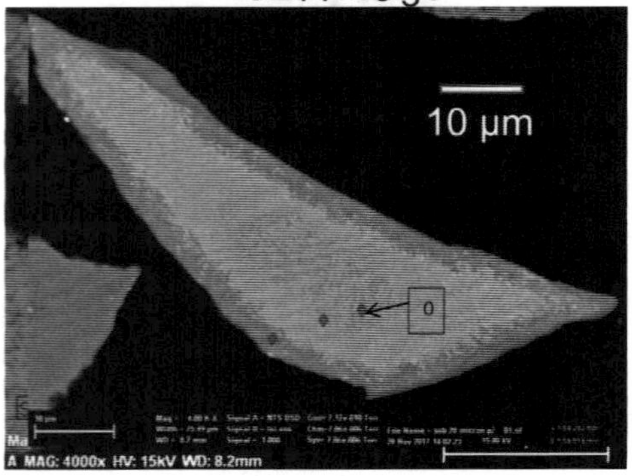

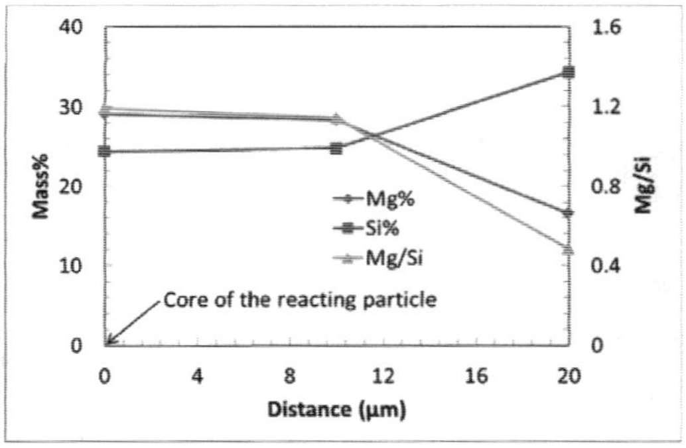

▶ 18

Thickness of Si passivating layer for all size fractions

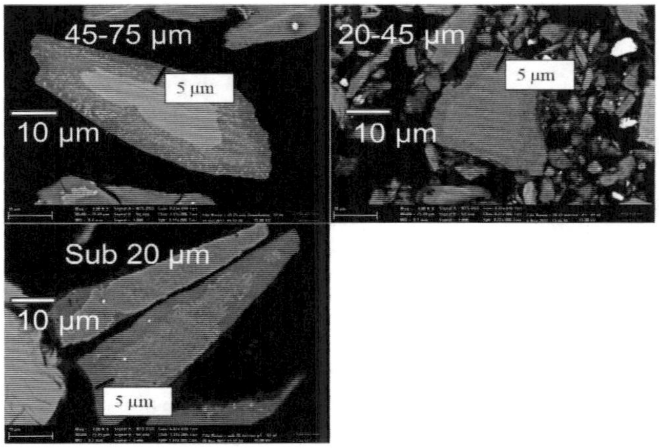

19

Comparison between carbonation and dissolution samples

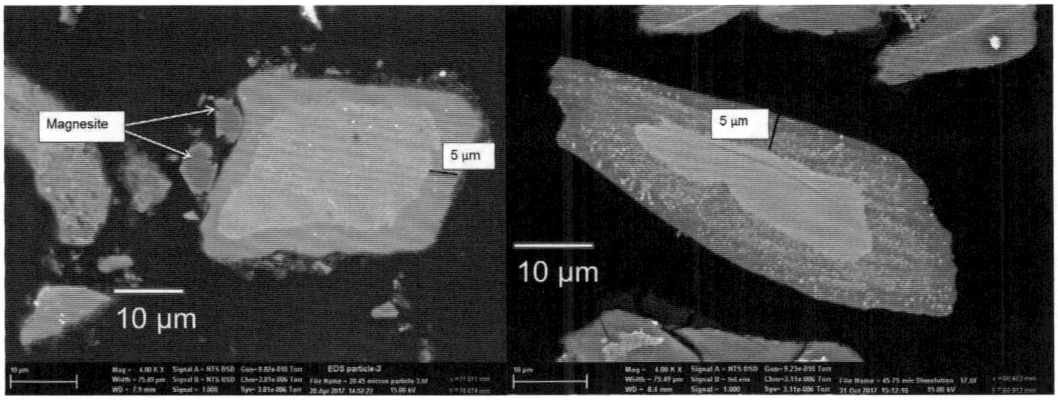

Carbonated sample at 130 bar, 150 °C & 0.64M $NaHCO_3$ (ARC conditions)

Dissolved sample at 100 bar, 40 °C & no $NaHCO_3$

20

Conclusions

- The rate of Mg extraction decreased with time as the particle dissolving
- Same thickness of the silica layer for all PSD which explains the highest Mg extraction for the fine particles.
- **Significant concentration of Mg still in the silica layer and even there are very small particles it is impossible to have 100% extraction.**
- The dissolved particles look very similar to the carbonated particles (ARC conditions).

21

Acknowledgements

I would like to sincerely acknowledge:

- The University of Newcastle for post graduate scholarship.
- Funding from MCi;

http://mineralcarbonation.com

22

Thank You!

Questions/comments welcome

Integrating MEA-based CO_2 absorption and fly ash-based amine regeneration for energy-efficient CO_2 sequestration from flue gas

Long Ji[1, 2, 3] | 12 March 2018

CSIRO ENERGY
www.csiro.au

1 CSIRO Energy, Newcastle, Australia
2 China University of Mining & Technology Beijing, Beijing, China
3 Macquarie University, North Ryde, Australia

Who we are

People	~5500
Sites	55
Flagships	9
Budget	$1B+

64% of our people hold university degrees over **2000** hold doctorates over **500** hold masters

We develop **832** postgraduate research students with our university partners

Commonwealth Scientific and Industrial Research Organization (CSIRO)

Post-combustion CO_2 capture (PCC) team

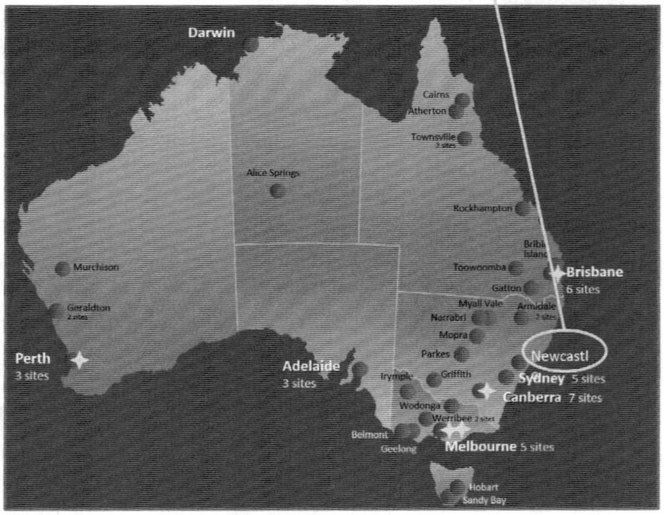

Absorbents-based CO_2 capture

Absorbents-based CO_2 in integrated absorber-stripper column systems is regarded as the most mature and technically ready for commercial deployment .

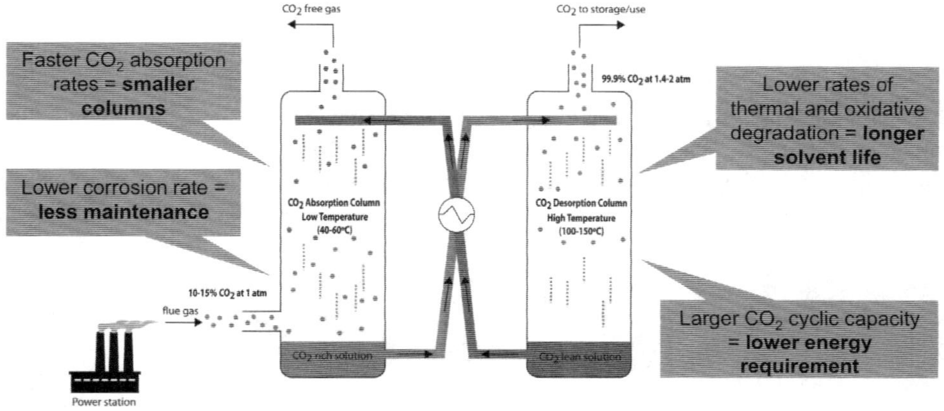

Puxty, G., Rowland, R., Allport, A., Yang, Q., Bown, M., Burns, R., ... & Attalla, M. (2009). Carbon dioxide post combustion capture: a novel screening study of the carbon dioxide absorption performance of 76 amines. Environmental science & technology, 43(16), 6427-6433.

Absorbents-based CO_2 capture – Pilot plant summary

Plant	Loy Yang	Munmorah -> Vales Point	Tarong	Newcastle PDF
Solvent			mine	
Flue gas source				
Scale				
Focus			ptimisation	
Other activities			olar thermal ntegration	cutting edge processes

Direct CO_2 mineralization by fly ash

Fly ash Elemental Quantification (XRF) in this study

FA	Compositions							
	SiO_2	Al_2O_3	Fe_2O_3	CaO	MgO	Na_2O	K_2O	SO_3
BJ	42.80	19.15	9.06	16.41	1.23	1.68	1.50	1.91
LY	42.80	16.20	9.53	3.60	7.13	4.95	0.34	9.4
HW	5.82	3.01	14.00	32.40	29.30	0.20	0.17	12.8
YA	8.92	5.92	42.30	9.40	27.90	0.56	0.16	2.84

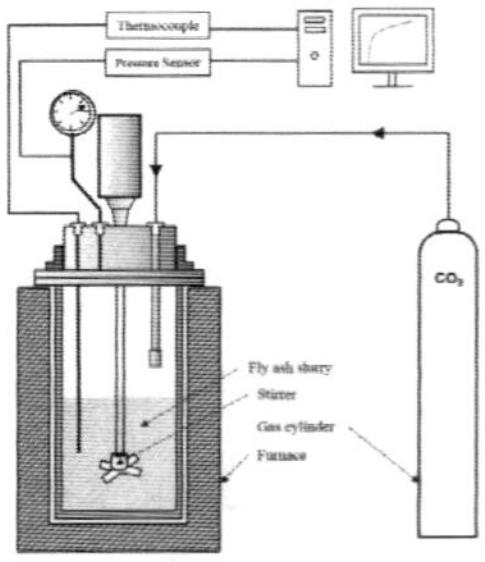

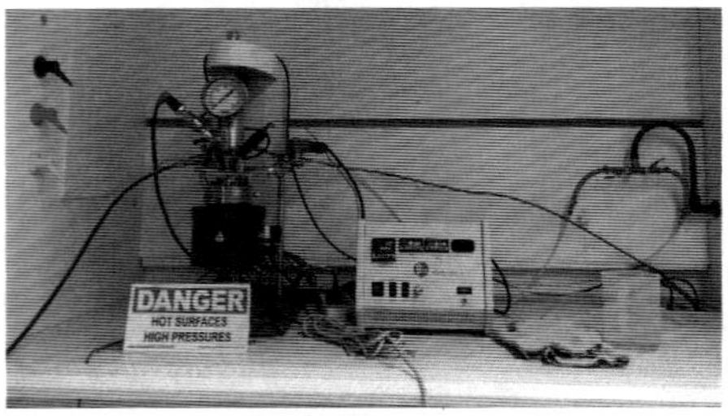

Parr reactor

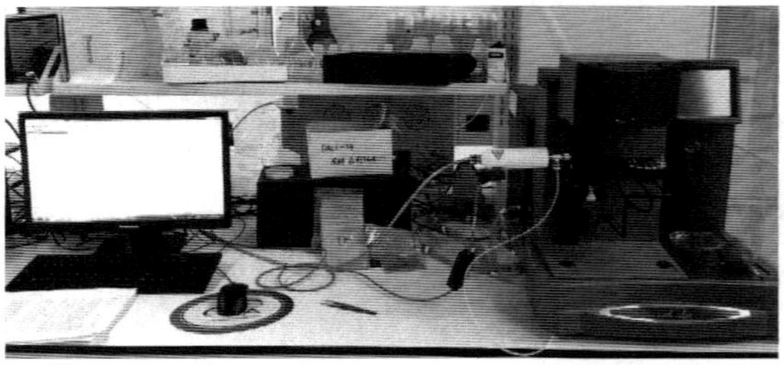

TGA-MS

Carbonation performance of Beijing fly ash

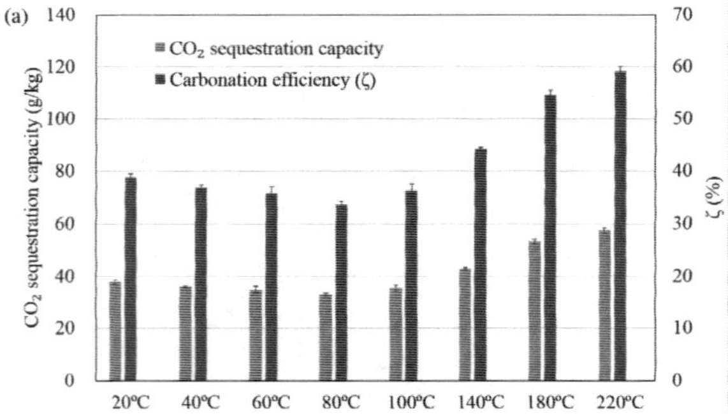

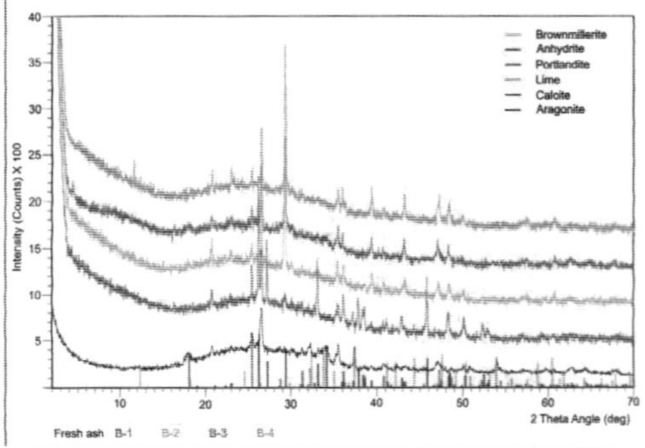

Id.	Chemical formula	BJ ash	B4 (40°C)	B3 (80°C)	B2 (180°C)	B1 (220°C)
Quartz	SiO_2	3.3	2.5	2.4	2.5	2.2
Mullite	$Al_6Si_2O_{13}$	1.4	1.2	1.2	1.0	1.3
Hematite	Fe_2O_3	1.4	0.4	2.0	2.1	2.8
Maghemite	Fe_2O_3	1.8	1.7	2.2	1.9	1.9
Brownmillerite	$Ca_2(Al,Fe)_2O_5$	0.7				
Anhydrite	$CaSO_4$	2.1	0.6	0.7	0.8	1.5
Portlandite	$Ca(OH)_2$	3.5				
Lime	CaO	1.2				
Calcite	$CaCO_3$		7.6	6.1	5.8	1.0
Aragonite	$CaCO_3$				1.7	10.3
Amorphous		84.6	85.9	85.4	84.2	79.1

Carbonation performance of Yallourn fly ash

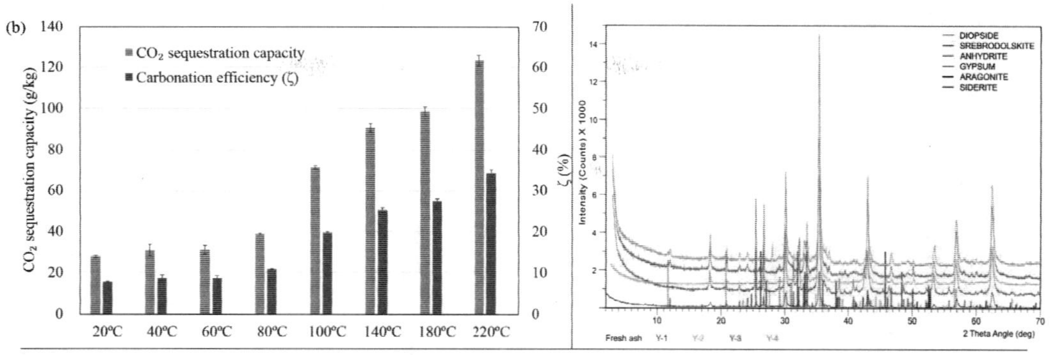

Id.	Chemical formula	YA ash	Y4 (40°C)	Y3 (80°C)	Y2 (180°C)	Y1 (220°C)
Quartz	SiO_2	3.5	2.6	3.3	1.0	3.0
Albite	$NaAlSi_3O_8$	0.8	1.5			
Diopside	$CaMgSi_2O_6$	2.3	0.1	0.6		0.8
Magnesioferrite	$MgFe_2O_3$	50.7	55.9	54.4	39.4	50.2
Ferrian spinel	$MgFe_{0.23}Al_{1.77}O_4$	3.8	3.2	3.4	1.7	3.0
Srebrodolskite	$Ca_2Fe_2O_5$	12.3	9.9	9.1	4.2	
Anhydrite	$CaSO_4$	3.1	1.9	1.1	0.3	9.6
Gypsum	$CaSO_4 \cdot 2H_2O$		0.7	1.0		
Aragonite	$CaCO_3$		1.8	1.6	4.0	2.3
Siderite	$Ca_{0.10}Mg_{0.33}Fe_{0.57}CO_3$				4.3	2.4
	$Mg_{0.92}Ca_{0.08}CO_3 \cdot 3H_2O$					6.7
Amorphous		23.5	22.4	25.5	45.0	22.0

Carbonation performance of Loy Yang fly ash

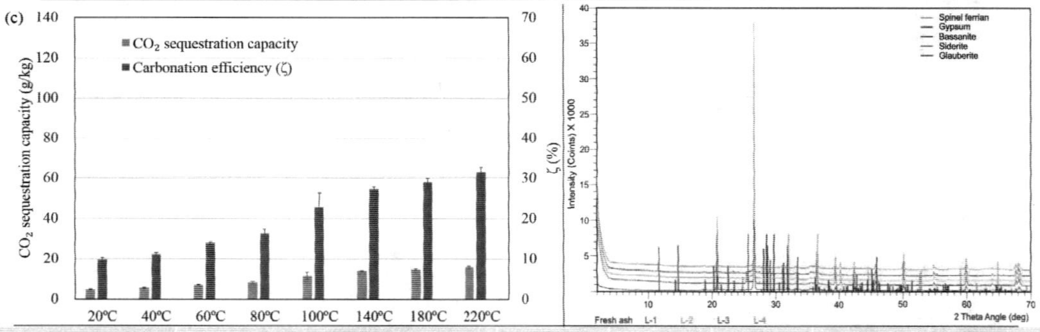

Id.	Chemical formula	LY ash	L4 (40°C)	L3 (80°C)	L2 (180°C)	L1 (220°C)
Quartz	SiO_2	37.6	53.3	47.9	54.0	55.8
Microcline	$KAlSi_3O_8$	1.9	3.0	3.4	3.2	3.3
Albite	$NaAlSi_3O_8$	1.7		1.2	1.1	1.3
Mullite	$Al_6Si_2O_{13}$		3.8	3.8	4.3	4.7
Hematite	Fe_2O_3	2.4	1.4	0.9	0.5	0.5
Maghemite	Fe_2O_3	11.7	7.0	5.4	3.5	4.4
Ferrian spinel	$MgFe_{0.23}Al_{1.77}O_4$	9.5	10.0	6.6	4.0	3.5
Gypsum	$CaSO_4 \cdot 2H_2O$		3.3	4.1	0.8	6.0
Bassanite	$CaSO_4 \cdot 0.5H_2O$	2.1			1.6	
Siderite	$FeCO_3$					0.7
Glauberite	$Na_2Ca(SO_4)_2$	7.9				
Thenardite	Na_2SO_4	3.0				
Polyhalite	$K_2Ca_2Mg(SO_4)_4 \cdot 2H_2O$	0.6				
Amorphous		21.8	18.0	26.7	27.0	19.7

Carbonation performance of Hazelwood fly ash

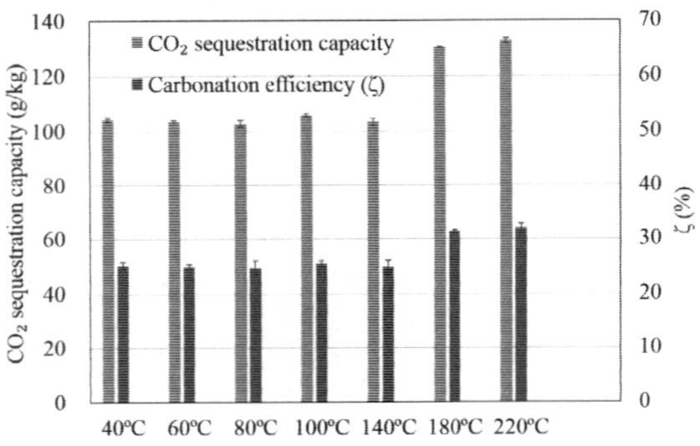

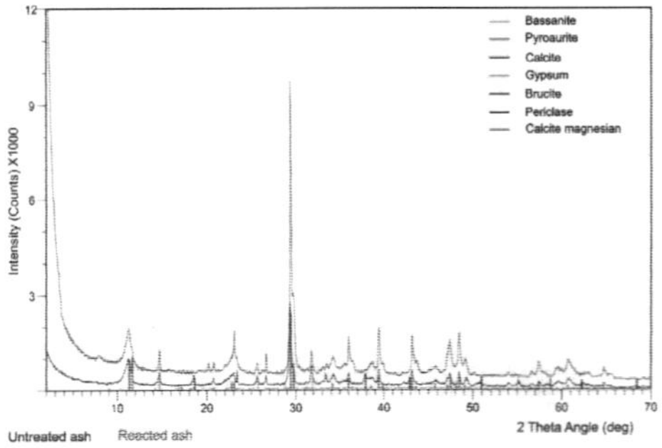

Id.	Chemical formula	Fresh ash	Carbonated ash
Quartz	SiO_2	1.3	1.0
Augite	$(Ca,Na)(Mg,Fe,Al,Ti)(Si,Al)_2O_6$	2.2	5.3
Srebrodolskite	$Ca_2Fe_2O_5$	3.0	2.2
Gypsum	$CaSO_4 \cdot 2H_2O$	3.0	0.4
Bassanite	$CaSO_4 \cdot 0.5H_2O$	10.7	4.8
Calcite	$CaCO_3$	23.0	36.8
Aragonite	$CaCO_3$	-	2.6
Mg-calcite	$MgCO_3$	-	10.5
Brucite	$Mg(OH)_2$	4.8	-
Periclase	MgO	3.2	-
Hexahydrite	$MgSO_4 \cdot 6H_2O$	-	2.1
Pyroaurite	$Mg_6Fe_2CO_3(OH)_{16} \cdot 4H_2O$	25.6	11.6
Amorphous	-	23.2	22.4

Opportunity for Integrated CO_2 Absorption–Mineralisation (IAM)

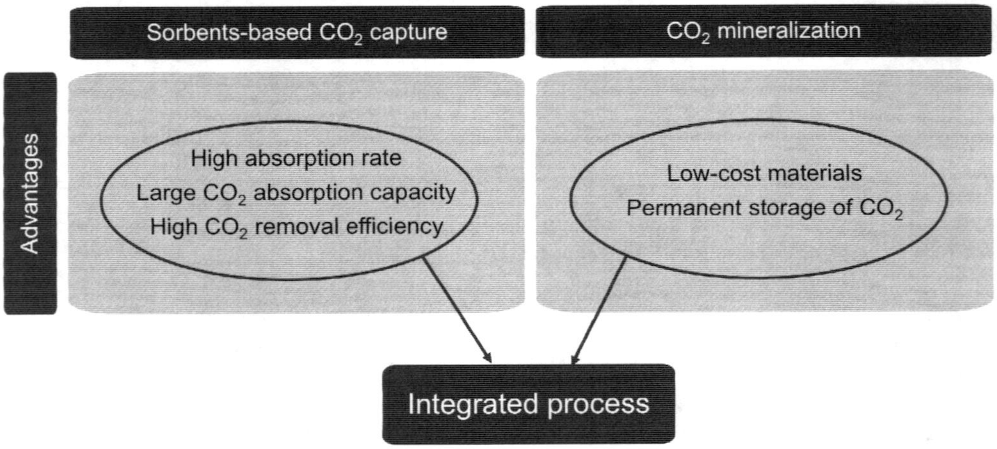

Integrated CO_2 Absorption–Mineralisation (IAM)

Integrated CO_2 Absorption–Mineralisation (IAM)

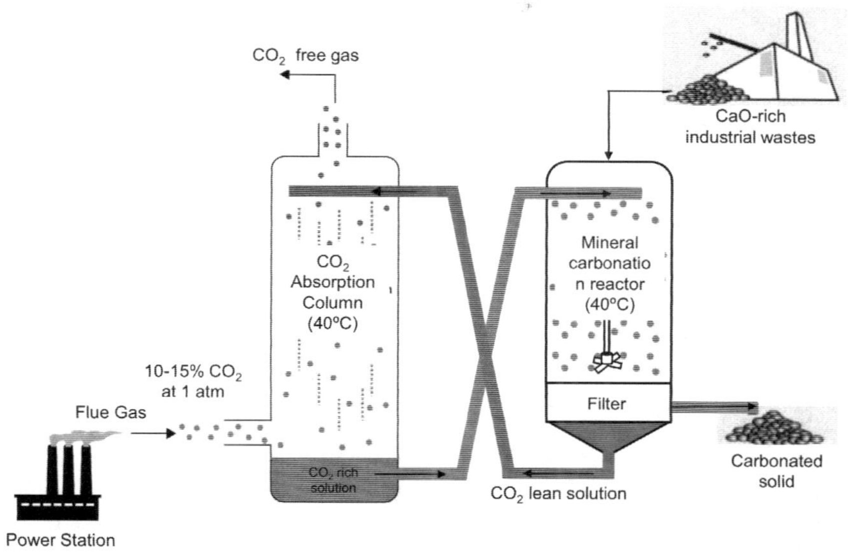

Integrated CO_2 Absorption–Mineralisation (IAM)

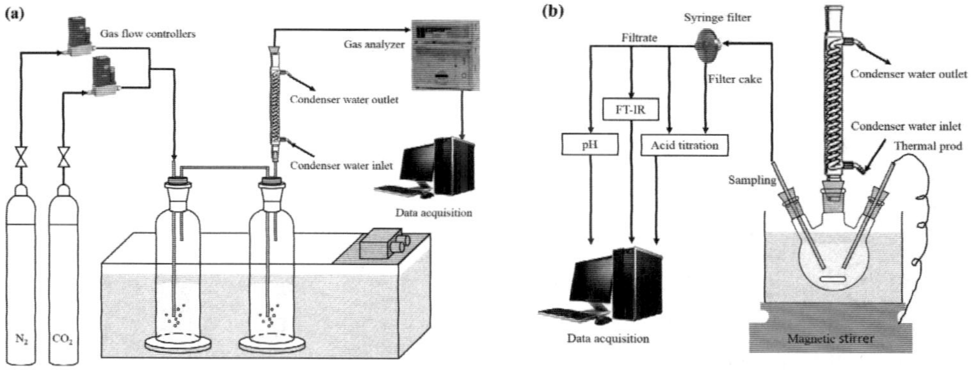

Schematic diagram of CO_2 absorption (a) and regeneration (b) experiments

IAM – Mechanism study by CaO

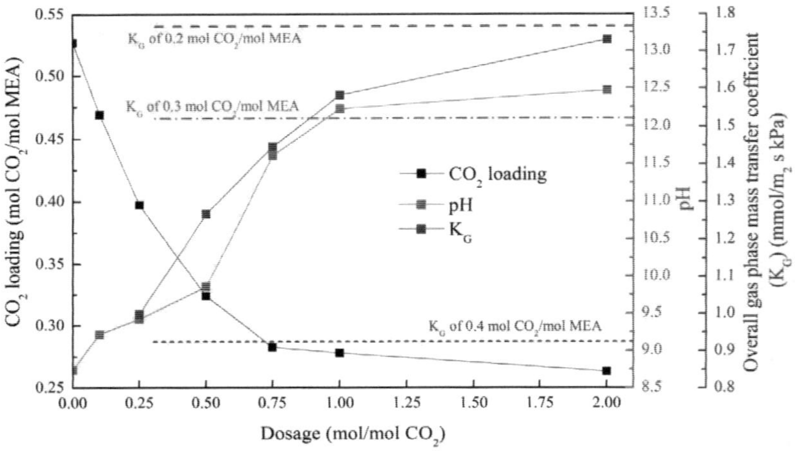

CO_2-loading, pH value and overall CO_2 mass transfer co-efficiencts (KG) of the MEA solutions after 15 min reaction as a function of dosages of calcium oxide

IAM – Mechanism study by CaO

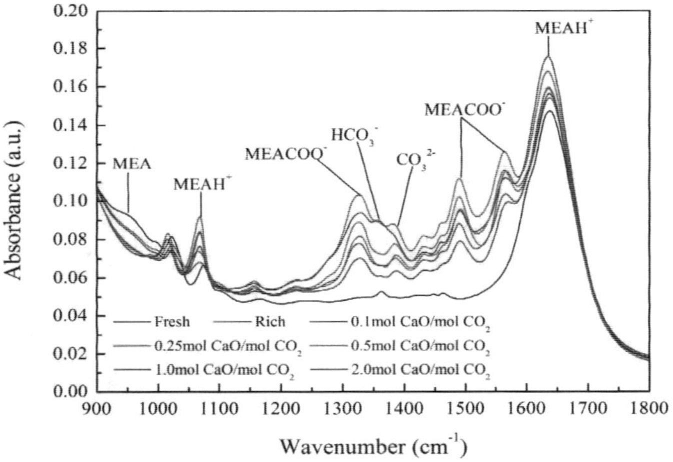

Corresponding FT-IR spectra of the MEA solutions

IAM – Mechanism study by CaO

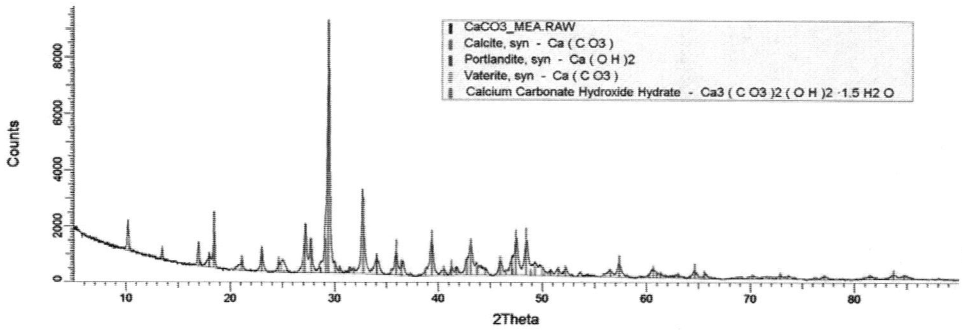

X-ray diffraction patterns of solid sample after reaction

IAM – Mechanism study by CaO

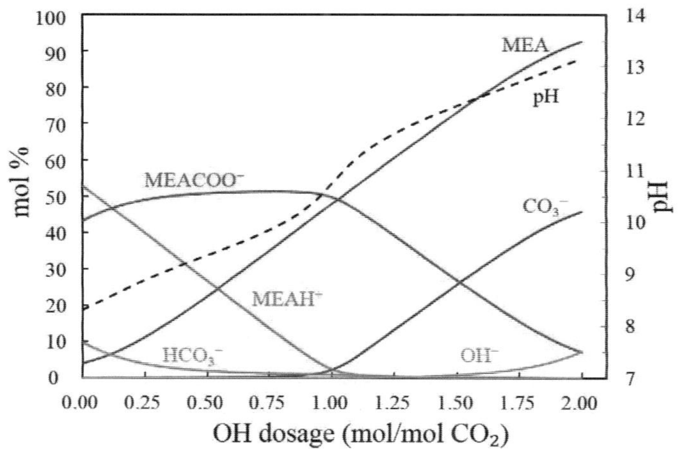

Calculated speciation profile of the MEA solutions with 0.53 mol/mol CO2-loading as a function of OH dosage

Chemistry of IAM

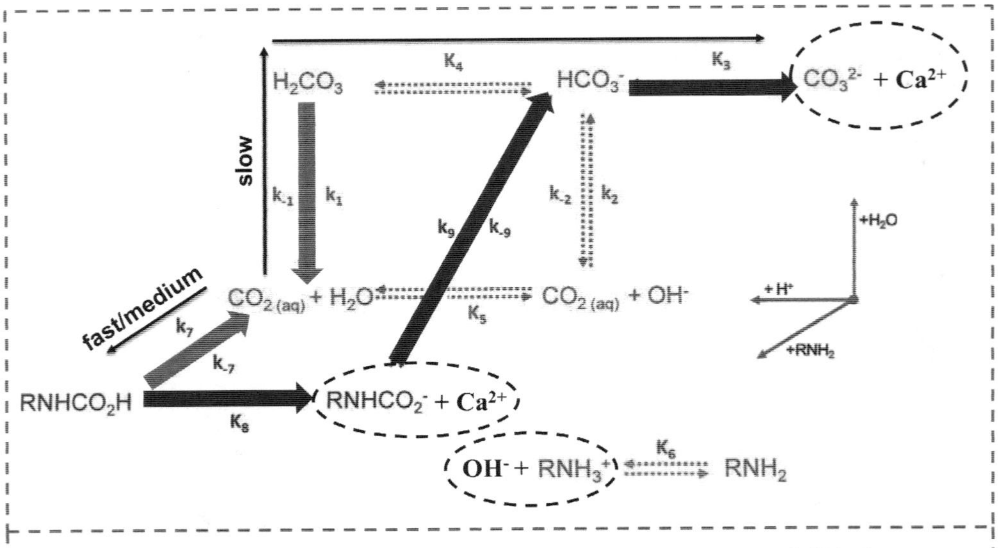

$CaO(s)+H_2O(l) \rightarrow Ca(OH)_2(aq)$, $Ca(OH)_2(aq) \leftrightarrow 2OH^-(aq)+Ca^{2+}(aq)$

IAM – Performance by BJ fly ash

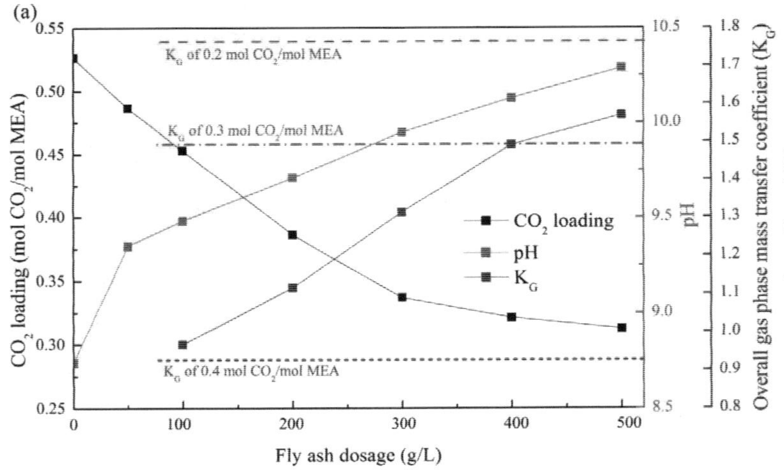

CO_2-loading, pH value and overall CO_2 mass transfer co-efficiencts (KG) of 2M MEA solutions as a function of dosages of fly ash

IAM – Performance by BJ fly ash

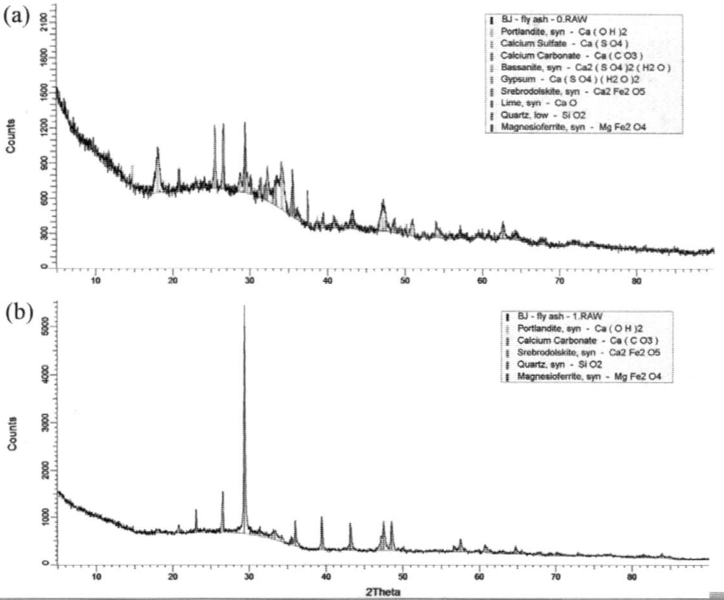

X-ray diffraction patterns of (a) fresh fly ash and (b) carbonated ash

IAM – Performance comparison

CO_2 capture technology	Chemical regeneration (CaO)	Chemical regeneration (fly ash)	Conventional amine scrubbing
Solvent concentration, mol/L	2	2	2
Flue gas CO_2, kPa	9	9	9
Absorption temperature, °C	40	40	40
Regeneration pressure, atm	1	1	2
Temperature approach of heat exchanger, K	–	–	10
CO_2-loading after absorption, mol/mol	0.51	0.53	0.51
CO_2-loading after regeneration, mol/mol	0.30	0.32	0.30
Cyclic CO_2-loading, mol/mol	0.21	0.21	0.21
Regeneration temperature, °C	40	40	116
Regeneration duty, MJ/kg CO_2	0	0	4.7
Product	$CaCO_3$	Fly ash rich with $CaCO_3$	CO_2

Bigger picture and future work

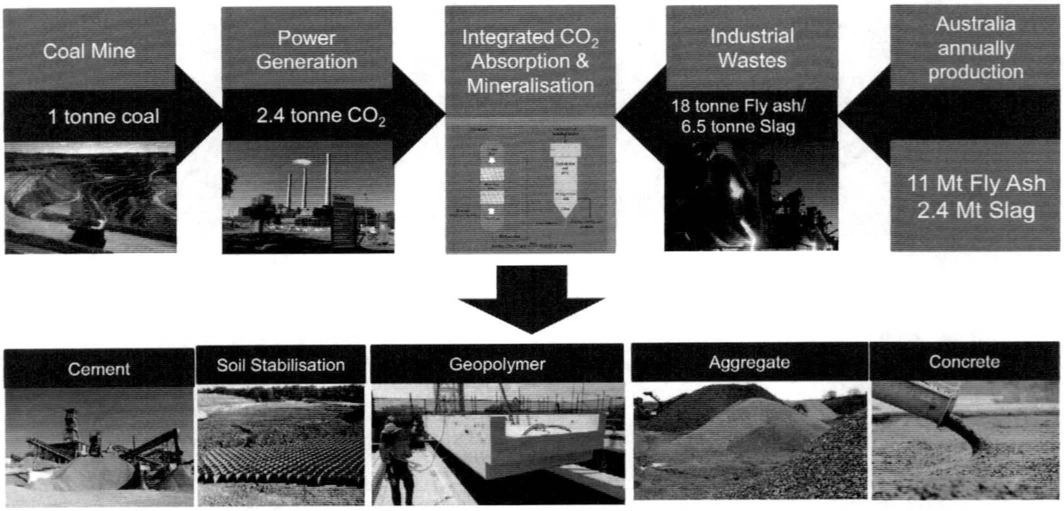

Acknowledgements

My supervisors:
Hai Yu
Ming Zeng
Damian Gore
Shuaifei Zhao

My colleagues:
Kangkang Li
Bing Yu
Will Conway
Mihaela Grigore

People who helped a lot:
David French
Xiaolong Wang

CSIRO
MACQUARIE University
中国华能集团清洁能源技术研究院有限公司
CHINA HUANENG CLEAN ENERGY RESEARCH INSTITUTE

Thank you !

Mineralogical Study of Thermally Activated Serpentine Dissolution Process for Direct Flue Gas CO2 Abatement By Mineral Carbonation

Authors:

Du Breuil, C., INRS-ETE, University of Quebec
Pasquier, L. C., INRS-ETE, University of Quebec
Mercier, G., INRS-ETE, University of Quebec
Dipple, G. M., The University of British Columbia
De Baere, B., The University of British Columbia
Carrol, K., The University of British Columbia
Blais, J. F., INRS-ETE, University of Quebec
Iliuta, M. C., Université Laval

Direct flue gas mineral carbonation is an efficient way to reduce industrial greenhouse gas emissions. Along with other silicates, serpentine minerals show great potential for carbonation. Most of all, serpentine minerals represent a great interest due to their large availability and accessibility. With more than 2Gt of serpentine mining residues closed to industrial emitters, the abatement potential is very important in the Province of Quebec. Magnesium leaching is identified as carbonation limiting factor. However, magnesium is strongly bonded to hydroxyl groups inside of the serpentine structure. Thereby, serpentine needs to be heat activated to undergo a dehydroxylation process to reach sufficient carbonation rates. Such step is responsible for high energy demand and is very sensitive for the process application. Temperature and duration of the activation are the parameters to be controlled during the dehydroxylation process. This study proposes a mineralogical approach by using XRD and Rietveld Refinement method combined to flow through time resolved analysis (FT-TRA) and aqueous-phase mineral carbonation reaction in flue gas conditions experiments to understand and apprehend the impacts of the thermal treatment on the dissolution of serpentine and furthermore on the carbonation process.

Metal Mobility during Passive and Accelerated Carbonation of Ultramafic Mineral Wastes

Authors:

Wilson, S. A., Monash University
Hamilton, J. L., Monash University
Morgan, B., The University of Sydney
Turvey, C. C., Monash University
McCutcheon, J., The University of Queensland
Paterson, D. J., Australian Synchrotron
Tait, A. W., Monash University
Southam, G., The University of Queensland

Ultramafic rocks and mineral wastes rank amongst the most plentiful and most reactive feedstocks for carbon mineralisation. Ultramafic rocks commonly host Niâ€"Cuâ€"PGE and chromite deposits and they can be enriched in gold, silver and cobalt. Recovery of metal resources during ore processing is not 100% efficient and ore processing circuits are typically designed to only recover one or two key elements or minerals. As such, ultramafic mineral wastes can contain valuable, and potentially toxic, first row transition metals at abundances of several percent by weight. We have undertaken a series of synchrotron, laboratory and field experiments to better understand the mobility â€" and ultimate fate â€" of transition metals during passive and accelerated carbonation of ultramafic mineral wastes.

Our laboratory experiments demonstrate that first row transition metals (Cr, Mn, Fe, Co, Ni, Cu) are rapidly incorporated into the structures of hydrated Mg-carbonate and Fe-oxyhydroxide minerals under conditions relevant to accelerated carbonation of ultramafic rocks. Greater than 99 wt.% of aqueous transition metals are sorbed to colloidal carbonate and Fe-oxyhydroxide precipitates within 5 minutes. These are retained during recrystallisation to form mm-scale, euhedral crystals of nesquehonite ($MgCO_3$â‹…$3H_2O$) indicating that metal sequestration is likely to be stable during Mg-carbonate phase transitions.

We see the same association of first row transition metals with Mg-carbonate and Fe-oxyhydroxide minerals during passive carbonation in the mineral waste at the Woodsreef Chrysotile Mine, New South Wales, Australia. Our synchrotron X-ray Fluorescence Microscopy data indicate that mobility of transition metals is limited during passive carbonation of ultramafic landscapes. Transition metals are immobilised on lengthscales of tens of micrometres, they are retained by alteration minerals at reaction fronts, and their concentrations are generally below detection in mine pit waters.

Our accelerated carbonation experiments employ weak sulfuric acid leaches to enhance the availability of Mg^{2+} for carbonation reactions. In field-based mesocosm experiments and laboratory-based column experiments, we find that periodic addition of acid can be used to leach Mg from both brucite and serpentine minerals, producing an increasingly deep Mg-depleted horizon. An Fe-oxyhydroxide-rich horizon occurs at the neutralisation front between leached and unreacted ultramafic mineral wastes. Our X-ray fluorescence data confirm that first row transition metals are associated with, and immobilised by, this Fe-oxyhydroxide layer.

These observations suggest there is low risk of transition metal release into process waters (for industrial-style carbon mineralisation) or surface waters (for enhanced weathering of ultramafic landscapes) so long as the neutralisation potential of mineral wastes is not exceeded. The ability of Mg-carbonate minerals to scavenge potentially hazardous transition metals may limit their use by some industries (e.g., pharmaceuticals). Ultimately, understanding the deportment of transition metals during carbonation reactions provides a valuable opportunity to use CO_2 sequestration as a technology to recover increasingly scarce metal resources.

TECHNISCHE UNIVERSITÄT DRESDEN

DRESDEN concept

Bastian Alexander Sakowski, M. Sc.

Reduction of CO_2 and H_2S for Biogas Upgrading using accelerated Mineralization

Institute of Process Engineering and Environmental Technology, TU Dresden

Newcastle, 13.05.2018

Research Question

General Questions

- Is it possible to **reduce carbon dioxide** and **hydrogen sulfide** from biogas using accelerated **mineralization** processes?
- Minerals come from **fly ash** from **waste combustion** power plants
- Target values are **hydrogen sulfide and carbon dioxide separation efficiency**, absolute **amount of stored H_2S and CO_2** from biogas
- **Influence of different factors** like ash, temperature, hydrogen sulfide partial pressure, and process water volume flow

Mineralization for Biogas Upgrading

Capture and Storage Carbon Dioxide

- Solving CO_2 in Water (≈ Capture)

  $CO_2 + H_2O \leftrightarrow HCO_3^- + H^+ \leftrightarrow CO_3^{2-} + 2\ H^+$

- Solving of Metal Oxide in Water

  $CaO + H_2O \leftrightarrow Ca(OH)_2 \leftrightarrow Ca_2^+ + 2\ OH^-$

- Forming Carbonates (≈ Storage)

  $Ca^{2+} + CO_3^{2-} \leftrightarrow CaCO_3$

$Ca^{2+} + 2\ HCO_3^- \leftrightarrow Ca(HCO_3)$

$2\ H^+ + 2\ OH^- \leftrightarrow 2\ H_2O$

Mineralization for Biogas Upgrading

Capture and Storage Hydrogen Sulfide

- Solving H_2S in water (≈ Capture)

  $H_2S + H_2O \leftrightarrow HS^- + H^+ \leftrightarrow S^{2-} + 2\ H^+$

- Solving of metals (Primarily Zinc and Lead) in water

- Forming sulfides (≈ Storage)

  $Pb^{2+} + S^{2-} \leftrightarrow PbS$

$Zn^{2+} + S^{2-} \leftrightarrow ZnS$

Test Station

Piping and Instrumentation Diagram

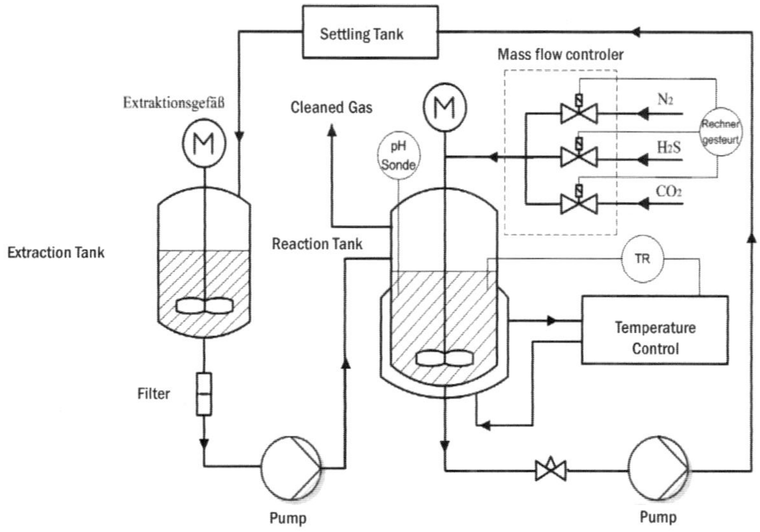

Test Station

Testfield in the Laboratoy

Test Station

Boundary Conditions for Experiments

- **3 liters of process water** in extraction tank, **2 liters of process water** in the reaction tank
- Ash: **300g fly ash** from waste combustion power plants in Germany
- **Ash separated** from process water using a **filter**
- **Gas: 200 liters/h**
- **Minimum 2% H_2S** in gaseous phase
- **Stopping criterion**: pH value < 8.5 in reaction tank
- **Analysis of precipitation product** for CO_2 and H_2S separation with ICP

Test Station

Ashes

- Both ashes from waste incineration plant in Germany

Species	Ash 1 [g/kg]	Ash 1 [mmol/kg]	Ash 2 [g/kg]	Ash 2 [mmol/kg]
Calcium	298	**7,435.5**	256	**6,387.5**
Lead	2.99	**14.4**	1.32	**6.37**
Zinc	14	**214**	4.46	**68.2**

Research Question

Detailed Questions

- Does the **presence** of carbon dioxide **support** the hydrogen sulfide sequestration?
- Reaction temperature & time: **kinetics** vs. **absorption capacity**
- What is the **limitation for carbon dioxide** capture and storage?
- What is the **limitation for hydrogen sulfide** sequestration?
- **Ideas** for compensation

Presence of Carbon Dioxide in the Gas Phase

Sequestrated Carbon Dioxide and Hydrogen Sulfide

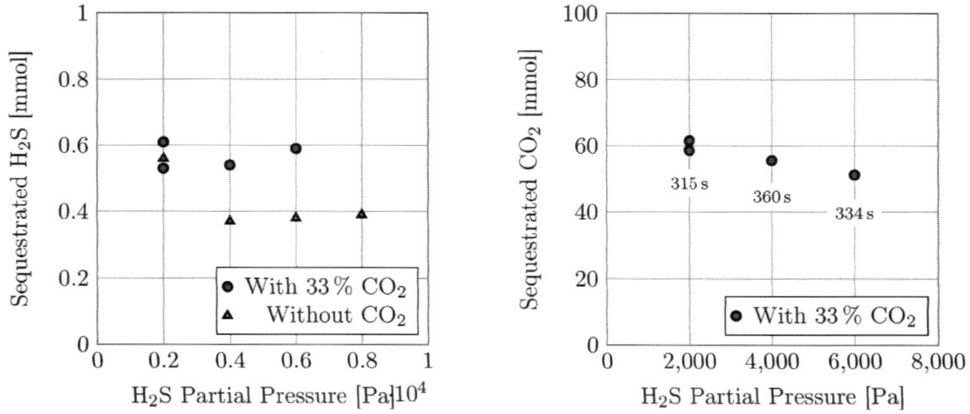

Presence of Carbon Dioxide in the Gas Phase

Separation Efficiency

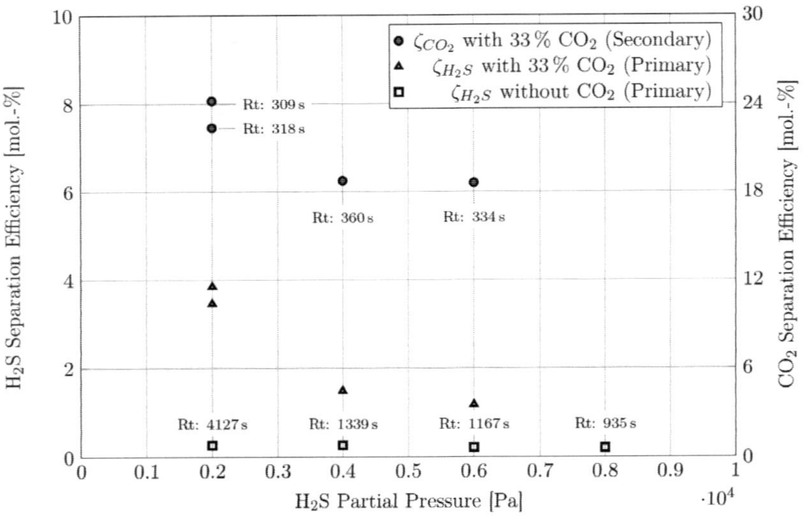

TECHNISCHE UNIVERSITÄT DRESDEN

Slide 12

>sustainable energy systems> Boysen-TUD-Graduiertenkolleg

Influence of Reaction Time

Sequestrated Carbon Dioxide and Hydrogen Sulfide

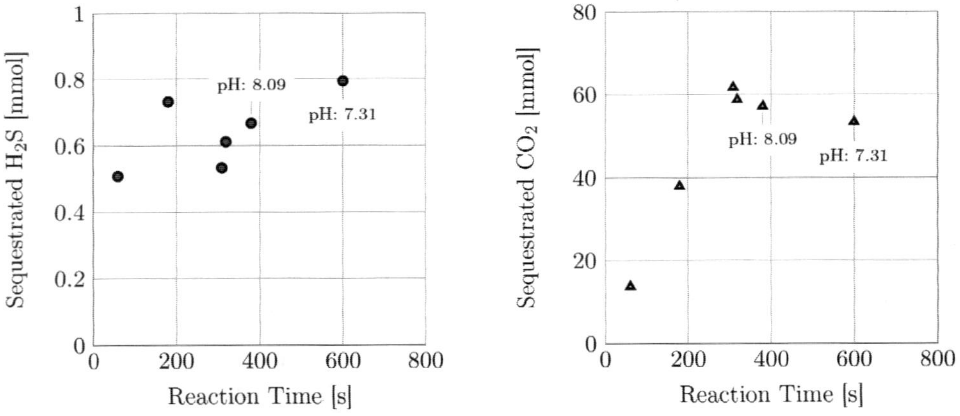

TECHNISCHE UNIVERSITÄT DRESDEN

Slide 14

>sustainable energy systems> Boysen-TUD-Graduiertenkolleg

Kinetics

Adding Hydrogen Sulfide to the Gas Phase

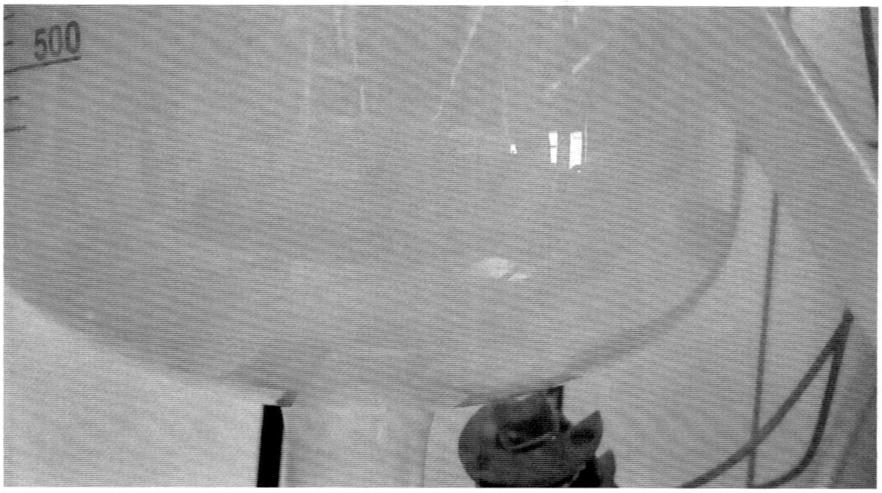

TECHNISCHE UNIVERSITÄT DRESDEN

Slide 15

›sustainable energy systems›
Boysen-TUD-Graduiertenkolleg

Compensation

Higher Temperatures for better Kinetics

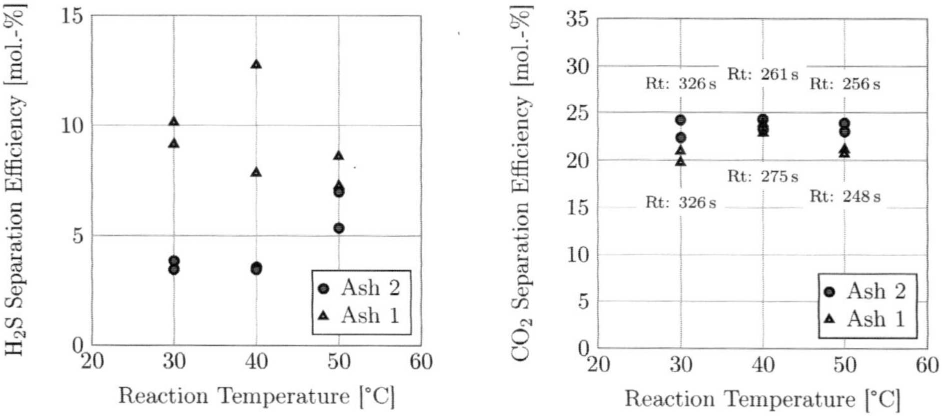

TECHNISCHE UNIVERSITÄT DRESDEN

Slide 18

›sustainable energy systems›
Boysen-TUD-Graduiertenkolleg

Results

Higher Process Water Volume Flow

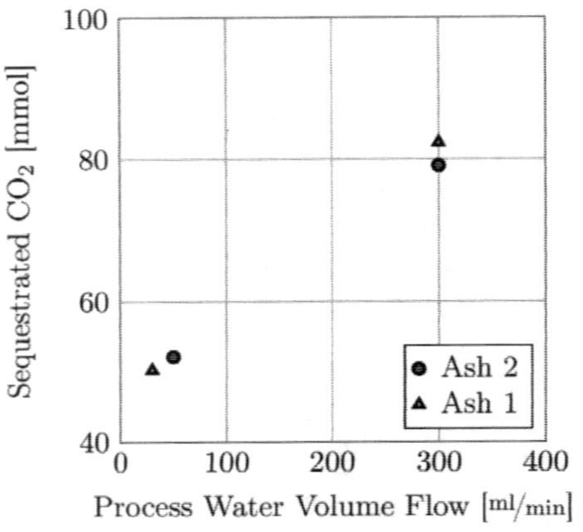

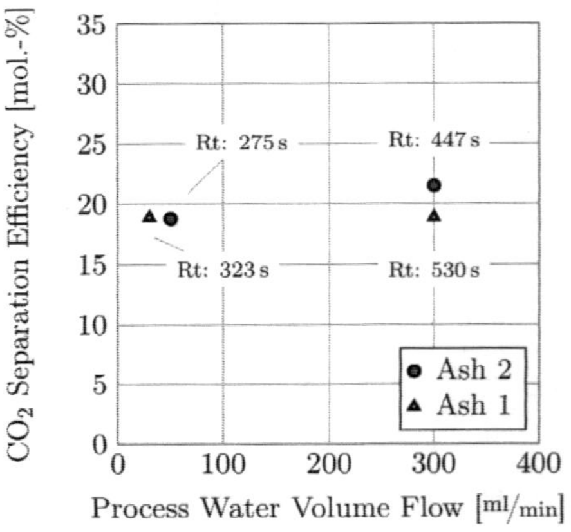

Conclusion

Ideas for future Tests

- Accelerated mineralization is a **promising technology for biogas upgrading** in the future
- Hydrogen sulfide **separation limitation: Amount of reactants** in the process water
- Carbon dioxide separation limitation: **Kinetics first**, amount of reactants in the process water second
- Solution 1: New test station with more process water volume flows (no filters)
- Solution 2: **New test station** with better kinetics for carbon dioxide separation (packed column)

Thank you very much!

TECHNISCHE UNIVERSITÄT DRESDEN

Slide 21

> sustainable energy systems >
Boysen-TUD-Graduiertenkolleg

ETH *zürich*

Carbonation kinetics of waste concrete

Johannes Tiefenthaler, Subrahmaniam Hariharan and Marco Mazzotti
ACEME 2018, Newcastle

ETH Zurich, Switzerland

ETH *zürich*

Lifecycle of cement – current state

- 900 kg CO_2/t cement
- 5% of global CO_2 emissions

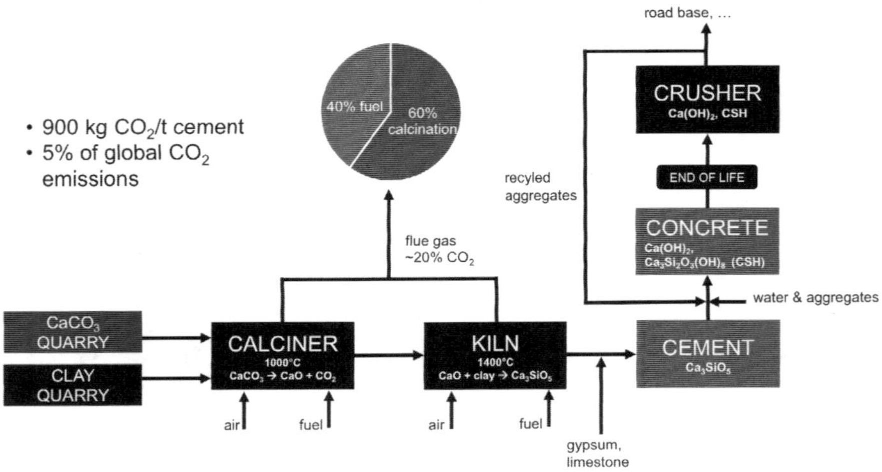

ETH zürich

Lifecycle of cement – future scenario

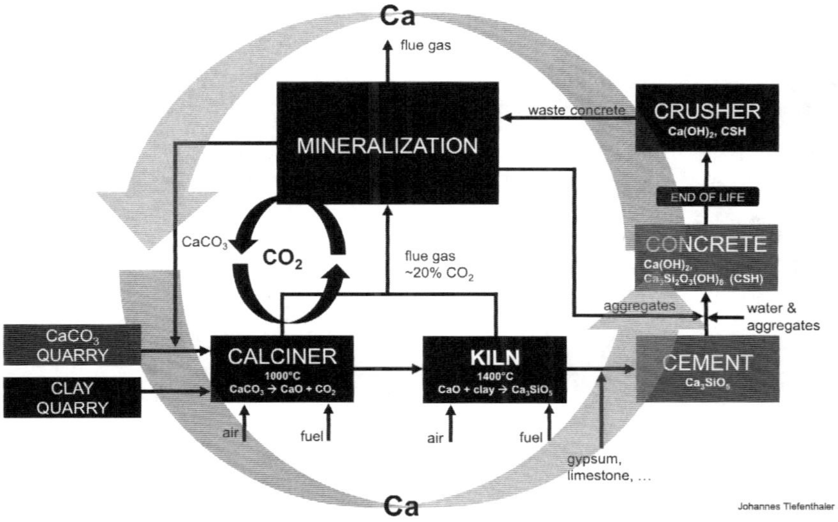

ETH zürich

Outline

Waste concrete characterization

Calcium leaching experiments

Carbonation experiments

Summary and conclusion

ETH zürich

Outline

Waste concrete characterization

Calcium leaching experiments

Carbonation experiments

Summary and conclusion

ETH zürich

Waste concrete characterization

concrete

Sand: 70 wt.%
Ca – phases: 30 wt.%

Ca-rich phases

40 % $Ca(OH)_2$
40 % Ca-silicate hydrate (CSH)
10 % $CaCO_3$
10 % other

↑ Solubility in $H_2O - CO_{2(aq)}$

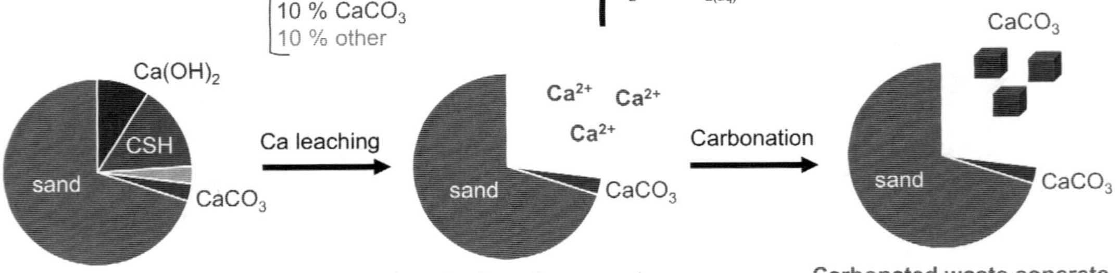

Leached waste concrete

Carbonated waste concrete

ETH zürich

Outline

Waste concrete characterization

Calcium leaching experiments

Carbonation experiments

Summary and conclusion

ETH zürich

Calcium leaching experiments

Experimental setup

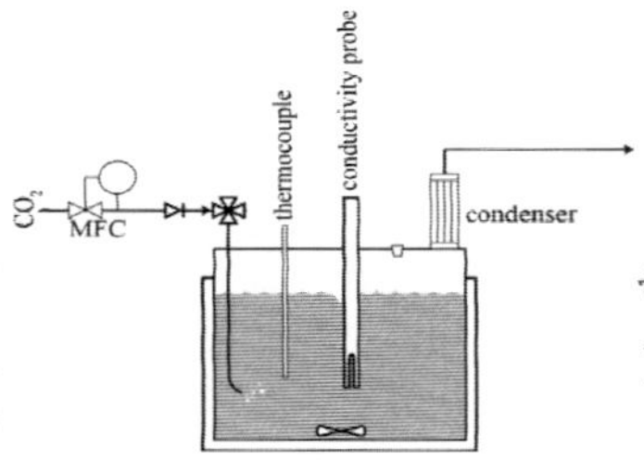

Ambient pressure
100 g aqueous solution

- Conductivity probe calibration
- CO_2 sorption rate measurement
- Batch dissolution experiments
- Controlled atmosphere

Reproducibility

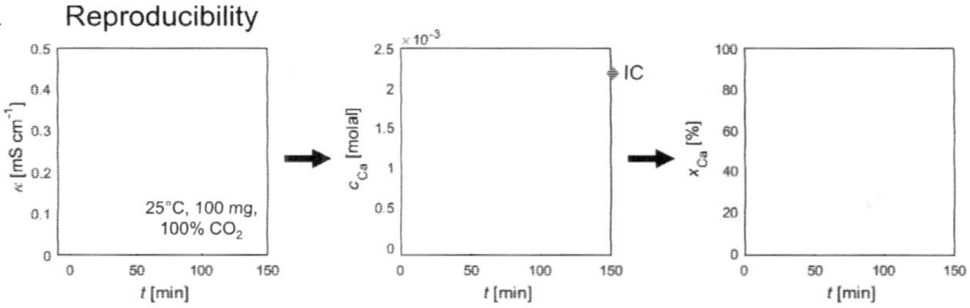

Calcium leaching experiments

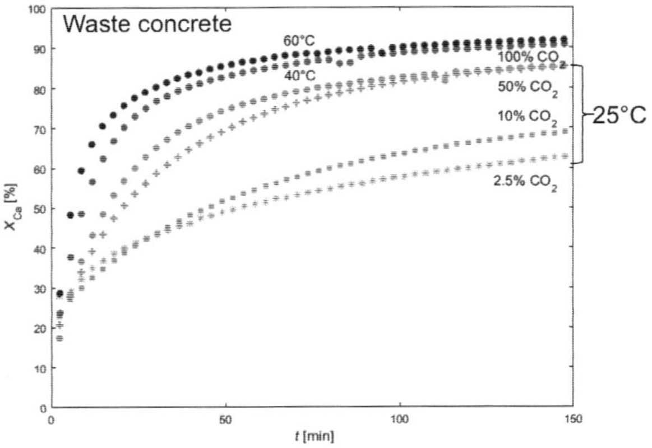

ETH zürich

Calcium leaching experiments

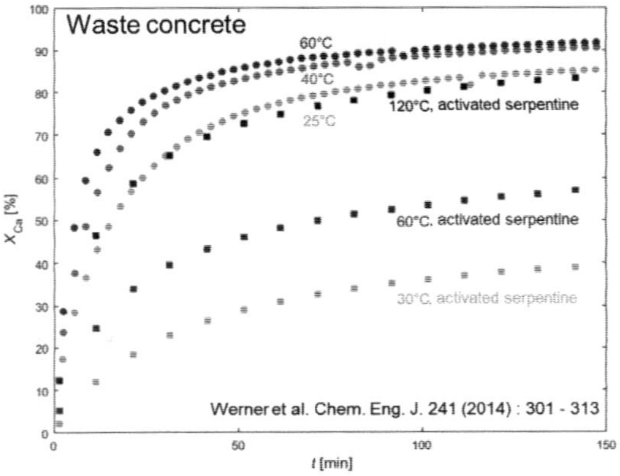

More than 80% of the calcium of waste concrete could be leached at mild operating conditions.

T and p_{CO2} accelerate the dissolution of calcium.

Waste concrete exhibits significantly faster dissolution kinetics than activated serpentine at similar operating conditions.

ETH zürich

Outline

Waste concrete characterization

Calcium leaching experiments

Carbonation experiments

Summary and conclusion

Separation Processes Laboratory

Johannes Tiefenthaler | 3/13/2018 | 11

ETH zürich

Carbonation experiments

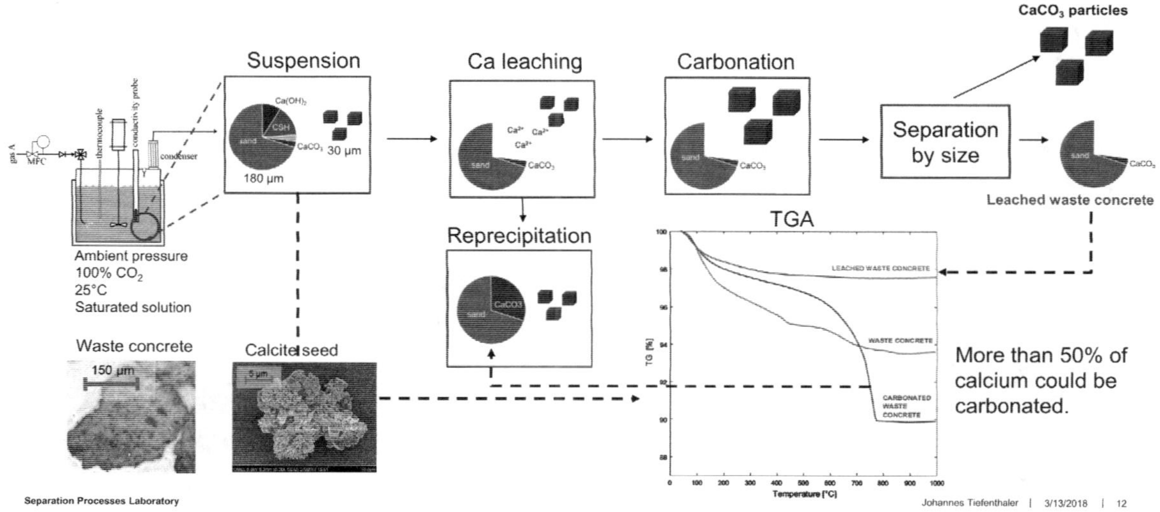

Separation Processes Laboratory

Johannes Tiefenthaler | 3/13/2018 | 12

Carbonation experiments

Summary and conclusions

- More than 80% of the Ca in waste concrete can be extracted at 25°C, 1 atm and 100% CO_2.
- First carbonation experiments show that more than 50% of the Ca in waste concrete can be carbonated in a single step.
- Waste concrete shows at lean operating conditions significantly faster dissolution kinetics and a higher extent of carbonation than activated serpentine.
- While promising dissolution and carbonation results were obtained, the separation of the carbonates from aggregate still has to be solved to be able to recycle waste concrete.

Backup – crystalline structure of seeds

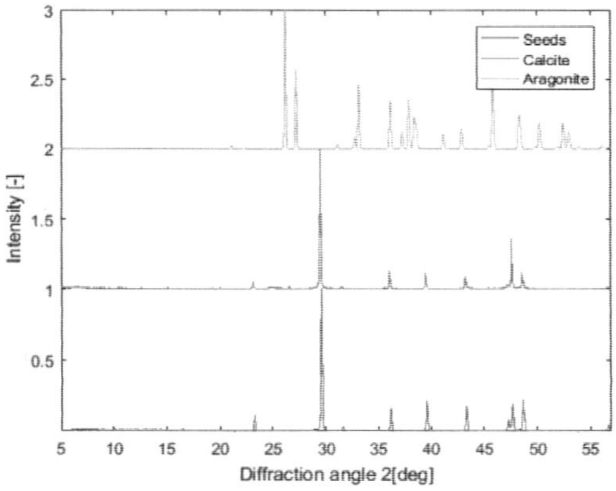

Vaterite

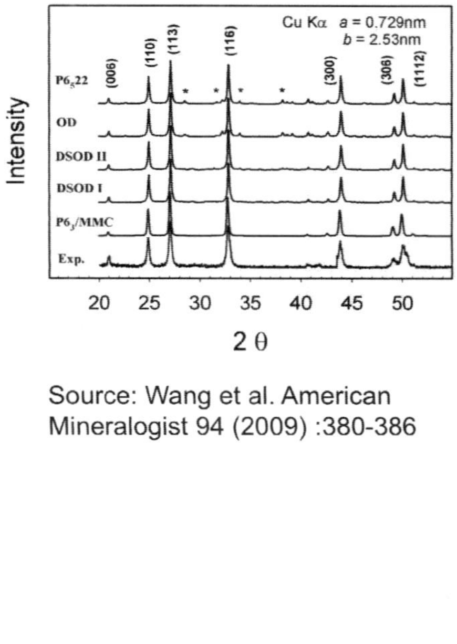

Source: Wang et al. American Mineralogist 94 (2009) :380-386

National Taiwan University

The 6th International Conference on Accelerated Carbonation for Environmental and Material Engineering
Newcastle City Hall, Newcastle, New South Wales, Australia

Integrated Carbon Mineralization and Air Pollutant Emission Reduction via a High-Gravity Process using Alkaline Wastes

Tse-Lun Chen[1,2], Yun-Ke Fan[1], Si-Lu Pei[1], Shu-Yuan Pan[1],
Yi-Hung Chen[3], Pen-Chi Chiang[1,2]

[1]Carbon Cycle Research Center, National Taiwan University, Taipei, Taiwan
[2]Graduate Institute of Environmental Engineering, National Taiwan University, Taipei, Taiwan
[3]Department of Chemical Engineering and Biotechnology, Taipei, Taiwan

2018.03.13

Outline

- Introduction
- Significance and Importance
- Materials and Methods
- Results and Discussion
- Conclusion and Recommendation

2

I. Introduction

3

1.1 Accelerated Carbonation of Alkaline Wastes

- **Industrial residues such as BOFS (Basic Oxygen Furnace Slag) are alkaline:**
  - CaO, MgO, SiO_2, Al_2O_3, Cr_2O_3, TiO_2, MnO, iron oxides
- **Advantages of carbonation using alkaline wastes**
  - Thermodynamically stable, exothermic reaction **(limits the costs)**
  - High available deposits, no on-site transport required **(cost -effective)**
  - Use in construction materials, improved environ. quality **(decrease heavy metal leaching)**

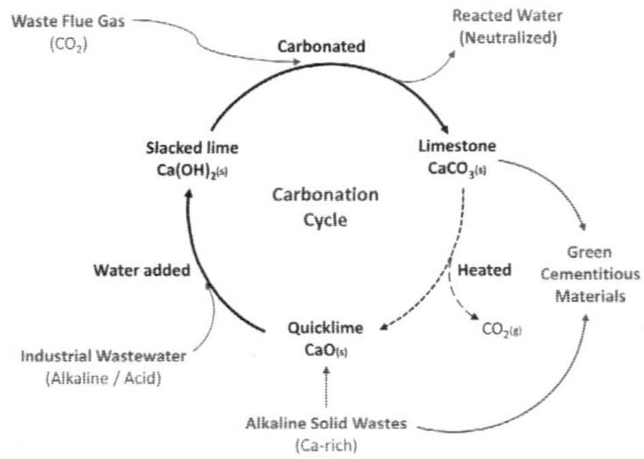

Coupling industrial carbon mineralization and waste product utilization: an integrated approach to application of accelerated carbonation process (Pan et al., 2015)

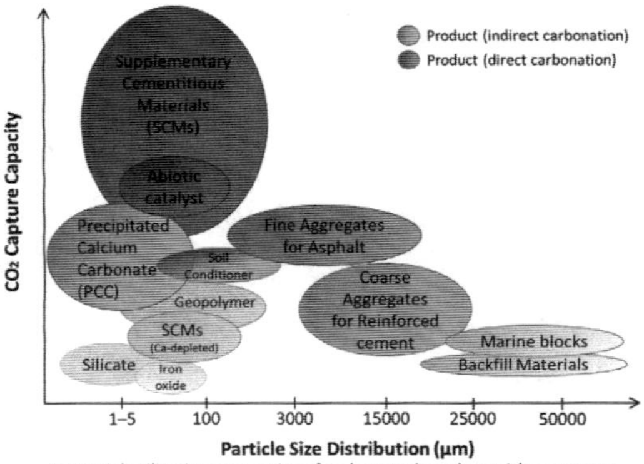

Potential utilization categories of carbonated product with respect to particle size distribution and CO_2 capture capacity (Pan et al., 2016)

4

1.2 Types of Alkaline Solid Wastes for CCUM

Alkaline Solid Waste	Example
Slag	Steelmaking slag (BOFS, EAFS) Coal slag Blast furnace slag (UF, FA, BFS)
Air pollution control (APC) residue	Municipal solid waste incinerator (MSWI) APC residue Cyclone dust
Fly ash	MSWI fly ash Coal fly ash Oil shale ash
Bottom ash	MSWI bottom ash
Cement wastes	Cement kiln dust (CKD) Cement bypass dust (CBD) Construction and demolition waste Cement / Concrete waste Blended hydraulic slag cement (BHC)
Mining and mineral processing waste	Asbestos tailings Nickel tailings Red mud (Bauxite)
Sludge (incinerator) ash	Sewage sludge incinerator ash (SSA) Steel wastewater sludge (SWS) Paper sludge incinerator ash (PSIA)
Paper pulping and mill waste	Paper mill waste (calcium mud) Green sludge dreg Lime mud Lime slaker grits

(Ref: Pan et al., 2012)

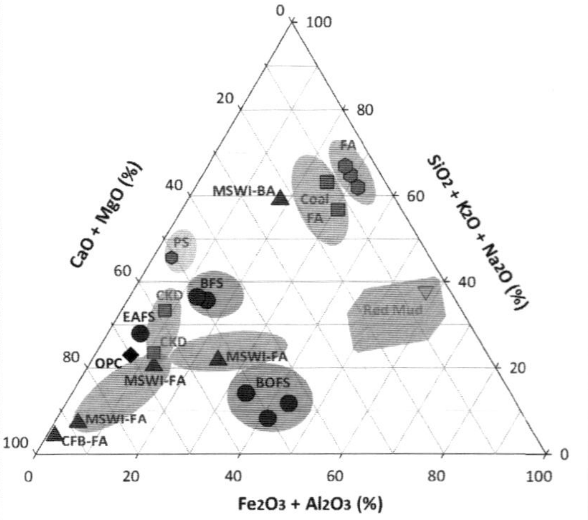

Normalized $CaO(MgO)-SiO_2(Na_2O,K_2O)-Al_2O_3(Fe_2O_3)$ phase diagram of various types of alkaline waste

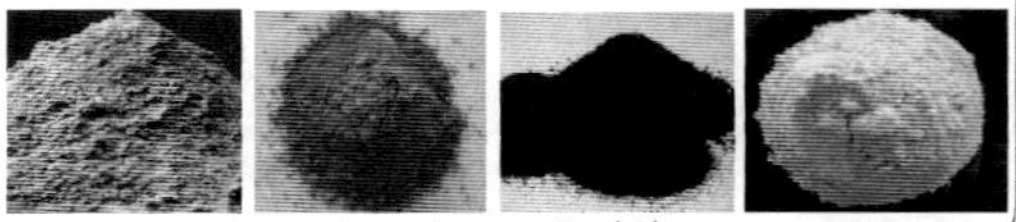

Incinerator Ash Red Mud Coal Slag BF Slag

1.3 High-Gravity Rotating Packed Bed (RPB)

- Generating **high acceleration** via **centrifugal force** to form **thin liquid films** and **micro-** or **nano-droplets**
- **Advantages**:
  - High mass transfer (an order of magnitude higher than those in conventional)
  - Low liquid hold-up
  - High micro-mixing efficiency (mixing at the molecular level)
  - High hydraulic and flooding capacity
  - Small reactor size
  - Better self-cleaning of the rotor

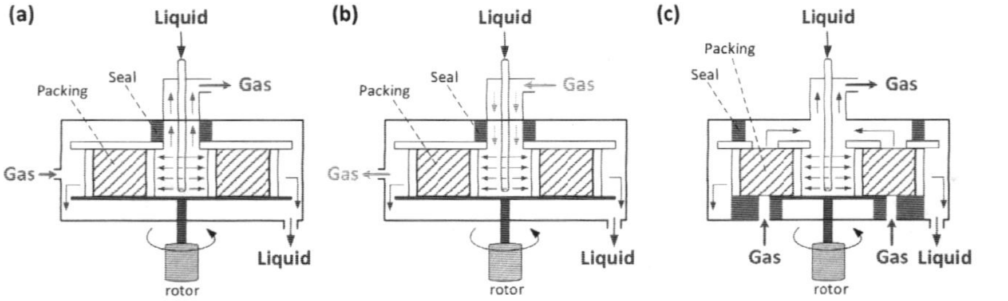

Various types of HIGEE reactors for reaction between gas and liquid phases: (a) countercurrent flow, (b) co-current flow and (c) cross flow.
(Pan et al., 2017)

II. Significance and Importance

7

2.1 Benefits of High-Gravity Carbonation

◆ Application of a rotating packed bed (RPB) for accelerated carbonation of high-gravity carbonation (HiGCarb) should be a promising approach to overcoming alkaline waste, wastewater and flue gas treatment.

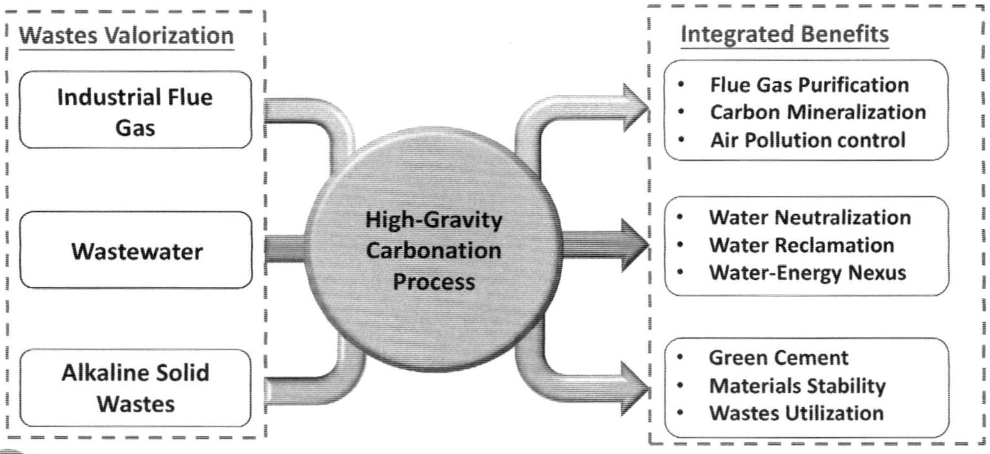

8

2.2 Technology Roadmap for HiGCarb

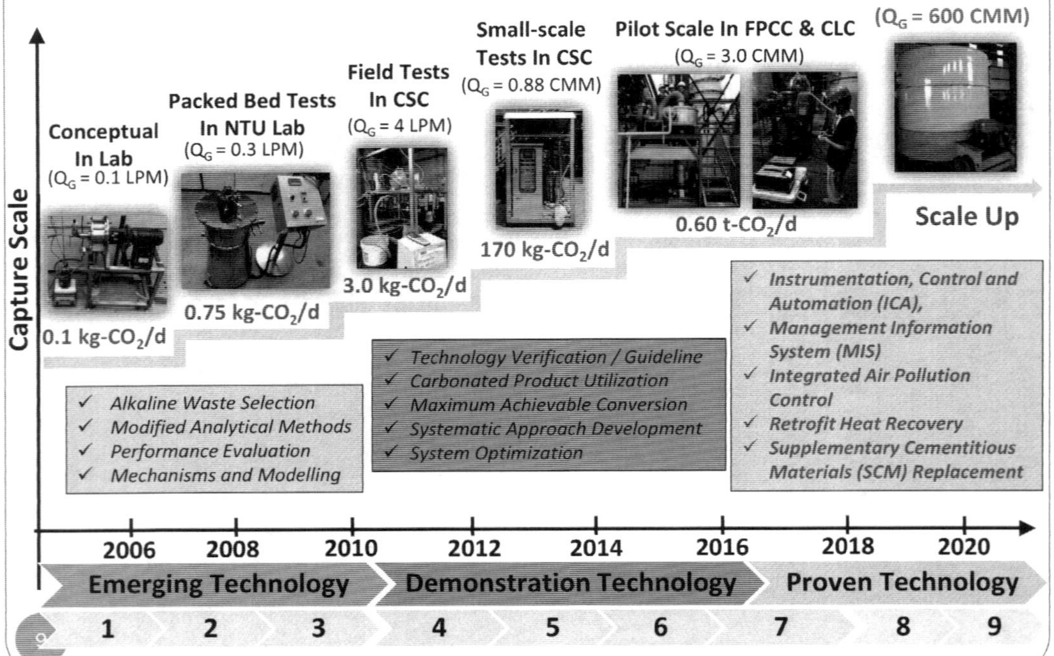

9

2.3 Objectives

- Evaluate the performance of HiGCarb process for CO_2 fixation using coal fly ash in a paper mill plant.
- Investigate the simultaneous removal efficiency for CO_2, SO_2, NOx and particle matters (PM) from exhausted flue gas via pilot-scale HiGCarb process .
- Assess the feasibility and potency of stabilized and carbonated coal fly ash utilization in cement substitution.
- Determine the optimal on-site operation conditions from environmental, economic and engineering (3E analysis) point of view.

10

III. Materials and Methods

11

3.1 Establishment of Pilot-Scale HiGCarb Process

- The flue gas properties from RDF boiler of paper mill plant:
  - Temperature of exhausted gas：105 °C
  - Alkaline wastes feedstocks: Coal fly ash
  - Humidity：14%-20%
  - The concentration of exhausted gas :
    - **CO_2: 8%vol-15%vol**
    - **SO_2: 40-300 ppm**
    - **NOx: 140-220 ppm**
    - **TSP: 4.18 mg/m^3 to 4.47 mg/m^3**

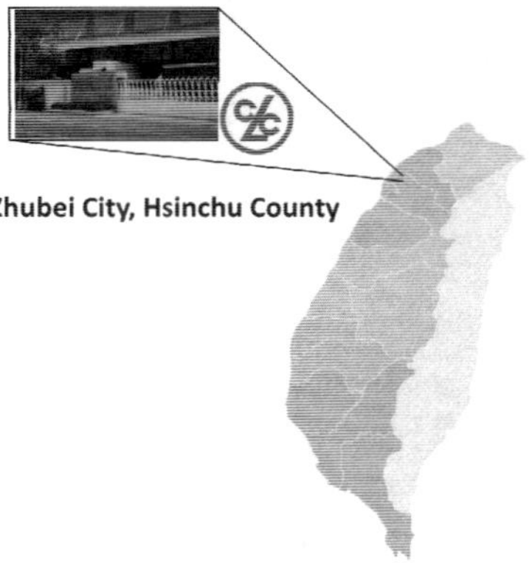

Chemical Composition and physical characteristics of coal fly ash via XRF analysis

Item (%)	SiO_2	Al_2O_3	Fe_2O_3	CaO	MgO	SO_3	K_2O	Na_2O	MnO_2	Free-CaO	BET area (m^2/kg)	Density (g/cm^3)
Coal Fly Ash	18.36	14.92	2.53	51.46	2.54	7.5	0.31	0.38	-	4.72	426.9	2.01

12

3.2 HiGCarb Process: On-site Experiment

◆ The fresh ash was firstly mixed with tap water for leaching process with the various liquid-to-solid (L/S) ratio then the leachate will be pumped into the RPB.

◆ The ozone was premixed with flue gas for NOx removal. The CO_2 could be reacted with alkaline leachate and the SO_2 and TSP could be removed simultaneously.

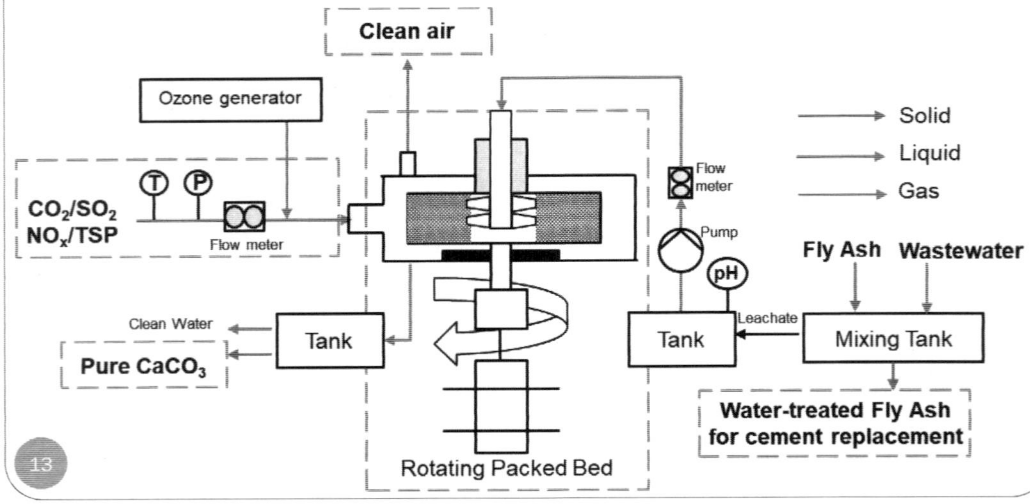

13

3.2 HiGCarb Process: On-site Experiment

◆ The range of inlet gas flow rate are 1.39 to 11.16 CMM.

◆ The range of inlet liquid flow rate are 0 to 32 L/min.

◆ The auxiliary equipment include pump, control box, storage tank and heat exchanger.

Orientation	Horizontal
Rotating speed	0~1100 rpm
Inner diameter	0.279 m
Outer diameter	0.292 m
Axial height	0.286 m
Volume	7.0×10^{-2} m^3

High Gravity Carbonation Equipment on Site.

14

3.3 Performance Evaluation: Product Analysis

1. Carbonation reaction

$$Ca^{2+}\ removal\ efficiency(\%) = \frac{M_{Ca-carbonation}(g)}{M_{Ca-extracted}(g)} \times 100\% \qquad \text{.......... (1)}$$

$$Carbonation\ efficiency(\%) = \frac{M_{Ca-carbonation}(g)}{M_t(g) \times \frac{C_{CaO}(\%)}{100} \times \frac{MW_{Ca}(\frac{g}{mol})}{MW_{CaO}(\frac{g}{mol})}} \times 100\% \qquad \text{.......... (2)}$$

STA 6000 TG Analyzer

$$CO_2\ capture\ capacity\left(\frac{g-CO_2}{kg-BOFs}\right) = \frac{M_{Ca-carbonation}(g) \times \frac{MW_{CO_2}(\frac{g}{mol})}{MW_{Ca}(\frac{g}{mol})}}{M_t(kg)} \qquad \text{.......... (3)}$$

$$\mathrm{ThCO_2(wt.\%)} = 0.785(\mathrm{CaO} - 0.56\mathrm{CaCO_3} - 0.7\mathrm{SO_3}) \qquad \text{.......... (4)}$$

$$\zeta_{Ca}(\%) = \frac{CO_2(wt.\%)}{\mathrm{Th}CO_2} \qquad \text{.......... (5)}$$

FT-IR Analyzer

2. Flue gas purification

$$\text{Removal Efficiency} = \frac{Q_{in} \times C_{in} - Q_{out} \times C_{out}}{Q_{in} \times C_{in}} \qquad \text{.......... (6)}$$

$$k_g a_e^{avg} = \frac{Q_g}{\pi h(r_o^2 - r_i^2)} * \ln\frac{C_b^g(r_i)}{C_b^g(r_o)} \qquad \text{.......... (7)}$$

$$E_{overall} = \int_{-\infty}^{+\infty} \phi_1(d'_p)d[\phi_2(d_p)] \qquad \text{.......... (8)}$$

$$E_{overall} = \sum_{i=1}^{12} E_i \times f_i \qquad \text{.......... (9)}$$

Gas Analyzer

Atomic Absorption Spectroscopy, AAS

15

3.4 Alkaline Product Utilization as SCMs

Supplementary Cementitious Materials (SCMs)

- the fresh FAS, water-treated FAS and carbonated product were utilized as SCM to substitute the clinker in cement mortar.
- The flow test, drying shrinkage test, sulfur resistance test and compressive strength of blended cement are investigated.

Items	Values	ASTM Regulations
Setting Time	• Initial setting no shorter than 45 min • Final setting time no longer than 375 min	C191
Sulfate resistance	• No more than 0.04% after 14 days	C452
Autoclave expansion	• No more than 0.8%	C151
Compressive Strength	• 12 MPa for 3 day • 19 MPa for 7 day • 28 MPa for 28 day (recommended)	C109/C109M

16

3.5 Engineering–Environment–Economy Analysis

(1) Data analysis

$$LCEI_{yi} = (e_{yi}) = \begin{bmatrix} e_{11} & \cdots & e_{1m} \\ \vdots & \ddots & \vdots \\ e_{s1} & \cdots & e_{sm} \end{bmatrix} \quad ....(17)$$

$$EP_{yj} = (p_{yj}) = \begin{bmatrix} p_{11} & \cdots & p_{1m} \\ \vdots & \ddots & \vdots \\ p_{s1} & \cdots & p_{sm} \end{bmatrix} \quad ....(18)$$

$$LCC_{yk} = (c_{yk}) = \begin{bmatrix} c_{11} & \cdots & c_{1m} \\ \vdots & \ddots & \vdots \\ c_{s1} & \cdots & c_{sm} \end{bmatrix} \quad ....(19)$$

(2) Standardization

$$X'_{yj} = (x'_{yj}) = \left(\frac{x_{yj} - \min\limits_{y=1}(x_{yj})}{\max\limits_{y=1}(x_{yj}) - \min\limits_{y=1}(x_{yj})} \right) = \begin{bmatrix} x'_{11} & \cdots & x'_{1m} \\ \vdots & \ddots & \vdots \\ x'_{s1} & \cdots & x'_{sm} \end{bmatrix} \quad .....(20)$$

$$(y = 1, 2, \ldots, s; j = 1, 2, \ldots, n)$$

$$T'_{yi} = (t'_{yi}) = \left(\frac{\max\limits_{y=1}(t_{yi}) - t_{yi}}{\max\limits_{y=1}(t_{yi}) - \min\limits_{y=1}(t_{yi})} \right) = \begin{bmatrix} t'_{11} & \cdots & t'_{1m} \\ \vdots & \ddots & \vdots \\ t'_{s1} & \cdots & t'_{sm} \end{bmatrix} \quad .....(21)$$

$$(y = 1, 2, \ldots, s; i = 1, 2, \ldots, m)$$

(3) Weighting Factor

$$KPI_y = \sum_{i=1}^{m} (KPI'_{yi} \cdot WF_{yi}) \quad ....(22)$$

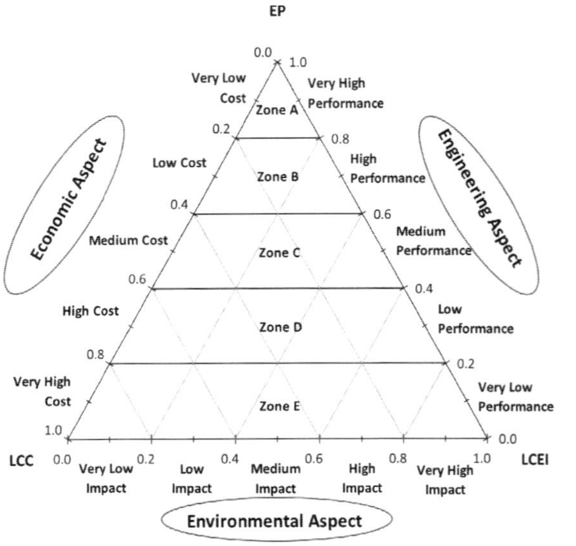

3E (Engineering–Environment–Economy) triangle model for evaluating high-gravity carbonation process. LCEI (X axis): life-cycle environmental impact; EP (Y axis): engineering performance; LCC (Z axis): life-cycle cost

(Ref: Pan et al., 2016, Applied Energy)

17

3.6 Determination of Key Performance Indicators

- 8 indicators in Environmental Impact (EI)
- 6 indicators in Engineering Performance (EP)
- 2 indicators in Economic Cost (EC)
- Weighting factor (W_i): determined by expert-consulting (Delphi method)
- Experts: academic (70%), industrial (20%), and government officials (10%).

Key performance indicators (KPIs) for 3E triangle model and their corresponding weighting factors determined by Delphi method

Aspects		Key performance indicators (KPI) [b]	Units	W_i [c]	Remarks
Engineering	EP_1	Practical CO2 capture capacity	t CO_2/t-FA	0.40	Related to technology risk
	EP_2	Attached removal quantity of SO_2	m³/t-FA	0.15	Related to technology risk
	EP_3	Attached removal quantity of NOx	t/t-FA	0.15	Related to technology risk
	EP_4	Attached removal quantity of PM	kg-PMtsp/t-FA	0.15	Related to technology risk
	EP_5	End-product production for cement substitution	t /t-FA	0.1	Related to commercialization risk
	EP_6	Calcium carbonate produced	t-$CaCO_3$/t-FA	0.05	Related to commercialization risk
Environmental [a]	$LCEI_1$	Global warming potential (GWP100) reduction	kg CO_2-Eq/t-FA	0.30	Related to ecosystem risk
	$LCEI_2$	Freshwater ecotoxicity reduction	kg 1,4-DCB/t-FA	0.10	Related to ecosystem risk
	$LCEI_3$	Particulate matter formation reduction	kg PM_{10}-eq/t-FA	0.10	Related to human health risk
	$LCEI_4$	Marine eutrophication potential reduction	kg N-eq/t-FA	0.10	Related to ecosystem risk
	$LCEI_5$	Urban land occupation avoidance	m²a/t-FA	0.10	Related to ecosystem risk
	$LCEI_6$	Ecosystem quality improvement	points/t-FA	0.10	Related to ecosystem risk
	$LCEI_7$	Human health protection	points/t-FA	0.10	Related to human health risk
	$LCEI_8$	Resource depletion reduction	points/t-FA	0.10	Related to ecosystem risk
Economic	EC_1	Operating & maintenance costs for capturing	US$/t-FA	0.70	Related to economic risk
	EC_2	Capital costs for facility installation	US$/t-FA	0.20	Related to economic risk

[a] Functional unit = per ton of fly ash input for HiGCarb process. [b] LCEI: life-cycle environmental impact; EP: engineering performance; EC: economic cost. [c] W_i =Weighting factor.

18

IV. Results and Discussion

- Changes in Chemical Properties of fly ash via Indirect Carbonation Process
- Performance Evaluation of Air Emission Reduction from Exhausted Flue Gas
- Utilization of fly ash in Cement Mortars
- Engineering, Environmental and Economic Analysis for HiGCarb Process

19

4.1-1 TG-DTG and FT-IR Results for Carbonated Product

◆ The TG-DTG curves of the carbonated product show the weight loss of 40% at a temperature region between 600°C to 900°C.

◆ The FT-IR analysis results presented that the carbonated product and pure calcium carbonate both at the highest peak at 1400 wave length (cm^{-1}) with the similar wave trend.

◆ The purity of carbonated product could be reached at 97.6%.

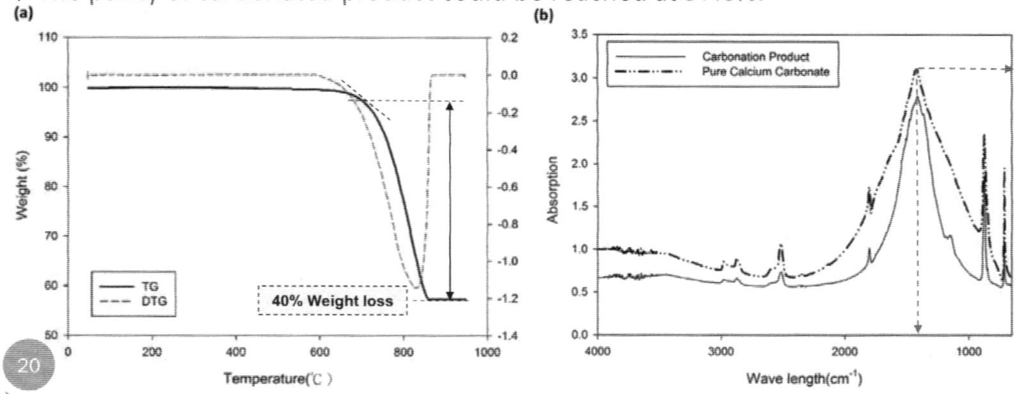

20

4.1-2 SEM Results of Fly Ash

Surface characteristics of (a) and (b) carbonated product and (c) unreacted fly ash.

- The result showed that lots of rhombohedral crystals were found on the surface of carbonated product, with small particles coated on the surface
- The unreacted fly ash was composed of heterogeneous grains and has a smooth surface.

4.1-3 Changes in Chemical Properties

◆ The CaO and free-CaO content of fly ash could be decreased via leaching process.

◆ According to TCLP, Barium (Ba^{2+}), lead (Pb^{2+}), and cadmium (Cd) can be potentially leached out from the solid matrix of the fresh FA.

Chemical properties analysis of fly ash and OPC.

Analysis	Items	Units	OPC	Unreacted FA	Leachate 1	Leachate 2
X-ray(XRF)	SiO_2	%	18.36	18.36	21.93	20.26
	Al_2O_3	%	14.92	14.92	16.23	15.47
	Fe_2O_3	%	2.53	2.53	3.08	3.42
	CaO	%	51.46	51.46	45.96	46.80
	MgO	%	2.45	2.45	3.05	2.87
	SO_3	%	7.50	7.50	5.88	5.69
	K_2O	%	0.31	0.31	0.28	0.33
	Na_2O	%	0.38	0.38	0.12	0.45
	Total	%	97.91	97.91	96.53	95.29
Chemical and thermal	Lime(free-CaO)	%	0.92	4.72	0.73	0.90
	CO_2 capture capacity	Kg/t	-		186.6	183.8
TCLP	Pb	mg/L	-	0.045	N.D.	N.D.
	Cd	mg/L	-	0.008	N.D.	N.D.
	Ag	mg/L	-	0.033	0.028	0.035
	Se	mg/L	-	0.002	0.008	0.003
	Ba	mg/L	-	3.58	0.113	0.065

4.2-1 Effect of Calcium leaching for Carbonation

- The maximal total carbon capture amount reaches 186.6 g-CO_2/kg-FA under L/S of 30 and leaching with tap water.
- The carbonation conversion of the 1st leaching solution is 99%.
- The pH of the solution will decline as increasing the leaching times, result in decrease the carbonation conversion.
- After four times leaching, the extraction rate of Ca^{2+} are higher than 40%.

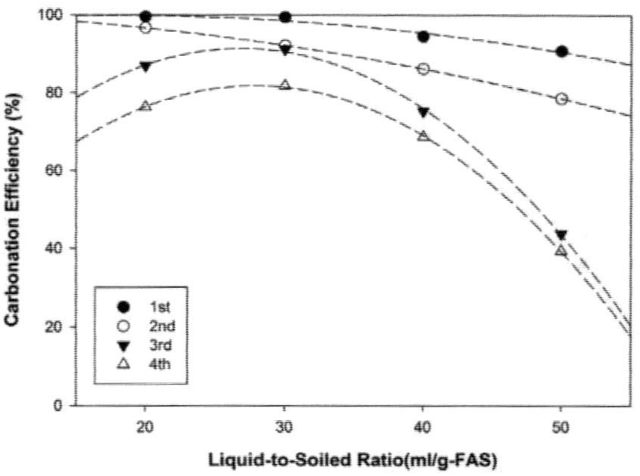

Effect of soiled-to-liquid ratio and times of leaching by tap water on carbon capture capacity of fly ash. (LFR : 1.0lpm; GFR : 0.5lpm; Rotating Speed: 600rpm)

23

4.2-2 Simultaneous Removal of SO_2, NOx and CO_2

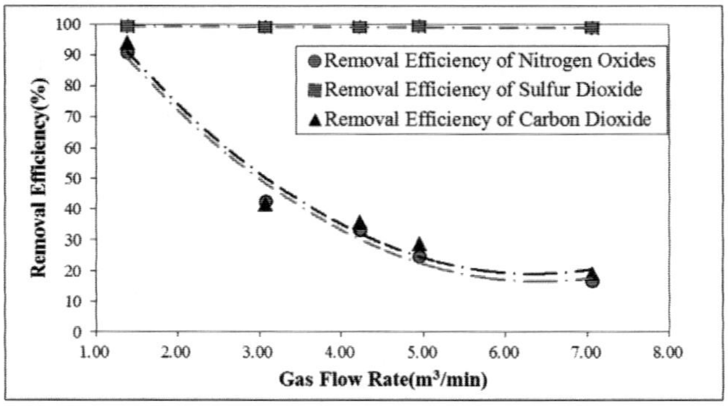

Simultaneous Removal of SO2, NOx and CO_2 under various gas flow rate (pH Value of Inflow Solution: 12.41; Gas flow rate: 1.39m^3/min; Rotating Speed: 600rpm)

- The removal efficiency of NOx and CO_2 decreased as increasing gas flow rate.
- While the capture capacity of NOx and CO_2 remain the same value.
- The removal efficiency of SO_2 are reached up to 99%.

Summary of Air Pollutants Removal Efficiency and Capture Capacity Under Different Operating Conditions.

	NOx(ppm)				SO_2(ppm)				CO_2(vol%)			
Gas Flow Rate (m^3/min)	Inlet	Outlet	Removal Efficiency (%)	Capture Capacity (kg/day)	Inlet	Outlet	Removal Efficiency (%)	Capture Capacity (kg/day)	Inlet	Outlet	Removal Efficiency (%)	Capture Capacity (kg/day)
1.39	194.2	18.1	90.7	0.365	51.8	0.3	99.3	1.1	10.9	0.7	93.8	442.32
3.08	145.8	84.1	42.4	0.377	46.1	0.4	99.2	2.5	8.1	4.7	41.5	433.76
4.23	139.1	92.8	33.3	0.407	161.2	1.4	99.1	3.5	11.8	7.6	35.6	510.61
4.94	161.8	121.6	24.8	0.355	98.2	0.7	99.3	4.0	12.9	9.2	28.9	484.46
7.06	181.8	151.4	16.7	0.342	149.6	1.9	98.7	5.7	12.5	10.1	19.0	454.71

24

4.2-3 The SO_2 Removal under Higher Gas Flow Rate

Observation

- The capture capacity of SO_2 are reached a high level at the highest gas flow rate of 11.16 m^3/min.
- The capture capacity of SO_2 are depended on the water input and rotating speed.
- The highest value can be reached to 9.1 kg/day.

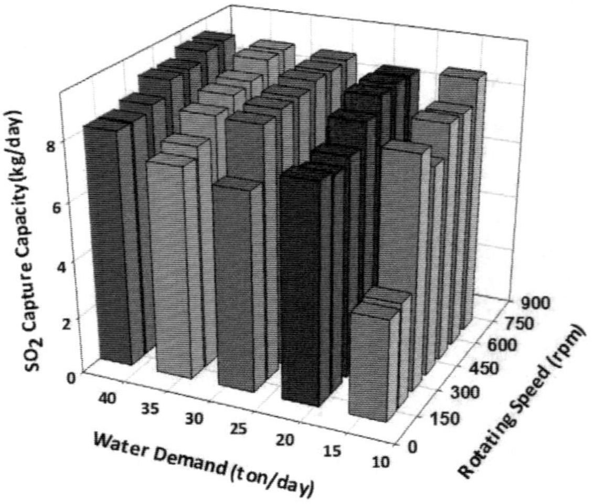

The SO_2 capture capacity and water demand with highest gas flow rate of 11.16 m^3/min under different rotating speed.

25

4.2-4 Simultaneous Removal of SO_2 and TSP

Observation

- The highest overall removal efficiency of TSP and capture capacity of SO_2 are respectively reached up to 80% and 9.7 kg/day under the rotating speed of 800rpm and liquid flow rate of 30 L/min.
- Those observation could be attributed to the change in velocity of water droplet and liquid holding in the RPB.

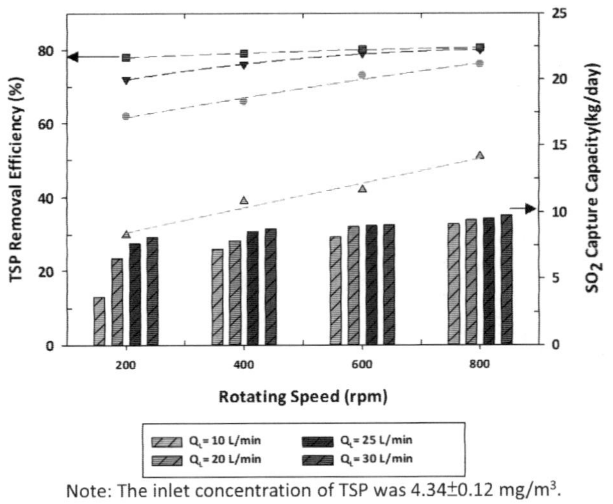

Note: The inlet concentration of TSP was 4.34±0.12 mg/m^3.

The simultaneous removal efficiency of SO_2 and TSP under various rotating speed and liquid flow rate.
(Operation condition: liquid agent: tap water; QG: 1.39 m^3/min)

26

4.2-5 Systematic Approach to Determination of Performance

Balancing engineering performance and energy consumption

- The energy consumption was decreased as increasing the rotating speed.
- The minimum energy consumption, i.e., 110 kWh/t-CO_2 (following the arrow 1), can determine the highest $CaCO_3$ production, i.e., 192 kg/day. (following the Arrow 2).
- The capture capacity of NOx and CO_2 were expected to 0.465 kg/day and 510 kg/day, respectively.

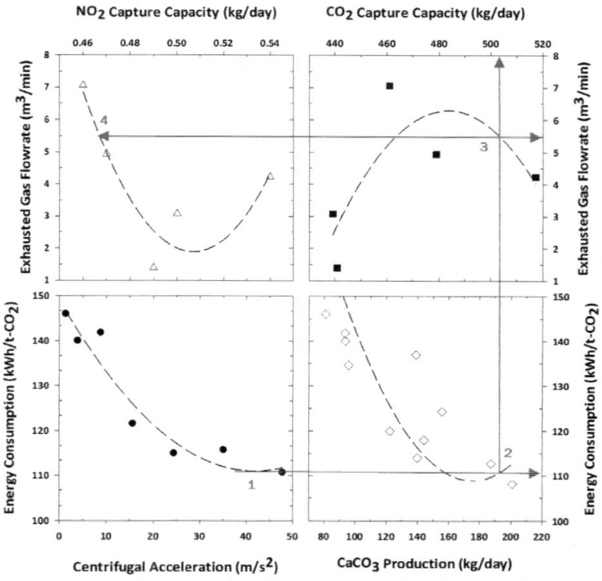

Systematic approach to determination of performance for simultaneous removal of NOx and CO2, energy consumption and CaCO3 production process. The system include rotating packed bed and pump.

27

4.3-1 Utilization of FA in blended cement

Durability

- Drying shrinkage test: Blended cement containing fresh fly ash presented the lowest drying shrinkage at 14th day.
- Sulfate Expansion: Blended cement containing carbonated fly ash performed good sulfate resistance at 14th day.

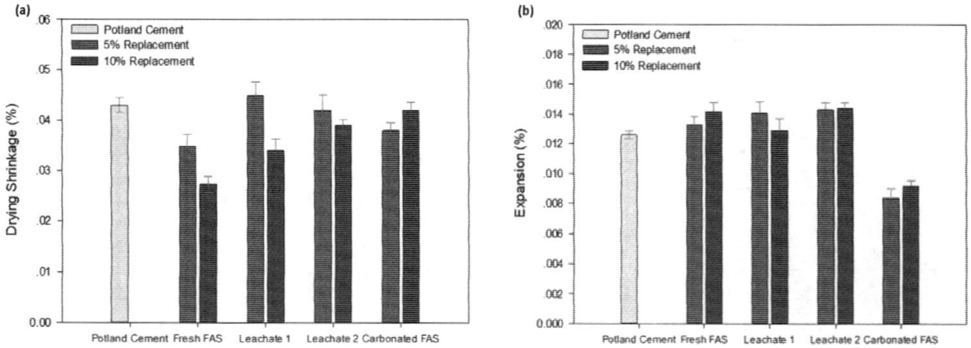

Effect under different substitution ratio and materials (a) 14th day drying shrinkage test results; and (b) 14th day potential expansion results.

28

4.3-2 Compressive Strength of Blended Cement FA

- The cement substituted with 5% carbonated fly ash exhibited superior compressive strength to others including the OPC.
- The minerals such as C_3S and C_3A could be activated by calcite, resulting in the acceleration of hydration process.

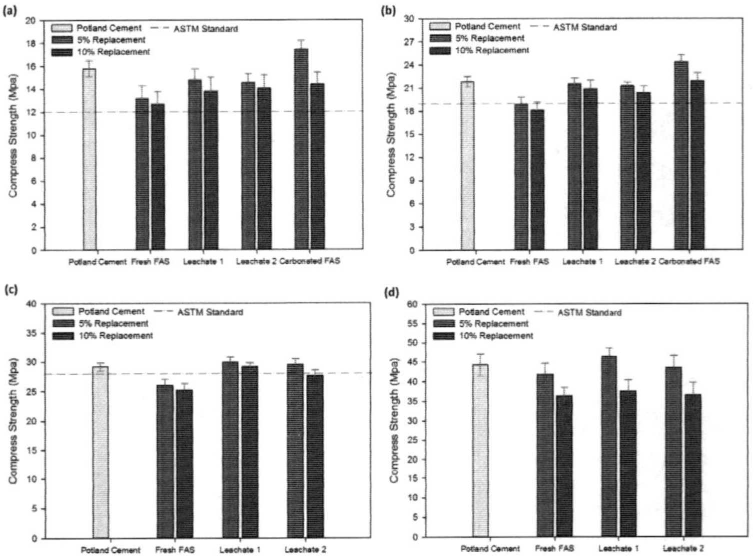

Compressive strength of blended cement with different substitution ratios of fresh, leachate and carbonated FAS at curing ages of. (a)3rd day (b) 7th day (c) 28th day (d)56th day.

29

4.4-1 3E Analysis for HiGCarb Process

- Summary of material, energy consumptions and engineer performance under different operation conditions.

Operation Condition				Input of Material and Energy (per t-fly ash consumed)		Performance (per t-fly ash consumed)					
Serial number	Q_G (m^3/min)	Rotating Speed(rpm)	L/S ratio (ml/g)	Tap water (t)	Electricity Consumption (kW)	CO_2 (t)	NOx (kg)	SO_2 (kg)	PM (kg)	Stabilized Fly Ash (t)	Calcium Carbonate (t)
Q1	1.39	600	40	80	16.2	0.14	0.17	0.37	1.72	0.92	0.16
Q2	3.08	600	40	80	16.3	0.15	0.18	0.83	3.82	0.92	0.17
Q3	4.23	600	40	80	17.2	0.16	0.19	1.15	4.37	0.92	0.18
Q4	4.94	600	40	80	17.4	0.14	0.16	1.34	4.08	0.92	0.16
Q5	7.06	600	40	80	17.8	0.13	0.15	1.91	4.38	0.92	0.15
R1	1.39	123	40	80	14.6	0.10	0.09	0.37	1.03	0.92	0.11
R2	1.39	200	40	80	15.4	0.11	0.12	0.37	1.22	0.92	0.13
R3	1.39	300	40	80	15.6	0.11	0.14	0.37	1.39	0.92	0.13
R4	1.39	400	40	80	15.8	0.13	0.15	0.37	1.43	0.92	0.15
R5	1.39	500	40	80	16.1	0.14	0.15	0.37	1.64	0.92	0.16
R6	1.39	600	40	80	16.2	0.14	0.17	0.37	1.72	0.92	0.16
R7	1.39	700	40	80	16.6	0.15	0.18	0.37	1.75	0.92	0.17
LS1	1.39	600	20	40	12.4	0.11	0.18	0.37	1.72	0.94	0.13
LS2	1.39	600	30	60	14.1	0.13	0.17	0.37	1.72	0.94	0.15
LS3	1.39	600	35	70	15.6	0.15	0.17	0.37	1.72	0.93	0.17
LS4	1.39	600	40	80	16.2	0.14	0.17	0.37	1.72	0.92	0.16
LS5	1.39	600	45	90	17.5	0.13	0.17	0.37	1.72	0.92	0.15
LS6	1.39	600	50	100	18.4	0.09	0.16	0.37	1.72	0.91	0.10

30

4.4-2 Environmental Aspect: Endpoint Impact Assessment

- Comparing with BAU, HiGCarb provided a **direct reduction in environmental impact via air pollutants emission reduction.**
- The utilization of **alkaline waste** and **calcium carbonate production processes** avoided impact points generated in production process of cement and limestone.

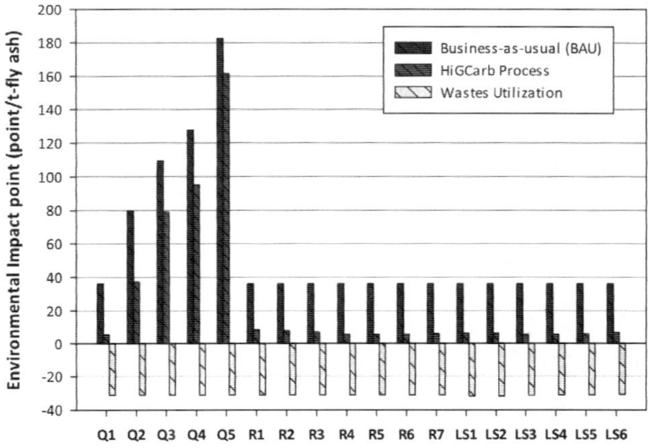

Result of endpoint of high gravity carbonation process.
(Function unit: per ton of fly ash slag)

31

4.4-3 Economic Aspect: Cost and Benefit Analysis

Cost Benefit = (carbon credit + avoided solid disposal fee+ product utilization)-(operation cost + capital cost)

- Carbon credit in ERU system (2016) was 6.25 USD/t-CO_2.
- Current treatment fees for NOx, SO_2 and fly ash were 0.33 US$/kg-NO and 0.30$/t-SO_2 and 25 US$/t-fly ash.
- Total benefits were estimated around **-5.05–22.22 USD** per ton of FA.

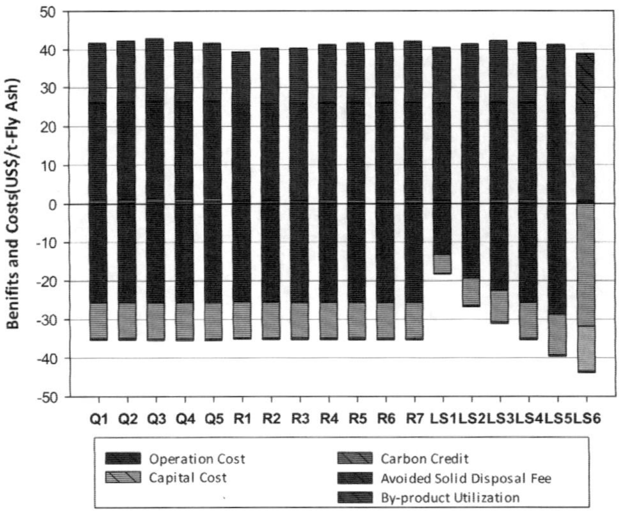

Cost benefits Accounting results. (Unit: US$/t-fly ash)

32

4.4-4 System Optimization using 3E Triangle Model

3E Performance Evaluation throughout Triangle Model

- Engineering, environmental and economic evaluations for these 18 operation conditions were summarized by standardization of unit.
- Scenario Q_3 exhibited a **superior engineering performance** (as indicated by 1) with a relatively **lower environmental impact** and a **relatively lower costs** .

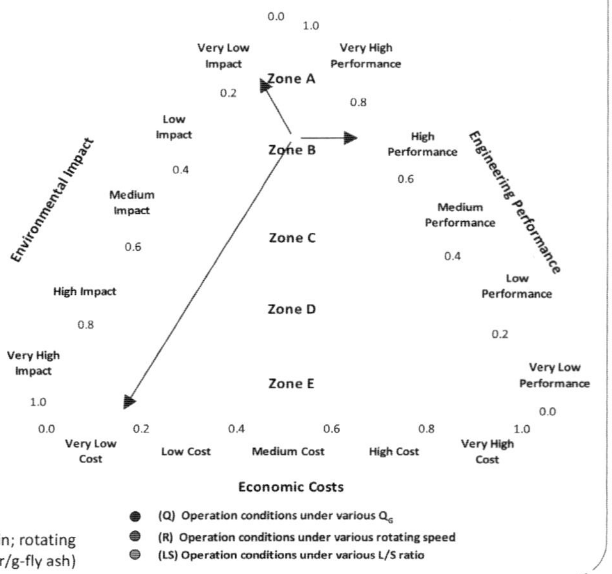

33

(Q3: operation condition: gas flow rate: 4.23 m3/min; rotating speed: 600 rpm; liquid-to-solid ratio: 40ml-tap water/g-fly ash)

V. Conclusion and Recommendation

34

5.1 Conclusion

- The TG-DTG and SEM results show the **existence of calcium carbonate.**
- The maximal total CO_2 capture amount reached at **186.6 g-CO_2/kg-FA** under L/S of 30 and leaching with tap water.
- The best simultaneous removal efficiency of **CO_2, SO_2 and NOx** can reach **96.3±2.1%, 99.4±0.3% and 95.9±2.1%** respectively. The **removal efficiency of TSP** reached **83.4±2.6%.**
- Cement mortar with FA exhibits **better compressive strength** and **durability.**
- The HiGCarb process can be comprehensively evaluated from engineering, environmental and economic points of views. Scenario Q3 exhibited a **better engineering performance** (as indicated by 1) with a relatively **lower environmental impact** (2) and a **relatively lower costs** (3).

35

5.2 Recommendation

- **Alkaline Wastes:** establishing sustainable Waste-to-Resource Supply Chains for CO_2 capture and waste treatment.
- **Process Integration:** Utilization of brine wastewater in the HiGCarb process to enhance CO_2 fixation capacity while purifying the water quality of brines.
- **Utilization Diversification:** Integration of HiGCarb with a slurry reactor to provide multiple pathways for carbonated product utilization, such as marine blocks and SCM.
- **Model Development:** Incorporation of process chemistry with reaction (leaching and carbonation) kinetics, hate transfer and mass transfer for the HiGCarb process.
- **System Optimization:** combining automation and monitoring, material-energy balance analysis and life cycle assessment (LCA), .

36

Thank you for your attention !!!

Questions and Comments ??

National Taiwan University

國立台灣大學
環境工程學研究所

CCRC/NTU
Carbon Cycle Research Center

科技部 Ministry of Science and Technology

第二期能源國家型科技計畫
National Energy Program-Phase II

減碳淨煤主軸中心
Carbon Reduction and Clean Coal Focus Center

37

TTÜ 100

UTILIZATION OF INDUSTRIAL WASTES AND CO_2 IN CONSTRUCTION MATERIALS - AN ESTONIAN PERSPECTIVE

Hakan Berber[1]
Kadriann Tamm[1]
Rein Kuusik[1]
Colin D Hills[2]
Paula Carey[3]
Mai Uibu[1]

(1) Tallinn University of Technology, Estonia
(2) University of Greenwich, UK
(3) Carbon8 Systems Limited, UK

International Conference on Accelerated Carbonation for Environmental and Material Engineering (ACEME)

March 11-14, 2018
Newcastle City Hall, Newcastle
New South Wales, Australia

TTÜ 100

OUTLINE

- Introduction
- Initial Materials
  - Chemical and phase composition
  - Physical properties
  - Leaching
- Methodology
- Results
  - CO_2 binding
  - Mechanical properties
  - Leaching
- Conclusions

2

Introduction

Estonian oil shale industry in 2016 [1]:

- Sales revenue: 742 million of EUR
- Oil shale used: 15.2 million of tonnes
- Energy content: 8 MJ/kg
- Estonia was only 8.9% dependent on energy imports in 2014: The lowest level in the whole European Union [2]

Oil shale ash deposits in Northeastern Estonia

Problem description:

- Ash produced: 6.9 million of tonnes
- 95% of the ash generated is not utilized
- CO_2 generated: 11.5 million of tonnes

The objective:

- **To study the applicability of producing light weight aggregates from the waste ashes and CO_2**
  - **Commercial product**
  - **CO_2 abatement**
  - **Utilization of industrial wastes**
  - **Stabilization of industrial wastes**

3

[1] - Eesti Energia (2016). Annual Report 2016

[2] - Eesti Energia (2015). Estonian oil shale industry yearbook 2015

Idea behind this study

From CO_2 to light-weight aggregates

- **Incinerator ashes**
- **By-pass dust**
- **Quarry fines**

http://www.c8s.co.uk/technology.php

+ CO_2 =

UK's Carbon8 wins permission to more than double production of carbon negative aggregates

First published on www.AggBusiness.com

Pioneering carbon negative aggregate specialist Carbon8 Aggregates is set to more than double production after gaining planning permission to build a second manufacturing facility at Avonmouth, near Bristol, southwest England.

Work on the €5.1 million (£4 million) project will begin later this year and the plant is expected to be fully operational by early 2016, more than doubling Carbon8's ability to meet rising construction industry demand for its high quality, lightweight aggregates.

Using an award-winning patented technique known as Accelerated Carbonation Technology

ZOOM

The size of this Carbon8 ball, pictured beside the Severn Bridge in southwest England, shows the equivalent of capturing 4,000tonnes of CO2 a year – the amount Carbon8's new Avonmouth facility will capture annually

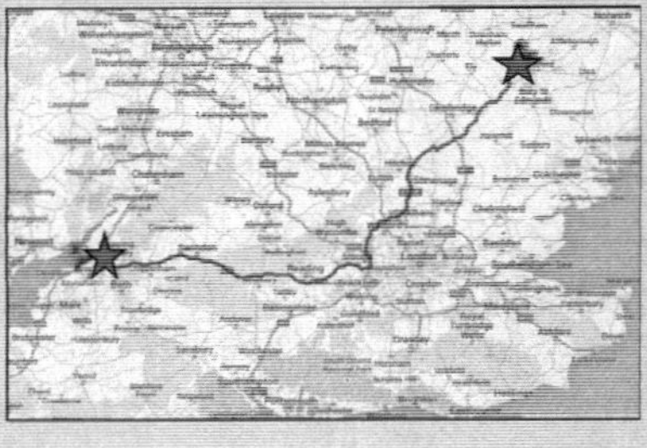

4

Initial Materials

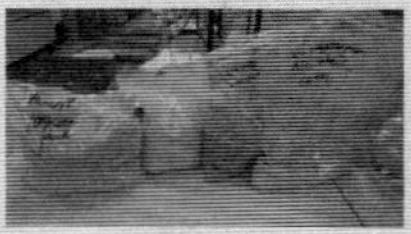

Oil shale ashes from Narva PP and Enefit280

Initial sample	d_{mean} [µm]	SSA [m²/g]	Free CaO [%]	CO_2 [%]	S_{SO4} [%]
Enefit 280 FA	36.48	4.24	0.36	16.,83	2.26
CFBC-FA1	20.99	4.75	12.00	5.31	1.93
CFBC-FA2	43.58	3.08	9.76	3.79	2.20
PF-FA	14.86	1.27	8.56	4.26	2.42

Cement kiln dust and cement from AS KNC

Initial sample	d_{mean} [µm]	SSA [m²/g]	Free CaO [%]	CO_2 [%]	S_{SO4} [%]
CKD	34.33	4.16	2.01	20.30	2.49
Cement	16.76	1.76	1.91	22.77	1.41

MSWI APC and FA Iru PP

Initial sample	d_{mean} [µm]	SSA [m²/g]	Free CaO [%]	CO_2 [%]	S_{SO4} [%]
APC	40.42	7.88	10.02	6.35	2.49
Fly ash	263.15	1.20	5.49	3.58	1.53

5

Initial Materials

Phase Composition (XRD)

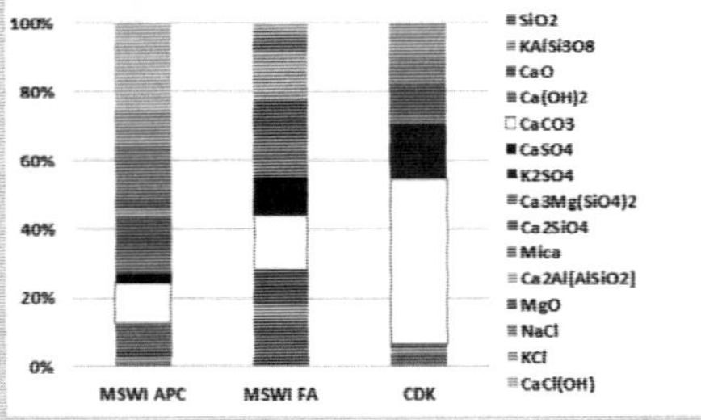

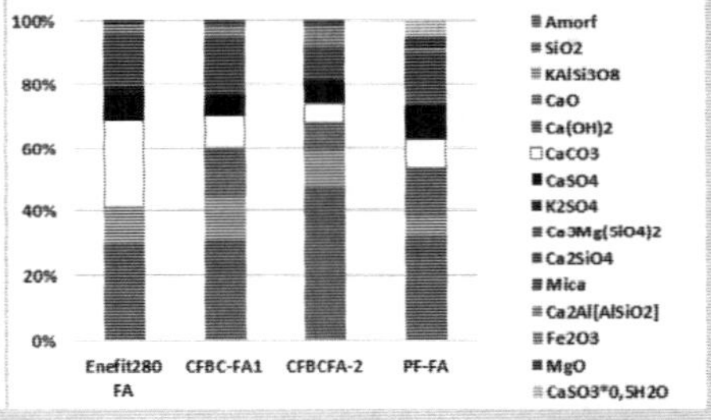

- Oil shale ashes (CFBC-FA1, CFBC-FA2) and MSWI ashes contain lime and/or portlandite
- MSWI ashes contain Cl-compounds NaCl, KCl and CaCl(OH)
- Carbonates are represented as calcite, mainly in CKD and Enefit 280 FA
- Silica is represented as quartz and K-feldspar, also as secondary silicates like belite etc
- Amorphous phase content is the highest in CFBC-FA2 (cannot be determined in MSWI ashes and CKD)

6

Initial Materials

Environmental hazardousness of the ashes

Parameter	Target value	Limit in industrial land	unit	MSWI APC	MSWI FA	CKD	CFBC-FA1
Dry matter			wt%	99.7	100	100	100
Ca			g/kg	271.4	204.2	252.3	238.5
K			g/kg	50.6	13.3	111.7	15.2
Na			g/kg	40	17	2.1	1.4
Mg			mg/kg	14580	17250	12530	20820
Mn			mg/kg	571	1410	696	344
Hg	0,5	10	mg/kg	20	0.31	0.18	<0.1
Cd	1	20	mg/kg	70	8	12	0,45
Pb	50	600	mg/kg	1900	206	1870	70
Cr	100	800	mg/kg	209	678	31	37
Ni	50	500	mg/kg	80	361	18	31
Zn	200	1000	mg/kg	8880	3030	773	<0.1
Cu	100	500	mg/kg	681	558	31	16
As	20	50	mg/kg	39	13	13	25
Mo	10	200	mg/kg	22	35	4,3	14
Ba	500	2000	mg/kg	751	1560	76	176

7

Initial Materials

Leaching properties of the ashes

Parameter	Limit for inert waste	Limit for hazardous waste	unit	MSWI APC	MSWI FA	CKD	CFBC-FA1
pH	6-9		-	>12 (13.3)	>12 (13.9)	>12 (13.9)	>12 (13.5)
EC			mS/cm	61.3	17.5	39.3	12.6
TDS	4	100	g/kg	306	87.9	196	63.1
Chloride	0,8	25	g/kg	192.2	16.8	40.8	2.1
Fluoride	0,01	0,5	g/kg	<0.15	0.05	0.26	<0.03
Sulfate	1	50	g/kg	14.6	10.8	53.2	0.016
Ca			g/kg	73.7	15.1	26.6	10.9
K			g/kg	46.6	7.1	97.8	2.8
Na			g/kg	36.4	6.6	1.7	0.1
Mg			mg/kg	<1	<1	<1	2.4
Mn			mg/kg	<0.1	<0.1	<0.1	<0.1
Hg	0,01	2	mg/kg	0.21	<0.002	<0.002	<0.002
Cd	0,04	5	mg/kg	<0.003	0.0093	<0.003	<0.003
Pb	0,5	50	mg/kg	700	4.7	22	0.22
Cr	0,5	70	mg/kg	2.6	17	4	0.07
Ni	0,4	40	mg/kg	<0.01	<0.01	<0.01	<0.01
Zn	4	200	mg/kg	130	6.2	0.5	0.09
Cu	2	100	mg/kg	4.1	0.2	<0.01	<0.01
As	0,5	25	mg/kg	<0.005	0.014	0.015	<0.005
Mo	0,5	30	mg/kg	4.5	1.8	2.5	0.3
Ba	20	300	mg/kg	29	2.3	5.3	27

8

Methodology

Carbonation and Granulation

- 1st step:
  - hydration (L/S=0.33;
  - carbonation (20% CO_2 in air for 30 minutes, 3000 rpm);
  - cement addition (0-10%);
- 2nd Step:
  - Granulation and carbonation (20% CO_2 in air for 20 min, -610 rpm);
- Both steps take place in the same Eirich-type mixer.

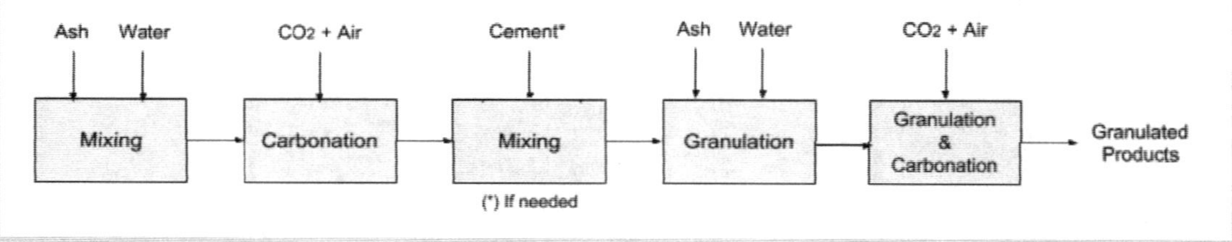

Methodology

Chemical Composition and Mechanical Properties

Chemical analysis determined from dried (3h, 105˚C) sample	Standard, method
TC, % (CO_2 binding degree)	ELTRA CS-580
Free CaO, %	
Mechanical properties are determined for the samples cured under in laboratory conditions (1 week)	
Fractional composition	CEN-EN 1097-11:2013
Apparent density, m³/g	EVS-EN 1097-6
Strength	CT3 25K, (CEN-EN 1097-11:2013)
BET SSA, porosity	KELVIN 1042, Poremaster 60-17

10

Methodology

Leaching Analysis

Leaching of ashes and granules	Standards, methods
Leaching tests	EN 12457-2(2002)
Cations, anions, heavy metals	DIN EN ISO 10304-1/-2(D19/20, E29)
pH, EC	pH (MT SevenGo pH), Electrical conductivity (MT SevenGo Duo Pro)
SO_4^{2-}, Cl^-	Lovibond
Cations, heavy metals	AAS (Carl Zeiss Jena AAS IN: Verian Spectra AA 55B)

11

European standard EN 12457-2:2002. Characterization of Waste – Leaching – Compliance test for leaching of granular waste materials and sludges – Part 2, European Committee for Standardization, Brussels, 2003.

Results

CO_2 binding in the first step (Hydration/Carbonation step)

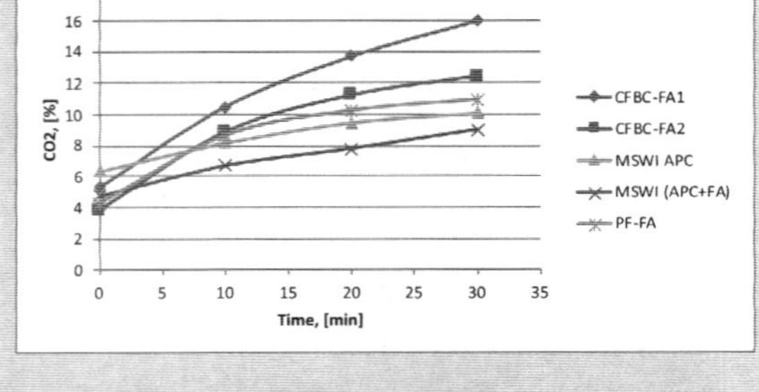

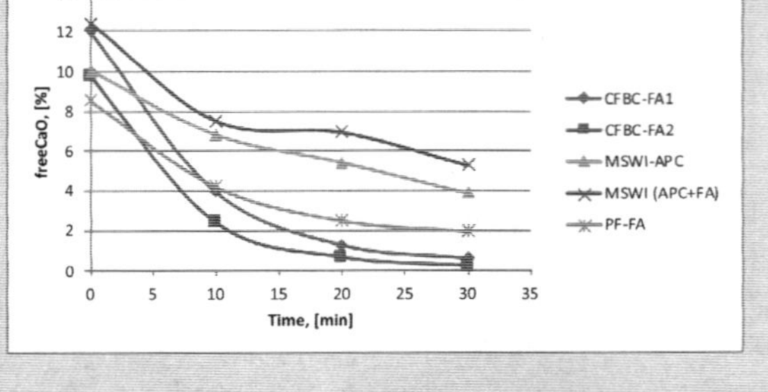

12

Results

Granules: CO_2 binding degree and free CaO (fCaO) content

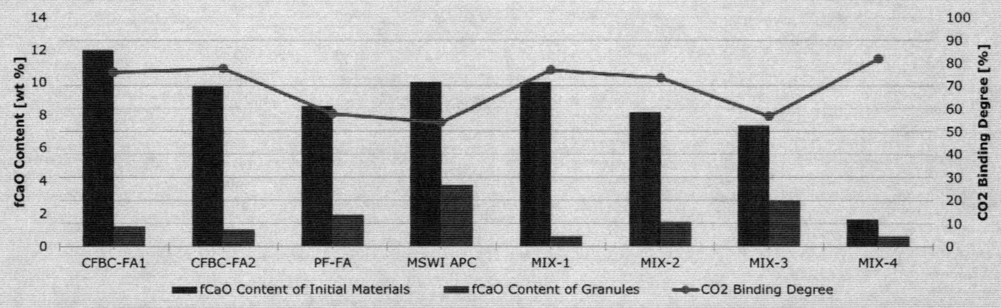

13

Results

Granules: Size distribution and compressive strength

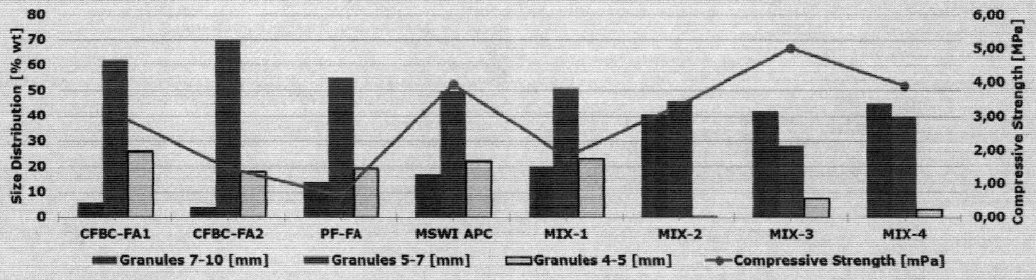

14

Results

Compressive strength and density of the granules produced before and after leaching

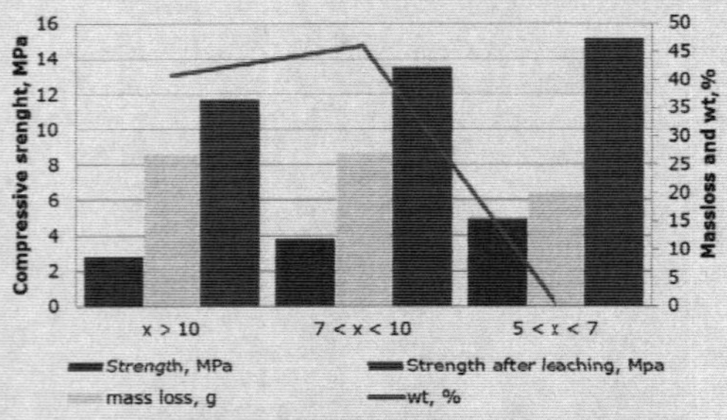

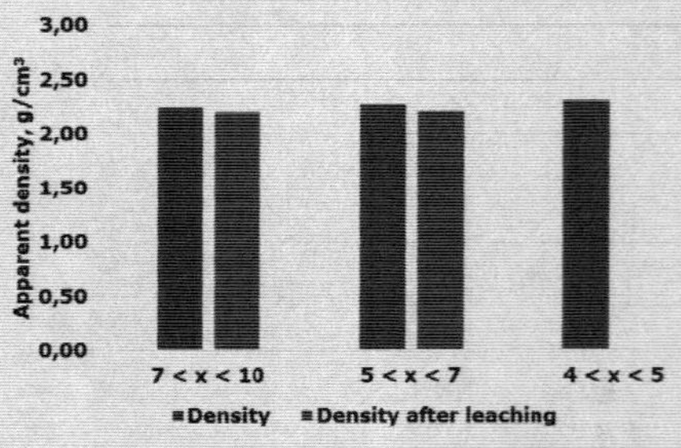

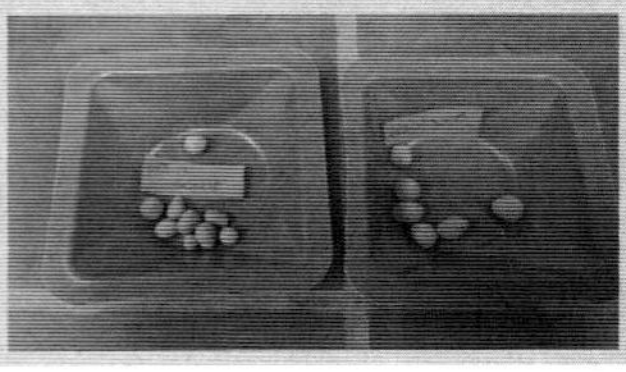

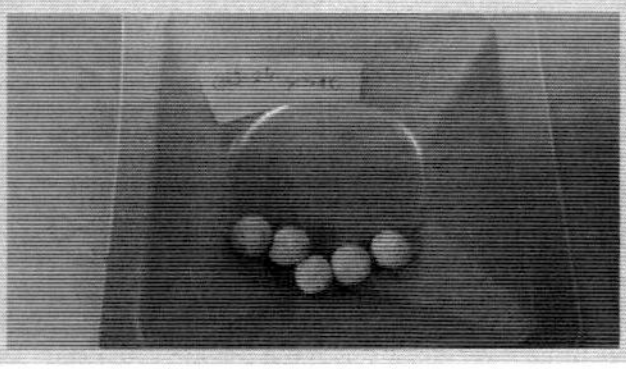

15

Results

Leaching – changes in the system pH and TDS

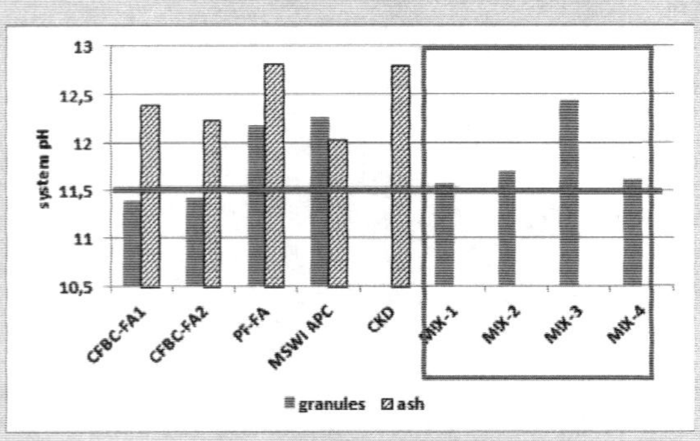

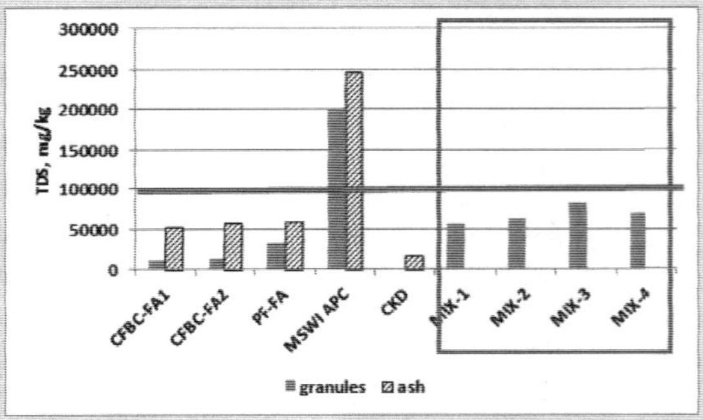

16

Results

Leaching of chlorides, sulphates and Zn

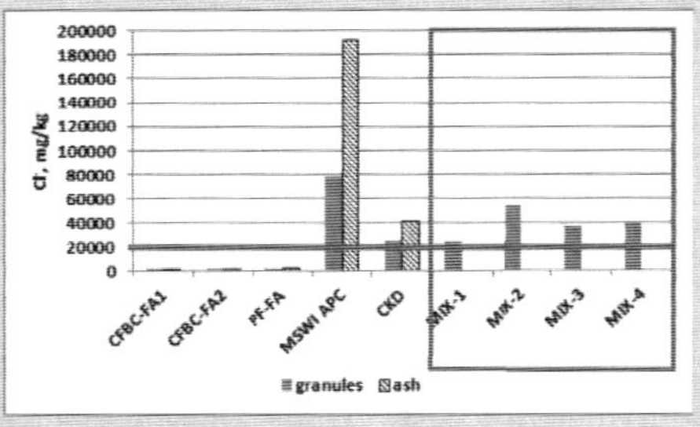

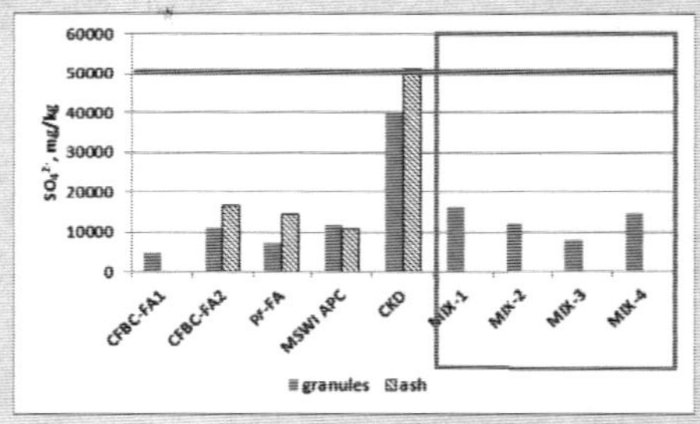

Limit for hazardous waste 200 mg/kg

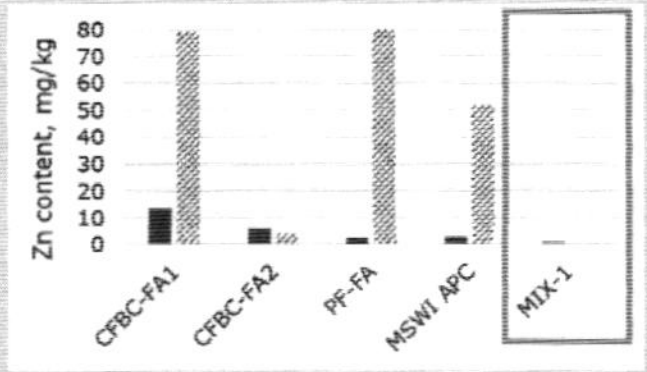

17

Conclusions

The applicability of utilizing different types of wastes from Estonian energy and cement sector as well as from municipal solid waste incinerators to bind CO_2 and produce light weight aggregates was studied.

- **Different types of oil shale ashes, cement kiln dust and municipal wastes were tested.**
- **All the process steps were carried out in one mixer-type apparatus.**
- **The CO_2 uptake was mainly attributed by the free lime content, which is relatively high (10-15%) in oil shale ashes and MSWI APC.**
- **The granules were characterized by 5 < x < 10 mm size range, compressive strength up to 6 MPa and apparent density close to 2.1 g/cm^3.**
- **Carbonation/granulation treatment led to significant decrease in the leachate pH and TDS values.**
- **In the case of hazardous MSWI APC ashes, the mobility of chloride and heavy metals could be decreased by optimizing mixtures.**
- **Leaching tests indicated that the granules suffered some weight loss, which indicates that at current cement additions (~5%), these granules are more suitable for using in building blocks than as a road base material. During leaching the compressive strength of the granules increased, however the density remained on the same level**

18

Acknowledgements

This study was supported by

- European Union Reginal Fund Dora plus
- Keskkonnainvesteeringute Keskus (KIK) (Project 11896 Uuring tööstusjäätmetest ehituslike täitematerjalide saamiseks koos CO2 sidumisega)
- Institutional research funding IUT (IUT33-19) of the Estonian Ministry of Education and Research.

International Conference on Accelerated Carbonation for Environmental and Material Engineering

Newcastle, March 13th, 2018

Analysis of the effects of slurry phase carbonation on the leaching behaviour of different types of waste incineration bottom ash

G.Costa*, F. Lombardi*, A. Polettini, R. Pomi** and R. Spagnuolo***

**DICII Department, University of Rome "Tor Vergata", Italy*

***DICEA Department, University of Rome "La Sapienza", Italy*

Waste incineration bottom ash management in Europe

WtE in a circular economy

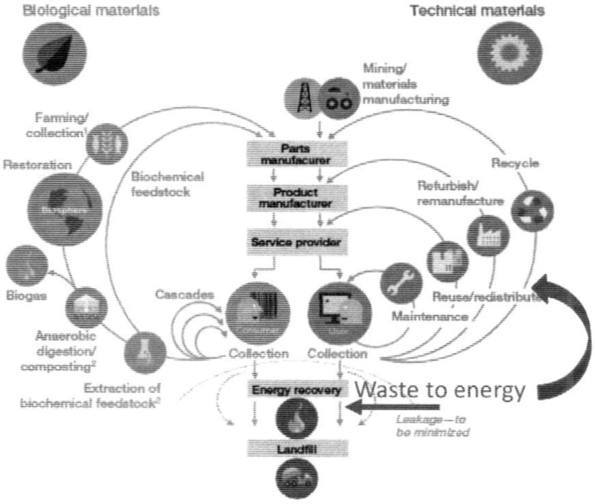

- Currently 400 WtE plants in operation in Europe
- 20 Mton/y Bottom ash generated (15-25% feed waste)
- Landfilling -> recovery and recycling
- Fe and Al recovery more and more applied at the plant or in specific facilities (≈200000 t/y)

Mineral fraction (80% wt) ?

Waste incineration bottom ash management in Europe

AEME 2018

Current management practices for the mineral fraction of BA

Country	Type of utilization
Austria	No intention to reuse except as landfill structure material
Czech Republic	Landfill construction
Denmark	Road sub-base and embankments, filler for marine structures, construction material
Finland	All bottom ash is landfilled
France	Use in road construction

Country	Type of utilization
Germany	Road sub-base, recovery in landfills
Italy	Recovery in cement kilns, road construction, landfill construction
The Netherlands	Road sub-base and embankments, concrete products
Sweden	Reuse as landfill covering material
Switzerland	Utilization only with waterproof cover

(CEWEP, 2010 e 2012)

- Before utilisation BA is typically weathered in heaps for 2 to 6 months
- Even when recycled restricted conditions are applied -> containment more than recycling

→ Dutch Green Deal: after January 1st 2020 BA cannot be recycled in restricted applications anymore only free application (just as primary construction materials) -> stricter requirements

Waste incineration bottom ash utilization

AEME 2018

Barriers for utilisation and treatments

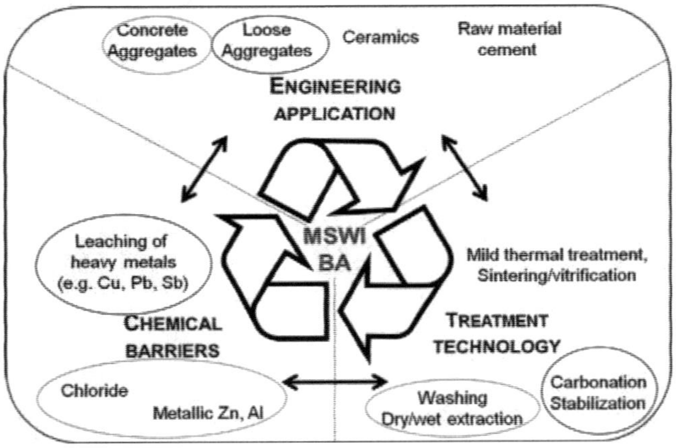

(Verbinnen et al 2017, Waste and Biomass Valor.)

G. Costa *Analysis of the effects of slurry phase carbonation on waste thermal treatment bottom ash*

Our research work on BA characterization and valorization

AC EME2018

➢ **Previous work on BA: characterization and leaching modelling of different types of BA (Tor Vergata)**

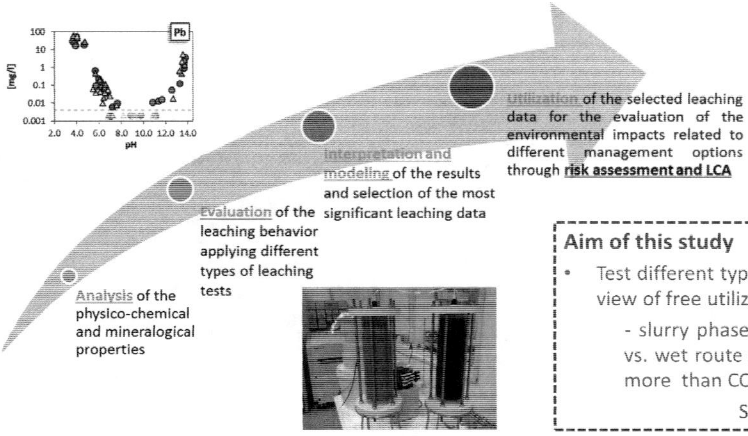

➢ **Previous work on BA: carbonation (Sapienza + Tor Vergata)**

- BA treatment (carbonation + others)
- Wet route carbonation aimed at CO_2 storage

Aim of this study

- Test different types of treatments to improve BA leaching in view of free utilization
  - slurry phase carbonation (Santos et al. J Env Manage) vs. wet route selecting operating conditions for leaching more than CO_2 uptake

  Storage -> utilization

Materials and Methods: *BA sampling and characterization*

AC EME2018

BA from a slag dry treatment unit

➢ Treatments: crushing, ferrous and non-ferrous metals removal

➢ Treated slag: refuse derived fuel (RDF) grate furnace incinerator

Treated all slag < 10 mm without additional grinding

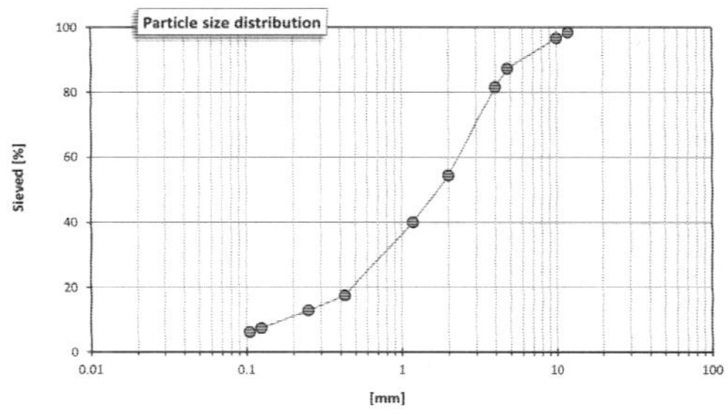

➢ *Total content analysis*

➢ *Compliance leaching test (EN 12457-2)*

➢ *pH-dependence test (CEN/TS 14429)*

➢ *Inorganic carbon content*

$$\%CO_{2,uptake} = \frac{\%CO_{2,fin} - \%CO_{2,in}}{100 - \%CO_{2,fin}} * 100$$

Materials and Methods: *Carbonation treatment tests*

AEME 2018

Slurry phase carbonation

- ***Experimental set-up***
  - pressurized stainless steel reactor with temperature control and magnetic stirring
- ***Operating conditions***
  - T= 30-50-100°C
  - L/S= 5 l/kg
  - 100% CO_2
  - P = 3-10 bar
  - 2 h

Wet carbonation

- ***Experimental set-up***
  - pressurized stainless steel reactor with temperature control
- ***Operating conditions***
  - T= 30-50°C
  - L/S= 0.3 l/kg
  - 100% CO_2
  - P = 3-10 bar
  - 2 h

Results: *BA characterization*

AEME 2018

Total content

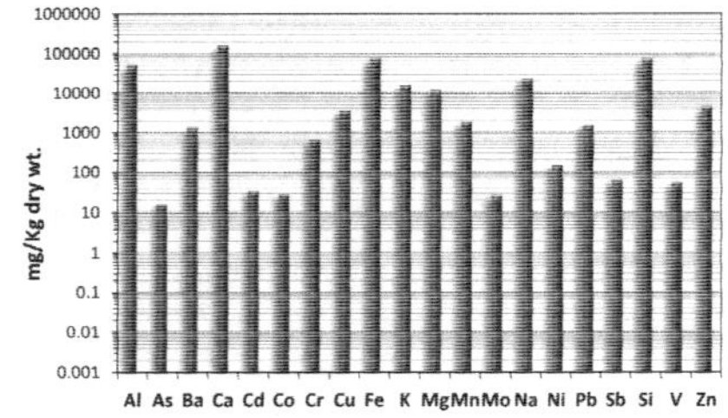

	IC [%]	$CaCO_3$ [%]
A.R.	2.16	18.04

Compliance test and limit values for unrestricted reuse (mg/kg)

Constituents	Leaching (As Received material)	The Netherlands (Cat 1)	Italy
Cl	2421	616	1000
SO_4	742	1730	2500
Ba	1.13	22	10
Pb	0.27	2.3	0.05
Cr	2.31	0.63	0.5
Cu	0.21	0.9	0.5
Zn	1.80	4.5	30
Sb	0.125	0.16	-
Mo	0.4	1	-
pH	12.45	-	5.5-12

Results: *Wet Carbonation*

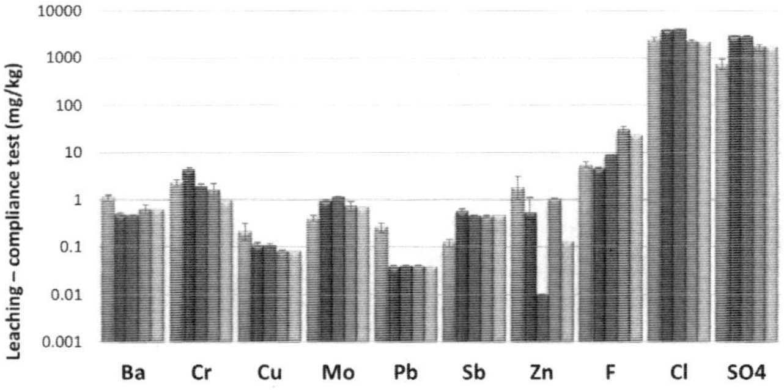

	pH	$CO_{2uptake}$(%)
A.R.	12.5	-
W30°C 3bar	10.7	7.95
W50°C 3bar	9.8	11.26
W30°C 10bar	11.0	11.32
W50°C 10bar	10.4	12.46

G. Costa *Analysis of the effects of slurry phase carbonation on waste thermal treatment bottom ash*

Results: *Slurry phase Carbonation*

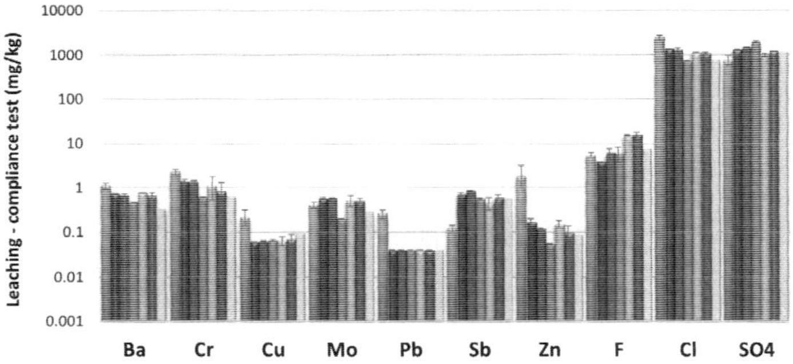

	pH	$CO_{2uptake}$(%)
A.R.	12.5	-
S30°C 3bar	10.7	9.00
S50°C 3bar	10.4	11.85
S100°C 3bar	10.7	9.96
S30°C 10bar	11.2	12.80
S50°C 10bar	11.0	9.39
S100°C 10bar	11.2	12.68

G. Costa *Analysis of the effects of slurry phase carbonation on waste thermal treatment bottom ash*

Results: *Wet vs. Slurry phase carbonation*

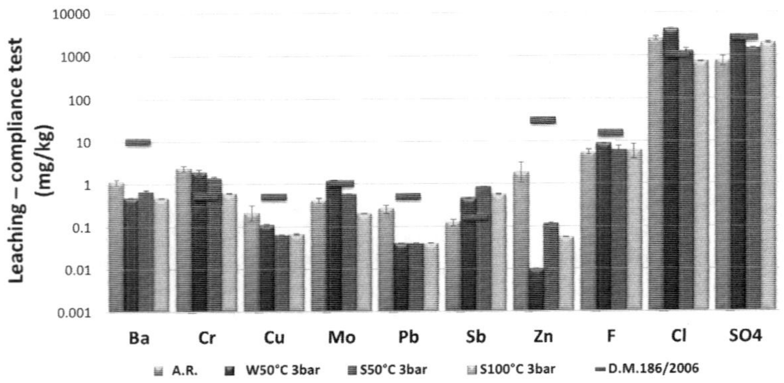

	pH	$CO_{2uptake}$(%)
A. R.	12.5	-
W50°C 3bar	9.8	11.31
S50°C 3bar	10.4	9.95
S100°C 3bar	10.7	9.38

Slurry carbonation at 100 °C

√ pH, Cr and Cl

X Sb

G. Costa *Analysis of the effects of slurry phase carbonation on waste thermal treatment bottom ash*

Results: *Acid/Base neutralization capacity*

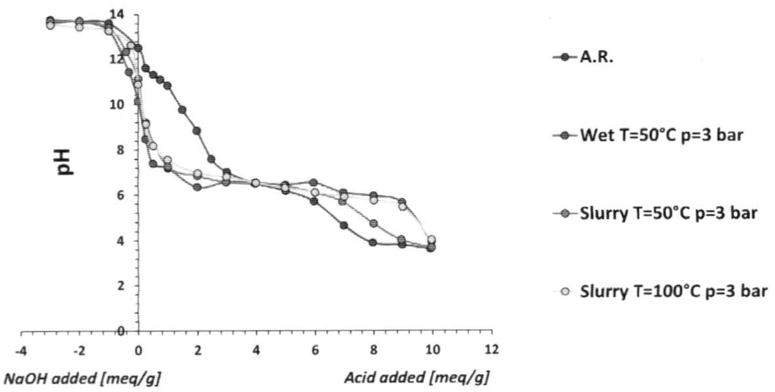

- As received sample: Acid buffering at pH 11, from 10 to 8 and at pH 7-6
- Carbonated samples: buffering capacity only for pH values between 7 and 6

G. Costa *Analysis of the effects of slurry phase carbonation on waste thermal treatment bottom ash*

Results: *Release as a function of pH*

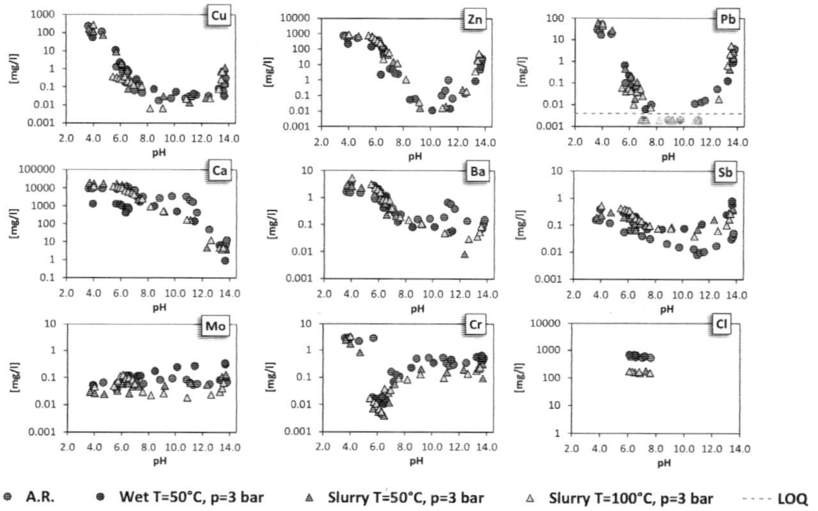

G. Costa — *Analysis of the effects of slurry phase carbonation on waste thermal treatment bottom ash*

Results: *Wet vs. Slurry phase carbonation for other types of BA*

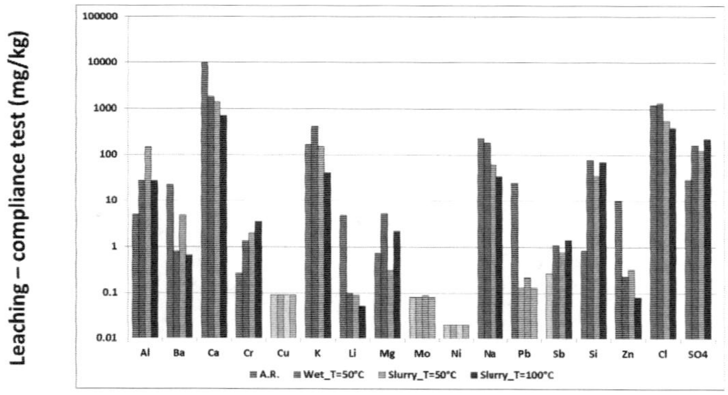

	pH	$CO_{2uptake}$(%)
A. R.	12.83	-
W50°C 3bar	11.33	2.8
S50°C 3bar	12.13	1.59
S100°C 3bar	11.54	1.09

Type 2 BA

BA from a fluidized bed plant treating RDF and extracted by dry treatment (no quenching)

Fine fraction d< 2 mm

Slurry carbonation at 100 °C

✓ pH, Ba and Cl

✗ Sb and Cr

G. Costa — *Analysis of the effects of slurry phase carbonation on waste thermal treatment bottom ash*

Results: *Wet vs. Slurry phase carbonation for other types of BA*

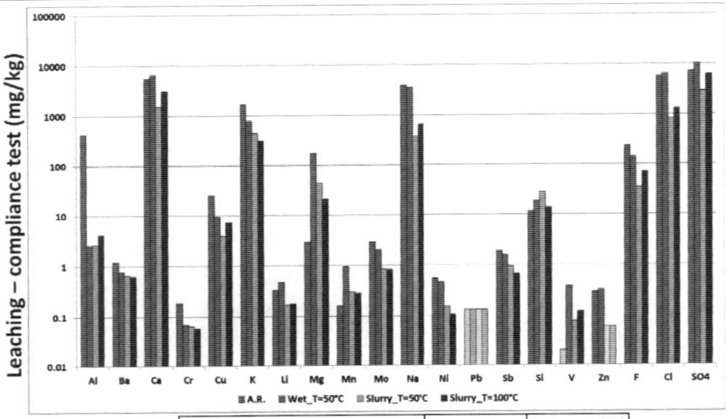

	pH	$CO_{2uptake}$(%)
A. R.	10.33	-
W50°C 3bar	8.45	1.81
S50°C 3bar	8.06	1.56
S100°C 3bar	8.46	1.33

Type 3 BA

BA from a large scale Dutch BA treatment plant before washing step (MSW grate fired plant with wet extraction)

Fine fraction d< 2 mm

DOC = 12.4 g/kg

Slurry carbonation at 50 °C

√ Mo, Cl and SO_4

✗ Sb (decreases), Cu, V and F

Open issues and work in progress

- *Mineralogy analysis*
- *Effect of 40% CO_2*
- *Analysis on the mechanisms controlling release by geochemical modelling* **ECN**
- *Treatments to reduce Sb leaching after carbonation (slag 1 and 2)* **ECN**
- *Effect of carbonation and/or washing on DOC leaching (Cu)*
- *Analyse other combined treatments*
- *Apply column leaching tests to evaluate leaching as a function of the L/S*
- *Compare the overall environmental sustainability of the most promising treatments by LCA*

Factors controlling the dissolution of thermally activated serpentinite under high pressure and temperature carbonic acid.

Jason P. Mann
Brian S. Haynes

THE UNIVERSITY OF SYDNEY

ACEME
Newcastle, NSW
11 – 14th March, 2018

Mineral Carbonation

Chemistry Overview

- Based on the natural carbonic acid equilibrium,

$$CO_{2\,(g)} + H_2O \leftrightarrow HCO_3^- + H^+$$

- Protons (H^+) generated from CO_2 hydration are used for Mg extraction.
  - e.g. periclase,

$$MgO_{\,(s)} + 2H^+ \leftrightarrow Mg^{2+} + 2H_2O$$

- $MgCO_3$ is formed directly through precipitation with CO_3^{2-},
  - Either by increasing pH (e.g. removing dissolved CO_2 or adding base) or temperature.

$$Mg^{2+} + CO_3^{2-} \leftrightarrow MgCO_{3\,(s)}$$

Mineral Carbonation

Serpentinite Activation

- Dissolution is extremely slow even at high temperatures & pressures,
  - e.g. 150 °C & 100 bar.
- Thermal activation is required to increase reactivity:

$$Mg_3Si_2O_5(OH)_4 \xrightarrow{T > 600\ °C} 3MgO.2SiO_2.(2\text{-}x)H_2O + xH_2O$$

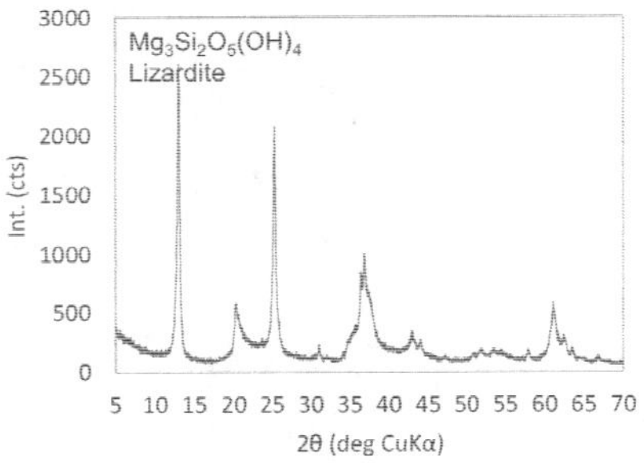

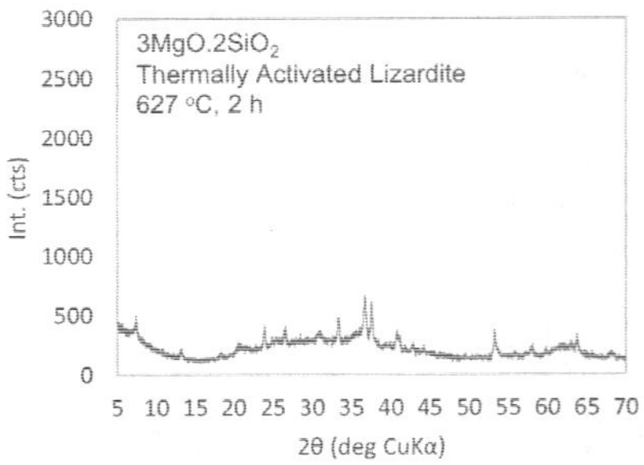

Mineral Carbonation

Deactivation

- A result of $3MgO.2SiO_2$ recrystallisation towards forsterite (Mg_2SiO_4).
  - Prominent at temperature above 650 °C.

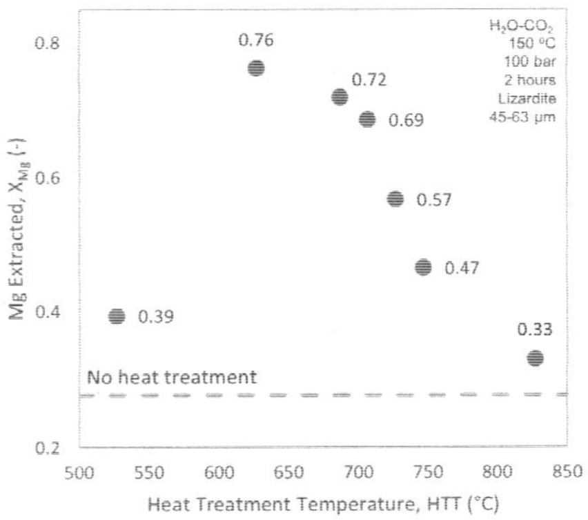

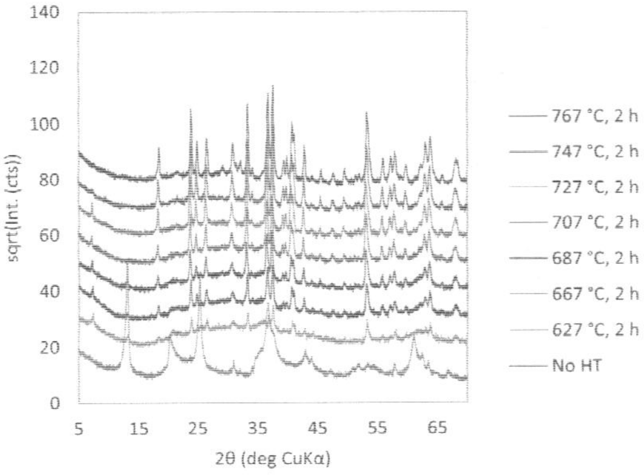

Research Questions

- What is the fundamental mechanism behind deactivation?
  - Is it chemically or physically induced?
    - Does the material become progressively homogenously inert, or
    - Does recrystallisation impart an external diffusion barrier of some type?
- And, can dissolution from deactivated material be easily modelled?

Experimental
How to achieve this?

- Exclusively probing the dissolution reaction,
  - Far from $MgCO_3$ & SiO_2 equilibrium,
  - Constant pH.
- Flow through reactor,
  - Semi-batch,
  - Micro-fluidized bed,
  - Upwards of 200 °C & 100 bar.
- Study Material,
  - Lizardite (SWOL)
  - 45-63 µm
  - 10 mg

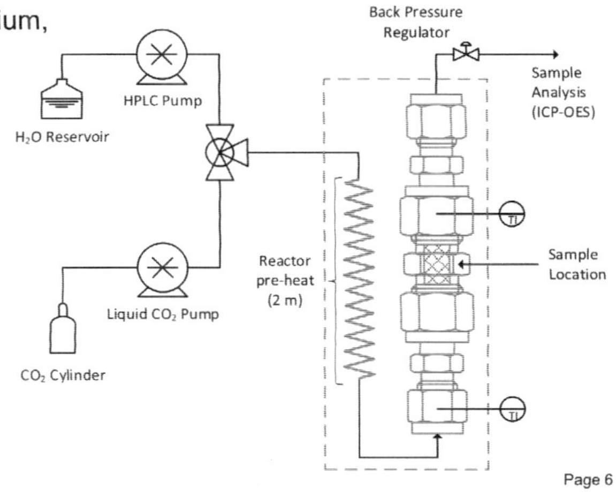

Experimental
Data Analysis

- Concentration profile converted to conversion rate profile,
  - i.e. $C_{Mg} \rightarrow \frac{d}{dt}X_{Mg} = \frac{C_{Mg}}{C_{Mg,0}}Q \rightarrow X_{Mg} = \int \frac{d}{dt}X_{Mg}.dt$

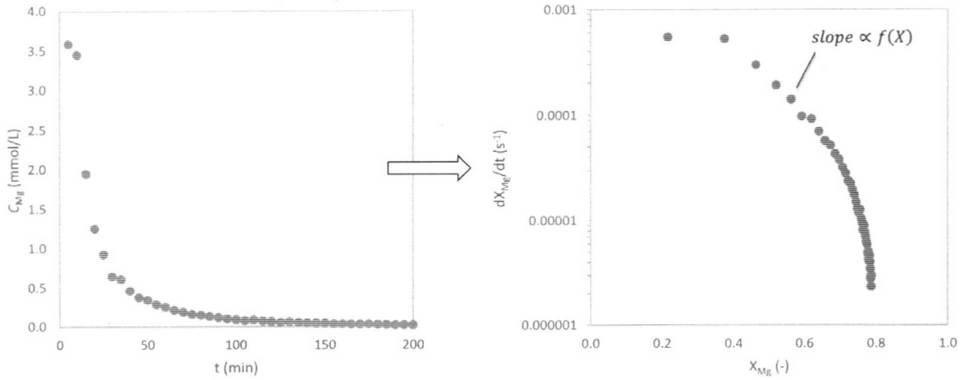

Dissolution Rate Profiles

- Reaction conditions:
  - 150 °C, 100 bar, H_2O-CO_2 (CO_2 saturated), 10 mg, Q = 1.2 cm^3/min.

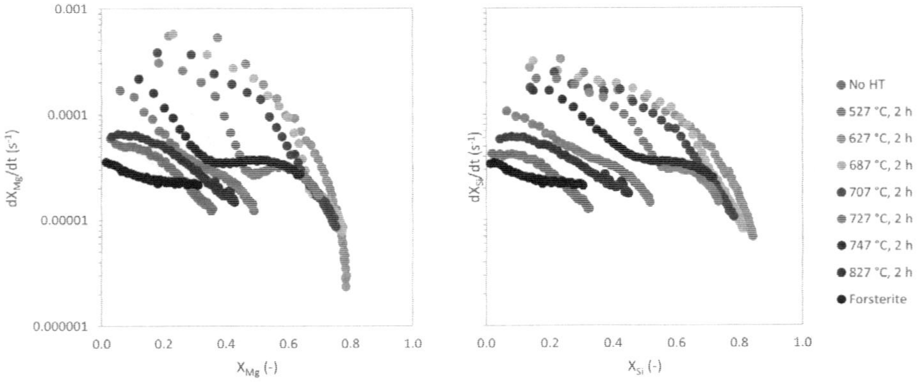

Conversion Rate Profiles

- Rate transition coincides with depletion of amorphous phase,
  - Determined through QPXRD (Rietveld refinement).

- Mg distribution

HHT (°C)	627	687	707	727	747
Forsterite	0.10	0.28	0.36	0.57	0.76
Amorphous	0.90	0.72	0.64	0.43	0.24

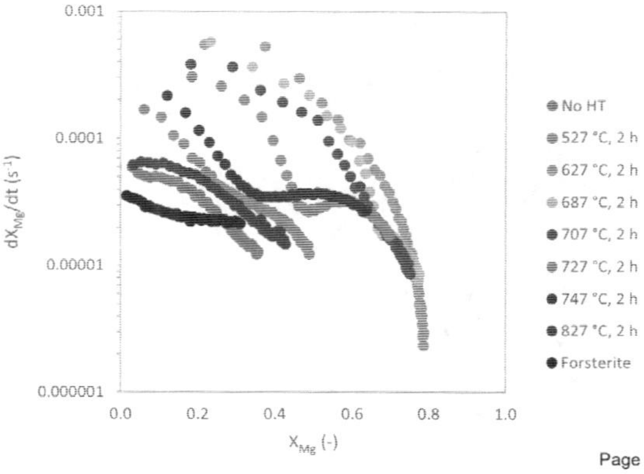

Residue Analysis
SEM-EDS

- Ultramicrotomy used to prepared electron transparent samples (~ 50 nm).
  - High resolution EDS analysis.

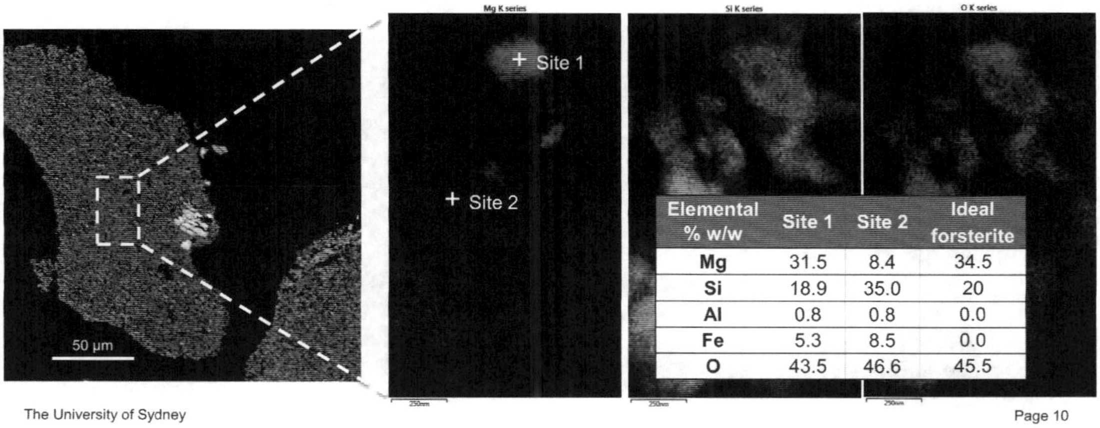

Elemental % w/w	Site 1	Site 2	Ideal forsterite
Mg	31.5	8.4	34.5
Si	18.9	35.0	20
Al	0.8	0.8	0.0
Fe	5.3	8.5	0.0
O	43.5	46.6	45.5

Residue Analysis
SEM-EDS/EBSD (TKD)

– Forsterite confirmed through EBSD-TKD.

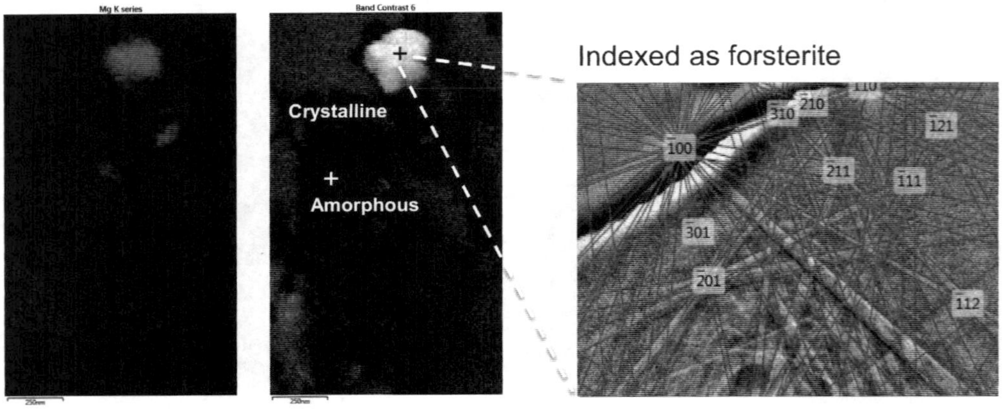

State of the residual Mg?
Higher heat treatment temperatures

– Residual material becomes increasingly forsteritic in nature,
  – Densification of forsteritic nodules,
  – Loss of amorphous background,
  – No change in forsterite grain size,

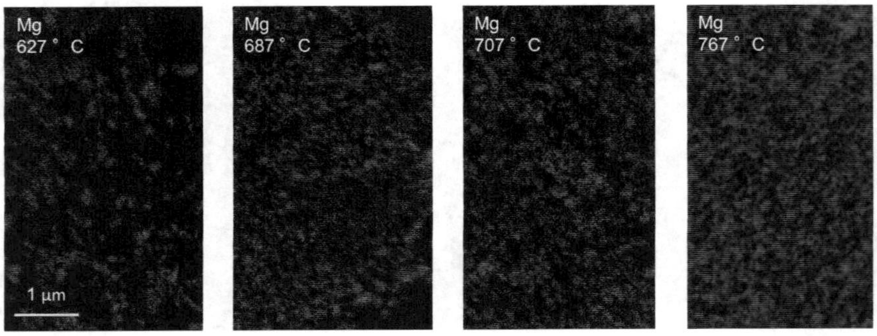

Dissolution Summary

- Dissolution proceeds through a two-stage mechanism.
- Combined QPXRD and SEM analysis indicates:
  - Remaining material consists of forsteritic Mg and residual amorphous Mg,
    - Residual amorphous Mg distributed in a siliceous rich matrix.
  - Two-stage behaviour arises from the sequential dissolution of the amorphous and forsteritic components,
  - No apparent physical barrier induced through recrystallisation,
    - N.B. at the current dissolution conditions.

- Deactivation is simply a result of the reduced amorphous content,
  - Forsterite dissolution is a more activated process.

- Can the reactions be decoupled?

Decoupling Dissolution

- $3MgO.2SiO_2 \rightarrow \frac{3}{2}(Y_{Fo})Mg_2SiO_4 + 3(1 - Y_{Fo})MgO + \left(2 - \frac{3}{2}Y_{Fo}\right)SiO_2$
  - Y_{Fo} determined through QXRD.
  - MgO and SiO_2 represent amorphous components.
- Mole balance gives,
  - Mg,

$$X_{Mg} = (1 - Y_{Fo})X_{Mg,am} + Y_{Fo}X_{Mg,Fo}$$

  - Si,

$$X_{Si} = \left(1 - \frac{3}{4}Y_{Fo}\right)X_{Si,am} + \frac{3}{4}Y_{Fo}X_{Si,Fo}$$

- Two potential pathways,
  - ~~Parallel, or~~
  - Series.

Decoupling Dissolution
Amorphous Component

– Forsterite initially remains untouched, i.e. $X_{Mg,Fo} = X_{Si,Fo} = 0$,

Mg: $X_{Mg,am} = \frac{X_{Mg}}{1-Y_{Fo}}$ Si: $X_{Si,am} = \frac{X_{Si}}{1-\frac{3}{4}Y_{Fo}}$

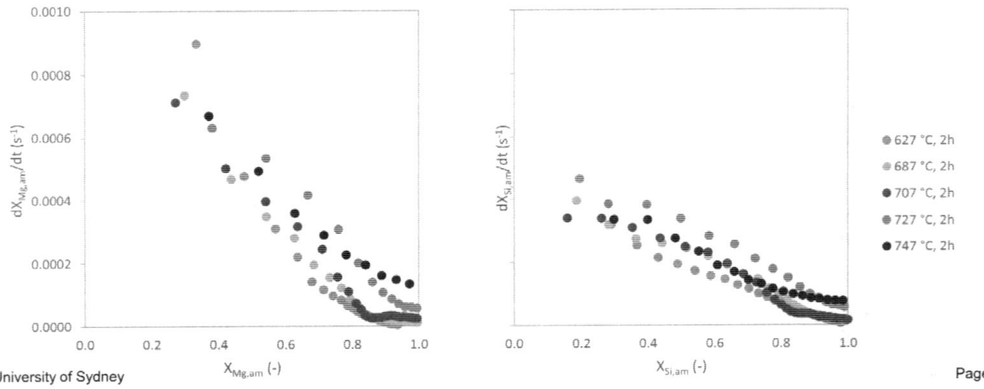

Decoupling Dissolution
Forsterite Component

– Amorphous material is consumed, i.e. $X_{Mg,am} = X_{Si,am} = 1$,

Mg: $X_{Mg,Fo} = 1 - \frac{(1-X_{Mg})}{Y_{Fo}}$ Si: $X_{Si,Fo} = 1 - \frac{4}{3}\frac{(1-X_{Si})}{Y_{Fo}}$

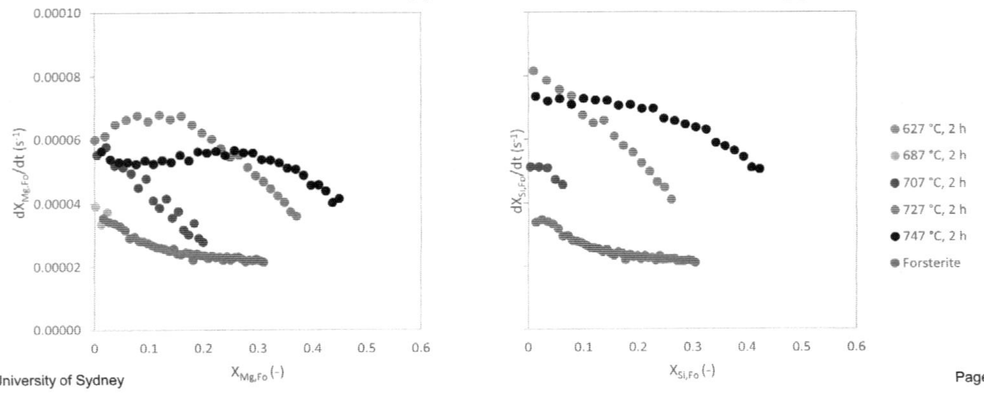

Conclusions

– Dissolution proceeds through a two stage mechanism,
  – Consistent with the amorphous and forsteritic fractions.
– Observed deactivation is a result of this characteristic,
  – Rate reduction occurs from the intrinsically slower rate of forsterite dissolution.
  – More forsterite, faster transition to this slower rate phase.
– Dissolution kinetics can be decoupled through stoichiometry alone,
  – Apparent first order dissolution behaviour observed for amorphous material.
  – Technique still evolving,
    - Role of Fe needs to be explored,
    - Better refinement for Y_{Fo}.

Acknowledgements

– Mineral Carbonation International (MCi)

ACEME2018

Presentation: ACEME2018
Date: 13th March 2018
Location: Newcastle Australia

ACEME2018
March 11-14
Newcastle City Hall, Newcastle, New South Wales, Australia

10-4-2018 2018 ACEME @ Newcastle 1 Green Minerals

Past, present

& future

Green Minerals

Introduction

CCS => CC**U**S (**U**tilization)

CO_2 utilization:	**Pro's**	**Cons**
1. CO_2 to fuels	Fit renewable energy	Storing time
2. CO_2 to chemicals	Added value	Amount
3. *CO_2 to minerals*	*Storing time*	***Reaction***

CCS ~~out~~, more precise moving from Utility
CCS => Steel, Cement, fertiliser
Utility sector moving to Wind / solar

Past: 1. Idea

- Ore mineralisation limited by:
  kinetics => residence time
  economics CO_2 price => products
- Exothermic reaction
- Gravity Pressure Vessel

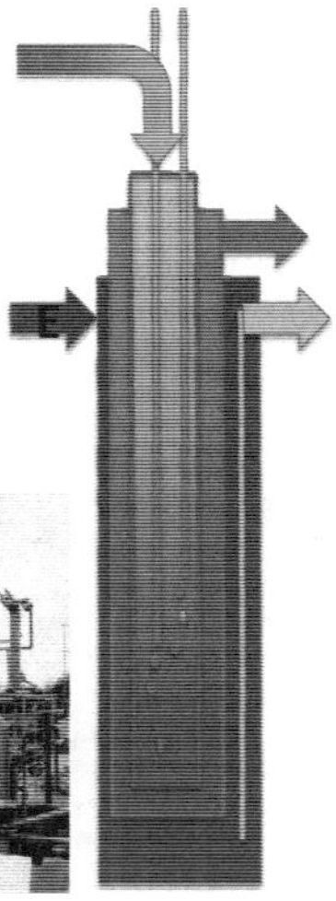

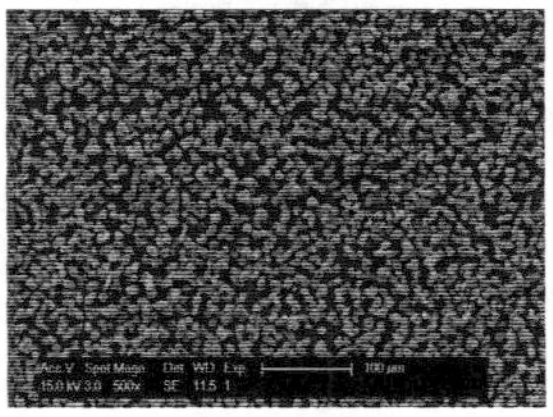

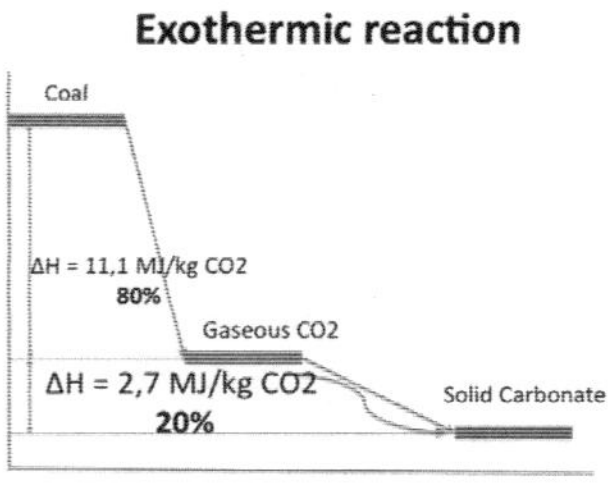

Past: 2. Academic

KUL (University of Louvain; prof. Tom van Gerven)

- 3 MSc thesis, ½ PhD-thesis (Dr. Rafael Santos)
- Various publications

University Nancy

- PhD-thesis Laura Turri

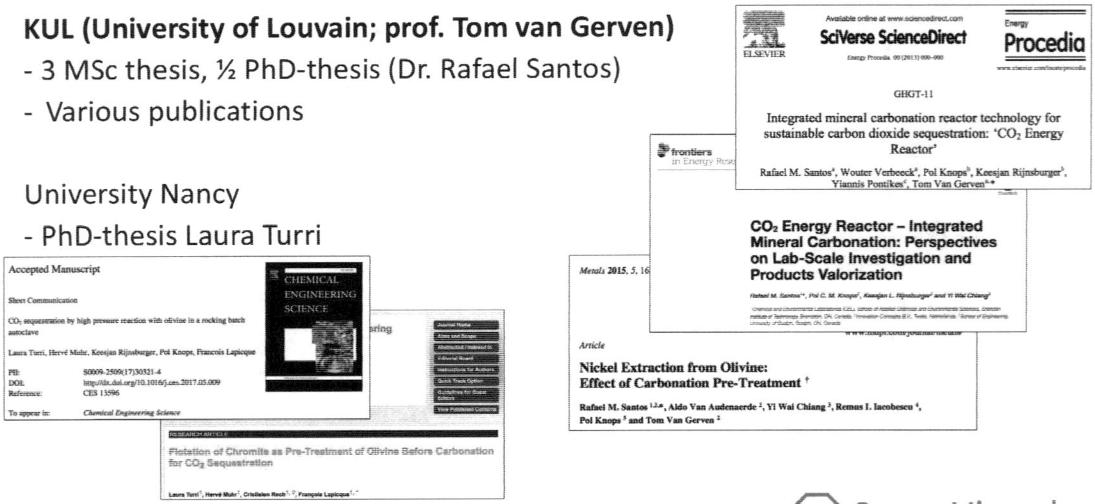

Past: 3. Industrial project

- Cooperation:
  - KU Leuven
  - ISPT
  - Arcelor Mittal
  - Sibelco
- Funding ± € 500k

- Stopped: Economic not feasible

Stopped: Economic not feasible		Focus
- CO_2 upgrading	- € 30/ton CO_2	Pure CO_2 sources, Exothermic
- CO_2 credits	€ 0	ETS including
- Products	+ € 30/ton	Higher value

Present 1: CO2MIN

- Partners:
  - RWTH (University of Aachen) 5
  - HeidelbergCement
  - IASS Potsdamm
  - Green Minerals

- Funding
  € 2.8 Million
  3 years

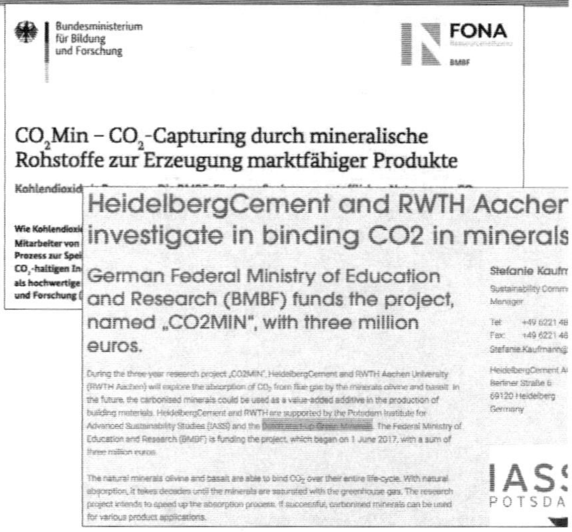
Bundesministerium für Bildung und Forschung

FONA BMBF

CO_2Min – CO_2-Capturing durch mineralische Rohstoffe zur Erzeugung marktfähiger Produkte

HeidelbergCement and RWTH Aachen investigate in binding CO2 in minerals

German Federal Ministry of Education and Research (BMBF) funds the project, named „CO2MIN", with three million euros.

During the three-year research project „CO2MIN", HeidelbergCement and RWTH Aachen University (RWTH Aachen) will explore the absorption of CO_2 from flue gas by the minerals olivine and basalt. In the future, the carbonised minerals could be used as a value-added additive in the production of building materials. HeidelbergCement and RWTH are supported by the Potsdam Institute for Advanced Sustainability Studies (IASS) and the Dutch start-up Green Minerals. The Federal Ministry of Education and Research (BMBF) is funding the project, which began on 1 June 2017, with a sum of three million euros.

The natural minerals olivine and basalt are able to bind CO_2 over their entire life-cycle. With natural absorption, it takes decades until the minerals are saturated with the greenhouse gas. The research project intends to speed up the absorption process. If successful, carbonised minerals can be used for various product applications.

Stefanie Kaufm
Sustainability Comm
Manager
Tel: +49 6221 48
Fax: +49 6221 48
Stefanie.Kaufmann@
HeidelbergCement A
Berliner Straße 6
69120 Heidelberg
Germany

IASS POTSDA

10-4-2018 2018 ACEME @ Newcastle 7 Green Minerals

Present 2: CO2MIN

- Partners:
  - 5 * RWTH
  - IASS
  - Heidelberg Cement
  - Green Minerals

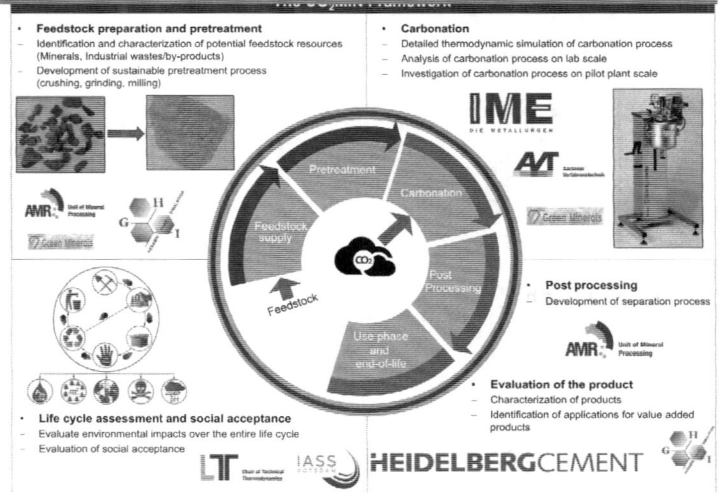

10-4-2018 2018 ACEME @ Newcastle 8

Present 3: Other projects

- Setting up Autoclave
  1.4 liter, 150 bar (max), 300 °C
- Validation beneficial use in polymer
  (replacement "functional fillers")
- Validation beneficial use in paper manufacturing
  (replacement PCC)

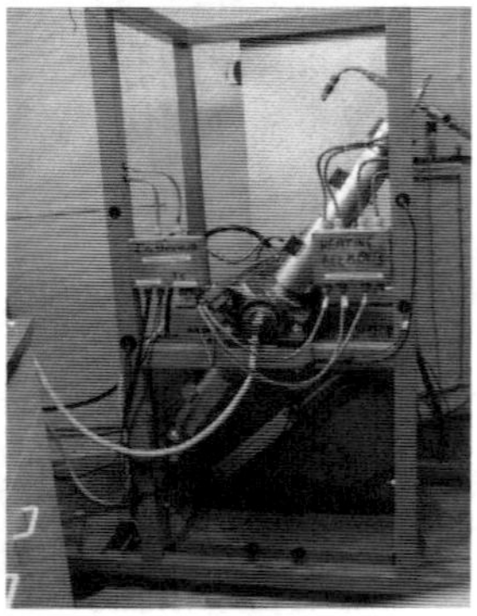

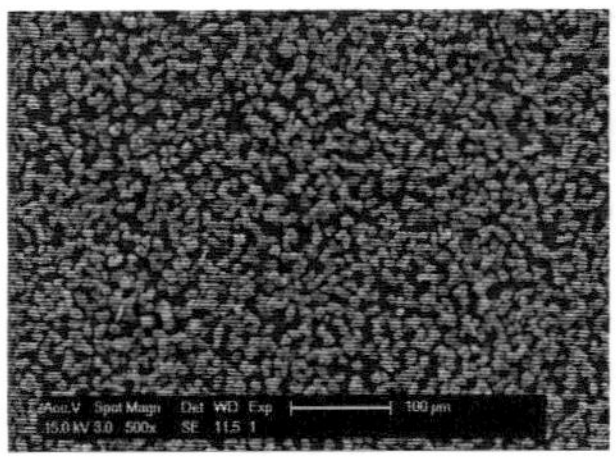

Future

• Participating CO2MIN project	Concrete	Heidelberg	☑☐
• Set-up project	Polymer	....	■
• Set-up project	Paper manufacturing		■

- Developing Continuous process

Afterword

- 1 Creating More awareness mineralization
  - CO2 "bound" forever.
  - Business cases

- 2. More attention towards academic - industrial cooperations
  - More funding research
  - More focus business cases

3. CCS => CCUS

Advantages: 1. Storing CO2 forever

2.Using CO2 as a feedstock => Products

10-4-2018 2018 ACEME @ Newcastle 11 Green Minerals

Turning CO_2 into Value

Green Minerals

Pol Knops
Tel: +31 6 5130 4842
Pol@green-minerals.nl
www.green-minerals.nl
@Greenolivine

10-4-2018 2018 ACEME @ Newcastle

Åbo Akademi University

2018 ACEME

Serpentinite carbonation using the Åbo Akademi routes – status update

Ron Zevenhoven , Martin Slotte§, Evelina Koivisto, Rickard Erlund

Åbo Akademi University, Thermal and Flow Engineering Laboratory, Turku, Finland

§ currently at Paroc Group Oy, Parainen Finland

Åbo Akademi University

Overview

- Background and scope
  - *The ÅA routes for stepwise serpentinite carbonation*
- Progress since ACEME-15
  - *Life cycle assessment (LCA) studies*
  - *Membrane separations for ammonium salt recovery*
  - *Thermal energy storage (TES) using hydrated magnesium carbonates*
  - *Other*
- Conclusions

Åbo Akademi University

Dr. theses so far at ÅA

- January 2017: **Martin Slotte** "Two process case studies on energy efficiency, life cycle assessment and process scale-up" https://www.doria.fi/handle/10024/130097
- January 2017: **Sonja Lavikka** (n. **Sjöblom**) "Geological and mineralogical aspects on mineral carbonation" http://www.doria.fi/handle/10024/130096
- December 2015: **Inês S. Soares Romão** "Production of magnesium carbonates from serpentinites for CO_2 mineral sequestration: optimisation towards industrial application" *double degree with Univ. of Coimbra, Portugal* http://www.doria.fi/handle/10024/117766
- December 2012: **Experience I. Nduagu** "Production of $Mg(OH)_2$ from Mg-silicate rock for CO_2 mineral sequestration" http://www.doria.fi/handle/10024/86170
- March 2012: **Johan Fagerlund** (n. **Sipilä**) "Carbonation of $Mg(OH)_2$ in a pressurized fluidized bed for CO_2 sequestration" http://www.doria.fi/handle/10024/74477
- see also S Eloneva (2010) and S Teir (2008) at Helsinki Univ. of Technol., nowadays Aalto Univ.

Åbo Akademi University

CCS / CCU & Finland

	CO_2 geological storage	CO_2 mineral sequestration	CO_2 utilisation
"CCS" according to EC directive 2009/31/EC	Yes	No	No
Potential inside Finland's borders	None	Several Gt CO_2	~ 0.3 Mt/year CO_2
CO_2 pre-separation ("capture") necessary	Yes	No	Preferable for product quality and purity
Final form of CO_2	Pressurised (super-critical)	Magnesium carbonate, hydrated magnesium carbonate	Precipitated calcium carbonate (PCC)
Primary interested industry sectors in Finland	All large-scale CO_2 producers; especially Power & Heat and Oil & Gas sectors	Mineral processing sector	Iron- & steel making, Pulp & paper sector, Mineral processing sector

Plan B for Finland according to some, for after 2030: ***BECCS**, forget fossil-carbon derived CO_2...*

Two "ÅA routes"

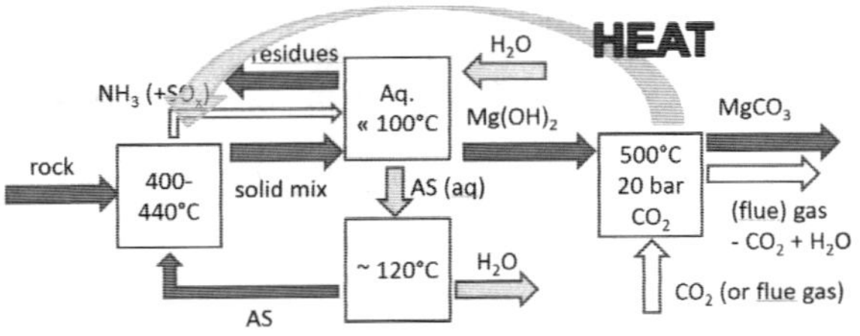

← *"Conventional" ÅA route*
Route I

↓ *"Alternative" ÅA route*
Route II

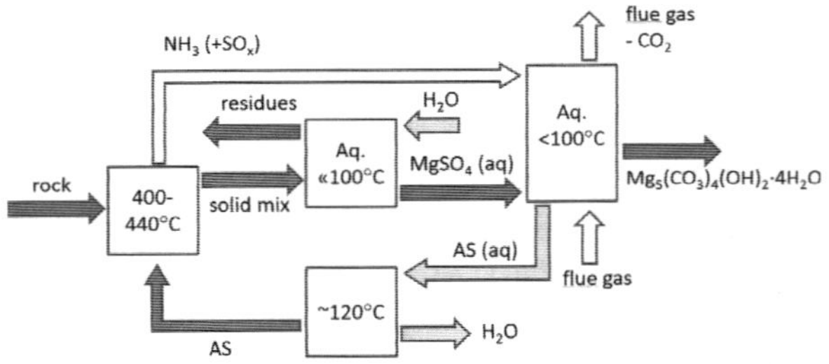

dark arrows: solid material

light: arrows aqueous

white arrows: gas

AS = ammonium sulphate

Two "ÅA routes"

See Aceme-08, -10, -13, -15, etc.

ute I

ute II

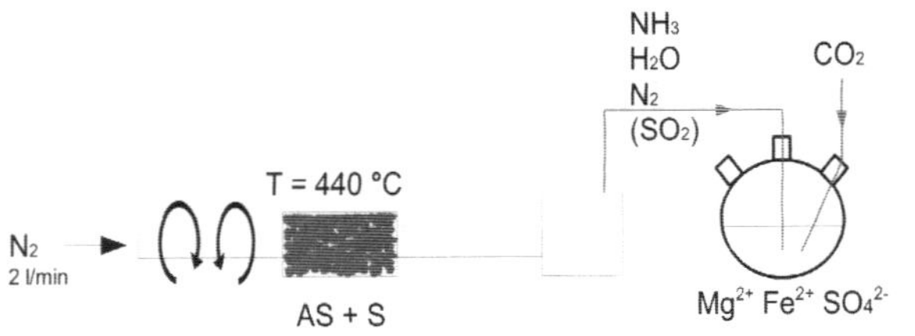

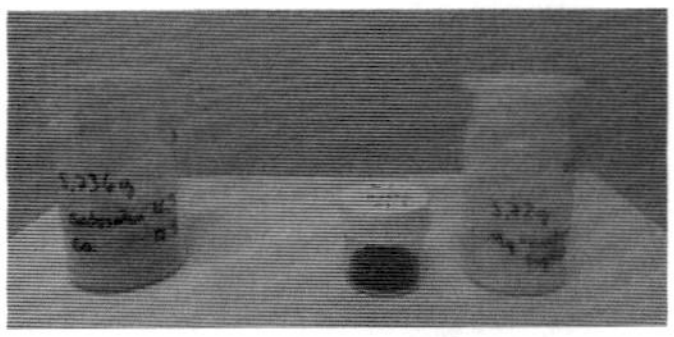

Interesting work from Canada

Contents lists available at ScienceDirect

International Journal of Greenhouse Gas Control

Technical & economic evaluation of a mineral carbonation process using southern Québec mining wastes for CO_2 sequestration of raw flue gas with by-product recovery

Louis-César Pasquier[a,*], Guy Mercier[a,*], Jean-François Blais[a], Emmanuelle Cecchi[a], Sandra Kentish[b]

L.-C. Pasquier et al. / International Journal of Greenhouse Gas Control 50 (2016) 147–157

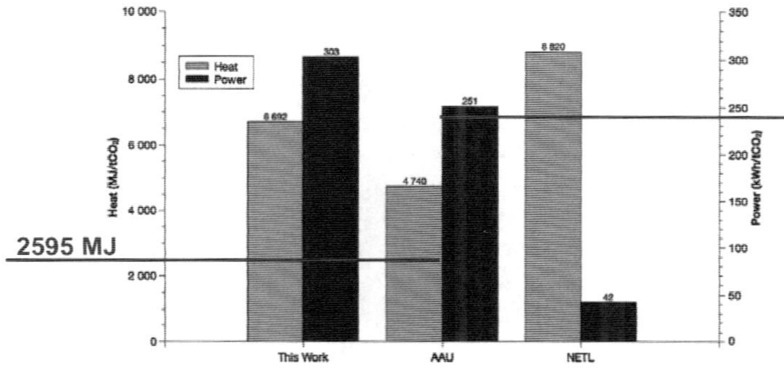

Fig. 8. Comparison of energy demands for different mineral carbonation processes using serpentine. Data for the AAU and NETL processes were taken from Nduagu et al. (2012).

Data
Slotte et al. 2013
doi 10.1016/j.energy.2013.07.009

Lime kiln
No CO_2 capture
Hitura serpentinite

LCA – system boundaries

ute I

ute II

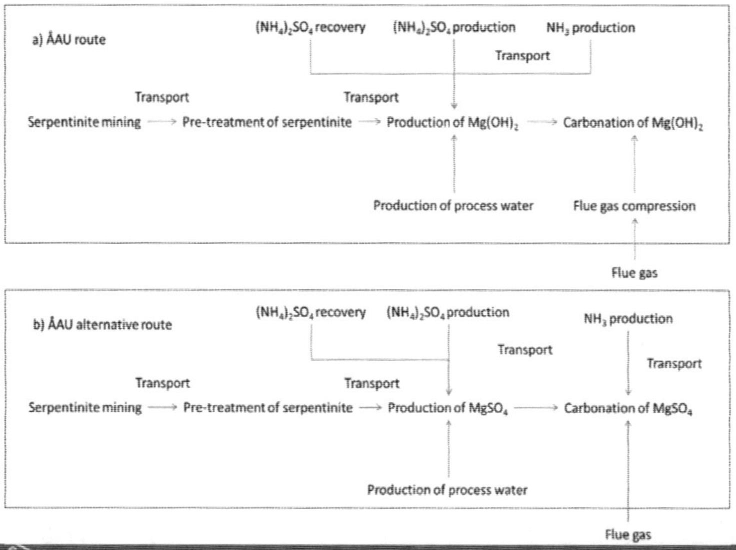

Note:
operation on flue gas, no CO_2 pre-capture

Calculation tool:

SimaPro 7.3
+
Ecoinvent 2.2 data

Dr. Thesis Slotte 2017
& doi: 10.1002/ente.201600702

kademi
versity

LCA – comparison two ÅA routes

LCI for "conventional" ÅA Route I (left) and ÅA Route II (right) for capture of CO_2 from a lime kiln per one kg of sequestered CO_2

	Human health	Ecosystem quality	Climate change	Resource depletion
	DALY [a]	PDF [b] m^2 yr	kg CO_2 eq	MJ primary
nation	0	0	-0.999	0
sium xide	$1.18 \cdot 10^{-7}$	0.014	0.194	3.16
onium te	$2.18 \cdot 10^{-7}$	0.002	0.051	1.09
ressed s	$1.00 \cdot 10^{-7}$	0.005	0.309	4.98
rocess	$4.36 \cdot 10^{-7}$	0.021	-0.444	9.24

[a] Disability adjusted life years
[b] Potentially disappeared fraction

	Human health	Ecosystem quality	Climate change	Resource depletion
	DALY	PDF m^2 yr	kg CO_2 eq	MJ primary
Serpentinite				
Mined	$2.26 \cdot 10^{-8}$	0.0027	0.02	0.31
Transport	$6.52 \cdot 10^{-8}$	0.0037	0.08	1.12
Amm. sulphate				
Recovered	$7.49 \cdot 10^{-9}$	0.0003	0.07	1.41
Make-up	$2.63 \cdot 10^{-8}$	0.0070	0.06	0.97
Dissolution H_2O	$9.08 \cdot 10^{-9}$	0.0005	0.03	0.40
Make-up NH_3	$7.40 \cdot 10^{-10}$	0.0001	0	0.04
Carbonation	0	0	-1.00	0
Total process	$1.31 \cdot 10^{-7}$	0.014	-0.74	4.24

Dr. Thesis Slotte 2017
& doi: 10.1002/ente.201600702

kademi
versity

LCA – comparison two ÅA routes

Effects of transportation distance of the magnesium containing rock on overall process LCI, for the "conventional" ÅA process Route I

	Human health	Ecosystem quality	Climate change	Resource depletion
	DALY	PDF m^2 yr	kg CO_2 eq	MJ primary
5 km	$1.78 \cdot 10^{-7}$	0.017	-0.519	8.17
150 km	$1.93 \cdot 10^{-7}$	0.018	-0.501	8.43
550 km	$2.40 \cdot 10^{-7}$	0.021	-0.445	9.24

[a] Disability adjusted life years
[b] Potentially disappeared fraction

Dr. Thesis Slotte 2017
& doi: 10.1002/ente.201600702

\kademi
versity

Separating ammonium bisulphate from sulphate

.queous leaching
ith ABS is much
›re effective than
with AS

BUT

50% AS / 50% ABS
;ives very good
ıction performance

01 6/j.hydromet.2016.07.005
01 6/j.hydromet.2016.07.004

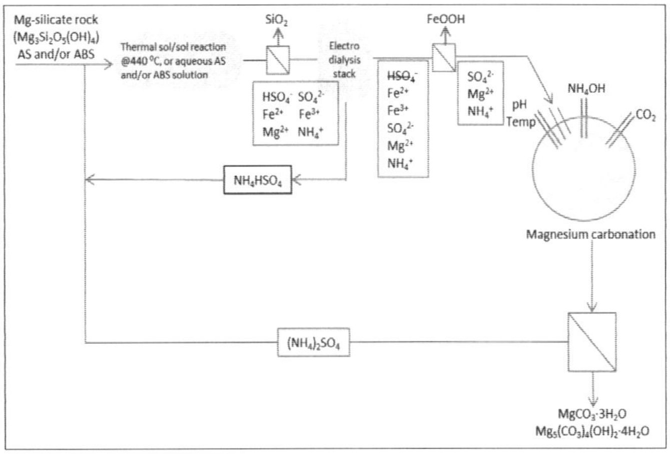

HSO_4^-
SO_4^{2-}
equilibrium

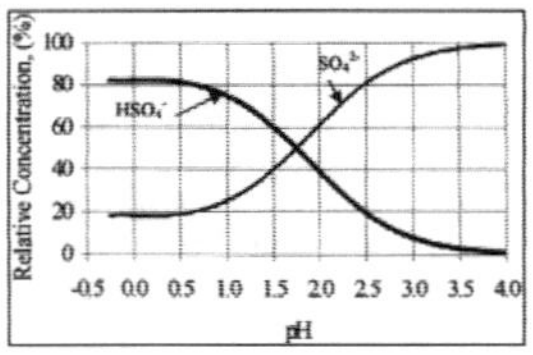

doi: 10.3390/geosciences8040123 (open source)

\kademi
versity

Separating ammonium bisulphate from sulphate

electrodialysis using monovalent ion-selective membranes

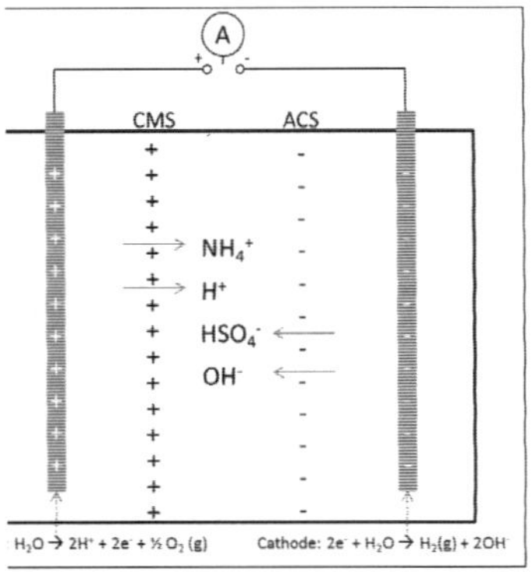

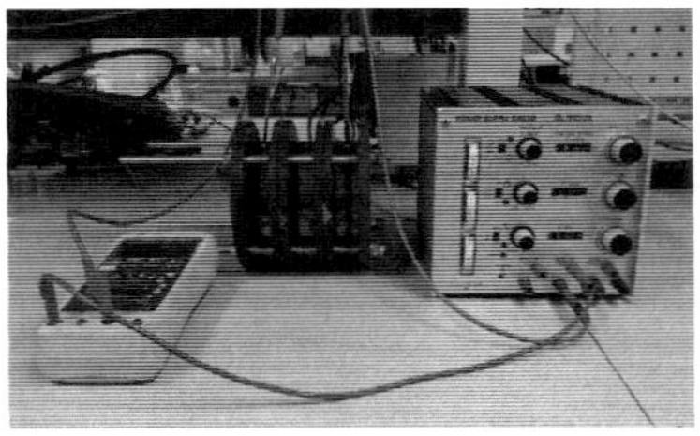

Three compartments
with cell width 40 mm

Test	Time (min)	Anode/Cathode	Voltage (V)	AS (mol/dm^3)	ABS (mol/dm^3)
1	120	Stainless steel (x2)	10	0.5	0.5
2	60	Stainless steel (x2)	20	0.03	0.03
3	25	Silver/Stainless steel	20	0.03	0.03
4	120	Stainless steel (x2)	20	0.1	0.1
5	120	Stainless steel (x2)	20	-	0.1
6	75	Stainless steel (x2)	20	0.5	0.5

doi: 10.3390/geosciences8040123 (open source)

Åbo Akademi University

Separating ammonium bisulphate from sulphate

NH_4^+ separation efficiency and pH variation vs. time: 3 cells

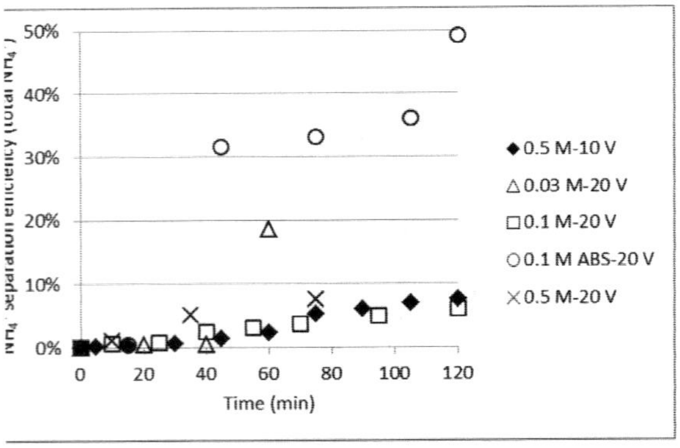

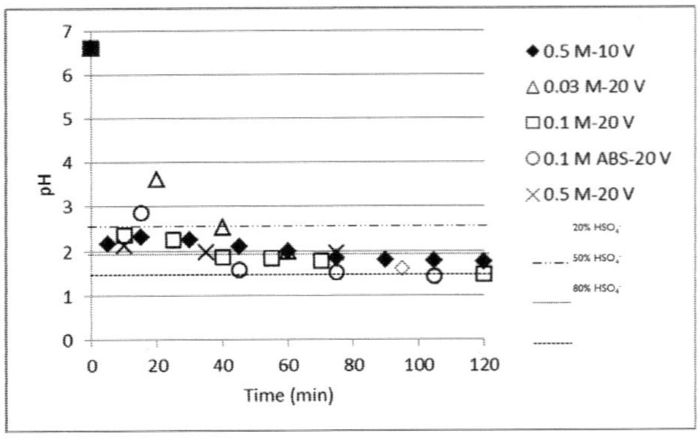

doi: 10.3390/geosciences8040123 (open source)

Åbo Akademi University

Separating ammonium bisulphate from sulphate

electric power use

gister current
ind voltage:

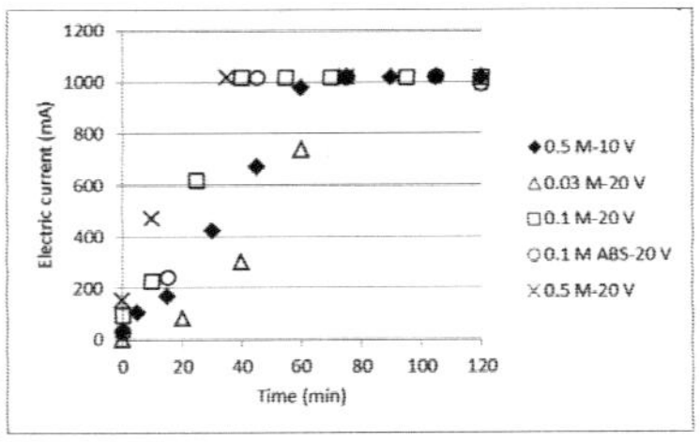

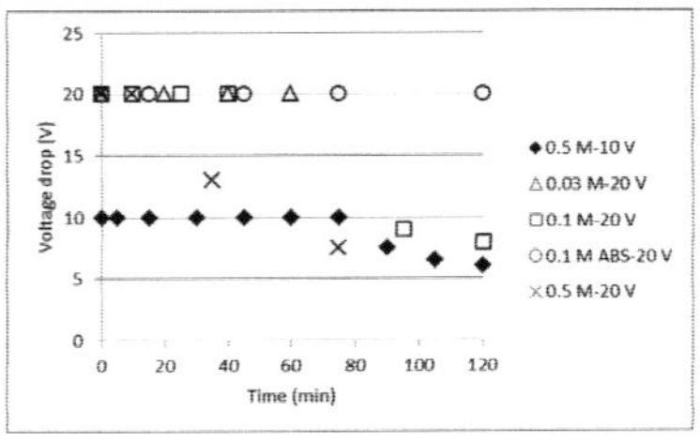

Calculate

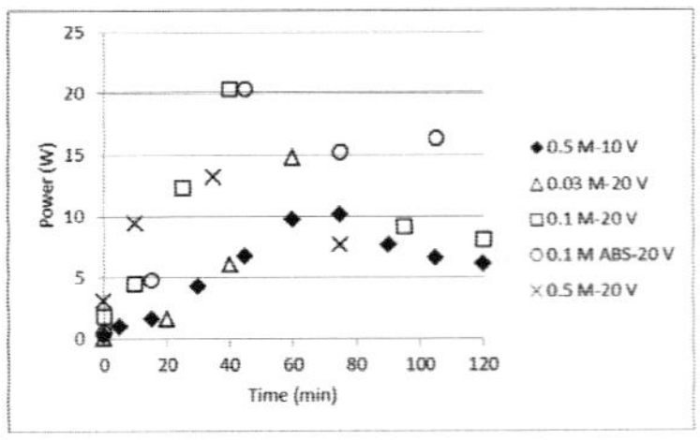

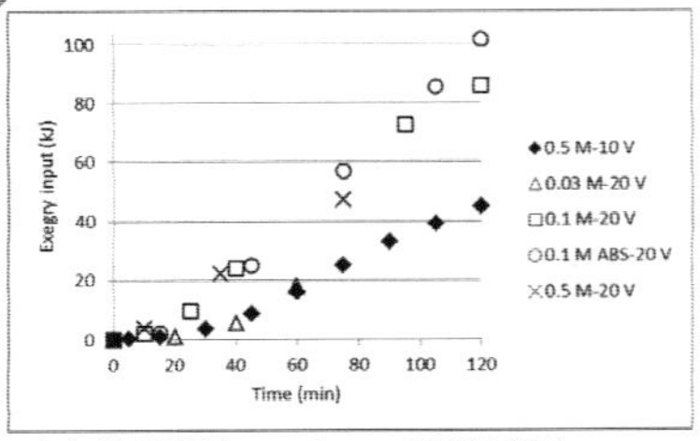

doi: 10.3390/geosciences8040123 (open source)

Separating ammonium bisulphate from sulphate

electrodialysis using monovalent ion-selective membranes: 5 cells

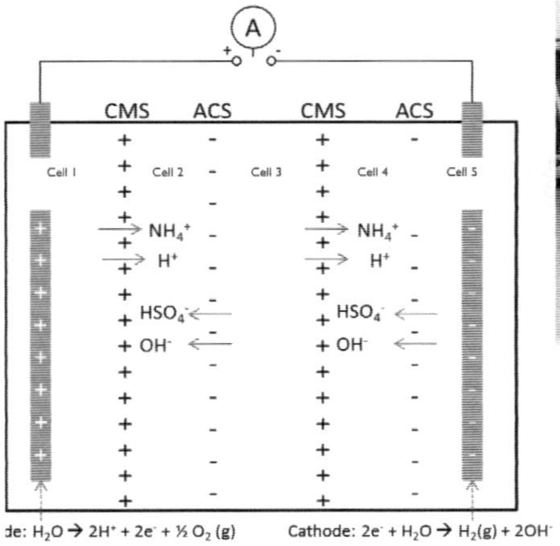

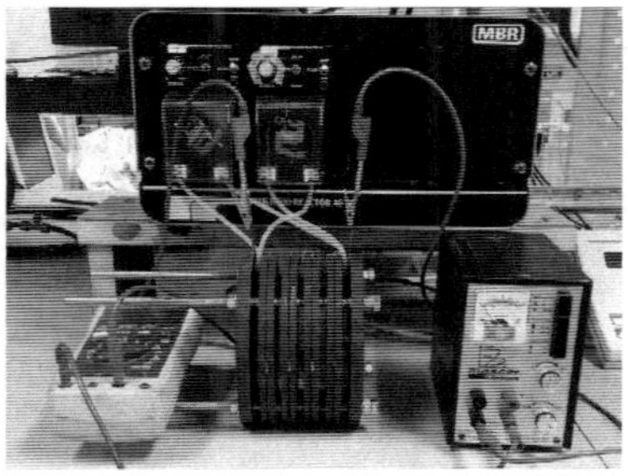

Five compartments with cell width 18 mm

Test	Time (min)	Anode/Cathode	Voltage (V)	AS/ABS (mol/dm^3)	Cell width (mm)	Recirculation (20 ml/min)
10	120	Stainless steel (x2)	20	0.05/0.05	40	Yes
11	120	Platinum/Stainless steel	20	0.05/0.05	18	No
12	120	Platinum/Stainless steel	20	0.05/0.05	18	Yes
13	120	Platinum/Stainless steel	10	0.05/0.05	18	Yes
14	120	Platinum/Stainless steel	15	0.05/0.05	18	Yes
15	120	Stainless steel (x2)	15	0.05/0.05	18	Yes

doi: 10.3390/geosciences8040123 (open source)

Separating ammonium bisulphate from sulphate

results: pH, current density, NH_4^+ removal efficiency, power, exergy input

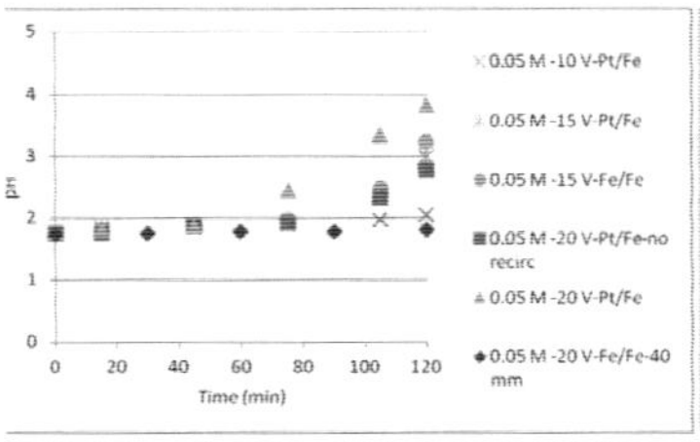

No voltage drop measured for tests done with cell width 18 mm

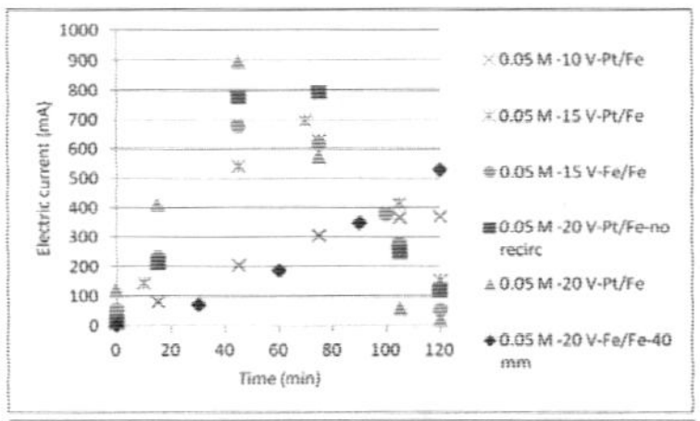

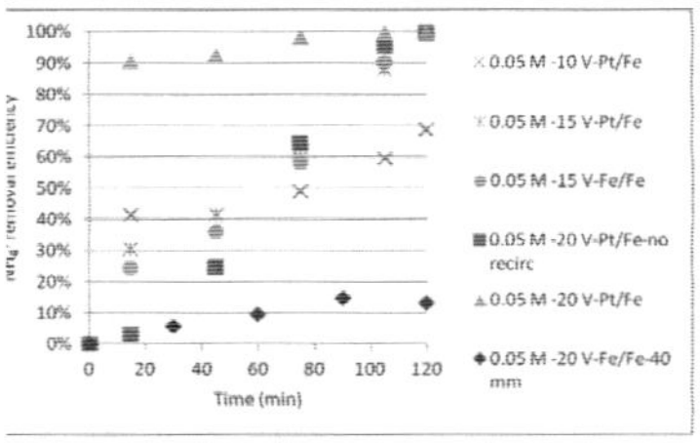

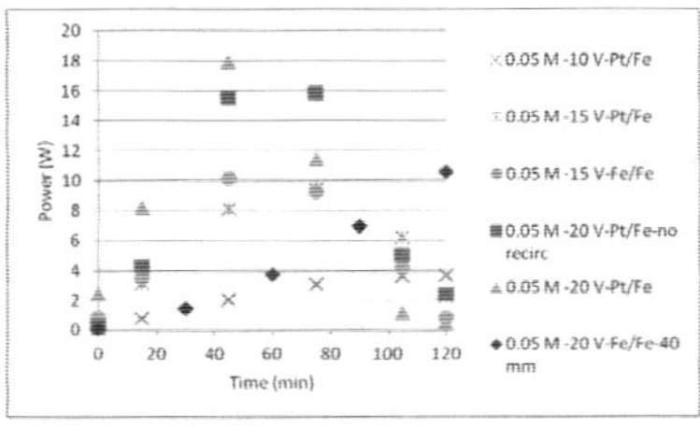

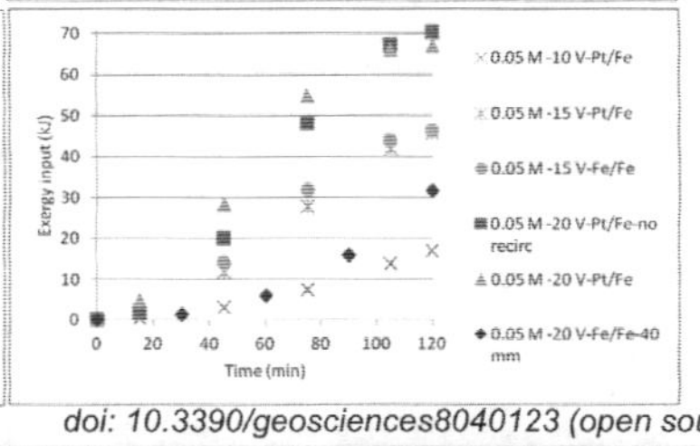

doi: 10.3390/geosciences8040123 (open source)

Thermal energy storage: Mg (hydro) carbonates

Reaction	Hydration temp. (°C)	T(ΔG=0) (°C)	90% dehydr. (°C)	Dehydr. temp. (°C)	Specific capacity (MJ/kg)	Specific capacity (GJ/m³)
$MgCO_3 + 3H_2O(g) = MgCO_3{\cdot}3H_2O$	~20	45	61	60-65	1,0	1,83
$MgCO_3 + 3H_2O(l) = MgCO_3{\cdot}3H_2O$		Irrev.	Irrev.			
$MgCO_3 + 5H_2O(g) = MgCO_3{\cdot}5H_2O$	5-10	55	66	65-70	1,41	2,54
$MgCO_3 + 5H_2O(l) = MgCO_3{\cdot}5H_2O$		Irrev.	Irrev.			
$MgSO_4 + 7H_2O(g) = MgSO_4{\cdot}7H_2O$	~20	145	153	122[1)]	1,70	2,9
$MgSO_4 + 7H_2O(l) = MgSO_4{\cdot}7H_2O$		Irrev.	Irrev.			
$5MgCO_3{\cdot}3H_2O$ g$_5(OH)_2(CO_3)_4{\cdot}4H_2O + CO_2(g) + 10H_2O(g)$	~20	72	75	70-80	0,82	1,46
$5MgCO_3{\cdot}3H_2O + 2OH^-(aq)$ g$_5(OH)_2(CO3)4{\cdot}4H_2O + CO_3(aq) + 11H_2O(l)$		Irrev.	Irrev.			
$\text{IgCO}_3{\cdot}3H_2O+2H_2O(g) = MgCO_3{*}5H_2O$	~20	74	97	95-100	0,63	1,06
$\text{IgCO}_3{\cdot}3H_2O+2H_2O(l) = MgCO_3{*}5H_2O$	~0	~8	86		0,15	0,25
$2H_2O(20\ °C) = 2H_2O(60\ °C)$	20[2)]	-	-	60[2)]	0,17	0,17
Silica Gel	~20	-	-	100	0,62	
Silica Gel dehydrated at 65 °C	~20	-	-	65	0,52	
lite HY (dehydr.) = Zeolite HY (hydr.) [4]	~20	-	~110	110	0,8	
te Na-Y (dehydr.) = Zeolite Na-Y (hydr.) [4]	~20	-	~140	140	1,0	
$CaSO_4 + 2H_2O(g) = CaSO_4{\cdot}2H_2O$	20	89	115	110-120		1,3

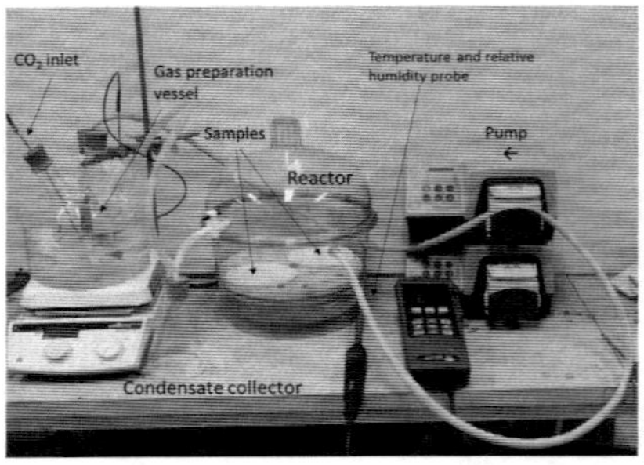

$MgCO_3 + 3H_2O(g) \leftrightarrow MgCO_3{\cdot}3H_2O$

$\Delta H = -1.0$ MJ/kg $MgCO_3{\cdot}3H_2O$, T=298K

$MgCO_3 + 5H_2O(g) \leftrightarrow MgCO_3{\cdot}5H_2O$

$\Delta H = -1{,}41$ MJ/kg $MgCO_3{\cdot}5H_2O$, T=298K

doi: 10.5541/ijot.5000209919 (open source)

Thermal energy storage: Mg (hydro) carbonates

material characterisation: mixtures with silica gel (SG)

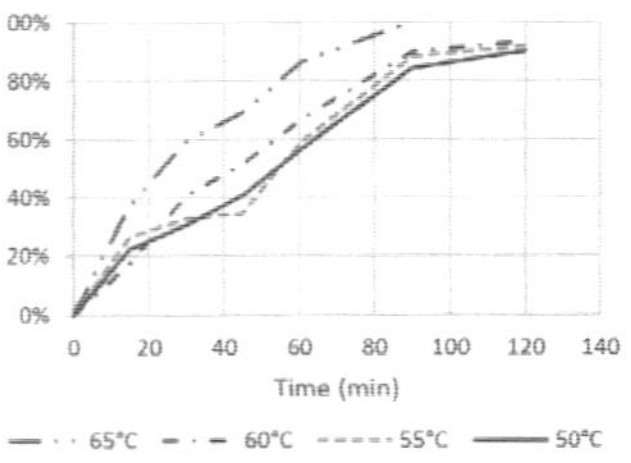

esquehonite dehydration rate

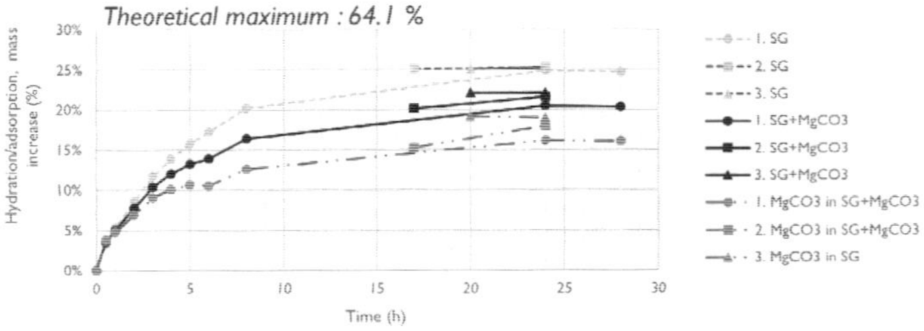

$MgCO_3$, SG and mixture rehydration rate

Nesquehonite / SG mixture granules

doi: 10.5541/ijot.5000209919 (open source)

Thermal energy storage: Mg (hydro) carbonates

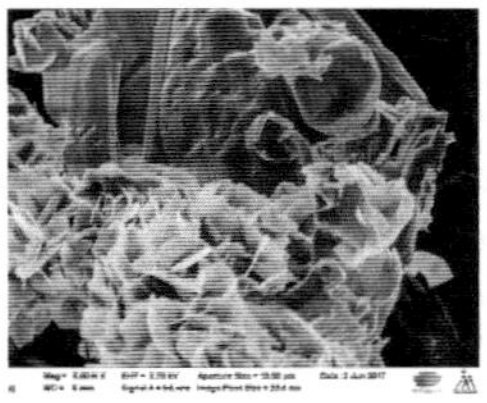

⊃ stored 2 month in air,
⊃ 20x dehydrated /re-hydrated.

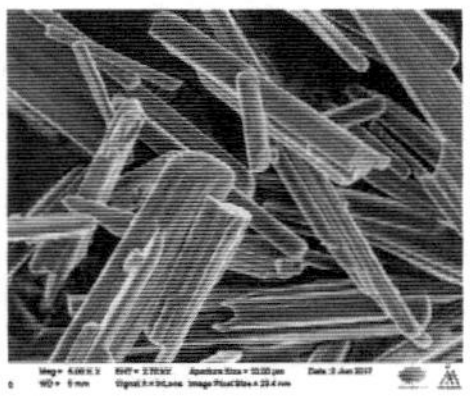

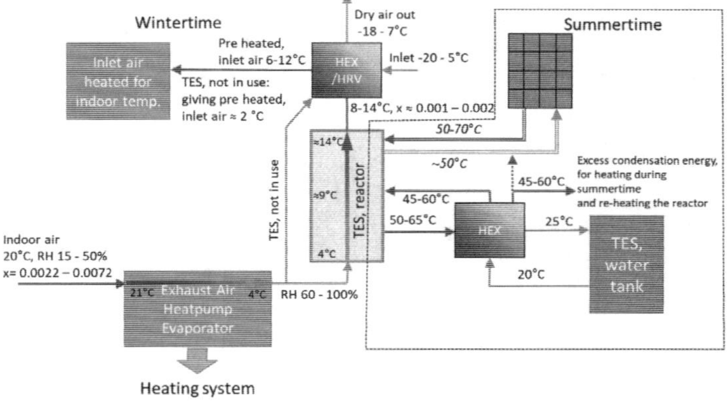

Scheme of thermal energy storage (TES) reactor combined with an auxiliary heat pump.
Single-lined arrows contain gas flows, double lined arrows fluid flows.

doi:10.3390/en11010170 open source

Conclusions

- CO_2 mineral sequestration offers a leakage-free alternative for CO_2 underground storage worldwide; for Finland the only option – may become essential for BECCS?
- CO_2 capture from oxygen containing gases isn't "taking off": capture is more expensive than economically viable CCS! → remove capture step from CCS chain by operating mineralisation directly on flue gas
- ÅA staged process route: ~ 80 % Mg extraction from serpentinite + fast (~10 min) $Mg(OH)_2$ carbonation ~70%, the wet route gives carbonation of $MgSO_4$ up to 88%
- ÅA Route I gives better energetic performance, while Route II has lower environmental footprint. Important: CO_2 concentration in flue gas, and waste heat.
- Depending on silicate rock, ABS may give better Mg extraction than AS
- Electrodialysis using ion-selective membranes allows for separating AS from ABS
- Hydrating / dehydrating nesquehonite, mixed with silica gel gives an interesting low temperature TES application with ~ 1MJ/kg storage capacity

Acknowledgements

- *Cleen / Clic CCSP (2011 - 2016)*
- *Academy of Finland graduate school for Energy Engineering and Systems (2012 - 2015)*
- *Ella & Georg Ehrnrooth foundation (2016 -)*
- *Walter & Lisi Wahl foundation (2016 -)*
- *Fortum foundation (2017 - 2018)*
- *Erkki Paasikivi foundation (2016 -)*
- *K V Lindholm foundation (2017 -)*
- *Foundation of Åbo Akademi Univ. (2017)*

Comparison two routes & two processes:

a lime kiln and a NG power plant: exergy analysis

	Serpentinite rock in, kg/h (T °C)	*Flue gas in, kg/h (T °C)*	*Mg-product out *, kg/h (T°C)*	*Flue gas out, kg/h (T °C)*	*CO_2 bound, kg/h*	*Exergy input heat kW*	*Exergy input power kW*	*Exergy input heat, MJ/kg CO_2 bound*	*Exergy input power, MJ/kg CO_2 bound*
Dry carbonation, lime kiln	550 (15)	620 (500)	386 (150)	489 (92)	187	135 $	46	2.6 §	0.89
Dry carbonation, natural gas PP (400 bar)	550 (15)	2427.4 500	385.8 (150)	2240.6 (18)	187	691 #	439	11.44 #	8.27
Wet carbonation, lime kiln	550 (15)	620 (500)	353 (30)	400 (30)	190	814 $	0	15.4 $	0
Wet carbonation, natural gas PP	550 (15)	2425.5 (500)	353.2 (30)	2294.2 (30)	189	635 #	0	12.1 #	0

* $MgCO_3$ for dry carbonation, $Mg_5(CO_3)_4(OH)_2 \cdot 4H_2O$ for wet carbonation.

§ This heat can be obtained from 28620 kg/h lime kiln flue gas, lowering the temperature from 500°C to 380°C.

$ This heat can be obtained from 39400 kg/h lime kiln flue gas, lowering the temperature from 500°C to 362°C.

This heat can be obtained from 28500 kg/h flue gas, lowering the temperature of the gas from 500°C to 380°C. (It is assumed that while 2427.4 kg/h flue gas is processed in the carbonation process, a larger flue gas of at least 28500 kg/h is produced by the power plant.)

Heat Q

$$Ex(Q) = Q \cdot \left(1 - \frac{T^{\circ}}{T}\right)$$

Power P

$$Ex(P) = P$$

doi: 10.1016/j.energy.2016.05.066

Formation of Magnesite and Hydromagnesite During Direct Aqueous Carbonation of Thermally Activated Lizardite

Supervisors: Profs. Kennedy and Stockenhuber

Advisors: Drs. Geoff Brent and Mark Rayson

Ammar Abu Fara 13th March 2018

1

Context

Looking for conditions favour magnesite

- Mixing (i.e. Impeller speed)
- Reaction conditions (i.e. T, P, $NaHCO_3$)

Scale up the factors in the pilot plant

2

Feed Characterisation

Lizardite
$Mg_3Si_2O_5(OH)_4$

Phases (Powder XRD):
Lizardite 95%
Clinochlore 4%
Magnetite 1%

Source: Great Serpentinite Belt, NSW, Australia

3

Lizardite Heat Treatment

- To remove chemically bound water to promote carbonation (to disrupt the long range order and create reactive amorphous phase)
- Lizardite was heat activated in an electrically heated rotary kiln at 630 °C for 4 hours.

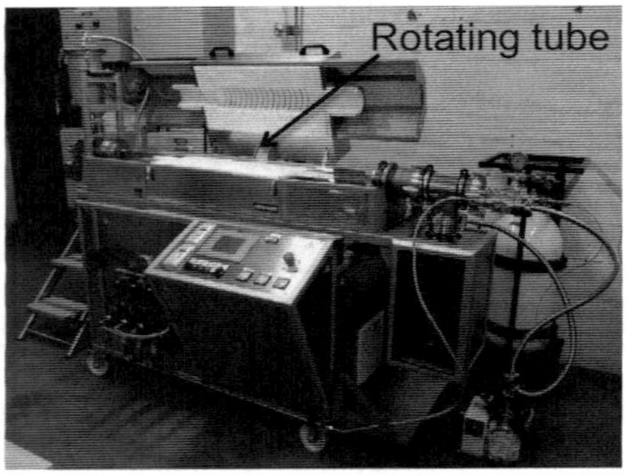

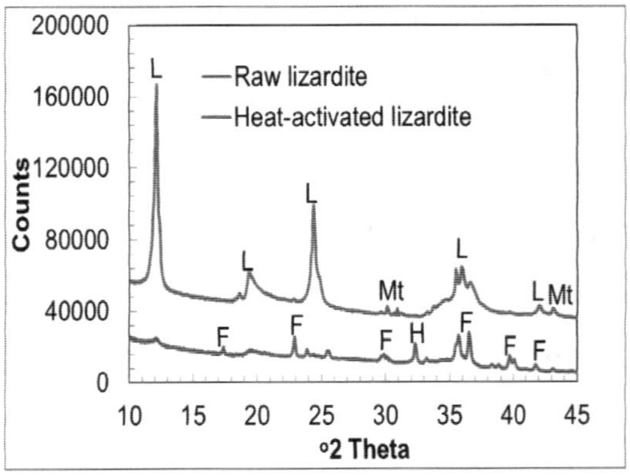

4

Materials and Methods

PARR reactor

Operating conditions (ARC process/No NaCl):

- ❑ T : 150°C, P: 130 bar
- ❑ Impeller speed: 100, 450 & 600 rpm
- ❑ Reaction time: 1 h (sampling t_0, t_{10}, t_{20}, t_{40}, and t_{60} min)
- ❑ Slurry concentration: 15% solid ratio
- ❑ Additive concentration ($NaHCO_3$): 0.64 M.

5

Thermodynamic Modelling

OLIAnalyzer 9.2 was used to simulate the effect of P, T and $NaHCO_3$ on the formation of Mg-carbonate phases

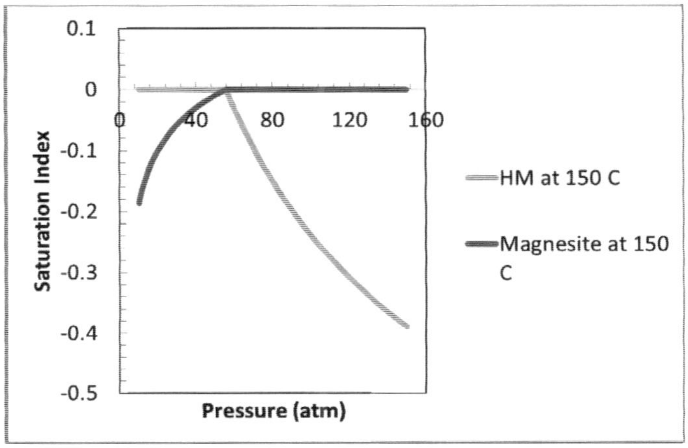

6

Effect of $NaHCO_3$

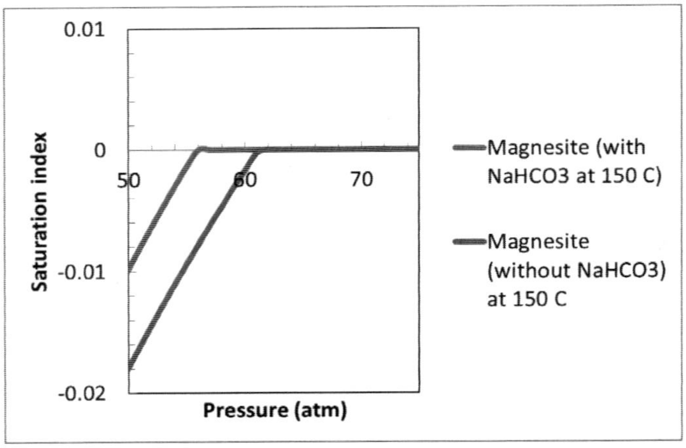

$$K_{sp-M} = [Mg^{2+}].[CO_3^{2-}]$$

$$K_{sp-HM} = [Mg^{2+}]^5.[CO_3^{2-}]^4.[OH^-]$$

▶ 7

Formation of magnesite (desired phase):

$Mg^{2+}_{(aq)} + CO_3^{2-}{}_{(aq)} \rightarrow MgCO_{3(s)}$

Formation of hydromagnesite (less carbonate ions are needed and the stoichiometric ratio of Mg to carbonate ion is 0.8)

$5\ Mg^{2+}{}_{(aq)} + 4CO_3^{2-}{}_{(aq)} + 2OH^-{}_{(aq)} + 4H_2O_{(l)} \leftrightarrow Mg_5(CO_3)_4(OH)_2.4H_2O_{(s)}$

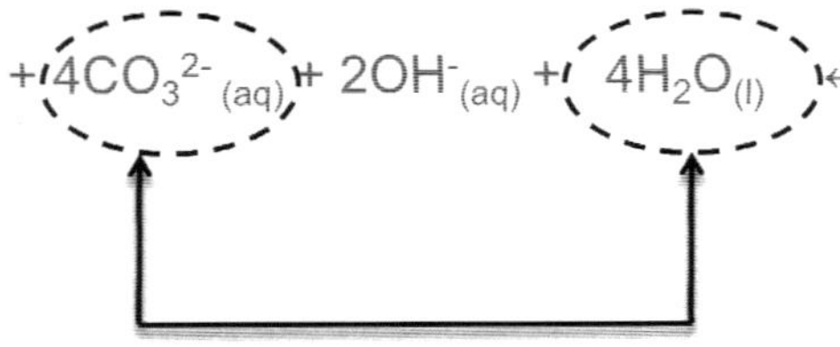

▶ 8

Magnesite is the most thermodynamically stable carbonate phase:

- less process water requirement
- chemical stability suitable for long-term storage
- Maximum CO_2 capacity (1:1)

Hydrated carbonate phases

- higher solubility (less effective for long-term storage of carbon).
- include added weight of water (transportation cost)
- High water demand for the process

9

Analysis Methodology of the Reaction System

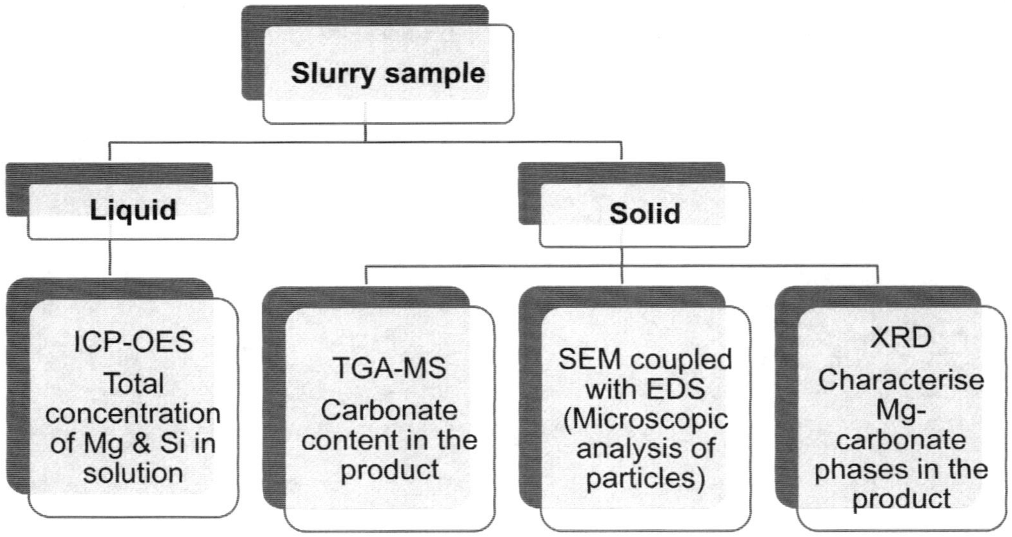

10

Liquid Analysis: ICP-OES Total concentration of Mg & Si in solution

Effect of impellor speeds on Mg concentration:

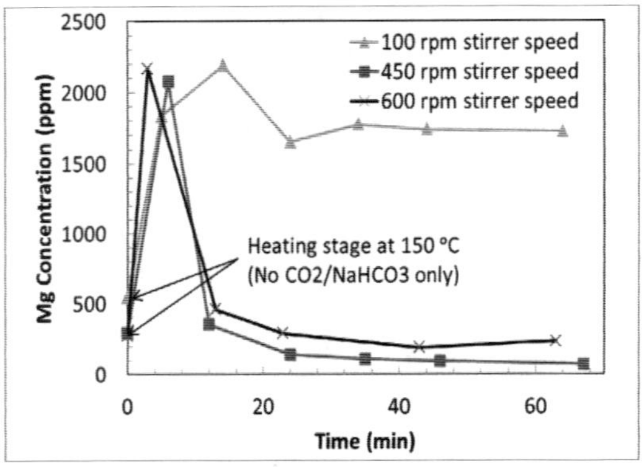

- Rapid extraction initially and Mg concentration reaches the supersaturation
- Dramatically decreases to a steady-state concentration in solution (100-200 ppm) (precipitation)
- Small extraction at the heating stage (no CO_2).
- Higher Mg concentration at the minimum impeller speed (100 rpm) (less precipitation??)

11

Effect of impellor speeds on Si concentration:

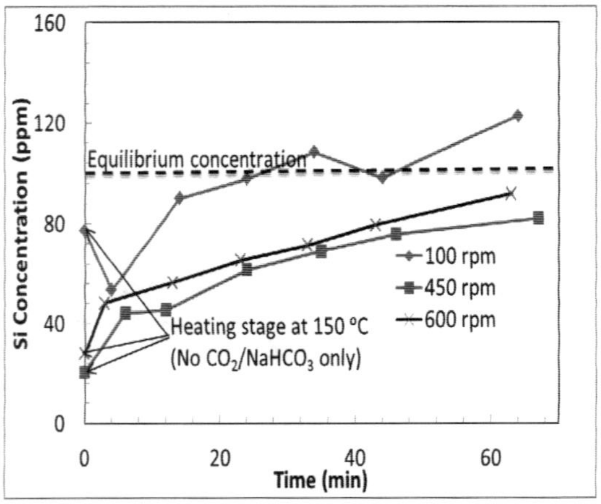

- Rapid rate of extraction of Si from the heat activated mineral initially (from t=0 to10 min).
- Si reaches the saturation limit for amorphous Si at the end of the reaction
- Small Si extraction at the heating stage (no CO_2).

12

Effect of different Impeller positions on Mg concentration :

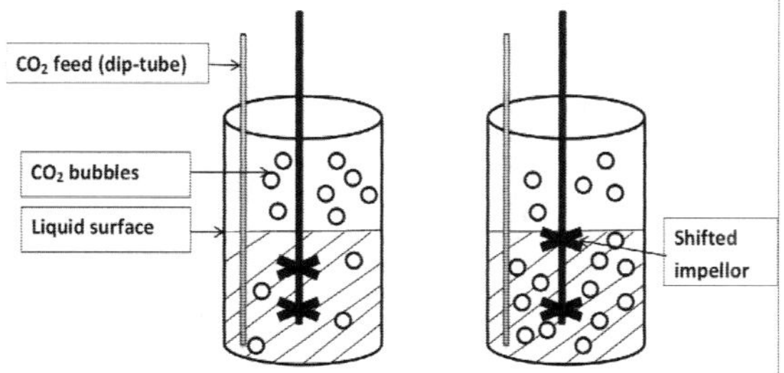

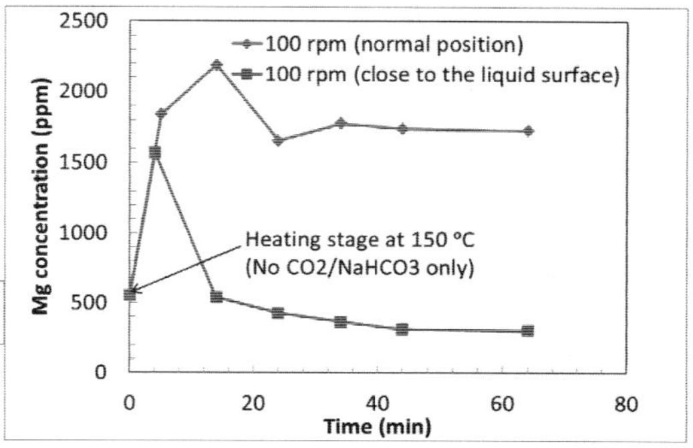

- The top impeller was located just below the surface of the slurry in an effort to increase the gas dispersion in the liquid and enhance the mass transfer of the reacting solution.

13

Effect of presence and absence of $NaHCO_3$ on Mg concentration:

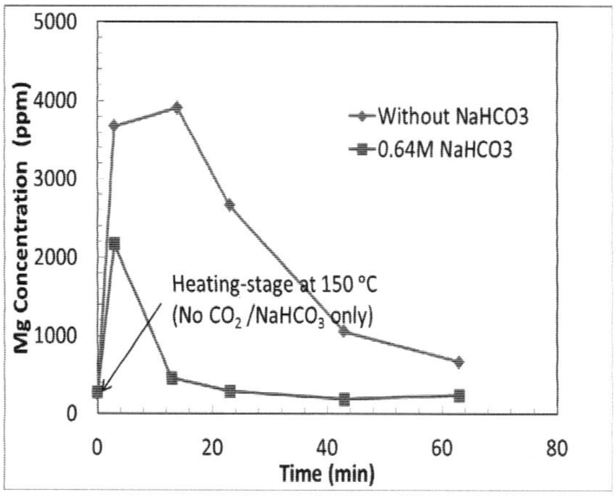

- Mg concentration was lower in the carbonation experiments undertaken with $NaHCO_3$, which indicates enhanced precipitation of Mg-carbonate phase.

- The presence of $NaHCO_3$ increased the concentration of carbonic ion (CO_3^{2-})

$$K_{sp-M} = [Mg^{2+}].[CO_3{}^{2-}]$$

14

Carbonate content in the product (TGA-MS)

- ❑ TGA-MS : Thermogravimetric and Mass Spectroscopic Analysis
- ❑ Analysis process (TGA-MS): the sample heated up to 150°C, then stable at 150 °C during 30 min, then ramp temperature up to 1000°C at 5°C/min
- ❑ Yield is measured based on the decomposition of magnesite

$$MgCO_3 \longrightarrow MgO + CO_2$$

Different Impellor speeds: Samples after 60 min

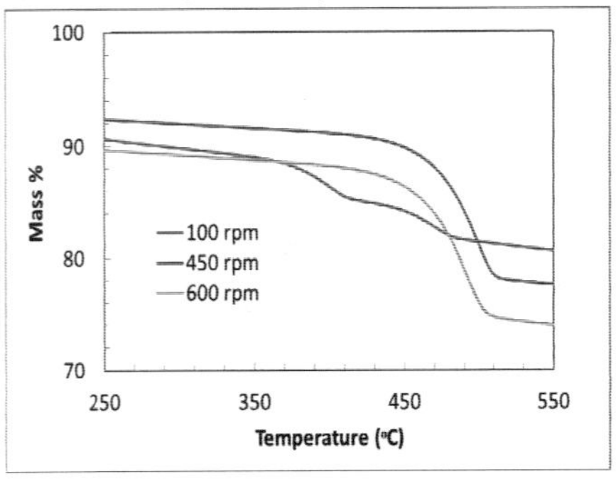

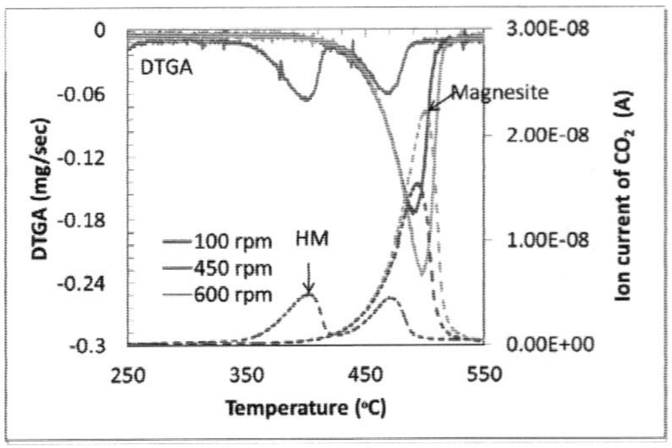

MS comparison between different Impeller positions:

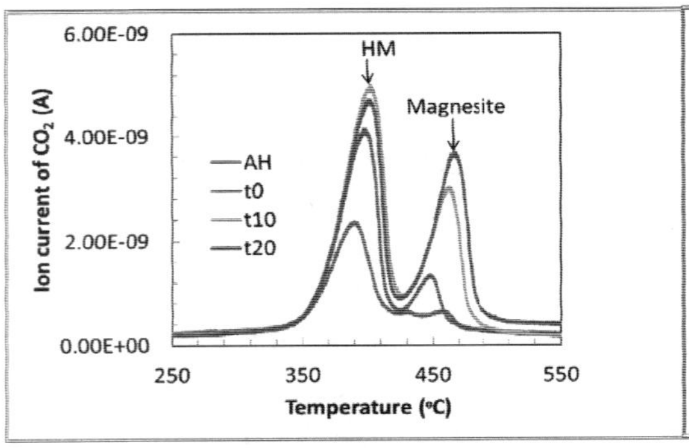

(a) normal impeller position

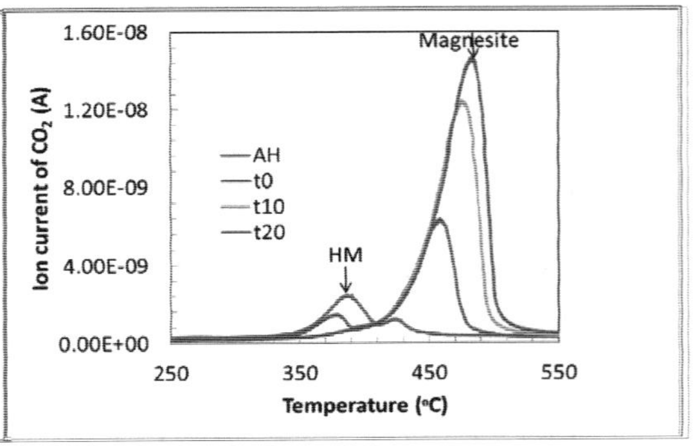

(b) shifted impeller position close to the liquid surface (high gas dispersion/promoted mass transfer)

MS comparison between presence and absence of $NaHCO_3$:

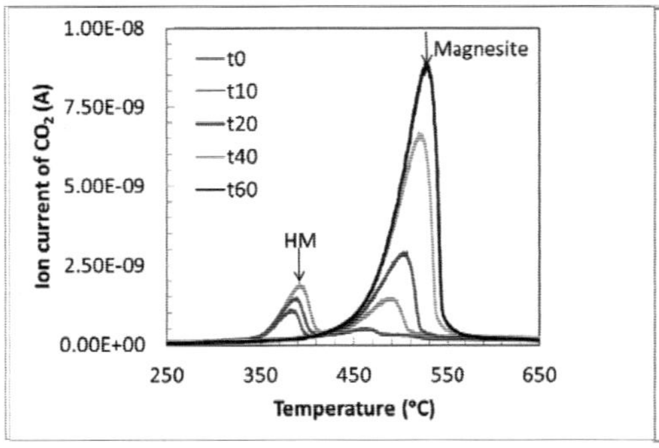

No $NaHCO_3$

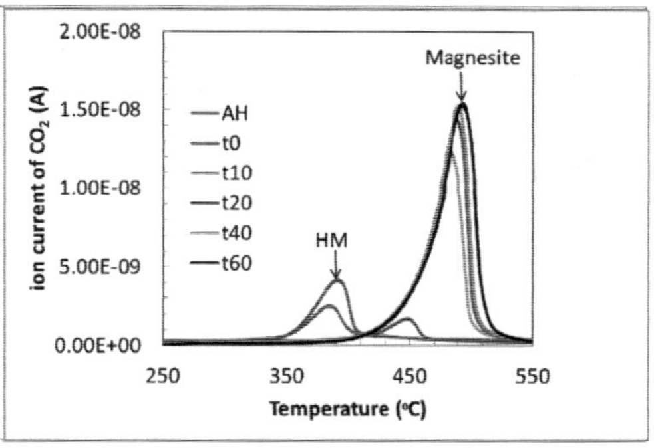

0.64M $NaHCO_3$

Direct aqueous carbonation: study of dissolution/precipitation kinetics

- ❑ Yield: magnesium in the feed converted into magnesium in the carbonate

Total yield% for 100 rpm impeller speed:

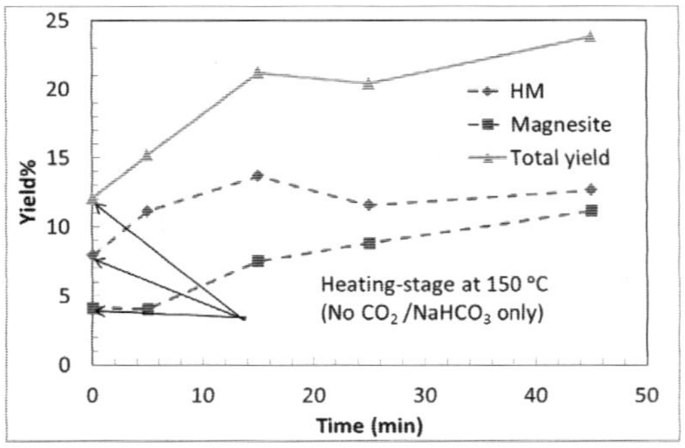

- ➢ magnesite and hydromagnesite co-existed
- ➢ Hydromagnesite is the most favoured phase
- ➢ Maximum yield was 24% (total)

Total yield% for 450 rpm impeller speed:

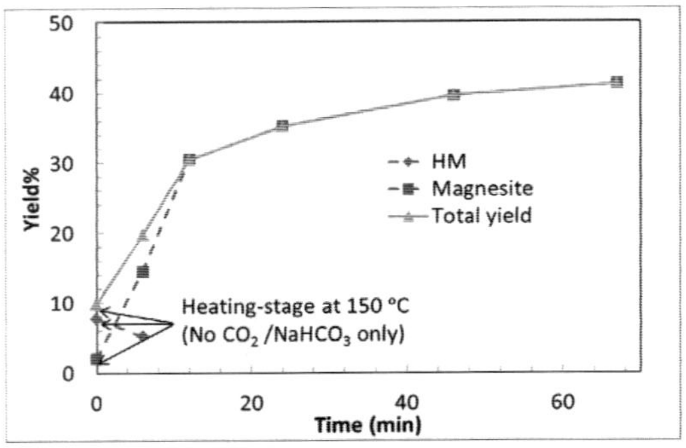

- Hydromagnesite co-existed only at AH sample and at time zero (t_0)
- Then magnesite is the favoured phase.
- Maximum yield was 41% (total)

Total yield% for 600 rpm impeller speed:

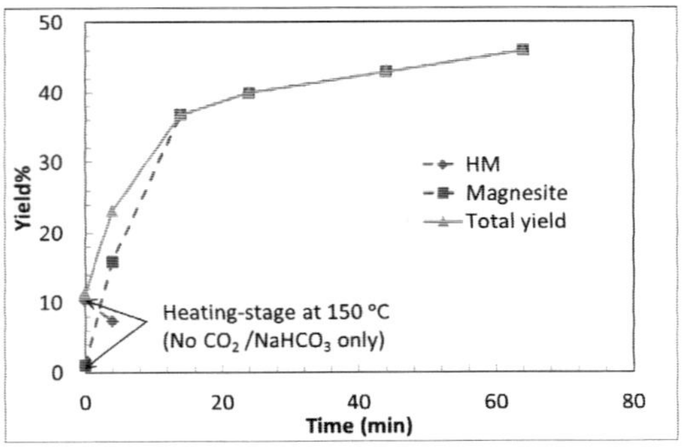

- Hydromagnesite at AH sample and at time zero (t_0)
- Then magnesite is the favoured phase.
- Maximum yield was 46% (total)

Total yield% comparison for all impeller speeds:

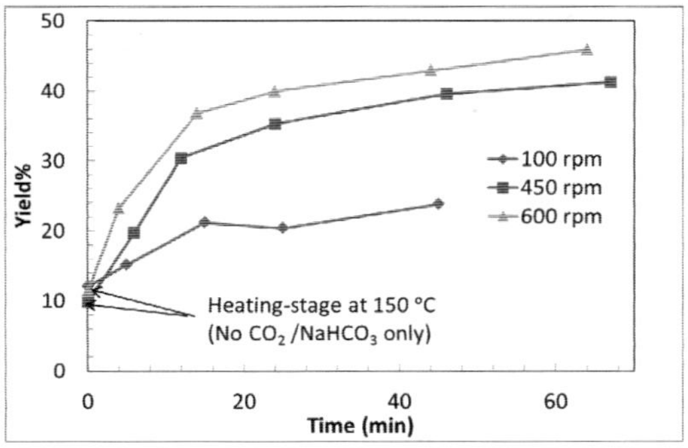

- The reaction is rapid for the first 20 min, achieving 40% yield (for 600 rpm experiment), after which point the reaction rate declines dramatically.
- Same yield% (12%) before introduction CO_2 to the system for all impeller speeds.

Total yield% comparison for different impeller positions:

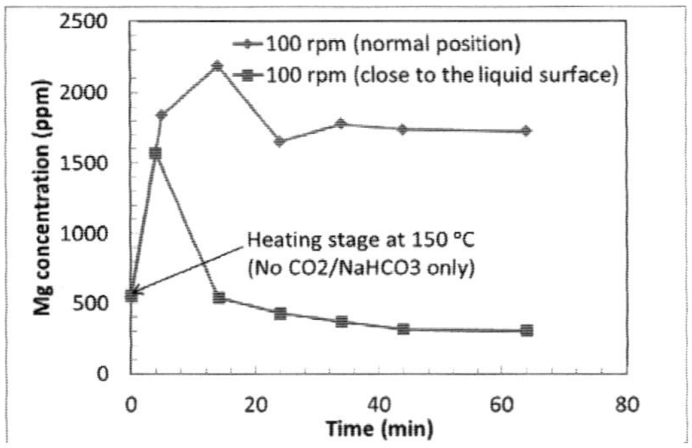

- The yield calculations based on the normal and new impellor position indicates a low precipitation compared with shifted impeller experiment with high vorticity.
- Referring to the ICP results, the concentration of Mg in solution at the standard impeller position achieved higher concentrations of Mg in solution.

Total yield% comparison for presence and absence of $NaHCO_3$:

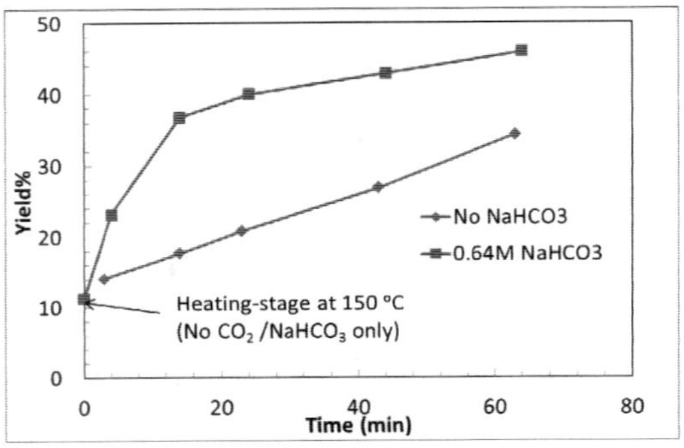

- The reaction rate with $NaHCO_3$ was higher than that for the experiment without $NaHCO_{3,}$
- A greater yield was achieved after 60 min (46%) compared to the experiment with no $NaHCO_3$ (34%).
- The yield due to $NaHCO_3$ was 12% (26% of the total yield).

23

Carbonate characterisation by XRD

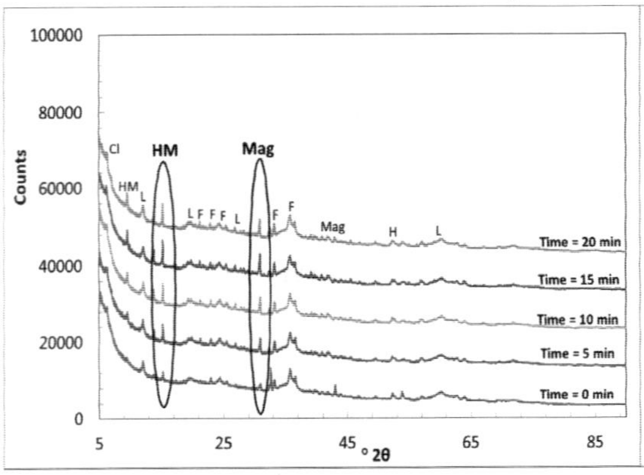

(a) No $NaHCO_3$

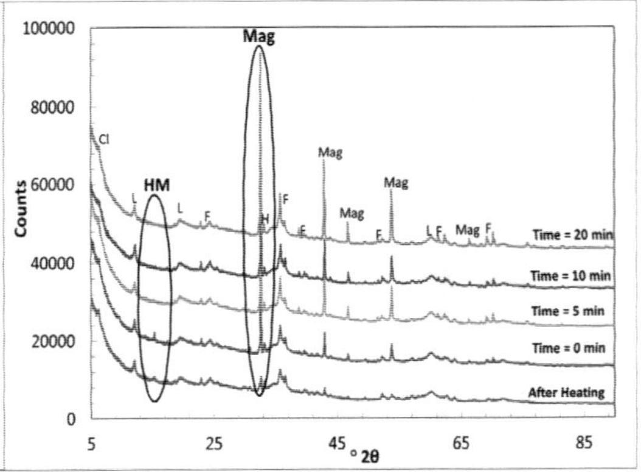

(a) 0.64M $NaHCO_3$

24

Microscopic analysis and direct observation for particles and their surfaces (SEM-EDS)

❑ SEM-EDS observation of the carbonated samples

(a) No $NaHCO_3$(after 20 min)

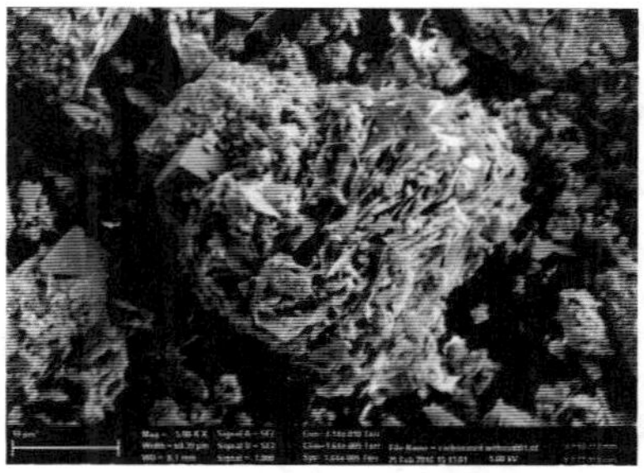

(b) 0.64M $NaHCO_3$(after heating/ no CO_2)

Conclusions

- Mass transfer will impact the results (increasing the stirring speed and selection of the impeller and its position to increase the gas dispersion in the liquid)
- $NaHCO_3$ generates an elevated concentration of carbonate ions which reacts with Mg cations with ratio 1:1 to produce magnesite instead of hydromagnesite which requires a lower carbonate to magnesium ratio (0.8).
- In the absence of $NaHCO_3$, hydromagnesite was grown with time and increased in parallel with magnesite.
- A significant contribution of $NaHCO_3$ on the product yield (up to 12%).

Acknowledgements

I would like to sincerely acknowledge:

- The University of Newcastle for post graduate scholarship.
- Funding from MCi;

http://mineralcarbonation.com

27

Thank You!

Questions/comments welcome

28

Development of Grinding Media for Aqueous Mineral Carbonation Applications

MUHAMMAD IMRAN RASHID (Presenter)

Co-authors:

E. Benhelal, L. Anderberg, F. Farhang, T.K. Oliver, M. Rayson. G. Brent, M. Stockenhuber, E. Kennedy

THE UNIVERSITY OF NEWCASTLE AUSTRALIA

ORICA

MCi Mineral Carbonation International

Aim

2

- Carbonate raw feedstocks (dunite rock, olivine, lizardite) without heat-activation
- Feedstocks
- Dunite, lizardite are collected from GSB, NSW, Australia
- Olivine is imported from Sibelco (Aheim plant), Norway

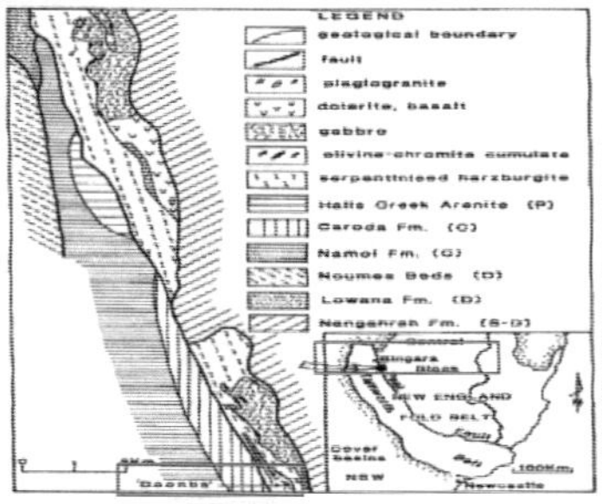

K. Yang, Canadian Mineralogist, 1993.

THE UNIVERSITY OF NEWCASTLE AUSTRALIA

Dunite Rock

3

- Dunite is source of magnesium (26% magnesium)
- Dunite is a mixture of lizardite[1] ,olivine, brucite

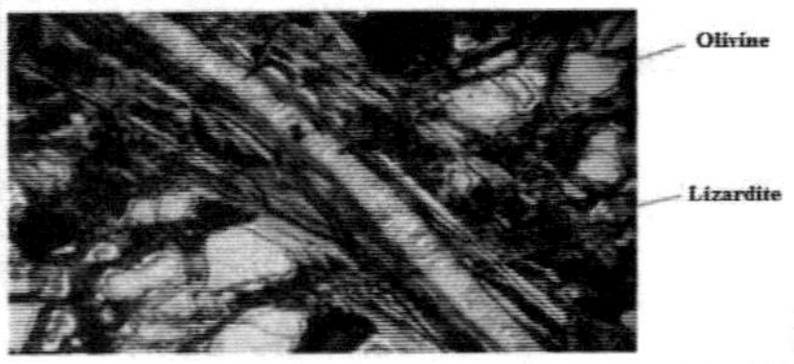

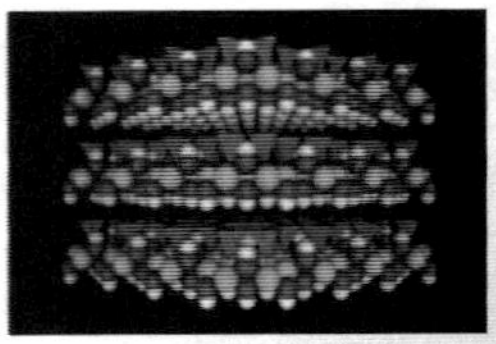

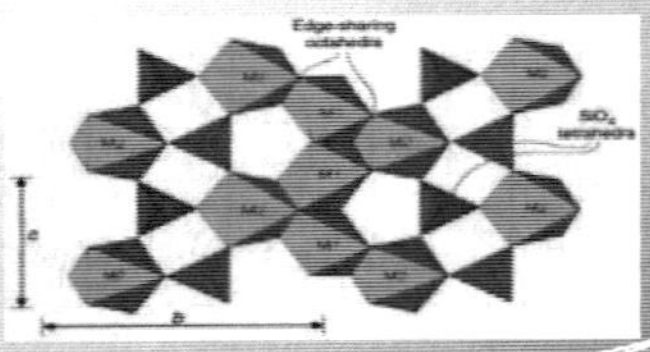

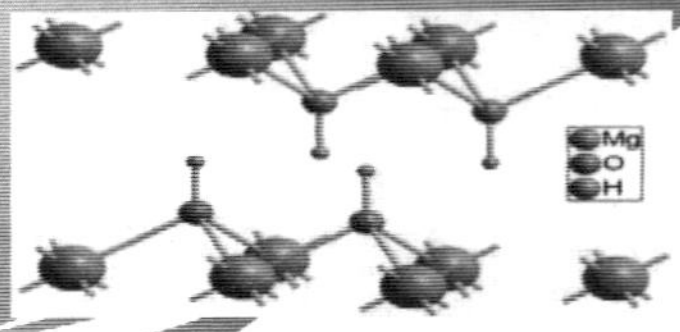

[1] Wicks. J. Fred, powder diffraction, 2000.

THE UNIVERSITY OF NEWCASTLE AUSTRALIA

Literature review (concurrent grinding)

4

Feed material	Grinding media	Media size	Yield (%)	Reference
Olivine sub 38 µm	Quartz	Sub 38 µm	68→84%	A.V.G. Chizmeshya et.al, 2007.
Olivine Harzburgite sub 100 µm	NA	NA	80% (24 h) 70% (24 h)	Bodenan et.al, 2014.
Olivine Sub 100 µm	Alumina Zirconia	1.3 mm 1 mm	80% (24 h) 35% (5 h) 38% (24 h), two-step	Julcour et.al, 2015.

Timescale (24 h), Feed PSD (fines), feed heat-treatment, Media size

THE UNIVERSITY OF NEWCASTLE AUSTRALIA

Dunite and olivine batches preparation

5

- Dunite and olivine are ground for 1-2 h
- Wet sieving performed for 1 h to remove fines
- Residual fines removed in ultra-sonic bath for 5 min

Grinding mill

Wet-sieving apparatus

Ultra-sonic bath

THE UNIVERSITY OF NEWCASTLE AUSTRALIA

Dunite batches PSD

6

- 20-45 µm dunite and 45-75 µm dunite batches PSD

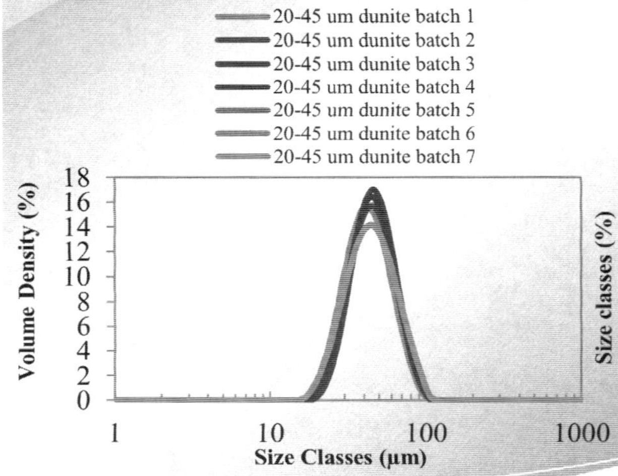

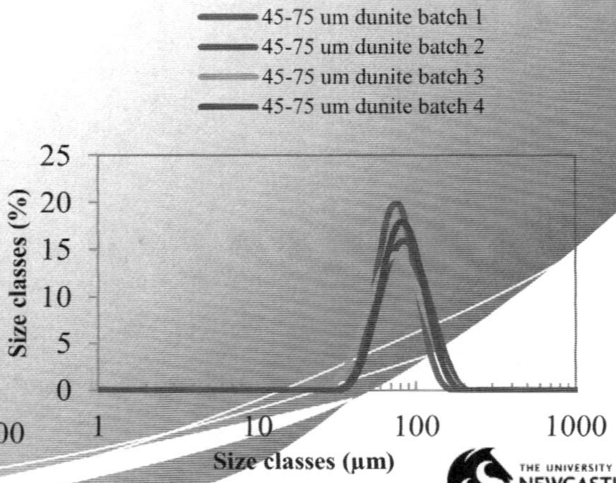

Fines are absent

THE UNIVERSITY OF NEWCASTLE AUSTRALIA

Dunite batches SEM

7

- 20-45 µm dunite and 45-75 µm dunite batches SEM

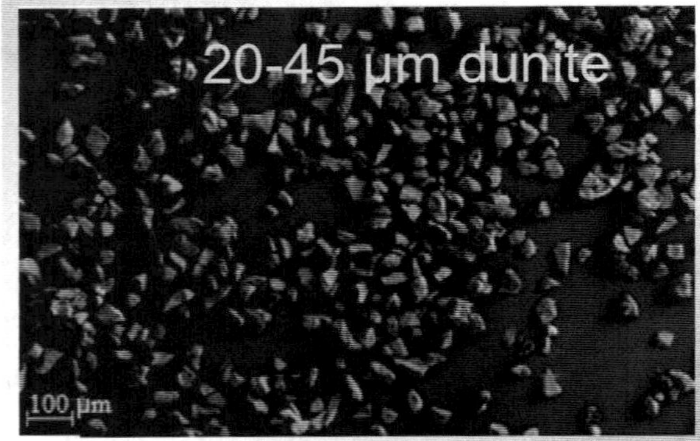

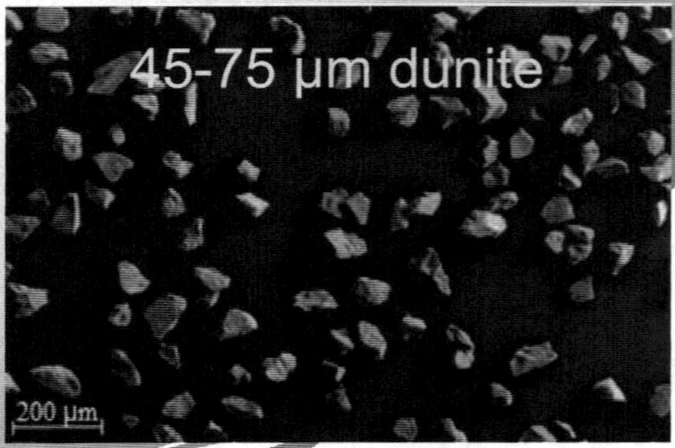

Fines are absent

THE UNIVERSITY OF NEWCASTLE AUSTRALIA

Olivine batches PSD

8

- 20-45 µm olivine and 45-75 µm olivine batches PSD

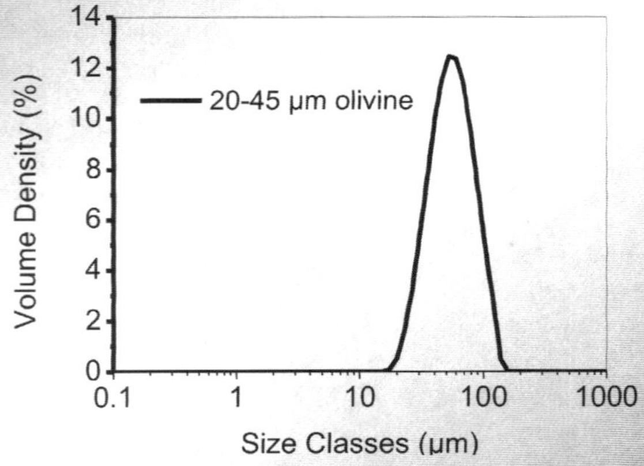

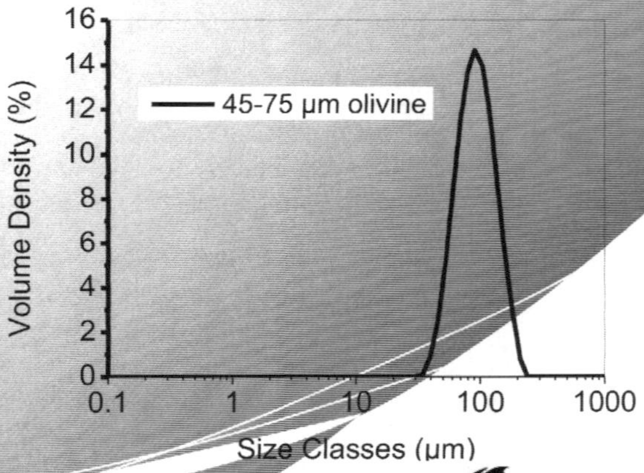

Fines are absent

THE UNIVERSITY OF NEWCASTLE AUSTRALIA

Olivine batches SEM

- 20-45 µm olivine and 45-75 µm olivine batches SEM

Fines are absent

Feed characterisation (TGA-MS)

- 20-45 µm dunite (7-8 mg) is heated in TGA furnace

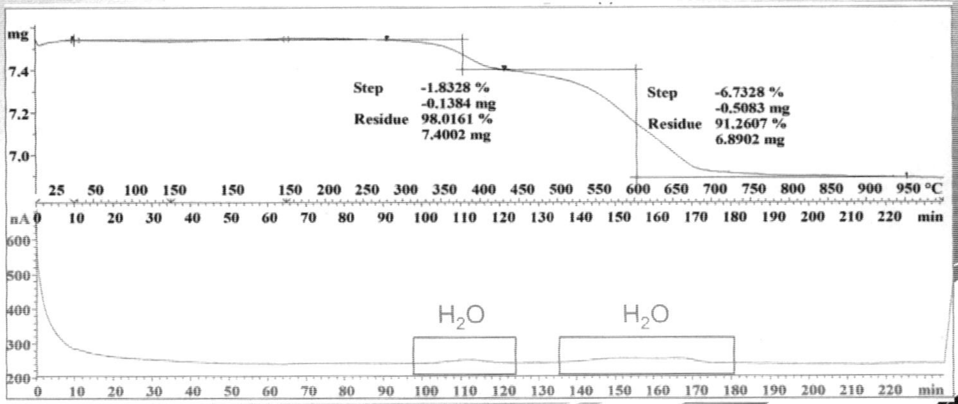

Brucite and lizardite dehydroxylation

Semi-quantitative PXRD Analysis

11

- Add silicon standard in solid sample
- Calculate peak areas from XRD pattern

$$C_i = A_x I_i RIR_x / (I_x RIR_i) \quad (1)$$

C_i=mass percentage of i phase

A_x=known amount of internal standard (silicon) added to original sample

I_i,I_x=Intensities (peak areas) of phases I and X

RIR_I, RIR_x= reference intensity ratio values of I and X respectively

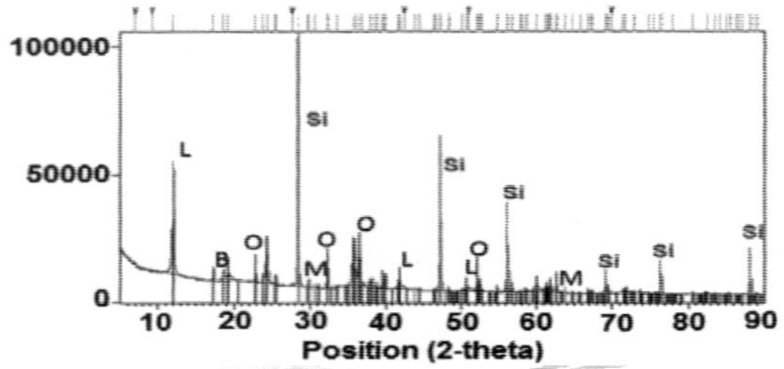

L=Lizardite B=Brucite O=Olivine
M=Magnetite Si=Silicon

THE UNIVERSITY OF NEWCASTLE AUSTRALIA

Dunite batches characterisation

12

- 20-45 μm dunite and 45-75 μm dunite are analysed by semi-quantitative XRD and TGA-MS

	Lizardite (%)	Olivine (%)	Brucite (%)	Magnetite (%)
20-45 μm dunite QXRD	52	42	5.4	0.46
20-45 μm dunite TGA-MS	52	42 (difference)	5.9	0.46 (XRD)
45-75 μm dunite QXRD	43	51	5.6	0.35
45-75 μm dunite TGA-MS	43	51 (difference)	5.5	0.35 (XRD)

THE UNIVERSITY OF NEWCASTLE AUSTRALIA

Carbonation test; Reactor setup

13

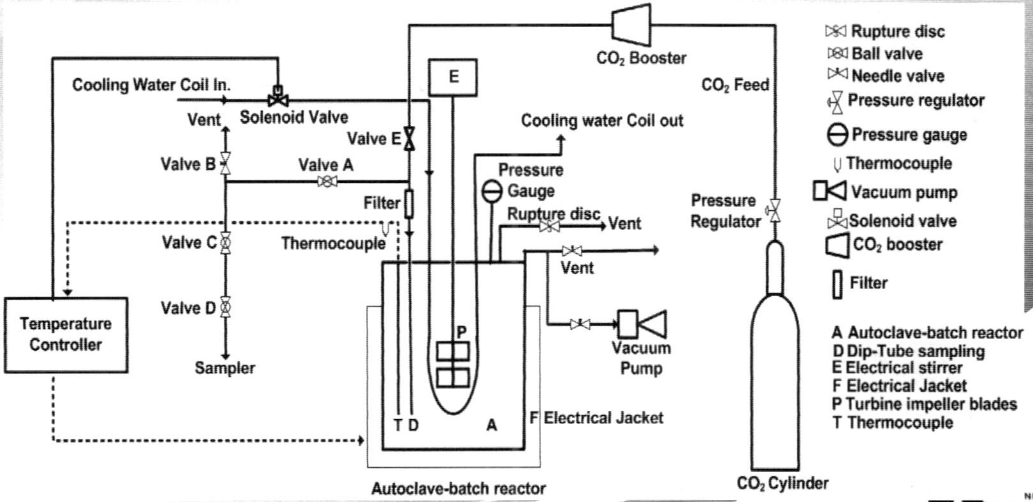

Calibration of Mass Spectrometer

14

- Sodium bicarbonate samples (different masses) were run in TGA coupled with MS from 0-300 °C
- Area under CO_2 peak and CO_2 mass was calculated.

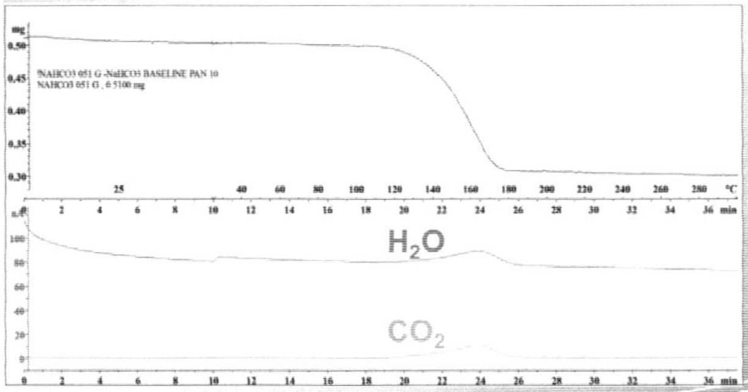

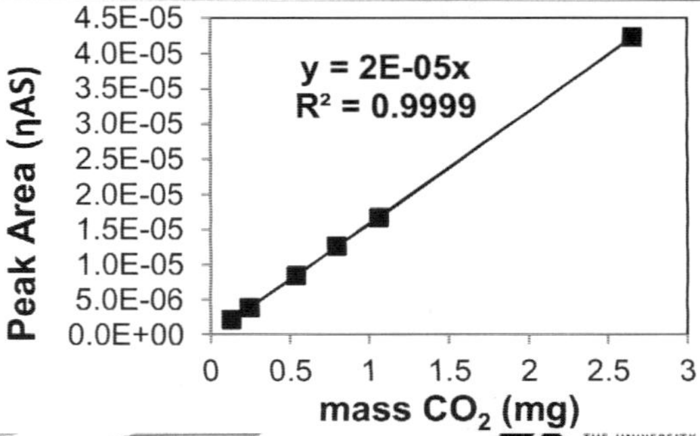

$2NaHCO_3 \rightarrow Na_2CO_3 + CO_2 + H_2O$

Magnesite ($MgCO_3$) yield measurement

15

- A product sample (7-8 mg) was heated 25-1000 °C in TGA
- Yield is amount of magnesium converted into magnesium carbonate

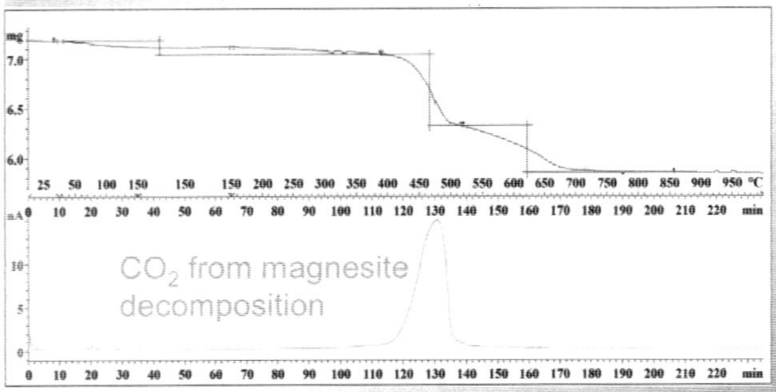

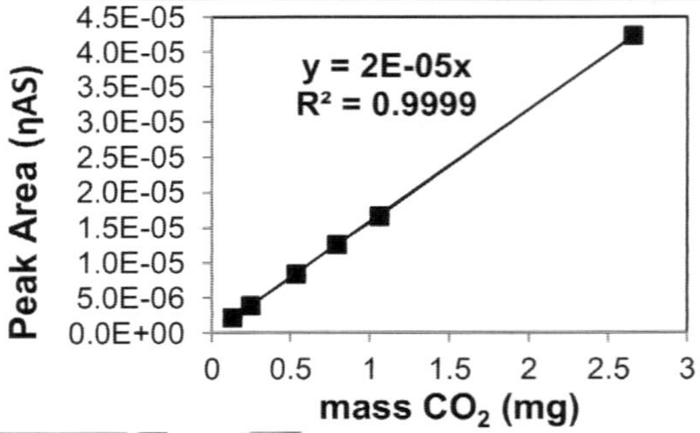

Yield is calculated using calibration curve

THE UNIVERSITY OF NEWCASTLE AUSTRALIA

Media efficacy during carbonation at constant volume

16

- 1 mm zirconia, 1 mm stainless steel and 1 mm alumina were compared based on constant media volume (10 cm^3)

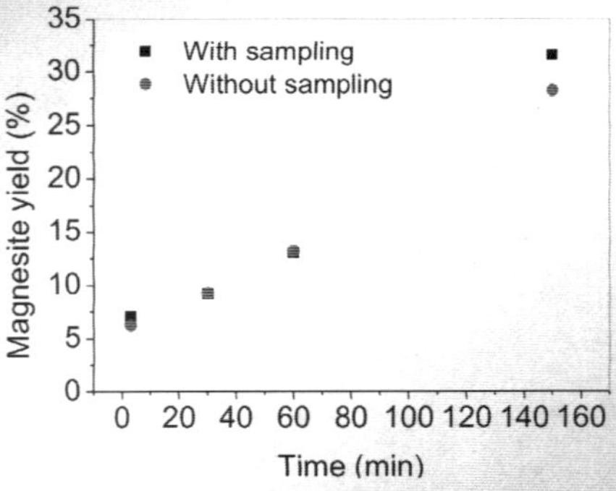

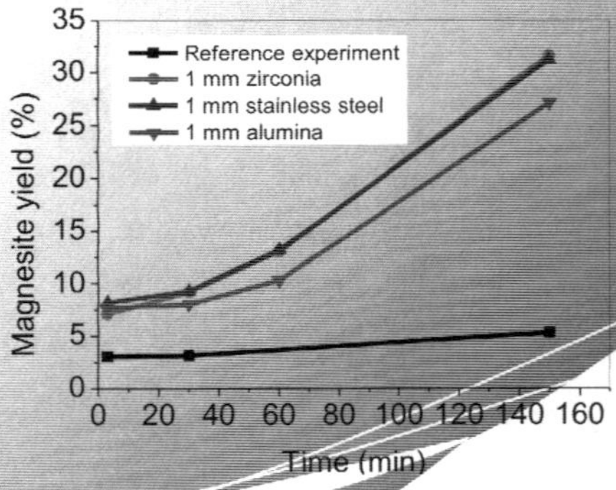

1 mm alumina is least effective

Sampling has negligible effect on yields

THE UNIVERSITY OF NEWCASTLE AUSTRALIA

Carbonation yield with grinding

17

- Statistical experimental design performed using Placket-Burman technique, 2.5 hr reaction (1 mm zirconia media)

Experiment	Slurry conc. (%)	Grinding media conc. (%)	Temperature (°C)	Pressure (bar)	Yield (%)
Reference	15	0	180	130	5.3
Final condition	15	50	180	130	32

Variable at Extreme	Yield
Grinding media conc. (50%)	6.3
Slurry conc. (30%)	-5
Pressure (130 bar)	2.3
Temperature (180 °C)	0.66

50% media, 15% solids optimum

Grinding media conc.>Slurry conc.>Pressure > Temperature

THE UNIVERSITY OF NEWCASTLE AUSTRALIA

Raw dunite carbonation (8 h)

18

- Reference and concurrent ground (1 mm zirconia) experiments performed for 8 hrs, 130 bar, 180 °C, 15% solids

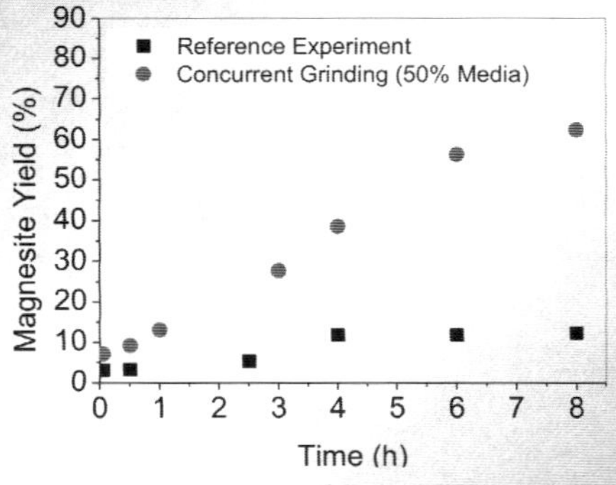

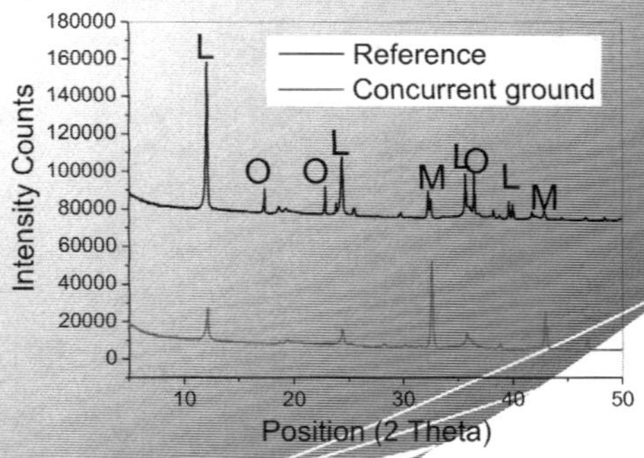

Olivine completely consumed
Lizardite partially consumed

62% yield obtained with raw dunite rock

THE UNIVERSITY OF NEWCASTLE AUSTRALIA

Concurrent (*in situ*) versus *In operando* grinding

19

- Concurrent (*in situ*) means grinding and carbonation simultaneously in presence of CO_2
- *In operando* means grinding in absence of CO_2 and carbonation

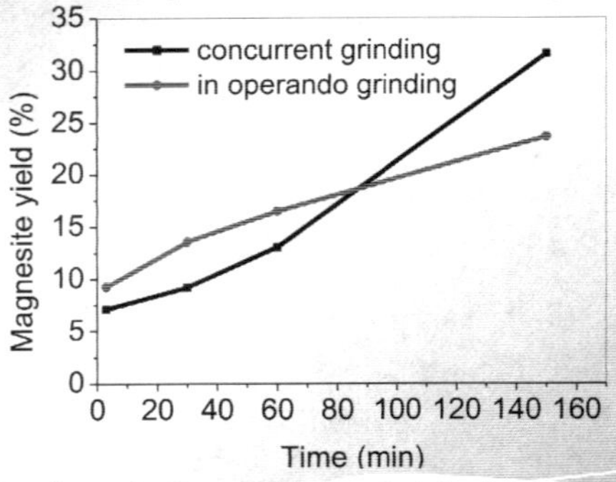

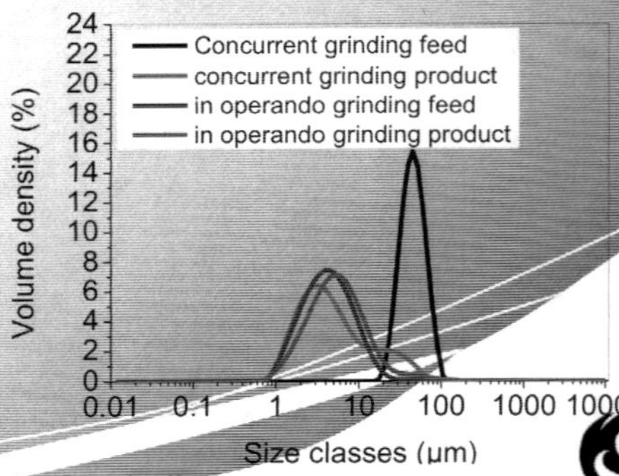

In situ grinding yield is slightly higher than *in operando* grinding

THE UNIVERSITY OF NEWCASTLE AUSTRALIA

Concurrent (*in situ*) versus *in operando* grinding

20

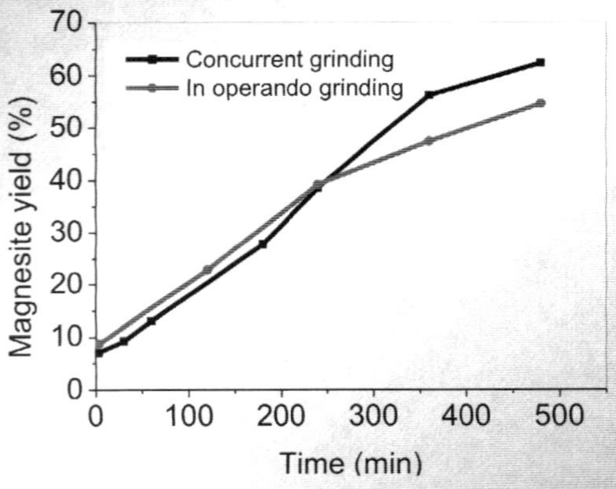

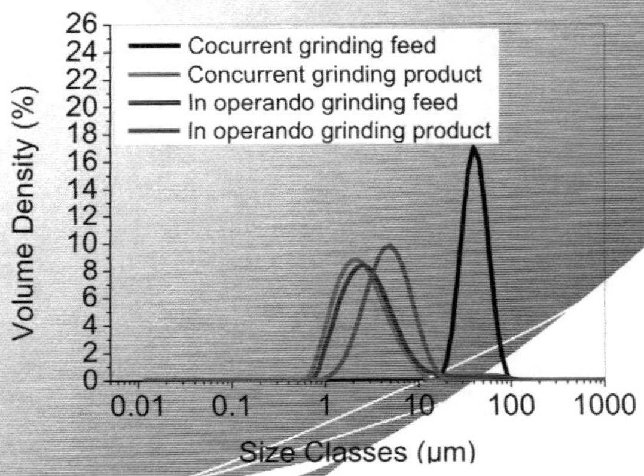

In situ grinding yield is slightly higher than in operando grinding

THE UNIVERSITY OF NEWCASTLE AUSTRALIA

Carbonation using 45-75 μm raw dunite

21

- Carbonation performed at 130 bar, 180 °C, 15% solids, 50% media (for concurrent ground)

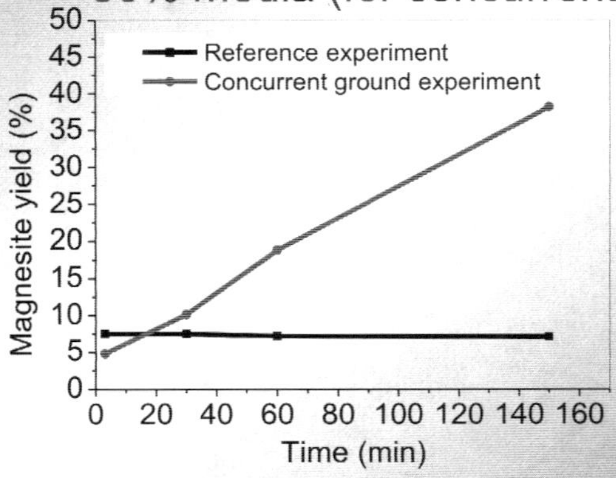

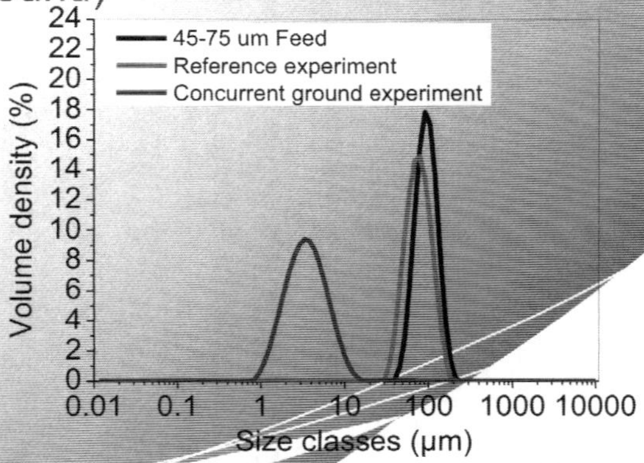

Reference experiment yield 7%
Concurrent grinding yield 38%

THE UNIVERSITY OF NEWCASTLE AUSTRALIA

Carbonation using 20-45 μm olivine

22

- Carbonation performed at 130 bar, 180 °C, 15% solids, 50% media (for concurrent ground)

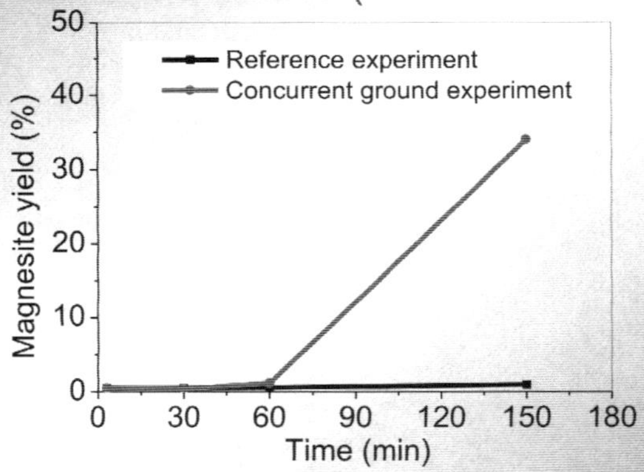

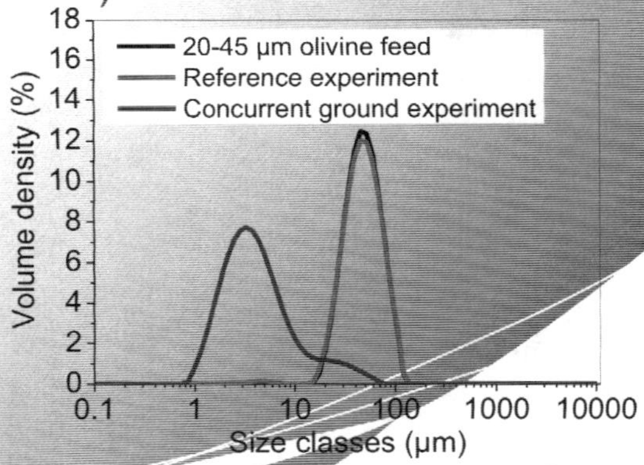

Reference experiment yield 1%
Concurrent grinding yield 34%

THE UNIVERSITY OF NEWCASTLE AUSTRALIA

Carbonation using sub 75 μm raw lizardite

23

- Carbonation performed at 130 bar, 180 °C, 15% solids, 50% media (for concurrent ground)

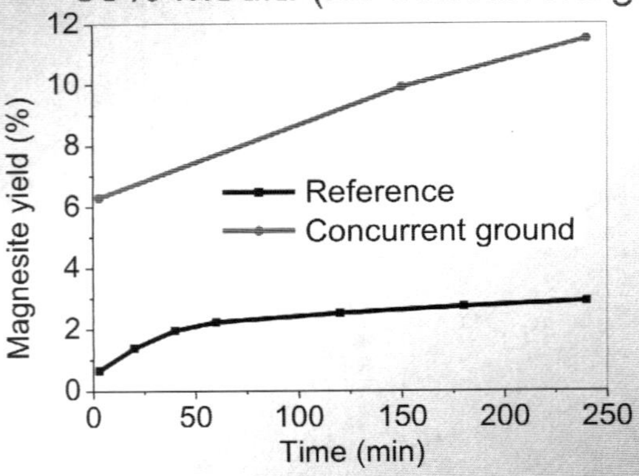

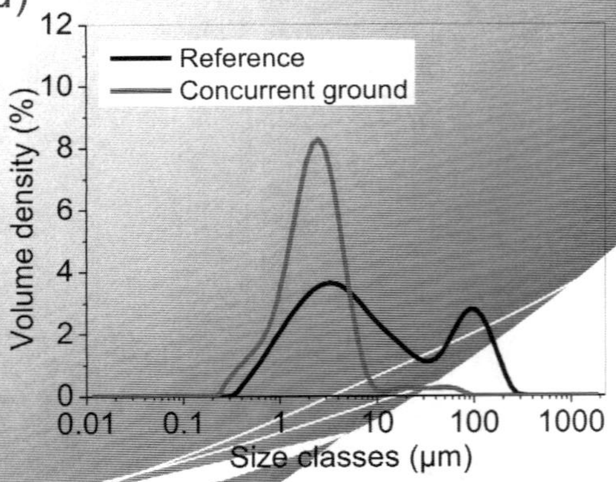

Reference experiment yield 3%
Concurrent grinding yield 12%

THE UNIVERSITY OF NEWCASTLE AUSTRALIA

Conclusions

24

- Alternative feedstocks (e.g. dunite, olivine, lizardite) can be utilised in mineral carbonation
- Magnesite yields 5-6 times higher when using grinding media
- 62% magnesite yield obtained using raw dunite (without heat-activation)
- Concurrent (*in situ*) grinding is better than *in operando* grinding
- Important variables, grinding media concentration>solid concentration>pressure>temperature

THE UNIVERSITY OF NEWCASTLE AUSTRALIA

Acknowledgments

- The University of Newcastle for postgraduate research scholarship
- Orica Limited and Mineral Carbonation International for funding
- Staff at EM-X-Ray unit (University of Newcastle) for XRD and SEM/EDS

25

THE UNIVERSITY OF NEWCASTLE AUSTRALIA

26

Thank You

Any Questions?

THE UNIVERSITY OF NEWCASTLE AUSTRALIA

Silica-rich layers formation

- During reference experiment (130 bar, 180 °C,8 h) silica-rich layers formation occurred
- 8.3 µm to 18 µm thick shell was observed

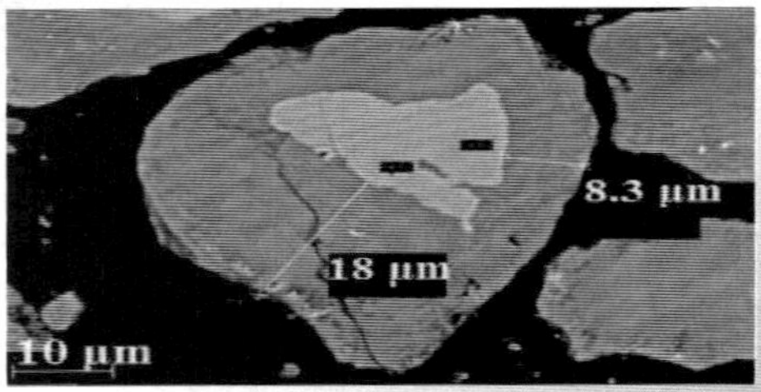

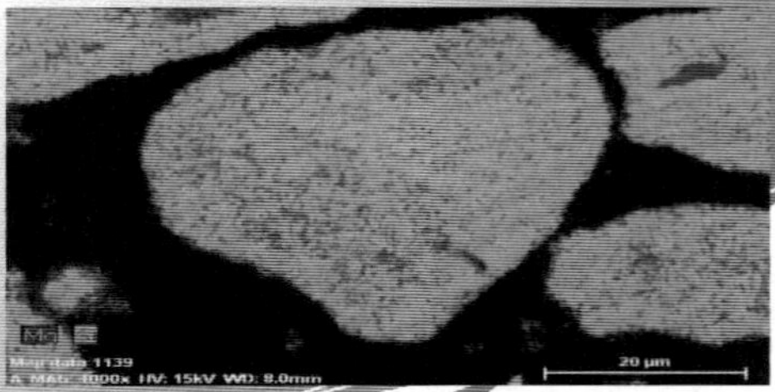

Inner core is rich with magnesium
Outer shell is rich with silicon

THE UNIVERSITY OF NEWCASTLE AUSTRALIA

Cont…

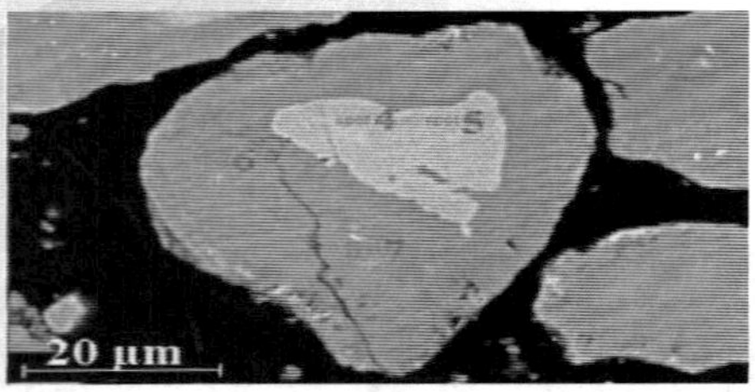

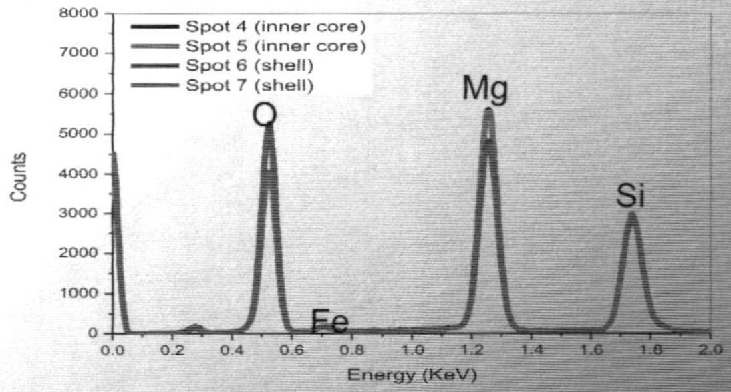

	Mg (wt%)	Si (wt%)	O (wt%)	Fe (wt%)	Mg/Si mass ratio
Spot 4 (Inner core)	31.7	19.6	40.3	8.86	1.61
Spot 5 (Inner core)	31.1	19.6	40.4	8.72	1.58
Spot 6 (shell)	26.6	18.3	47.9	7.23	1.45
Spot 7 (shell)	26.2	18.9	47.1	7.87	1.38

Mg/Si mass ratio low for shell

THE UNIVERSITY OF NEWCASTLE AUSTRALIA

Silica-rich layers disruption

29

- Silica-rich layers were disrupted during concurrent grinding (130 bar, 180 °C, 8 h)

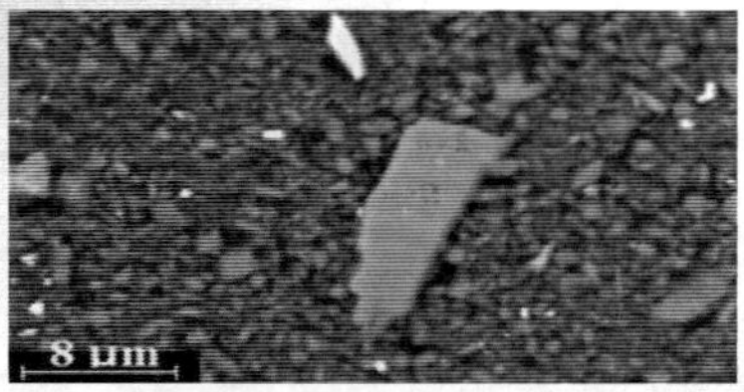

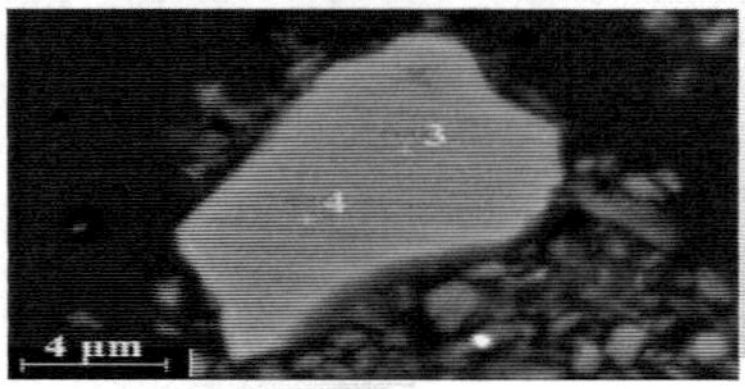

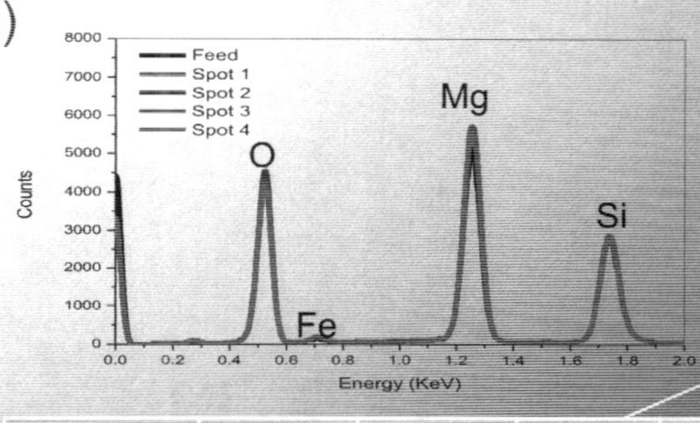

	Mg (wt%)	Si (wt%)	O (wt%)	Fe (wt%)	Mg/Si mass ratio
Feed	28.7	19.5	42.7	9.12	1.47
Spot 1	31.03	18.5	43.5	6.92	1.67
Spot 2	30.9	18.8	43.6	6.62	1.64
Spot 3	31.06	19.3	41.4	8.25	1.61
Spot 4	31.2	19.4	41.3	8.17	1.60

Fragments of disrupted layers

30

- Fragments of disrupted silica-rich layers observed

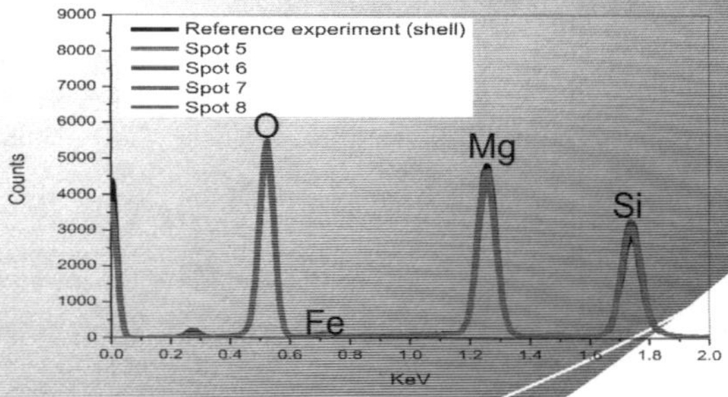

	Mg (wt%)	Si (wt%)	O (wt%)	Fe (wt%)	Mg/Si mass ratio
Reference (shell)	26.6	18.3	47.9	7.23	1.45
Spot 5	24.1	19.8	49.3	6.71	1.21
Spot 6	25.2	21.5	48.7	4.44	1.17
Spot 7	24.8	19.9	50.6	4.59	1.24
Spot 8	25.7	19.9	48.3	6.07	1.29

THE UNIVERSITY OF NEWCASTLE AUSTRALIA

Carbonation using sub 75 µm olivine

31

- Carbonation performed at 130 bar, 180 °C, 15% solids, 50% media (for concurrent ground)

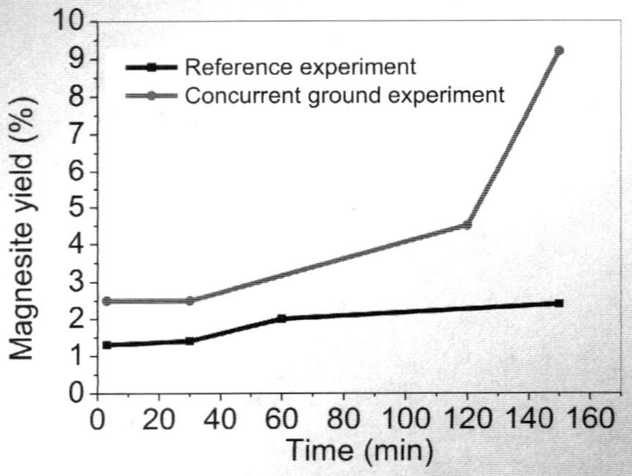

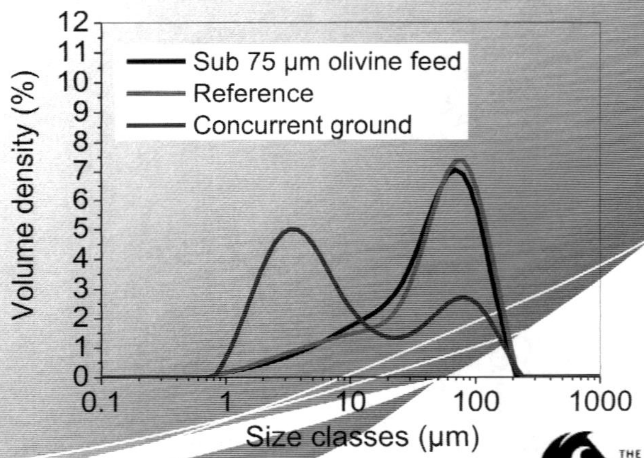

Some of the feed is not finely ground

THE UNIVERSITY OF NEWCASTLE AUSTRALIA

Exploring the Metastability Field for Precipitates Formed from Degassed Serpentinite Leachate

Timothy K. Oliver, Todd W. Hodgins, Mark S. Rayson, Geoff F. Brent, Michael Stockenhuber, Eric M. Kennedy

ACEME 2018

11-14 March, 2018, Newcastle

Mineral Carbonation International

About this research

2

- Desktop study to guide & inform experimental research
- Uses established thermodynamic data + thermodynamic data for a poorly crystalline antigorite phase (*p*-antigorite) evident from research for Mg-Si solutions (Gunnarsson et al., 2005)
- PHREEQC (Version 2.18) thermodynamic software (Parkhurst and Appelo, 1999) using the wateq4f thermodynamic database available in PHREEQC

Degassed serpentinite leachate

3

- CO_2 degassing = *controlled removal of CO_2 from solutions*
- Degassed serpentinite leachate = *Mg bicarbonate solutions, formed from carbonic acid dissolution of serpentinite, with (some) CO_2 removed*
- CO_2 degassing of aqueous mineral bicarbonate solutions ⟶ Inducement of a pH swing of predictable and measurable effect ⟶ supersaturation and precipitation
- Various metastable Mg carbonate phases can form due to slow rate of nucleation and growth of magnesite (e.g. nesquehonite, $MgCO_3 \cdot 3H_2O$ and hydromagnesite $Mg_5(CO_3)_4(OH)_2 \cdot 4H_2O$)

Methodology - Thermodynamic modelling

4

- Investigate the effect of T, solution composition, and the amount of CO_2 degassed from solution
- *Successive degassing of initial solution of given chemical composition & T*
- *Equilibrium-based calculations to assess most probable reactions & resultant solid phases*

Degassed serpentinite leachate

5

- **Initial solution** from the stoichiometric dissolution of a forsterite-like phase *(= fully dehydroxylated serpentinite)*
- **Various solutions:** Mg concentrations ranging from 0.05 M to 0.25 M & 25 to 65°C
- **Aqueous $H_4SiO_4^0$ (silicic acid) activity** chosen as that defined by the solubility of amorphous silica *(Ruiz-Agudo et al., 2012; Sposito, 1994)*

- **Approach adopted consistent with dissolution mechanisms showing precipitation of amorphous silica (on particle surface)**

THE UNIVERSITY OF NEWCASTLE AUSTRALIA | nier | GreenMag Group | NSW Resources & Energy | Australian Government Department of Industry, Innovation and Science | ORICA

Degassed serpentinite leachate

6

- **Delayed amorphous silica precipitation:** *Include cases where Si concentration of initial solution is increased x5 and x10 from the equilibrium concentration*
- **pH of the solution is fixed**, *equating to leachate exposed to a fixed partial pressure of CO_2 (pCO_2)*
- **Alkalinity** *defined by the amount of Mg in solution, equating to a prescribed extent of (forsterite-like) dissolution*

THE UNIVERSITY OF NEWCASTLE AUSTRALIA | nier | GreenMag Group | NSW Resources & Energy | Australian Government Department of Industry, Innovation and Science | ORICA

Methodology - SI & precipitation calculations

7

1\. **Calculate aqueous speciation of solution and saturation indices for a range of minerals**

Suppression of precipitation for all solid phases

2\. **Solution allowed to precipitate a select range of minerals**

- < 50°C : nesquehonite, *p*-antigorite, amorphous silica and silica gel
- > 50°C: hydromagnesite, *p*-antigorite, amorphous silica and silica gel

Nesquehonite, hydromagnesite and p-antigorite are acidity generating phases, so to maintain the solution at a fixed pH, additional CO_2 removed from the solution

$$S = \frac{IAP}{K} = MINERAL\ SATURATION$$

$$SI = log_{10}S = SATURATION\ INDEX$$

Metastable phases:

- Suitable carbonate phase selection, & magnesite excluded consistent with that observed in research *(e.g. Rheinheimer et al., 2017)*
- Mg silicate phases limited to the precipitation of a poorly crystalline antigorite phase (*p*-antigorite) evident from research of Mg-Si solutions *(Gunnarsson et al., 2005)*
- Si phases constrained to amorphous solids (amorphous silica and silica gel) because of the delayed formation of crystalline phases such as chalcedony and quartz *(Faimon and Blecha, 2008; Höllen, 2013)*

Saturation Indices (SIs) of various minerals

8

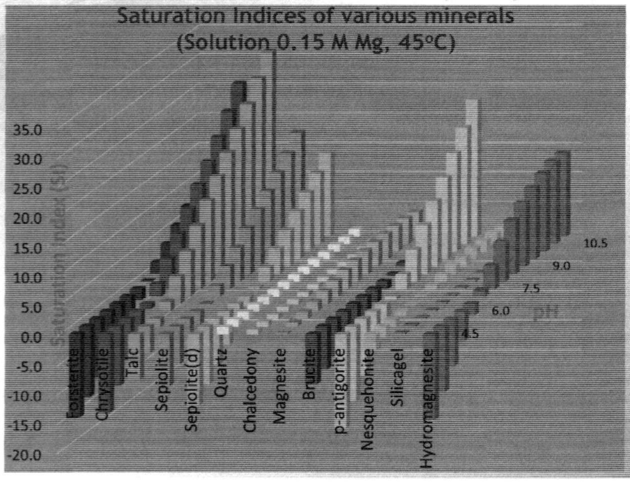

- Saturation Indices of various minerals (Solution 0.15 M Mg, 45°C)
- Typical of modelled trends for other Mg concentrations
- Thermodynamic data for *p*-antigorite provided by Gunnarsson et al. (Gunnarsson et al., 2005)

Saturation Indices (SIs) of select minerals with T

9

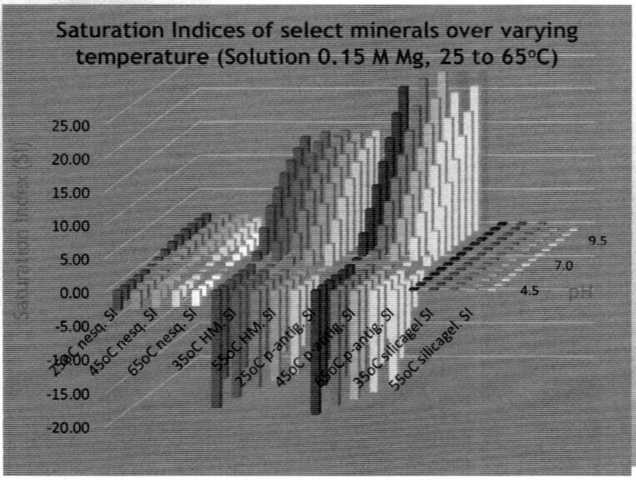

- Temperatures considered: 25°C, 35°C, 45°C (base case), 55°C, 65°C
- A general increase in saturation index (SI) is evident for the Mg carbonate phases and *p*-antigorite unlike that for silica gel

THE UNIVERSITY OF NEWCASTLE AUSTRALIA | nier | GreenMag Group | NSW Resources & Energy | Australian Government Department of Industry, Innovation and Science | ORICA

SIs for various minerals - Summary

10

- Mg silicate phases: well undersaturated at lower pH, at higher pH the SI for these solids increases markedly. pH dependency of reaction. E.g.

  $3Mg^{2+} + 2H_4SiO_4^0 + H_2O \rightleftharpoons Mg_3Si_2O_5(OH)_4 + 6H^+$
- Silica phases: uniform SI consistent with the assumption that the precipitation of amorphous silica is controlling.
- Mg carbonate phases: increase in SI with increasing pH, trend is more pronounced for hydromagnesite. pH dependency of reaction.

  $5Mg^{2+} + 4CO_3^{2-} + 6H_2O \rightleftharpoons Mg_5(CO_3)_4(OH)_2{\cdot}4H_2O + 6H^+$
- Other Mg carbonate metastable phases (such as barringtonite and dypingite) are possible. Thermodynamic data for these other metastable phases unavailable.
- Increase of SIs with T

THE UNIVERSITY OF NEWCASTLE AUSTRALIA | nier | GreenMag Group | NSW Resources & Energy | Australian Government Department of Industry, Innovation and Science | ORICA

Precipitates (with varying Mg)

11

- Mg concentrations considered: 0.05 M, 0.1 M, 0.15 M (base case), 0.20 M, 0.25 M
- Concentrated Mg solutions favour precipitation of nesquehonite over that of *p*-antigorite
- Tying Si concentration of bulk solution to amorphous silica solubility limits Si availability and precipitation of *p*-antigorite (till higher pH)

Precipitate moles nesquehonite and *p*-antigorite. Solution, 0.05 M to 0.25 M Mg, Si from 1.603E-03M (pH4) to 1.094E-01 M (pH11), 45°C

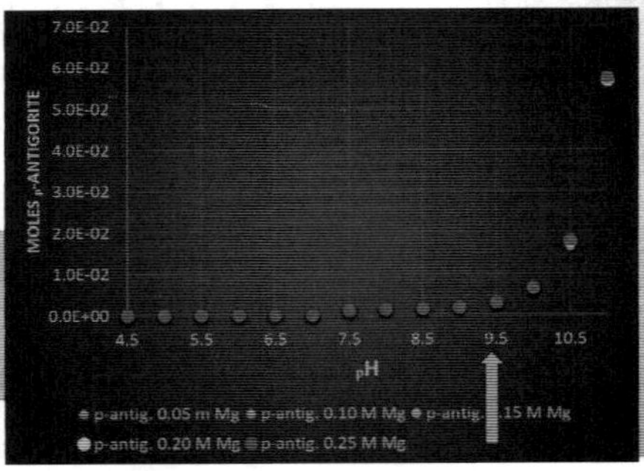

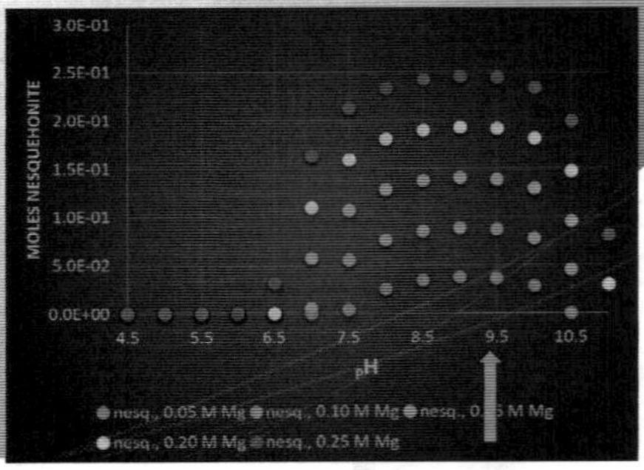

THE UNIVERSITY OF NEWCASTLE AUSTRALIA | nier | GreenMag Group | NSW Resources & Energy | Australian Government Department of Industry, Innovation and Science | ORICA

Precipitates (with varying Si)

12

- Mg concentrations considered: 0.05 M, 0.1 M, 0.15 M (base case), 0.20 M, 0.25 M, 45°C
- Si concentration of initial solution increased x5 and x10 from equilibrium Si concentration
- Higher T similar trends but with strong T dependence of the *p*-antigorite precipitation reaction (increasing amounts of *p*-antigorite with increasing T)

Precipitate moles nesquehonite and *p*-antigorite. Solution, 0.05 M to 0.25 M Mg, Si from (x1,x5,x10)1.603E-03 M (pH4) to (x1,x5,x10) 1.094E-01 M (pH11), 45°C

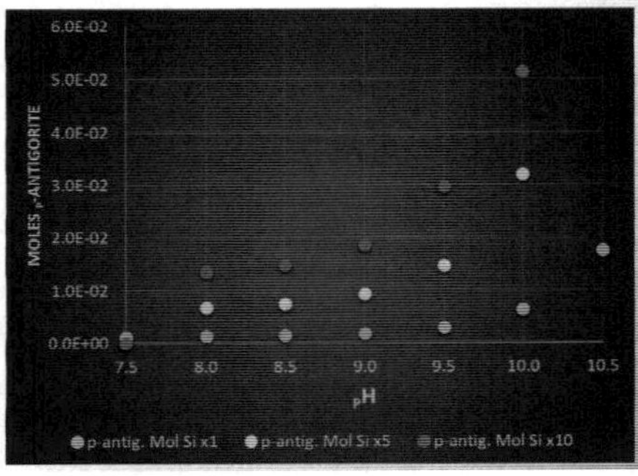

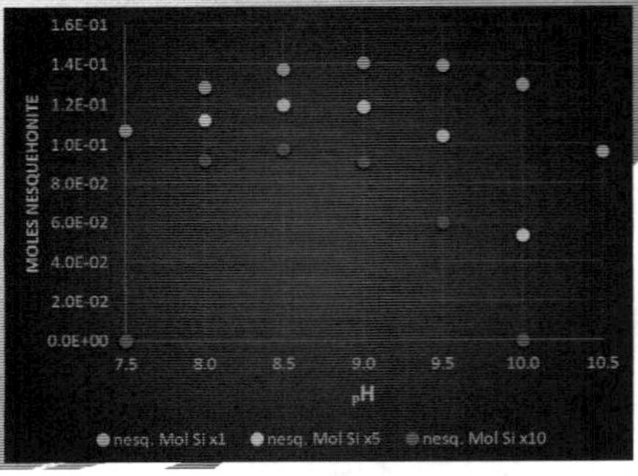

THE UNIVERSITY OF NEWCASTLE AUSTRALIA | nier | GreenMag Group | NSW Resources & Energy | Australian Government Department of Industry, Innovation and Science | ORICA

Activity Ratio Diagrams

13

- SI calculations are limited to consideration of the stability of only one phase at a time
- Activity Ratio Diagrams show;
  - phase relations among many related minerals
  - pH-dependence on the stability fields of minerals
- Give overall context to the solution-mineral system

Activity Ratio Diagrams

14

- *Activity of Mg ion, designated $\{Mg^{2+}\}$, is chosen as a reference state*
  - Lower solubility of silica than Mg related phases, so fix amount Si in solution
- *Activities of solid phases treated in the same way as the activities of soluble species*
- *Activities on a relative scale (relative to $\{Mg^{2+}\}$)*
- *Ordinate values = activity of the ith solid species, designated $\{A_i\}$ relative to $\{Mg^{2+}\}$ plotted (log scale) against pH*
- ***Solid with the largest activity ratio most stable and will form, if its SI also exceeds unity***

Activity Ratio Diagrams

15

Hydromagnesite:

$$Mg_5(CO_3)_4(OH)_2{\cdot}4H_2O_{(s)} + 2H^+_{(aq)} \rightleftharpoons 5Mg^{2+}_{(aq)} + 4CO_3^{2-}_{(aq)} + 6H_2O_{(l)}$$

$$\log\left|\frac{\{HM\}^{\frac{1}{5}}}{\{Mg^{2+}\}}\right| = \frac{1}{5}pK + \frac{4}{5}\log\{CO_3^{2-}\} + \frac{6}{5}\log\{H_2O\} + \frac{2}{5}pH$$

p-antigorite:

$$Mg_3Si_2O_5(OH)_{4\,(s)} + 6H^+_{(aq)} \rightleftharpoons 3Mg^{2+}_{(aq)} + 2H_4SiO_4^0{}_{(aq)} + H_2O_{(l)}$$

$$\log\left|\frac{\{p-antig.\}^{\frac{1}{3}}}{\{Mg^{2+}\}}\right| = \frac{1}{3}pK + \frac{2}{3}\log\{H_4SiO_4^0\} + \frac{1}{3}\log\{H_2O\} + 2pH$$

THE UNIVERSITY OF NEWCASTLE AUSTRALIA | nier | GreenMag Group | NSW Resources & Energy | Australian Government Department of Industry, Innovation and Science | ORICA

Activity Ratio Diagrams - Overall context

16

- **Formation and transformation governed by Gay-Lussac-Ostwald (GLO) Step Rule:**
- If initial composition is such that several solid phases can precipitate a given ion, the **solid phase that forms first** will be the accessible one for which the activity ratio is nearest above its initial value in the solution. Thereafter, the remaining accessible solid phases will form in order of increasing activity ratio. PROVIDING SOLUTION SUPERSATURATED/SATURATED

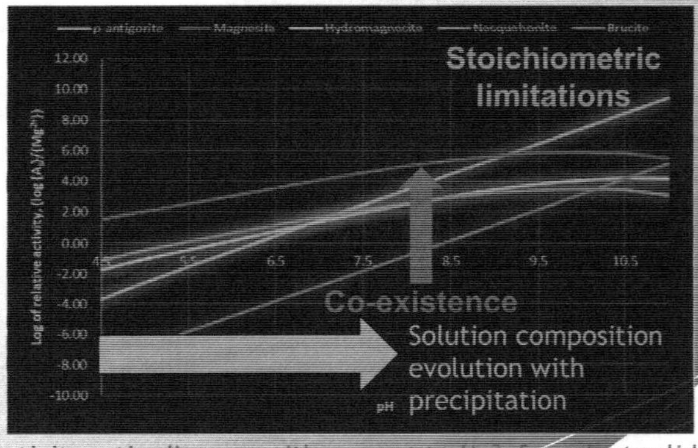

Activity ratio diagram with respect to Mg^{2+} for select solid phases in the $MgO\text{-}SiO_2\text{-}CO_2\text{-}H_2O$ system, conditions Mg 0.15M, Si from 1.603E-03M (pH4) to 1.094E-01M (pH11), 45°C

THE UNIVERSITY OF NEWCASTLE AUSTRALIA | nier | GreenMag Group | NSW Resources & Energy | Australian Government Department of Industry, Innovation and Science | ORICA

Conclusions

17

- Metastability field for MgO- SiO_2 -CO_2 - H_2O system for degassed serpentinite leachate:
  - < 50°C dominated by primary phases nesquehonite and *p*-antigorite. *p*-antigorite increasing solubility control to Mg^{2+} at pH values near neutral and above *(stoichiometric limitations apply!)*.
  - > 50°C (to at least 65°C) formation of less hydrated Mg carbonate phases (transitioning to hydromagnesite) along with *p*-antigorite expected *(stoichiometric limitations apply!)*.
- General: Value of Activity Ratio Diagrams and consideration of Gay-Lussac-Ostwald (GLO) Step Rule in appreciating dissolution and precipitation behaviour/phases. **These techniques complement thermodynamic SI and precipitation assessments.**

Acknowledgements

18

- The authors acknowledge and thank Mineral Carbonation International (MCi) for permission to publish this work and acknowledge financial support from Orica Ltd, Coal Innovation NSW and the Australian Commonwealth Government. This research is being conducted at the University of Newcastle and the MCi Research Pilot Plants are hosted on the TUNRA site.
- We would like to thank Dr Faezeh Farhang for helpful discussions.

References

19

1. Ball, J.W., Nordstrom, D.K., 1991. User's manual for wateq4f, with revised thermodynamic data. Base and test cases for calculating speciation of major, trace, and redox elements in natural waters, ed. U.S. Geological Survey, Menlo Park, California.
2. Faimon, J., Blecha, M., Interaction of freshly precipitated silica gel with aqueous silicic acid solutions under ambient and near neutral pH-conditions: A detailed analysis of linear rate law. Aquatic Geochemistry, 2008, 14(1), 1-40.
3. Gunnarsson, I., Arnórsson, S., Jakobsson, S., Precipitation of poorly crystalline antigorite under hydrothermal conditions. Geochimica et Cosmochimica Acta, 2005, 69(11), 2813-2828.
4. Höllen, D., 2013. Formation and use of amorphous silica. Graz University of Technology, Graz, Austria.
5. Krauskopf, K.B., Dissolution and precipitation of silica at low temperatures. Geochimica et Cosmochimica Acta, 1956, 10(1), 1-26.
6. Park, J., Norman, D., McLin, K., Moore, J., 2006. Modeling amorphous silica precipitation: a strategy to reduce precipitation near COSO injection wells, In Thirty-First Workshop on Geothermal Reservoir Engineering, Stanford University, Stanford, California.

THE UNIVERSITY OF NEWCASTLE AUSTRALIA | nier | GreenMag Group | NSW Resources & Energy | Australian Government Department of Industry, Innovation and Science | ORICA

References

20

7. Parkhurst, D.L., Appelo, C.A.J., 1999. User's guide to PHREEQC (Version 2) - A computer program for speciation, and inverse geochemical calculations, In U.S. Geological Survey Water-Resources Investigations Report 99-4259, p. 312.
8. Rheinheimer, V., Unluer, C., Liu, J., Ruan, S., Pan, J., Monteiro, P., XPS study on the stability and transformation of hydrate and carbonate phases within MgO systems. Materials, 2017, 10(1), 75.
9. Ruiz-Agudo, E., Putnis, C.V., Rodriguez-Navarro, C., Putnis, A., Mechanism of leached layer formation during chemical weathering of silicate minerals. Geology, 2012, 40(10), 947-950.
10. Sposito, G., Chemical equilibria and kinetics in soils 1994, Oxford University Press, New York.
11. Stumm, W., Morgan, J.J., 1996. Aquatic chemistry. Chemical equilibria in natural waters, 3 ed. J. Wiley & Sons, Inc., New York.

THE UNIVERSITY OF NEWCASTLE AUSTRALIA | nier | GreenMag Group | NSW Resources & Energy | Australian Government Department of Industry, Innovation and Science | ORICA

21
Thank you, any questions?
THE UNIVERSITY OF NEWCASTLE AUSTRALIA
nier
GreenMag Group
NSW GOVERNMENT
Resources & Energy
Australian Government
Department of Industry, Innovation and Science
ORICA

School of Earth and Ocean Sciences
Ysgol Gwyddorau'r Ddaear a'r Môr

CARDIFF UNIVERSITY
PRIFYSGOL CAERDYDD

GREENHOUSE GAS REMOVAL RESEARCH PROGRAMME

The potential of carbon storage in the ocean as bicarbonate

Phil Renforth

RenforthP@Cardiff.ac.uk

@phil_renforth

NERC SCIENCE OF THE ENVIRONMENT | EPSRC Investing in research for discovery and innovation | E·S·R·C ECONOMIC & SOCIAL RESEARCH COUNCIL | Department for Business, Energy & Industrial Strategy

ACEME 2018, Newcastle Australia

Outline

- Negative emissions, why bother?
- Ocean storage of carbon as bicarbonate
  - Environmental impact
  - Carbon sink stability
- Enhanced weathering
  - Weathering rates
  - Engineering requirements

CARDIFF UNIVERSITY PRIFYSGOL CAERDYDD

Negative Emissions

Anderson and Peters (2016), Science, 354, 6309

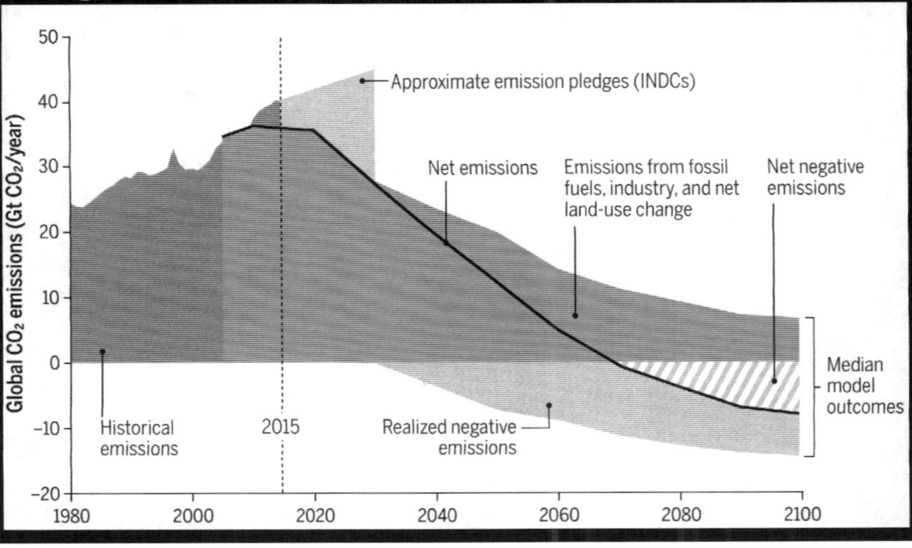

CARDIFF UNIVERSITY PRIFYSGOL CAERDYDD

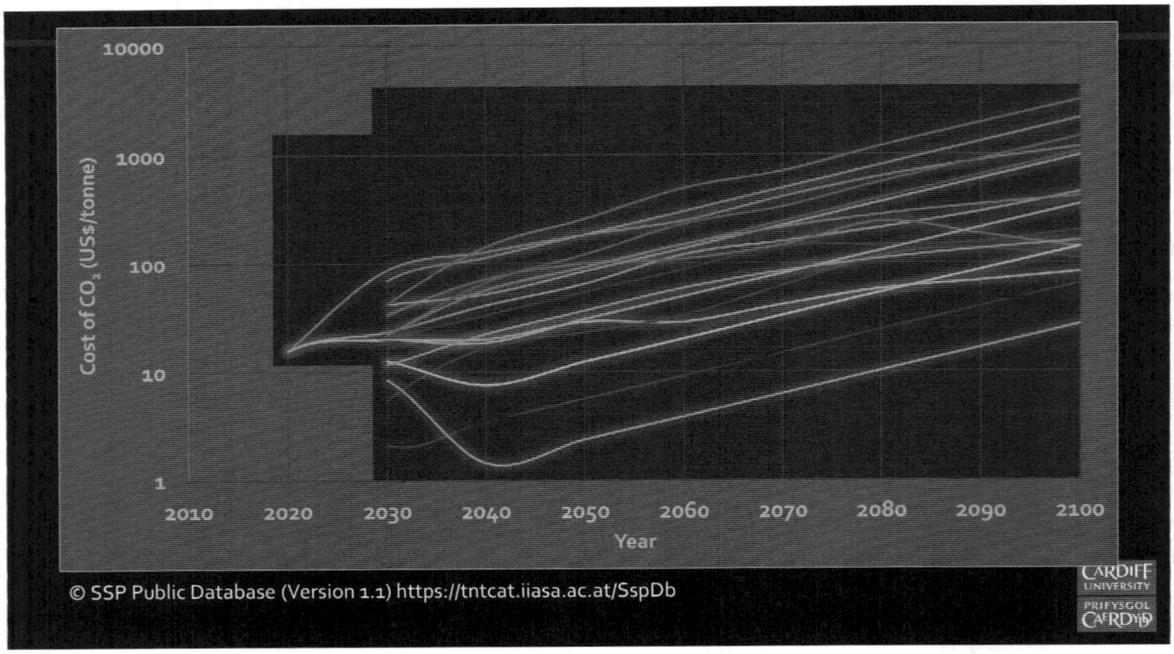

10000
1000
100
10
1
Cost of CO_2 (US$/tonne)
2010
2020
2030
2040
2050
2060
2070
2080
2090
2100
Year
© SSP Public Database (Version 1.1) https://tntcat.iiasa.ac.at/SspDb
CARDIFF UNIVERSITY
PRIFYSGOL CAERDYDD

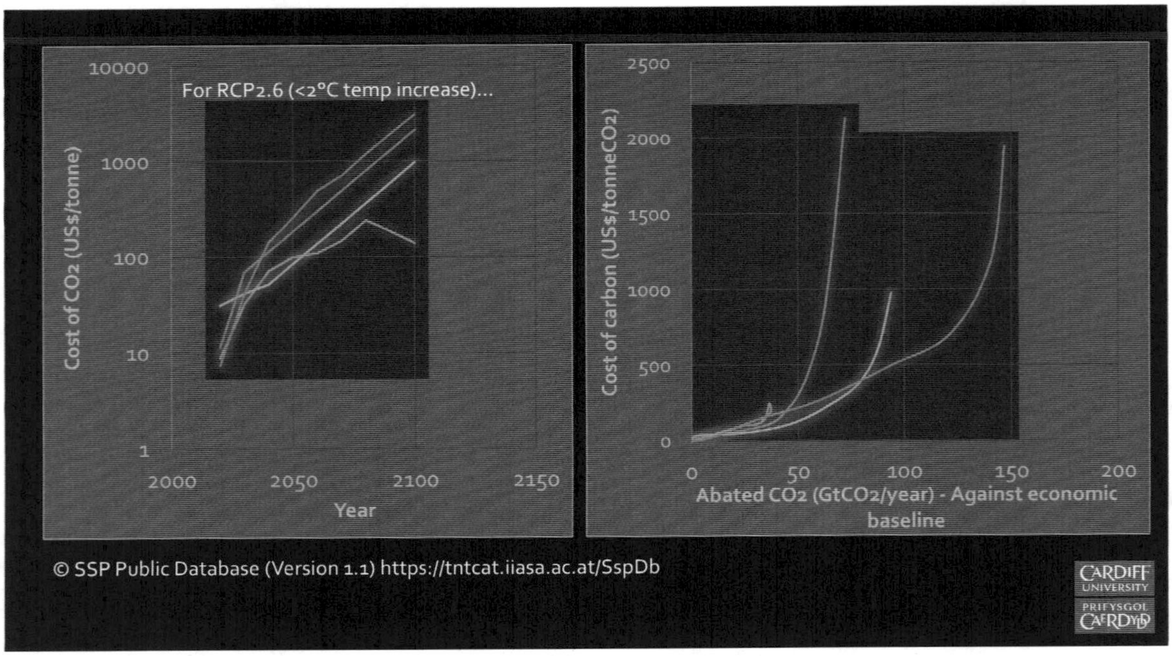

For RCP2.6 (<2°C temp increase)...
10000
1000
100
10
1
Cost of CO2 (US$/tonne)
2000
2050
2100
2150
Year
2500
2000
1500
1000
500
0
Cost of carbon (US$/tonneCO2)
0
50
100
150
200
Abated CO2 (GtCO2/year) - Against economic baseline
© SSP Public Database (Version 1.1) https://tntcat.iiasa.ac.at/SspDb
CARDIFF UNIVERSITY
PRIFYSGOL CAERDYDD

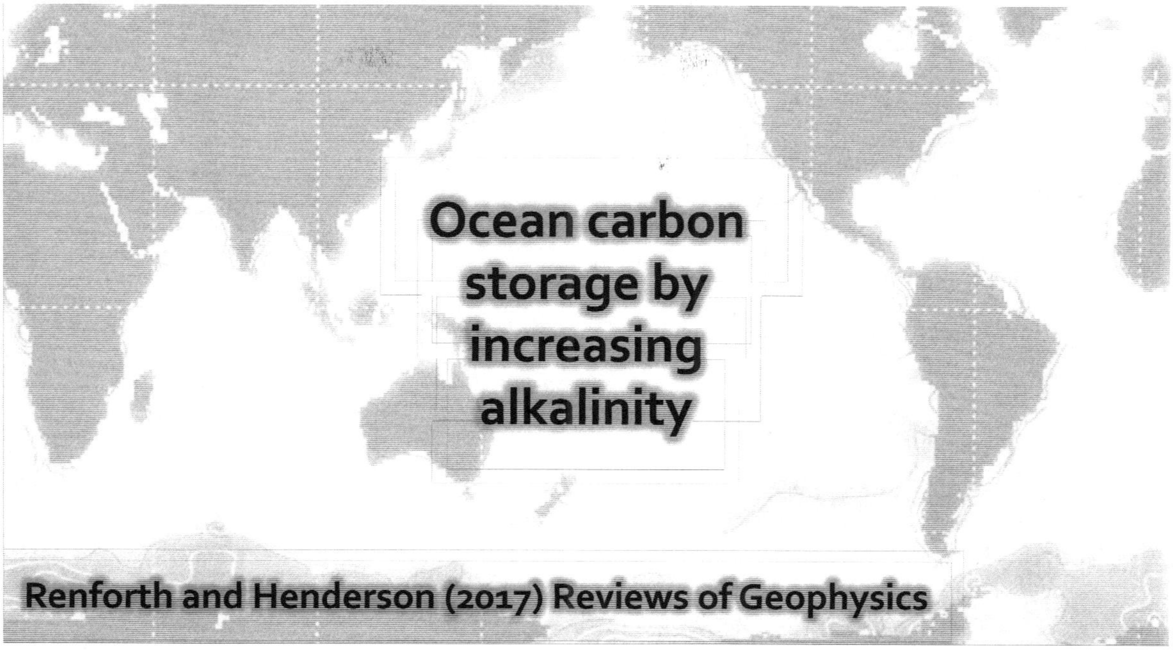

CO_2 storage by increasing ocean alkalinity

Forsterite:

$Mg_2SiO_4 + 2CO_2 + 5H_2O \rightarrow 2MgCO_3\ 3H_2O + H_4SiO_4$

$Mg_2SiO_4 + 4CO_2 + 4H_2O \rightarrow 2Mg^{2+} + 4HCO_3^- + H_4SiO_4$ (ΔG_r = -25.3 kJ/mole)

CARDIFF UNIVERSITY PRIFYSGOL CAERDYDD

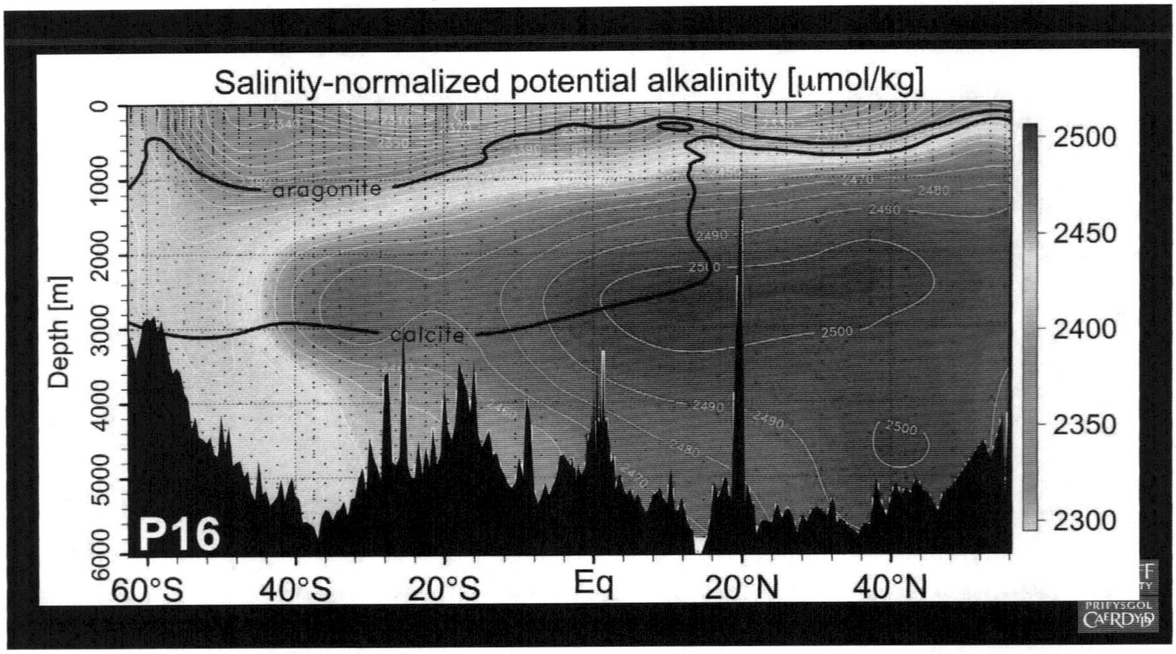
Salinity-normalized potential alkalinity [μmol/kg]
Depth [m]
0
1000
2000
3000
4000
5000
6000
aragonite
calcite
P16
60°S
40°S
20°S
Eq
20°N
40°N
2500
2450
2400
2350
2300

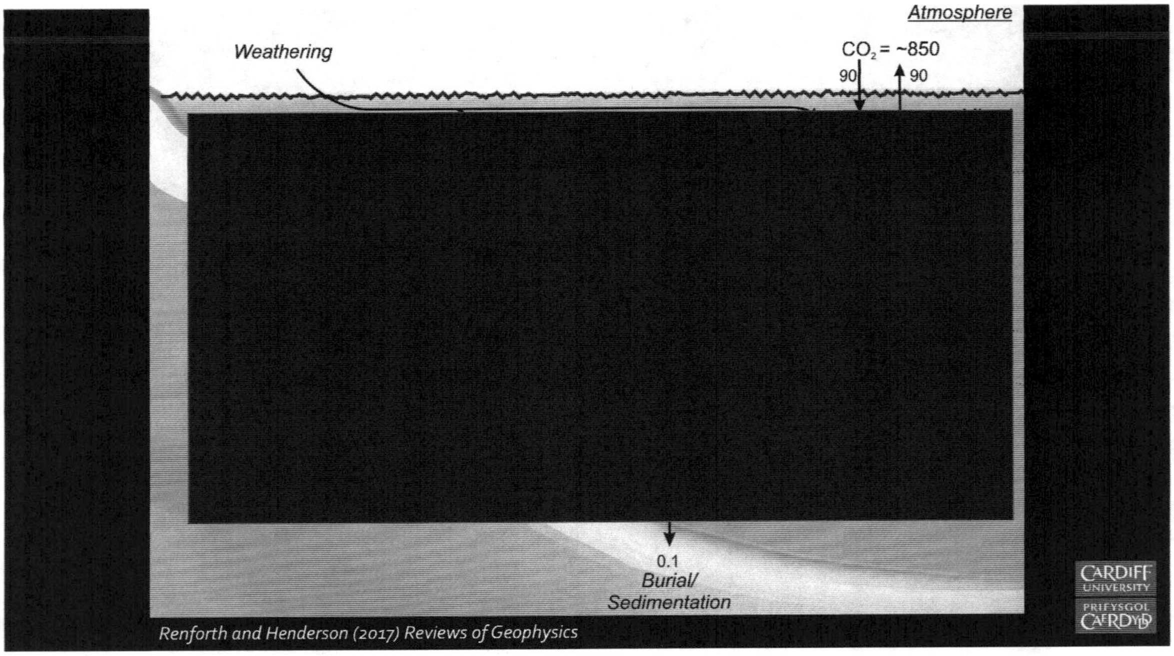
Atmosphere
Weathering
CO_2 = ~850
90
90
0.1
Burial/
Sedimentation
Renforth and Henderson (2017) Reviews of Geophysics

CO_2 storage by increasing ocean alkalinity

Forsterite:

$Mg_2SiO_4 + 4CO_2 + 4H_2O \rightarrow 2Mg^{2+} + 4HCO_3^- + H_4SiO_4$ (ΔG_r = -25.3 kJ/mole)

Calcite:

$CaCO_3 + CO_2 + H_2O \rightarrow Ca^{2+} + 2HCO_3^-$ (ΔG_r = 22.8 kJ/mole)

CO_2 storage by increasing ocean alkalinity

Forsterite:

$Mg_2SiO_4 + 4CO_2 + 4H_2O \rightarrow 2Mg^{2+} + 4HCO_3^- + H_4SiO_4$ (ΔG_r = -25.3 kJ/mole)

Calcite:

$CaCO_3 + CO_2 + H_2O \rightarrow Ca^{2+} + 2HCO_3^-$ (ΔG_r = 22.8 kJ/mole)

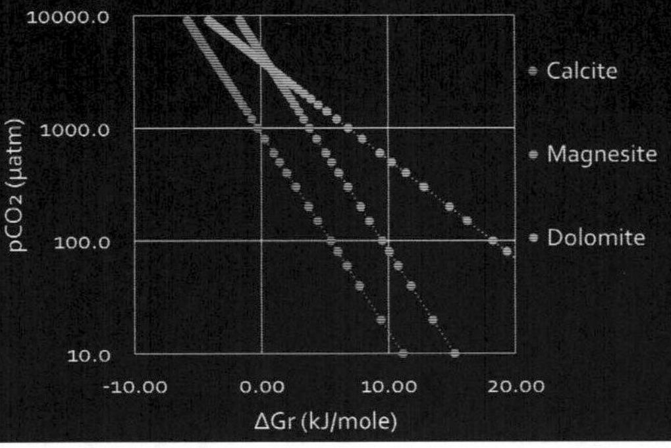

CO_2 storage by increasing ocean alkalinity

Forsterite:

$Mg_2SiO_4 + 4CO_2 + 4H_2O \longrightarrow 2Mg^{2+} + 4HCO_3^- + H_4SiO_4$ (ΔG_r = -25.3 kJ/mole)

Calcite:

$CaCO_3 + CO_2 + H_2O \longrightarrow Ca^{2+} + 2HCO_3^-$ (ΔG_r = 22.8 kJ/mole)

Portlandite:

$Ca(OH)_2 + 2CO_2 + H_2O \longrightarrow Ca^{2+} + 2HCO_3^-$ (ΔG_r = -60.1 kJ/mole)

...proposals for increasing ocean alkalinity

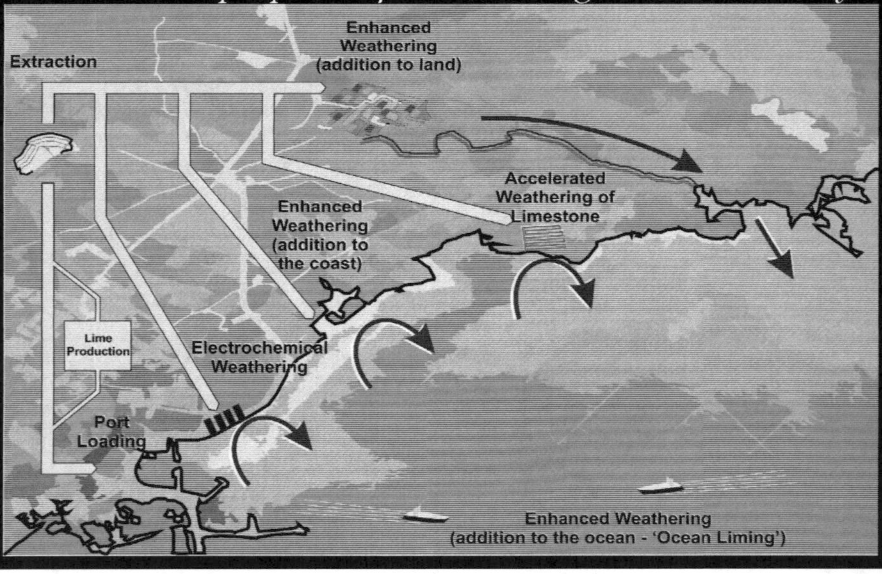

Table 3. Comparison of Electrical and Thermal Energy Requirements and Financial Costs of Ocean Alkalinity Carbon Storage Technologies[a]

Technology	GJ tCO_2^{-1} [b] Electricity	Thermal	US\$ tCO_2^{-1} [b]
Ocean liming (Oxy-fuel flash calciner: limestone)	1.3	4.8	126
Ocean liming (Endex CFC: limestone)	−0.1	5.5	100
Ocean liming (Oxy-fuel flash calciner dolomite)	0.7	3.2	95
Ocean liming (Endex CFC: dolomite)	−0.1	4.2	72
Ocean liming (Solar calciner: limestone)	0.4	0.6[c]	159
Electrochemical weathering (Mg-Silicate)	5[d]		86–154
Electrochemical weathering ($CaCO_3$)	5[d]		14–190
Electrochemical weathering (NaOH production)	3–18		–
Direct carbonate addition to upwelling regions	<0.1	3.6	–
Mineral carbonation/ocean liming	2.2	5.0	
Accelerated weathering of limestone			10–40
Enhanced weathering	0.1–8.4	0.8–4.2	20–600
Direct air capture	7.5–10		100–1000
Typical cost of "conventional" CCS	6.7		30–100
Energy cost of decommissioning coal	*3*	*10*	

[a][*Renforth et al.*, 2013; *Renforth and Kruger*, 2013 and references therein].
[b]Per net ton of carbon dioxide sequestered.
[c]Additional thermal requirements from fossil fuels.
[d]Per ton of CO_2 extracted rather than net sequestration.

Renforth and Henderson (2017) Reviews of Geophysics

CARDIFF UNIVERSITY PRIFYSGOL CAERDYDD

What about the environmental impact?

What is the stability of the carbon sink?

CARDIFF UNIVERSITY PRIFYSGOL CAERDYDD

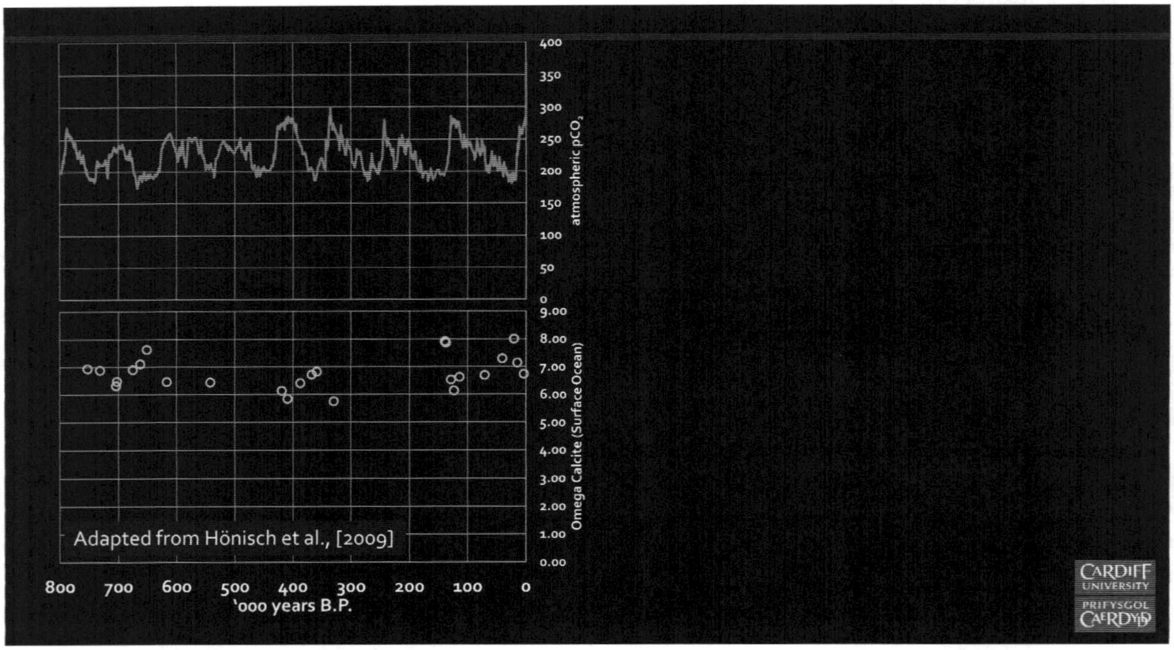
atmospheric pCO_2
400
350
300
250
200
150
100
50
0
9.00
8.00
7.00
6.00
5.00
4.00
3.00
2.00
1.00
0.00
Omega Calcite (Surface Ocean)
Adapted from Hönisch et al., [2009]
800 700 600 500 400 300 200 100 0
'000 years B.P.
CARDIFF UNIVERSITY
PRIFYSGOL CAERDYDD

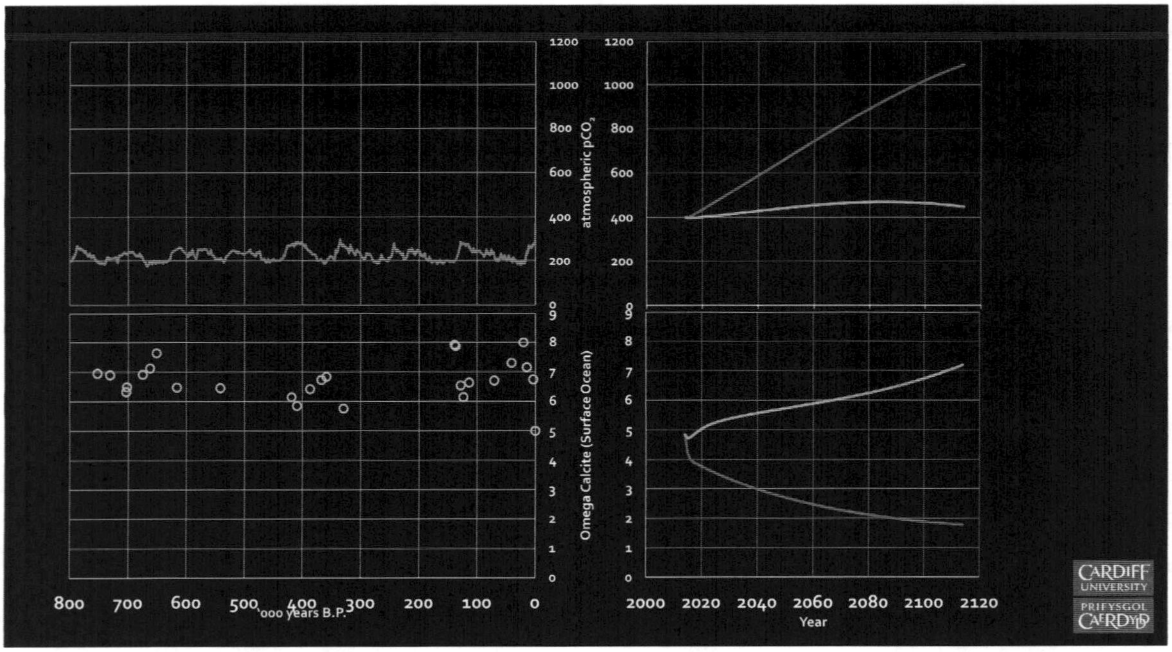
1200
1000
800
600
400
200
atmospheric pCO_2
9
8
7
6
5
4
3
2
1
0
Omega Calcite (Surface Ocean)
800 700 600 500 400 300 200 100 0
'000 years B.P.
2000 2020 2040 2060 2080 2100 2120
Year
CARDIFF UNIVERSITY
PRIFYSGOL CAERDYDD

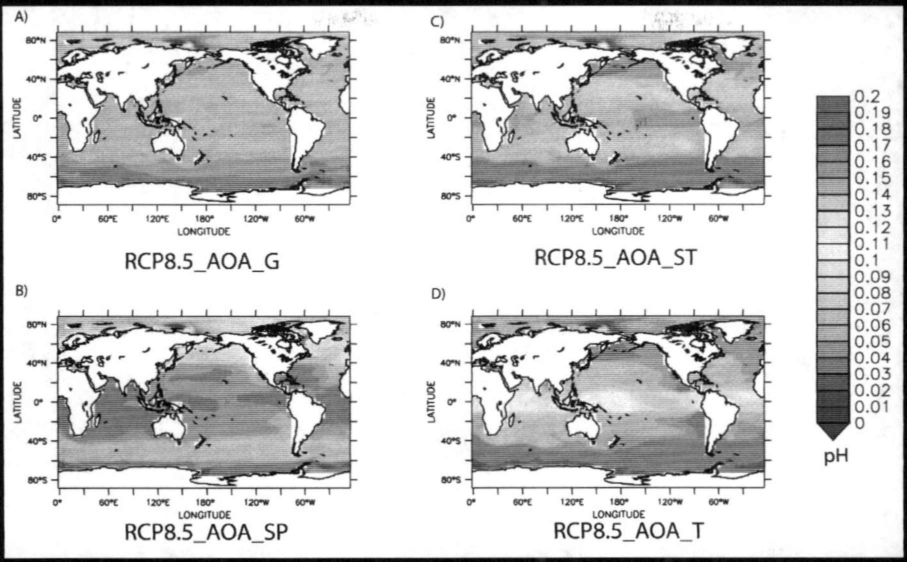

The ocean change in pH,
In 2100
For 650 $GtCO_2$ storage

[*Lenton et al.*, 2018 ESD]

The biological impact of changes in ocean chemistry

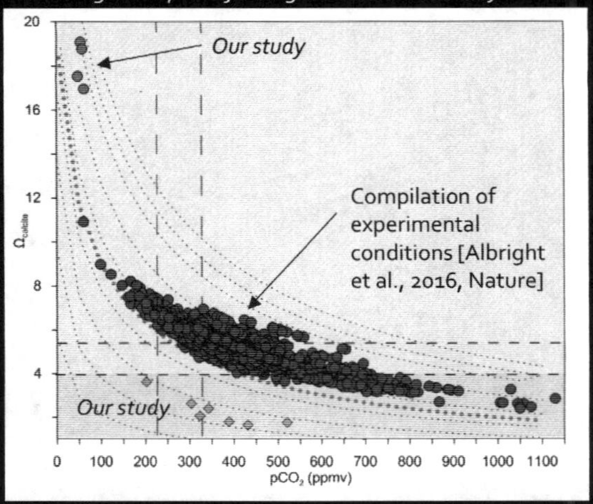

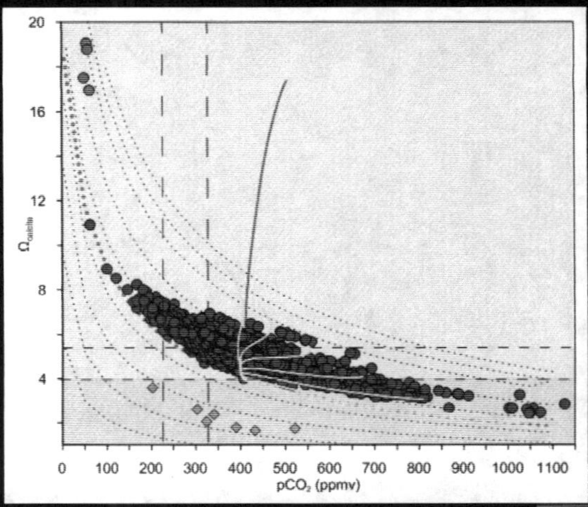

Slide courtesy S. Gore

Carbonation by Corallina algae

(Williamson et al., 2014)

- *Corallina* are calcifying organisms which are found globally.
- Potential 'canary in coal mine' for ocean chemistry changes.
- They are important ecologically because they create and maintain habitats, provide shelter from tidal processes and provide "mini-ecosystems" for microphytobentos.
- They are important $CaCO_3$ (high Mg calcite) producers in shallow temperature coastal waters.

Slide courtesy S. Gore

Methods: ex-situ study

Slide courtesy S. Gore

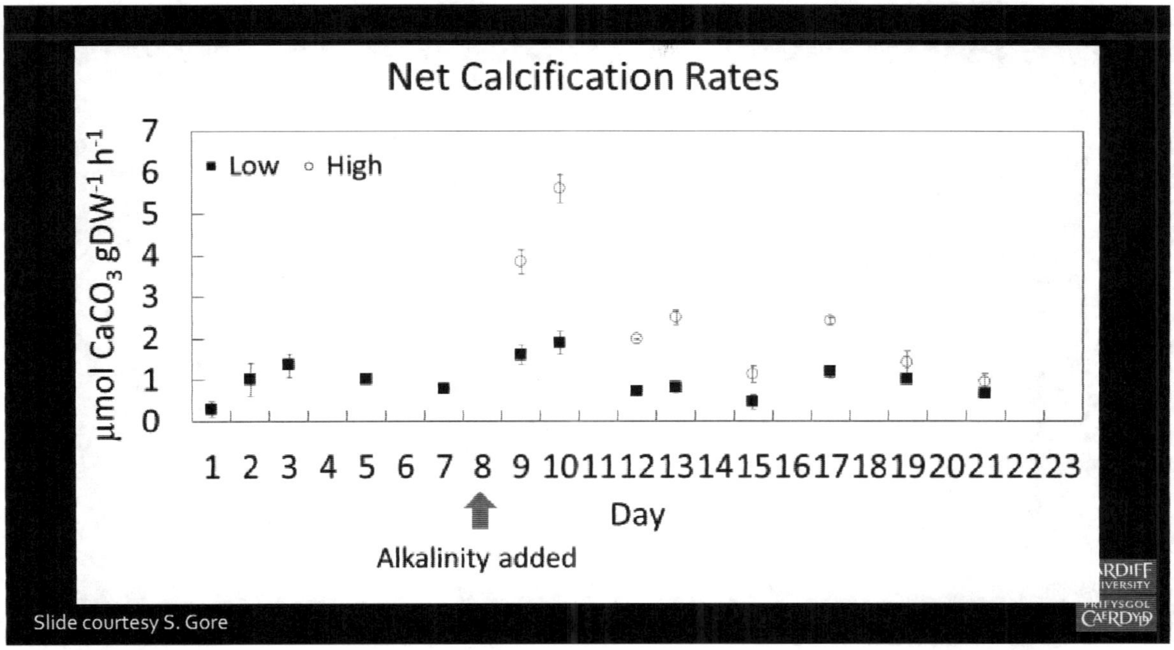
Net Calcification Rates
µmol CaCO$_3$ gDW^{-1} h^{-1}
Low
High
Day
Alkalinity added
Slide courtesy S. Gore

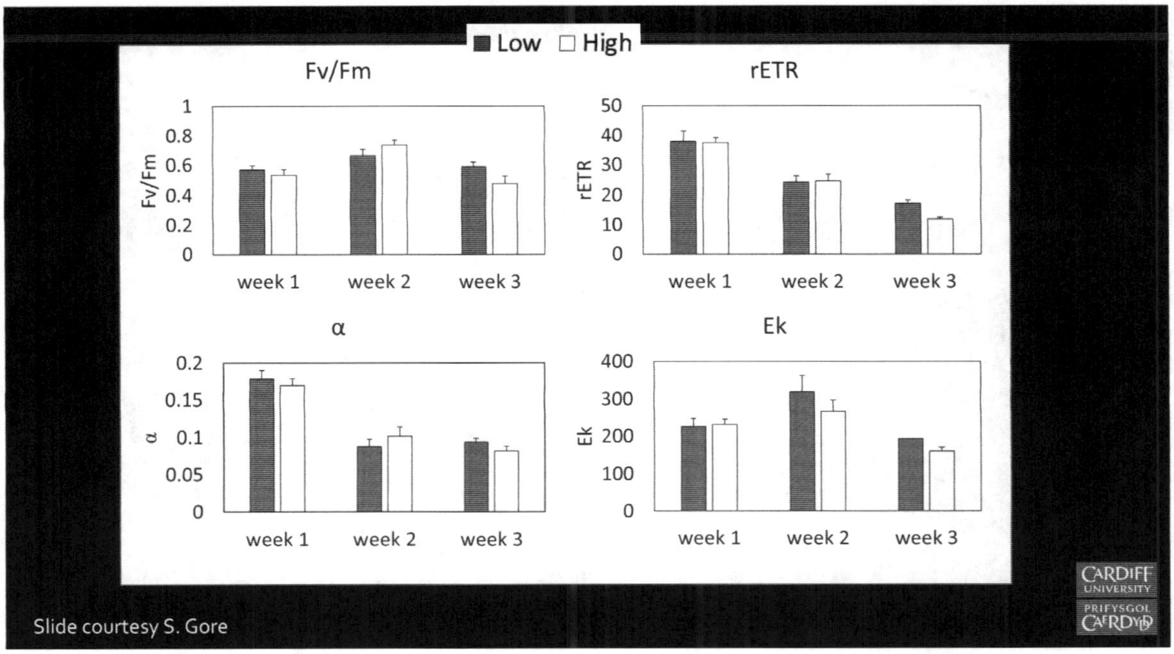
Low
High
Fv/Fm
rETR
α
Ek
week 1
week 2
week 3
Slide courtesy S. Gore

...proposals for increasing ocean alkalinity

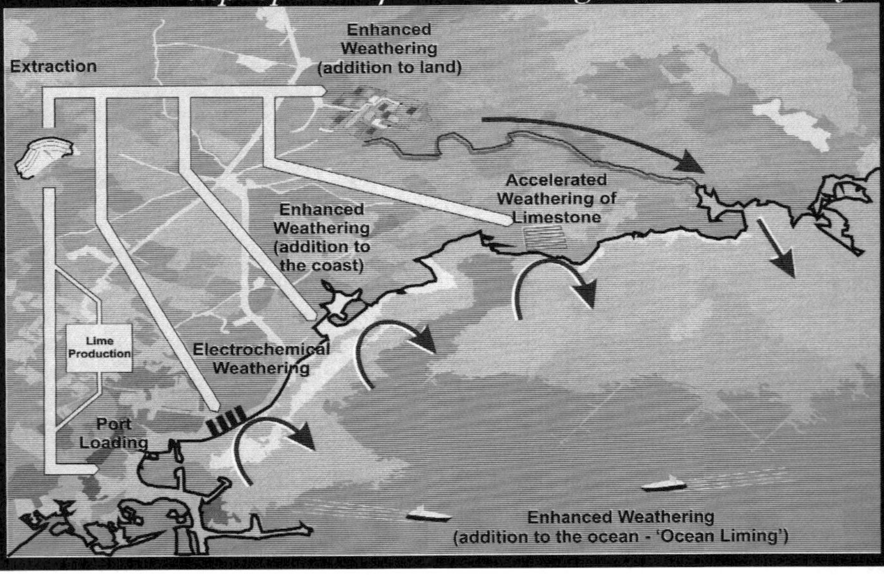

CARDIFF UNIVERSITY
PRIFYSGOL CAERDYDD

Enhanced Weathering

Possible at ambient conditions?

Hartmann et al., (2013) Reviews of Geophysics

-Reaction time ~ 10,000 hours

-Global cropland soil pore space ~ 250,000 m^3

-CO2 concentration ~5-20% at 1 atm

-0.5 – 4.0 mmol/l of various organic ligands (citrate, formate, oxalate, malate...)

-Free makeup water

CARDIFF UNIVERSITY
PRIFYSGOL CAERDYDD

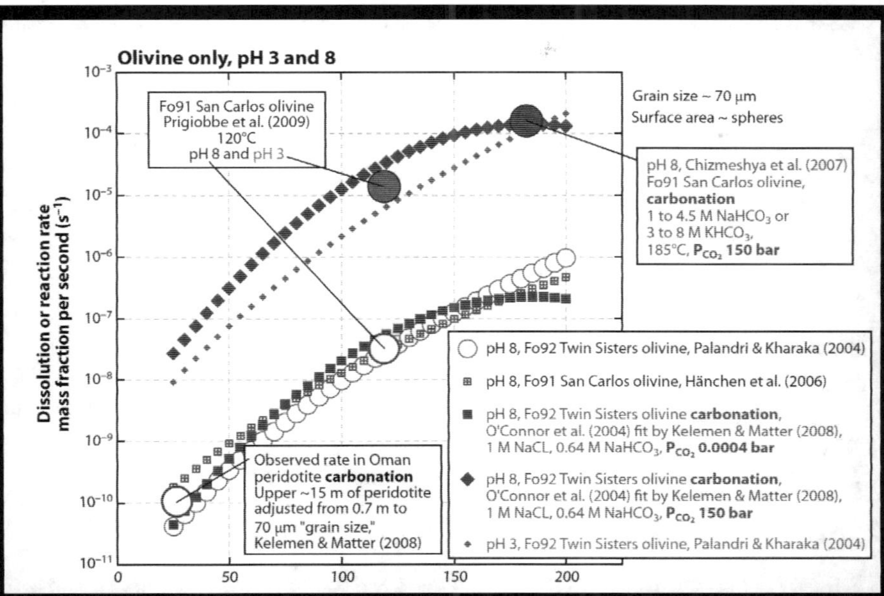

Kelemen et al., (2011)

Variations in weathering rate

Potential costs:
220 – 750 kWh/tCO_2 (ultrabasic)
650 – 3500 kWh/tCO_2 (basic)

24-578 US\$ tCO_2^{-1}
Renforth (2012)

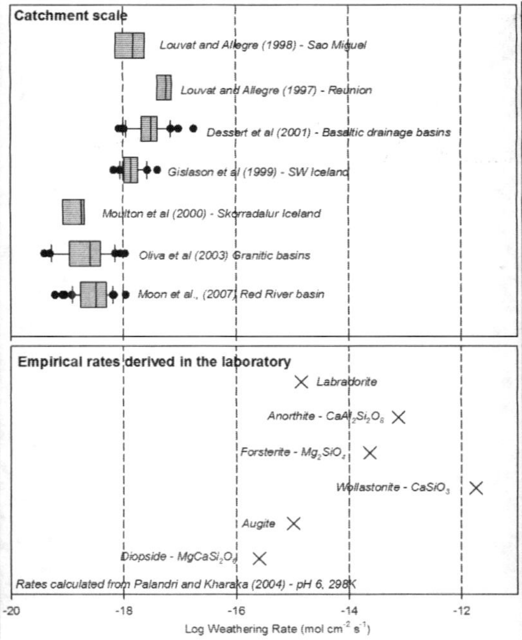

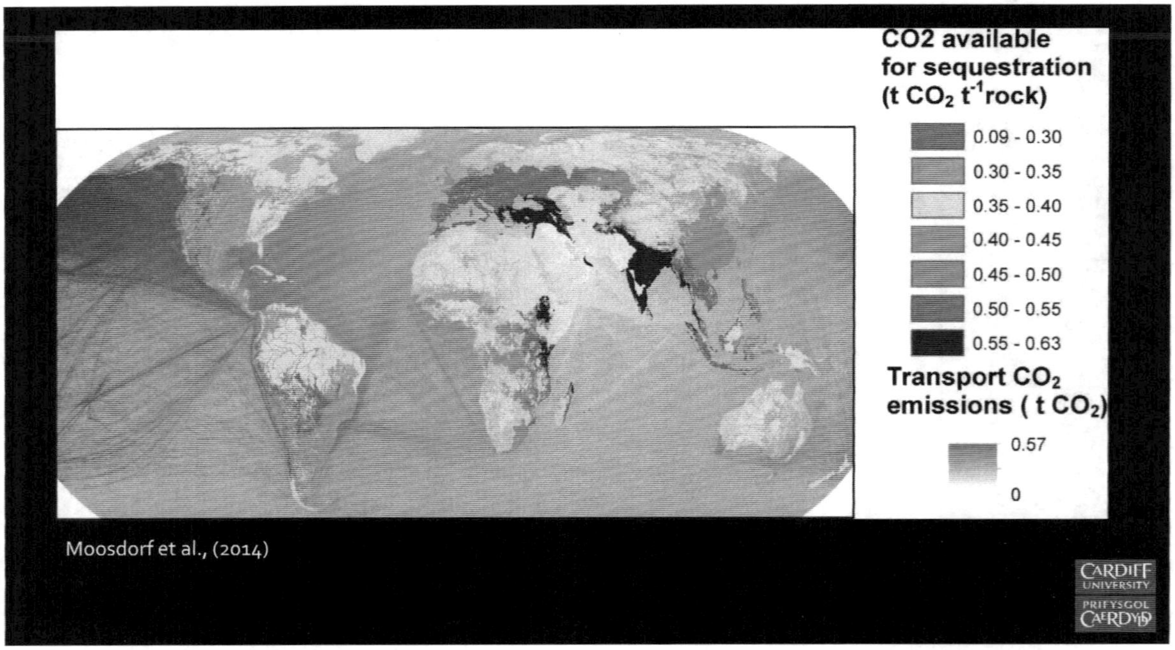
CO2 available for sequestration (t CO_2 t^{-1}rock)
0.09 - 0.30
0.30 - 0.35
0.35 - 0.40
0.40 - 0.45
0.45 - 0.50
0.50 - 0.55
0.55 - 0.63
Transport CO_2 emissions (t CO_2)
0.57
0
Moosdorf et al., (2014)
CARDIFF UNIVERSITY
PRIFYSGOL CAERDYDD

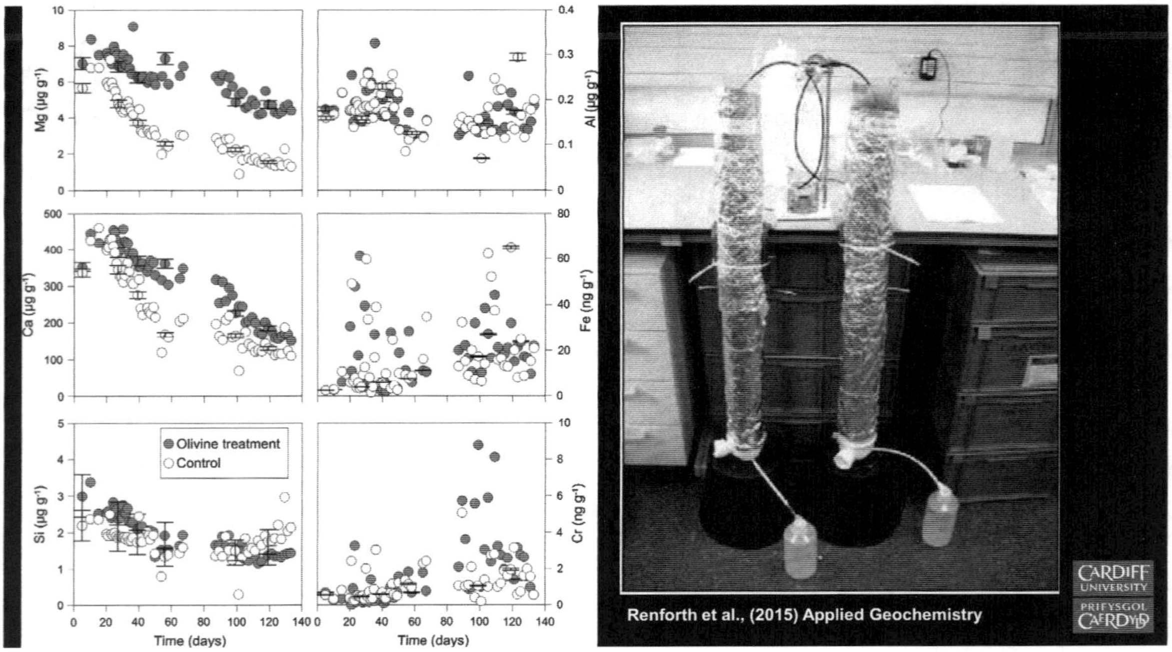
Mg (µg g⁻¹)
Ca (µg g⁻¹)
Si (µg g⁻¹)
Al (µg g⁻¹)
Fe (ng g⁻¹)
Cr (ng g⁻¹)
Olivine treatment
Control
Time (days)
Time (days)
Renforth et al., (2015) Applied Geochemistry
CARDIFF UNIVERSITY
PRIFYSGOL CAERDYDD

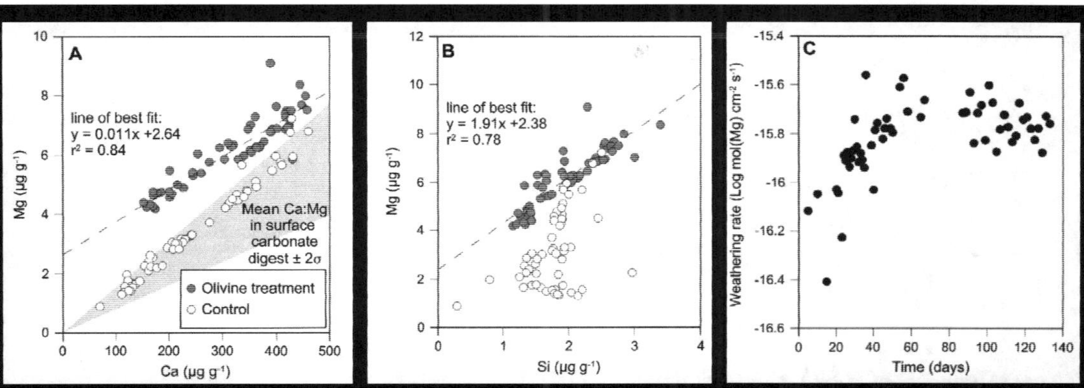

An Enhanced Weathering Experiment

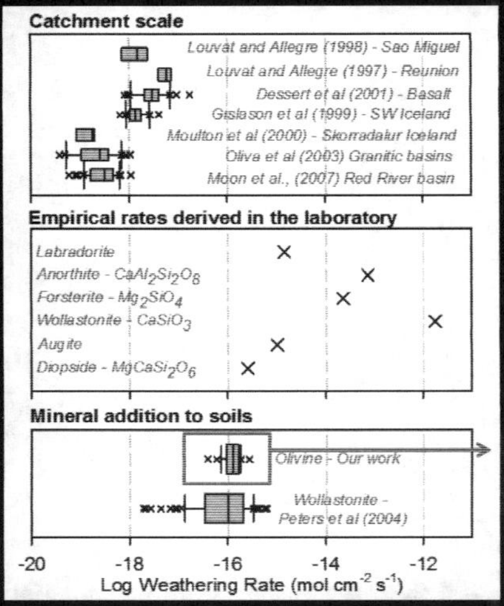

Initial particle diameter (P_{80})~ 1-10μm
Grinding energy ~ 1.5 GJ_e t(olivine)$^{-1}$

Renforth et al., (2015) Applied Geochemistry

Sufficient capacity to be able to supplement 'conventional' mitigation>

~1-10 kg m^{-2} yr^{-1} is typical agricultural lime addition rate

Potential co benefits:
- -Fertiliser (basalt)?
- -Crop resilience to pests?
- -Replacing soil erosion losses?

Uncertainties/risks
- -Fertilising terrestrial/coastal ecosystems(Si, Fe, trace metals)
- -Metal accumulation if using ultrabasic rock?
- -Dust generation at comminution and application site?
- -Monitoring and accounting?

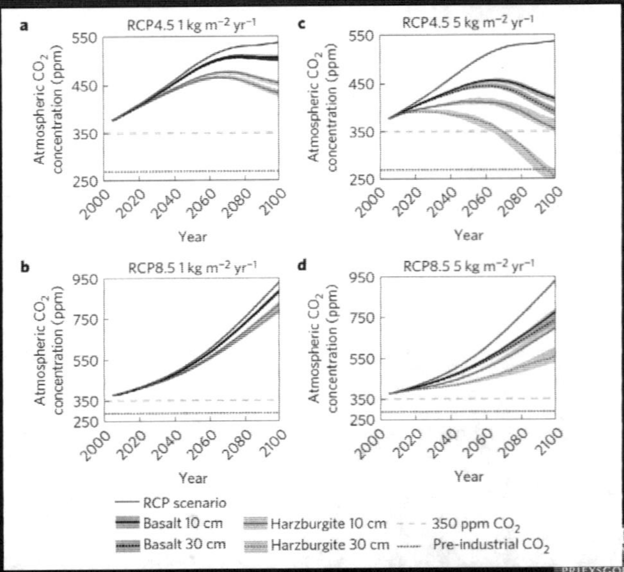

Taylor et al., (2016) Nature Climate Change

PRIFYSGOL CAERDYDD

School of Earth and Ocean Sciences
Ysgol Gwyddorau'r Ddaear a'r Môr

CARDIFF UNIVERSITY
PRIFYSGOL CAERDYDD

GREENHOUSE GAS REMOVAL RESEARCH PROGRAMME

The potential of carbon storage in the ocean as bicarbonate

Phil Renforth

RenforthP@Cardiff.ac.uk

@phil_renforth

ACEME 2018, Newcastle Australia

NERC SCIENCE OF THE ENVIRONMENT
EPSRC Investing in research for discovery and innovation
E·S·R·C ECONOMIC & SOCIAL RESEARCH COUNCIL
Department for Business, Energy & Industrial Strategy

Carbonate nucleation from seawater

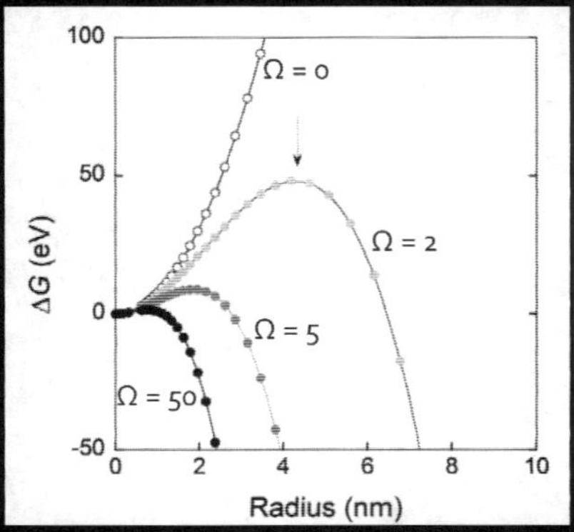

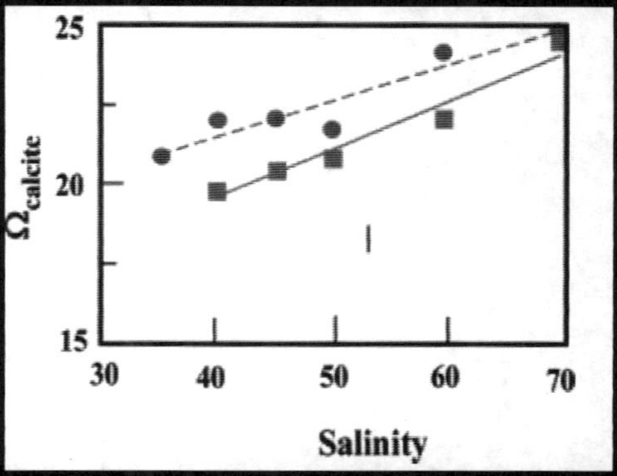

Morse et al., (2007), Chemical Reviews

Ocean Liming

Global lime prices: $50-150

Spare production capacity in US cement industry ~ 10 – 20 Mt/a

if similar globally between 0.5 - 1.2 Gt/a

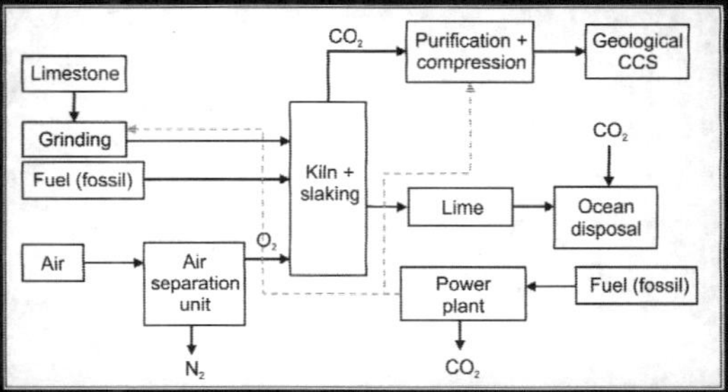

Kheshgi, [1995]; Renforth et al., [2013]

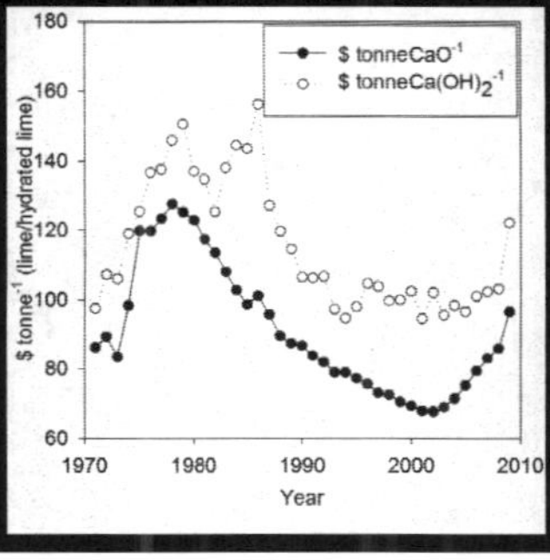

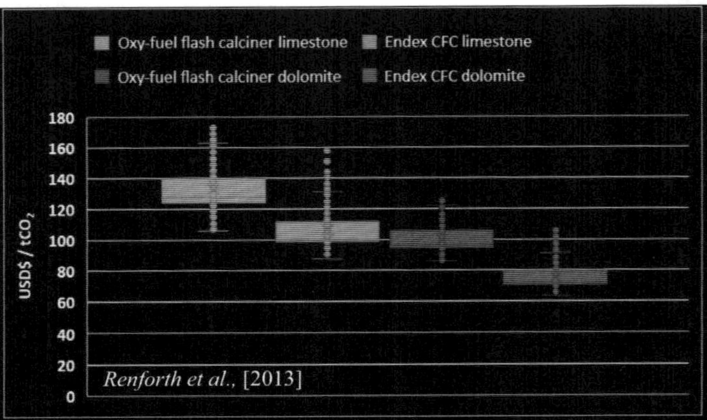

- Oxy-fuel flash calciner (limestone) 1.3_e 4.8_t GJ/tonne
- Endex CFC (limestone) -0.1_e 5.5_t GJ/tonne
- Oxy-fuel flash calciner (dolomite) 0.7_e 3.2_t GJ/tonne
- Endex CFC (dolomite) -0.1_e 4.2_t GJ/tonne
- solar calciner (limestone) 0.4_e 0.6_t GJ/tonne

$CaO + 2CO_2 + H_2O$ ----> $Ca^{2+} + 2HCO_3^-$ $\Delta Hr = -215$ kJ/mol

CARDIFF UNIVERSITY PRIFYSGOL CAERDYDD

Flash Calciner (limestone)

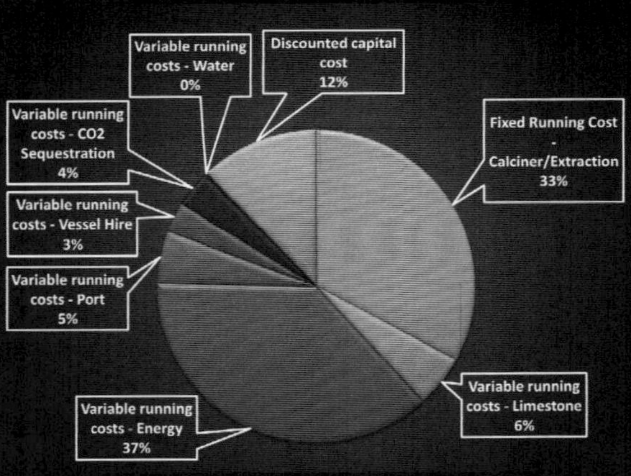

USD$ 126 / tonneCO_2
USD$ 100/ tonneCaO

CARDIFF UNIVERSITY PRIFYSGOL CAERDYDD

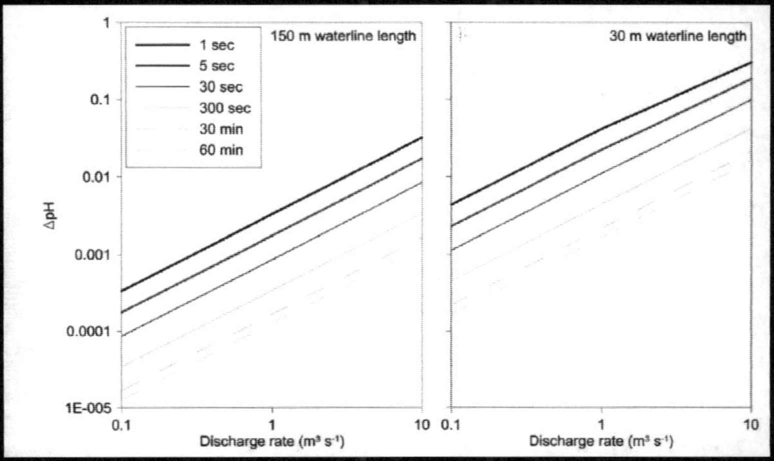

Renforth and Henderson (2017)

Scale limitations?

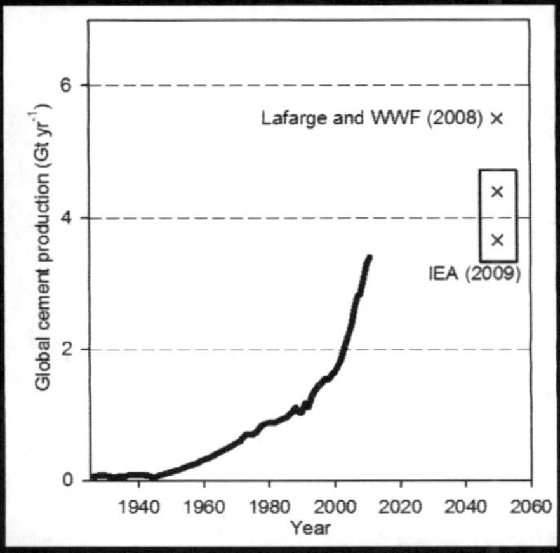

Current cement production is around **4 Gt/year**

Raw material for this represents around **8 Gt/year**

Use in concrete would suggest around >**30 Gt/year** of aggregate and sand

Total mass movement ~ **40 Gt/year**

13 Gt/year brought online since 2005

1

ACEME 2018 March 11-14

Scrubbing CO_2 via dissolution of heat-activated serpentinite

Todd Hodgins, Timothy Oliver, Faezeh Farhang, Eric Kennedy, Michael Stockenhuber, Mark Rayson, Geoff Brent

MCi · The University of Newcastle Australia · nier · GreenMag Group · NSW Government Resources & Energy · Australian Government Department of Industry, Innovation and Science · ORICA

2

Introduction

- Flue gas from coal-fired power station
- Heat activated (HA) Serpentinite
- Facilitate design of pilot plant

The University of Newcastle Australia

Serpentinite

3

- Serpentinite is Lizardite sourced from The Great Serpentinite Belt, NSW, Australia
- Milled & sieved to desired fraction
- Heat activation - 630°C, 4 hours, vacuum

THE UNIVERSITY OF NEWCASTLE AUSTRALIA

HA Lizardite PSD (Post HA)

4

- 3 hours grinding

Particle Size Distribution (PSD)

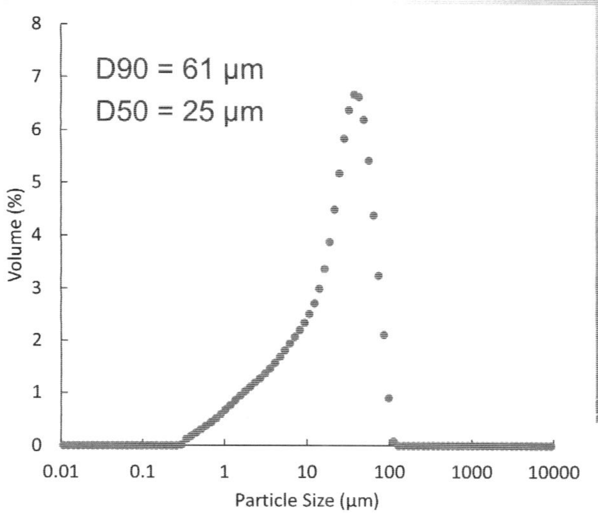

THE UNIVERSITY OF NEWCASTLE AUSTRALIA

HA Lizardite XRD

X-ray Diffraction (XRD)

Component	(%wt)
Forsterite	8.7
Lizardite	7.3
Clinochlore	5.4
Amorphous Phase	78.6

Determine amount of magnesium available

Reactor

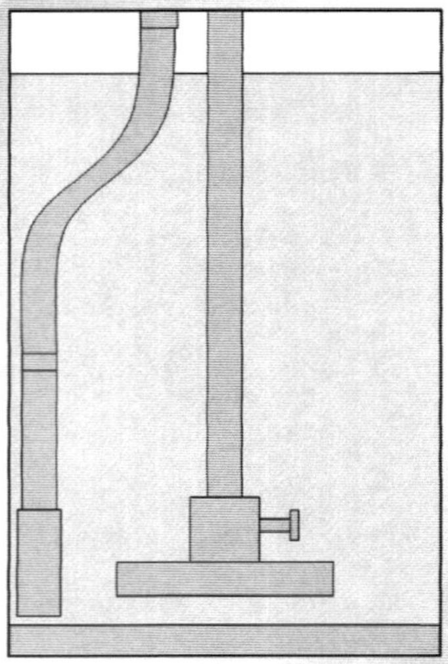

- Batch-style autoclave
- Filled with 900mL distilled water

Reactor

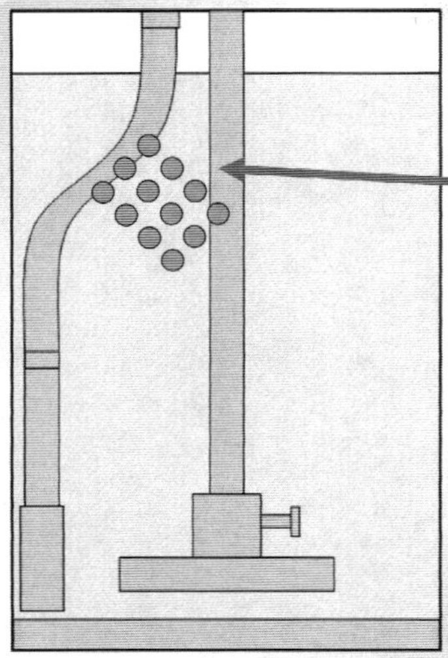

- Batch-style autoclave
- Heat activated mineral loaded into the reactor (900 mL H_2O) & reactor sealed

Reactor

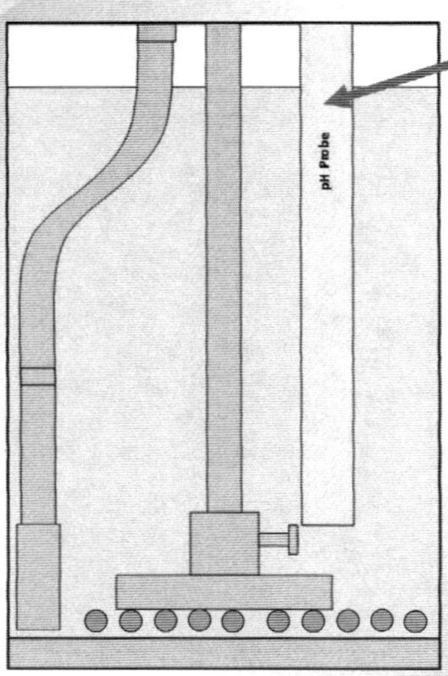

- pH probe inserted

Reactor

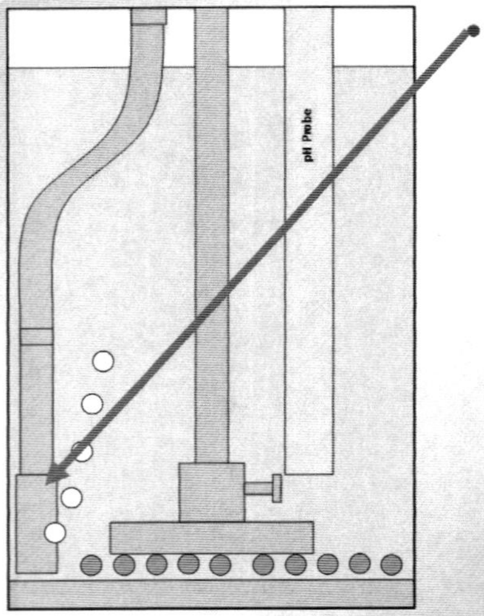

- Flue gas inlet flow is via SS 2 μm diffuser, at a constant rate

Reactor

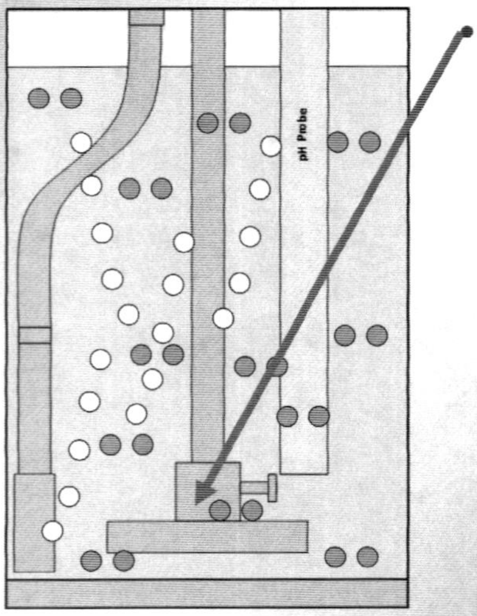

- Agitated via impellor

Reactor

11

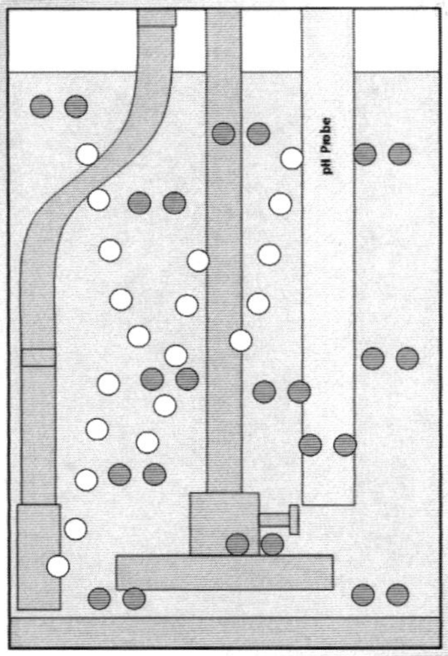

- Liquid samples drawn up via syringe
- Gas outlet leads to an NDIR sensor (detect CO_2)

THE UNIVERSITY OF NEWCASTLE AUSTRALIA

Fixed Conditions

12

Mineral

<75µm heat activated (HA) Lizardite

Reactor Temperature

45°C

Reactor Pressure

1atm

Flue Gas Composition

15% Vol. CO_2 bal. N_2

THE UNIVERSITY OF NEWCASTLE AUSTRALIA

Variables

13

Gas Flow Rate

Solids Loading

Mineral Substitution

Water (no additives) – Worst case

Magnesium Hydroxide – $Mg(OH)_2$ – Best case

THE UNIVERSITY OF NEWCASTLE AUSTRALIA

Analysis & Techniques

14

Online Logging

- pH (INPRO4800 pH probe)
- CO_2% in outlet gas (Ultramat 23, NDIR sensor)

Progressive Samples

- Conductivity (Oakton Con 6 portable meter)
- Elemental Mg (ICP-OES)

- Each experiment has 2 minute induction phase prior to gas flow

THE UNIVERSITY OF NEWCASTLE AUSTRALIA

Conductivity & Release of Magnesium

15

- Mazzotti et. al, 2016, ETH used conductivity measurements to indicate release of $Mg_{(aq)}$

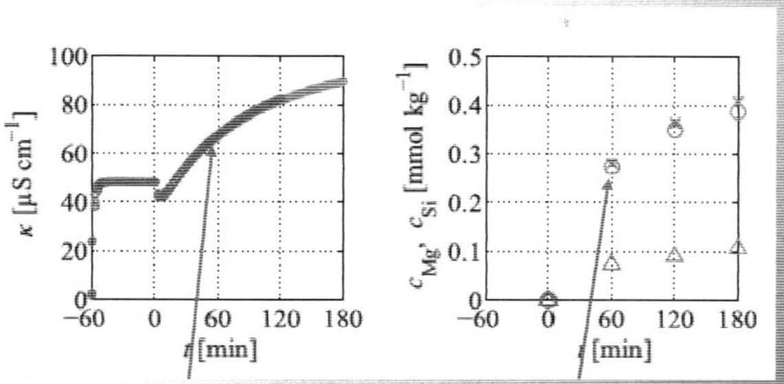

(Mazzotti et. al, 2016)

THE UNIVERSITY OF NEWCASTLE AUSTRALIA

Conductivity & Moles Mg in 2% HA Lizardite Solution

16

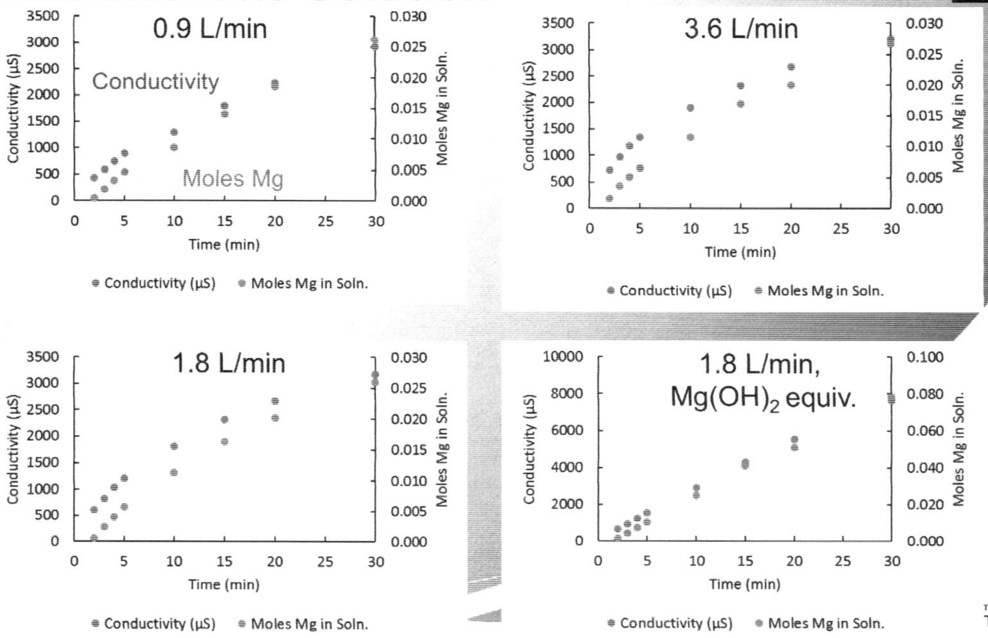

Solution pH, 2% HA Lizardite

17

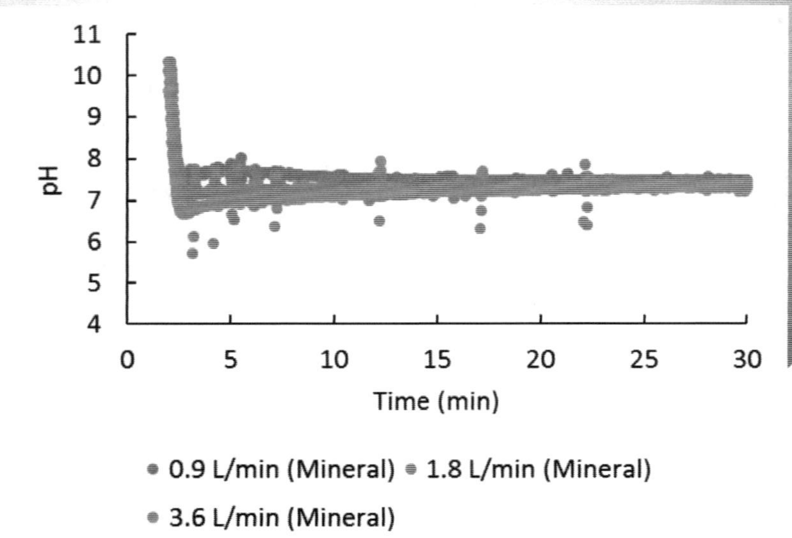

THE UNIVERSITY OF NEWCASTLE AUSTRALIA

Rate of CO_2 Capture – In Water

18

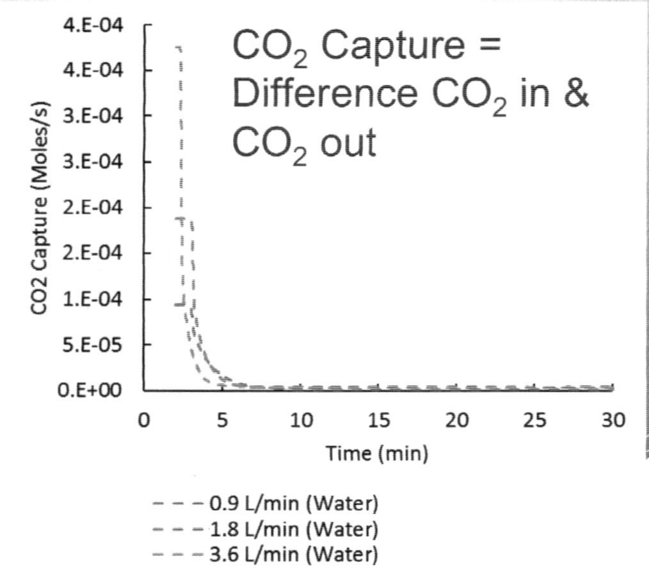

THE UNIVERSITY OF NEWCASTLE AUSTRALIA

Rate of CO_2 Capture – Lizardite

19

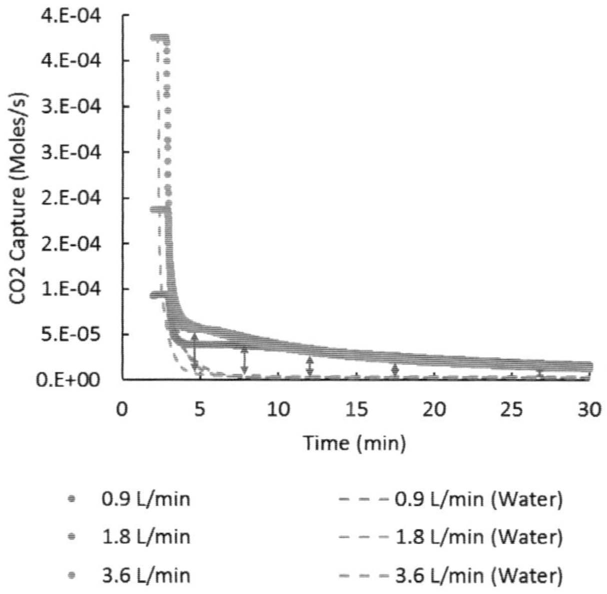

THE UNIVERSITY OF NEWCASTLE AUSTRALIA

Additional CO_2 Capture

20

- CO_2 capture capacity in addition to that captured by water alone

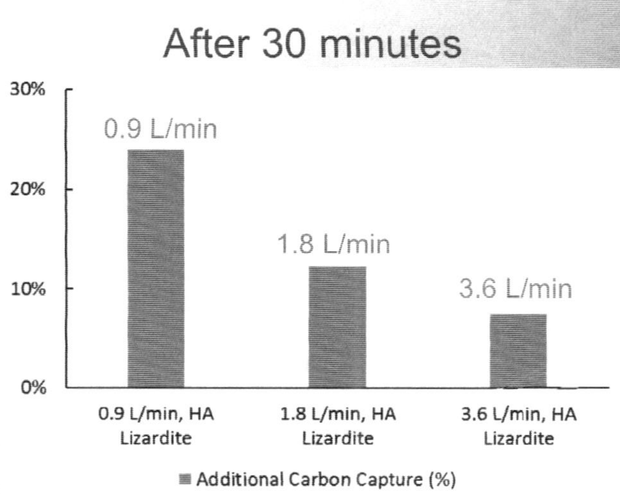

THE UNIVERSITY OF NEWCASTLE AUSTRALIA

Efficiency Vs. Inlet Flow

21

- Li, 2015 (University of Cincinatti)
- Low inlet flow rates – $Mg(OH)_2$
- Similar trend, more efficient at lower flows

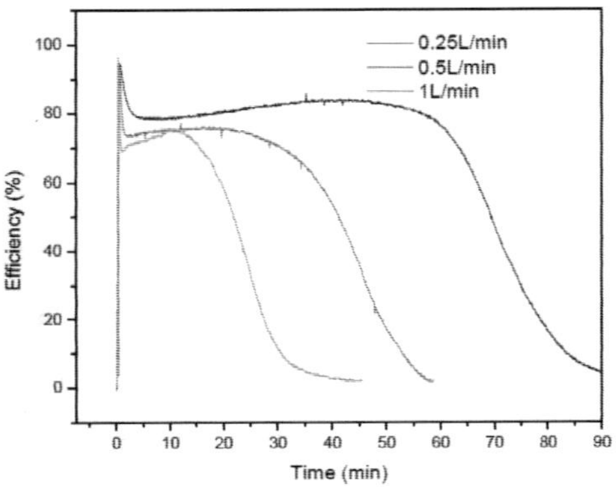

(Li, 2015)

THE UNIVERSITY OF NEWCASTLE AUSTRALIA

Surrogate Mineral, 1.8 L/min

22

- Surrogate mineral – $Mg(OH)_2$ with equivalent moles Mg to 2% solids loading Lizardite

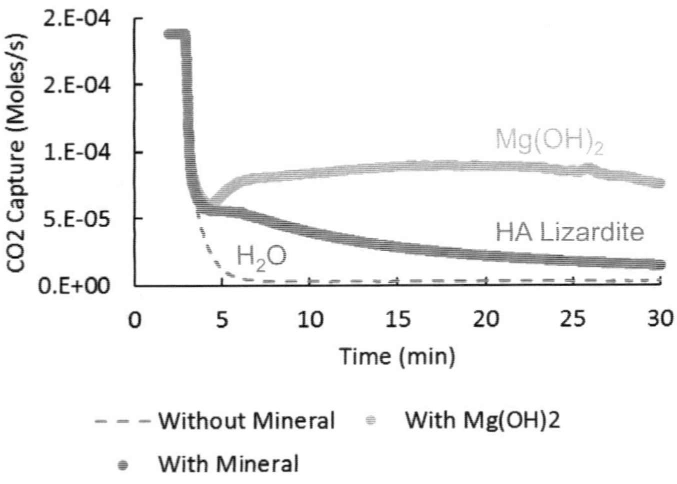

THE UNIVERSITY OF NEWCASTLE AUSTRALIA

Lizardite Vs. $Mg(OH)_2$

23

- CO_2 capture capacity in addition to that captured by water alone

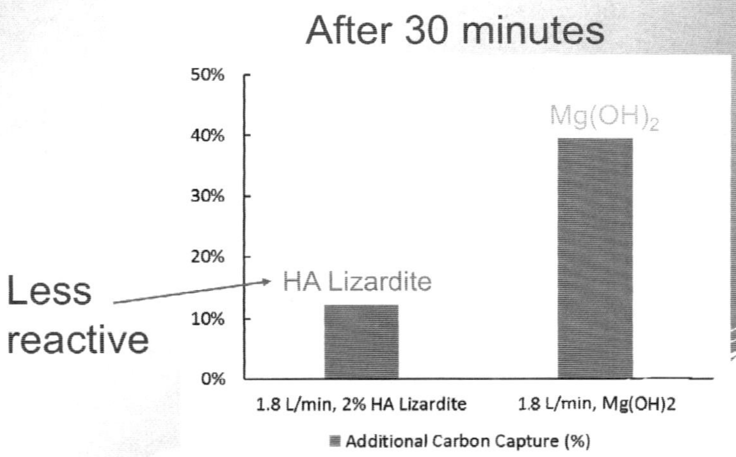

THE UNIVERSITY OF NEWCASTLE AUSTRALIA

Double (4%) Solids Loading

24

- Lizardite solids loading: 2% → 4%

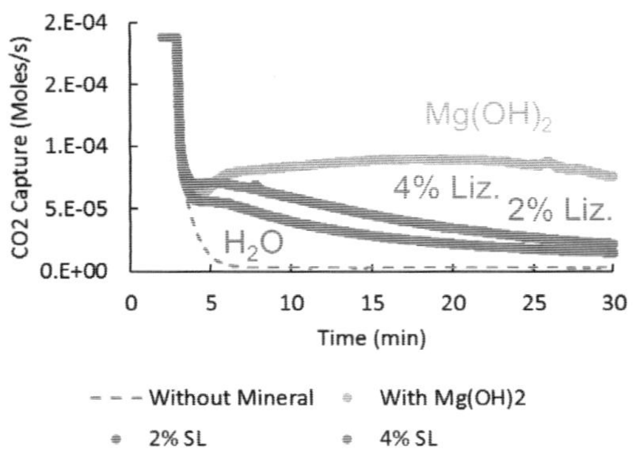

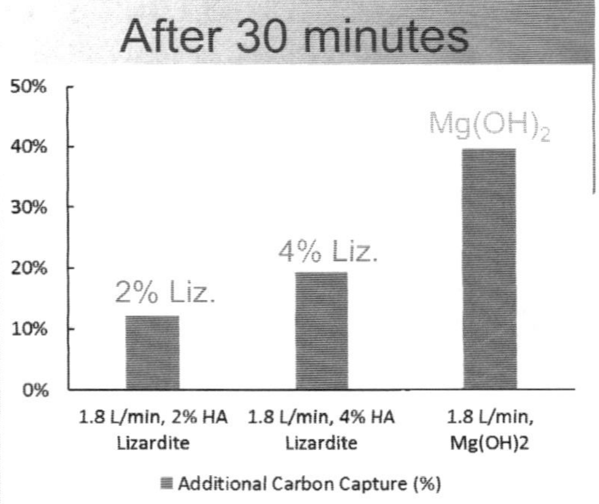

THE UNIVERSITY OF NEWCASTLE AUSTRALIA

Conclusions

25

- Addition of HA Lizardite increases capture efficiency over 30 minutes, trending downwards as inlet flow rate increased

- At identical inlet flows, HA Lizardite captures less CO_2 compared to equivalent moles Mg of $Mg(OH)_2$ – HA Lizardite less reactive than $Mg(OH)_2$

THE UNIVERSITY OF NEWCASTLE AUSTRALIA

Conclusions

26

- Doubling solids loading (HA Lizardite) further increases capture efficiency – however CO_2 capture is less effective per gram of feedstock

- Conductivity profiles matches that of the profile of moles Mg in the 2% HA Lizardite and equiv. $Mg(OH)_2$ solution

THE UNIVERSITY OF NEWCASTLE AUSTRALIA

Acknowledgements

27

- Co-authors
- Parties involved with our research

MG | THE UNIVERSITY OF NEWCASTLE AUSTRALIA | nier | GreenMag Group | NSW GOVERNMENT Resources & Energy | Australian Government Department of Industry, Innovation and Science | ORICA

THE UNIVERSITY OF NEWCASTLE AUSTRALIA

References

28

Li, T. (2015). *The Study of CO2 Removal in Slurries with Mg(OH)2 Suspended Particles and the Regeneration of Products.* University of Cincinnati.

Hariharan, S., Repmann-Werner, M., & Mazzotti, M. (2016). Dissolution of dehydroxylated lizardite at flue gas conditions: III. Near-equilibrium kinetics. *Chemical Engineering Journal, 298,* 44-54.

THE UNIVERSITY OF NEWCASTLE AUSTRALIA

Thank You

29

- Questions?

THE UNIVERSITY OF NEWCASTLE AUSTRALIA

Thank You

30

THE UNIVERSITY OF NEWCASTLE AUSTRALIA

Thank You

31

THE UNIVERSITY OF NEWCASTLE AUSTRALIA

Side View

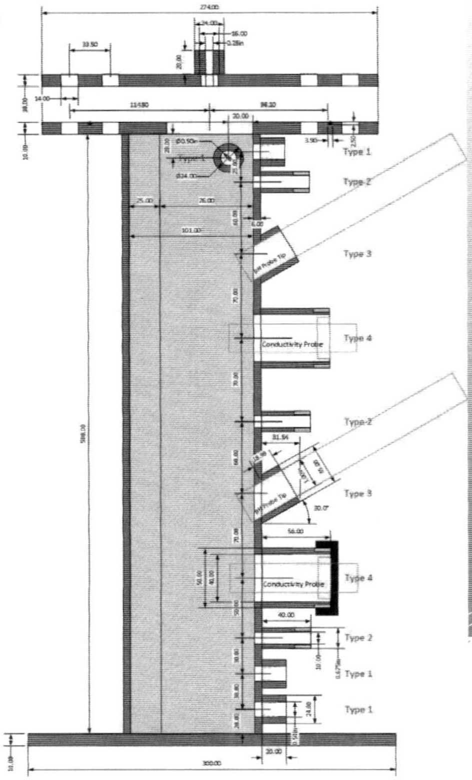

32

THE UNIVERSITY OF NEWCASTLE AUSTRALIA

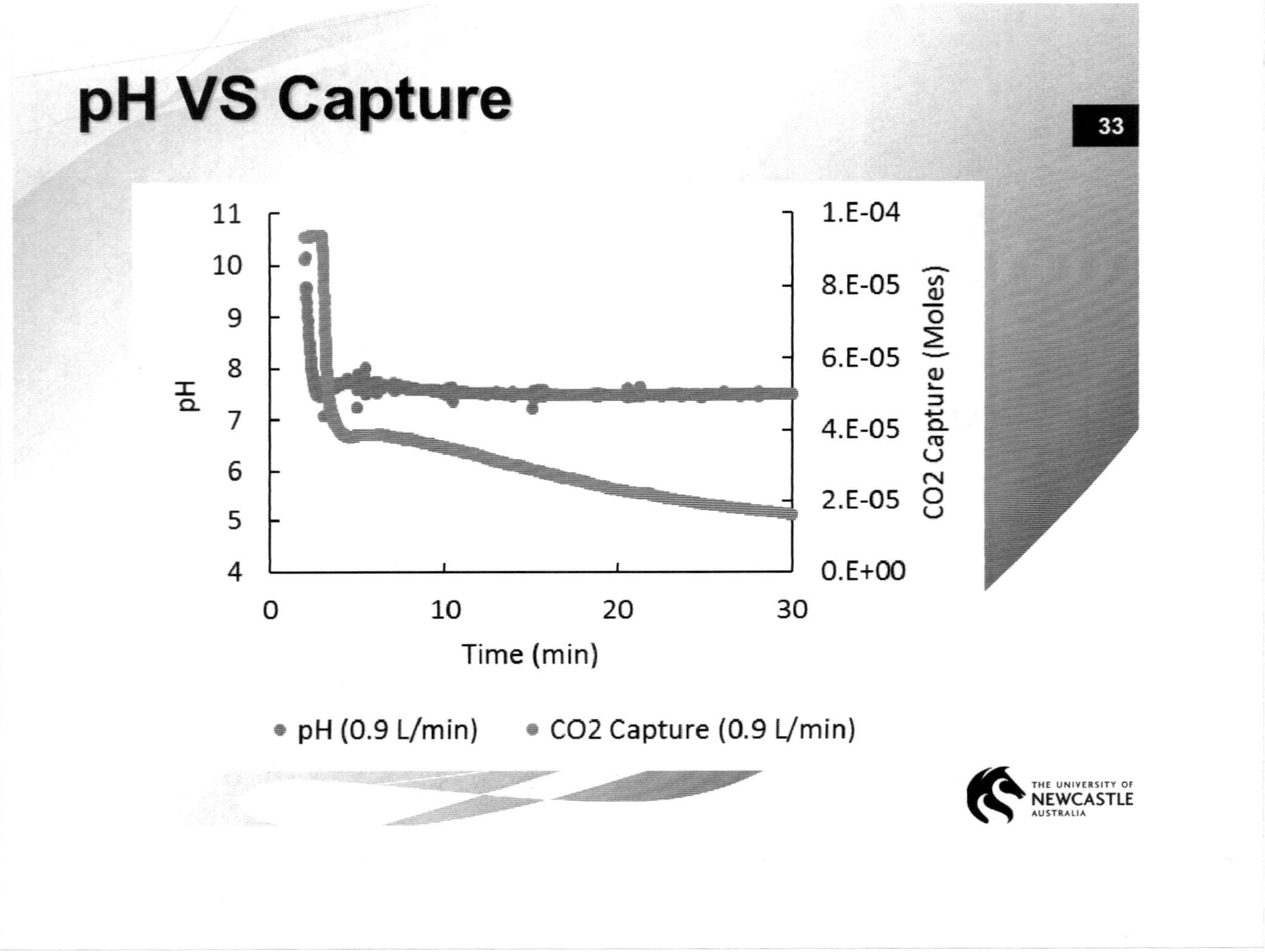
pH VS Capture
33
11
10
9
8
7
6
5
4
pH
1.E-04
8.E-05
6.E-05
4.E-05
2.E-05
0.E+00
CO2 Capture (Moles)
0
10
20
30
Time (min)
pH (0.9 L/min)
CO2 Capture (0.9 L/min)
THE UNIVERSITY OF NEWCASTLE AUSTRALIA

Augmenting the Magnesite Yield Produced during Aqueous Mineral Carbonation of Dunite Rock

MUHAMMAD IMRAN RASHID (Presenter)
Co-authors:
E. Benhelal, F. Farhang, O. Mowla, M. Rayson. G. Brent, M. Stockenhuber, E. Kennedy

THE UNIVERSITY OF NEWCASTLE AUSTRALIA ORICA MCi Mineral Carbonation International

Aim

2

- To utilise Mg rich (alternative) feedstocks (e.g. **dunite rock**) & widen feedstock viability
- NSW reserves
- Using serpentinised rocks in NSW will increase feedstock

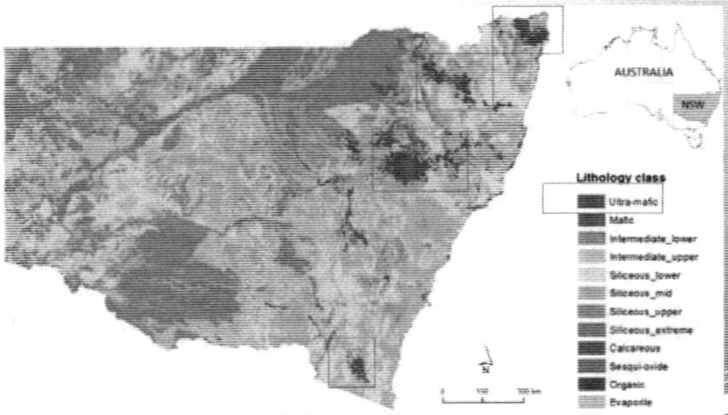

Ultra-mafic,
Dunite
Serpentinite
Peridotite

[1] J.M. Gray et.al, Catena, 2016.

THE UNIVERSITY OF NEWCASTLE AUSTRALIA

Dunite collection

3

- Dunite is collected from The Great Serpentine Belt, NSW, Australia
- Dunite "Doonba" deposit is located close to Bingara[1]

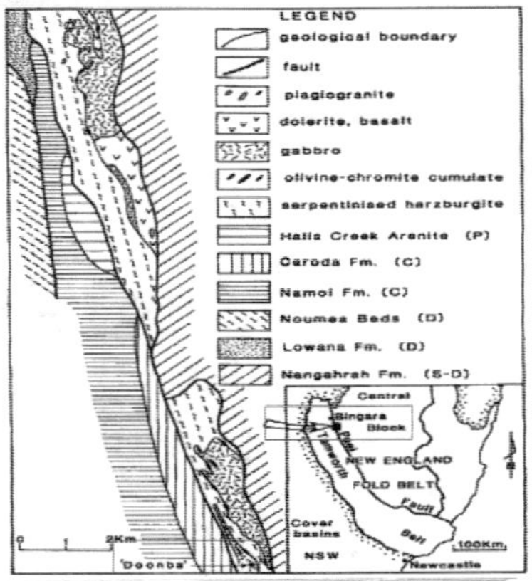

"Doonba" deposit

K. Yang, Canadian Mineralogist, 1993.

THE UNIVERSITY OF NEWCASTLE AUSTRALIA

Approaches for carbonation

4

- Dunite can be heat-activated like lizardite to get higher carbonation yields
- Heat-activated lizardite provide 42% yield in 1.5 h carbonation (140 bar, 185 °C, 0.64 M $NaHCO_3$)
- Regrinding to break silica-rich layers[1]

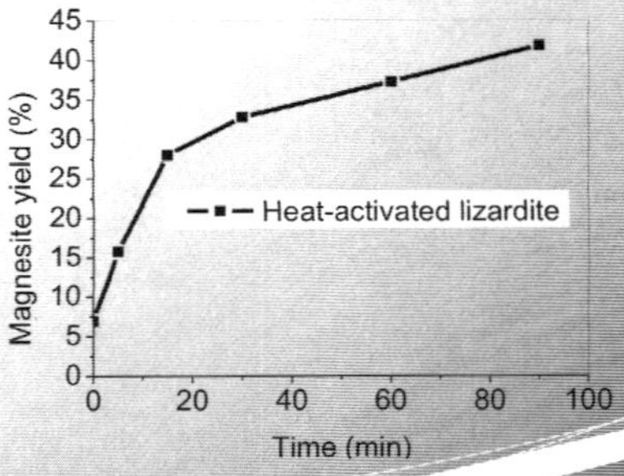

[1] Pokrovsky and Schott, Geochimica et Cosmochimica Acta, 2000.

THE UNIVERSITY OF NEWCASTLE AUSTRALIA

Semi-quantitative PXRD Analysis

5

- Add silicon standard in solid sample
- Calculate peak areas from XRD pattern

$$C_i=A_x\, I_i\, RIR_x\,/(I_x RIR_i) \ldots\ldots\ldots\ldots\ldots\ldots (1)$$

C_i=mass percentage of i phase

A_x=known amount of internal standard (silicon) added to original sample

I_i,I_x=Intensities (peak areas) of phases I and X

RIR_i, RIR_x= reference intensity ratio values of I and X respectively

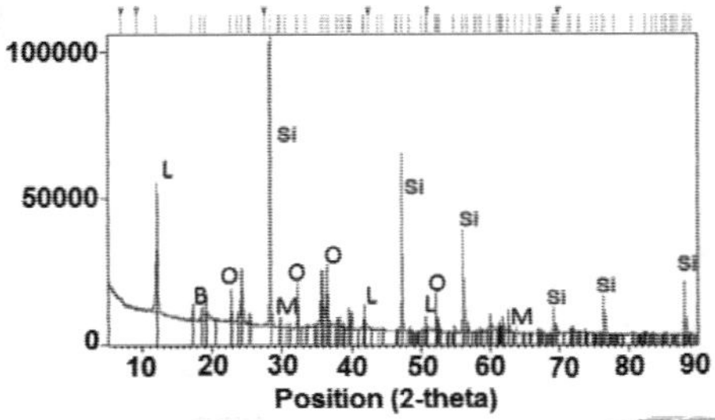

L=Lizardite B=Brucite O=Olivine
M=Magnetite Si=Silicon

THE UNIVERSITY OF NEWCASTLE AUSTRALIA

Dunite Rock Characterisation

6

- Dunite is source of magnesium (26% magnesium)
- Dunite is a mixture of lizardite[1] olivine brucite

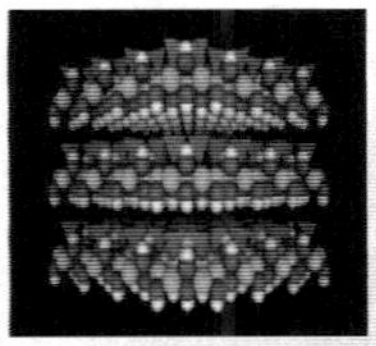

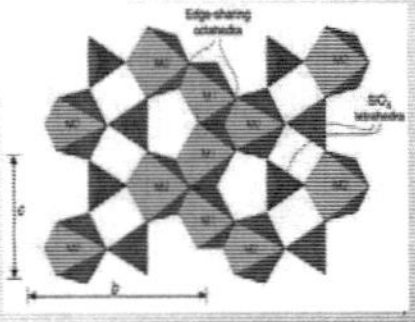

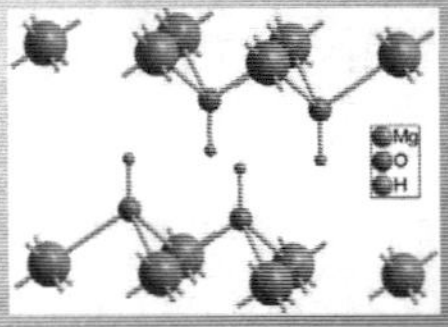

	Lizardite (%)	Olivine (%)	Brucite (%)	Magnetite (%)
Sub 75 μm dunite QXRD	62	33	3.8	0.5
Sub 75 μm dunite TGA-MS	61	30 (difference)	8.3	0.5 (XRD)

[1] Wicks. J. Fred, powder diffraction, 2000.

THE UNIVERSITY OF NEWCASTLE AUSTRALIA

Carbonation test; Reactor setup

7

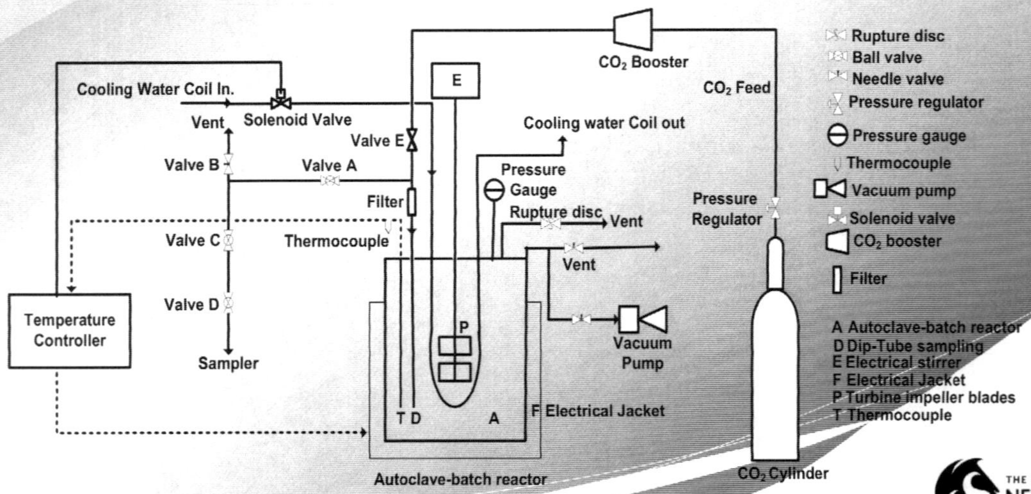

THE UNIVERSITY OF NEWCASTLE AUSTRALIA

Raw dunite carbonation

8

- Dunite was carbonated at 130 bar, 185 °C using 0.64 M $NaHCO_3$ for 1 h

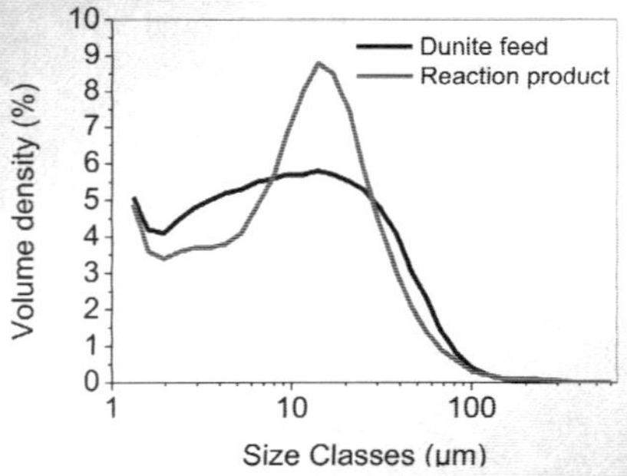

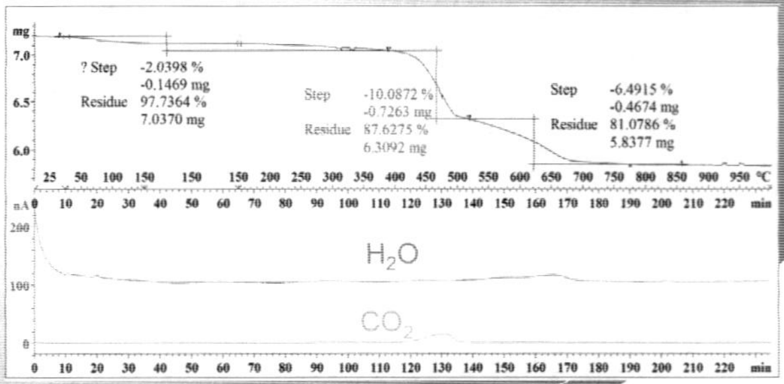

No mass loss for brucite in TGA-MS

Magnesite yield is 24%
(6% from brucite)
(18% from olivine)

THE UNIVERSITY OF NEWCASTLE AUSTRALIA

Reaction product SEM Analysis

9

- Magnesite and silica-rich phase identified
- PSD of medium size fraction increase due to precipitation of magnesite on feed particles

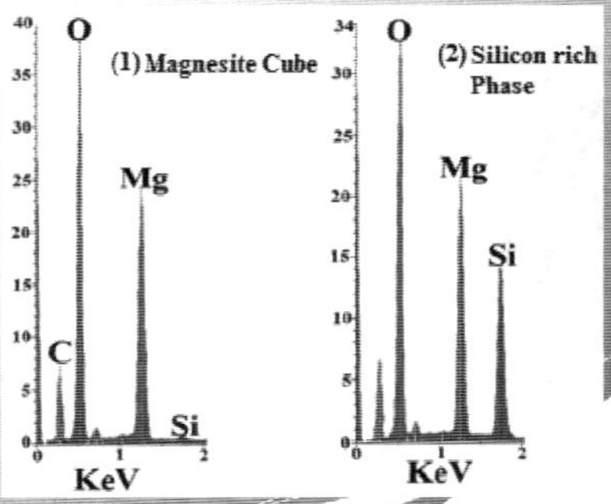

THE UNIVERSITY OF NEWCASTLE AUSTRALIA

Effectiveness of Regrinding

10

- Dunite was carbonated at 130 bar, 185 °C using 0.64 M $NaHCO_3$ for 90 min
- 20 min regrinding performed after 60 min reaction to break down silica-rich layers

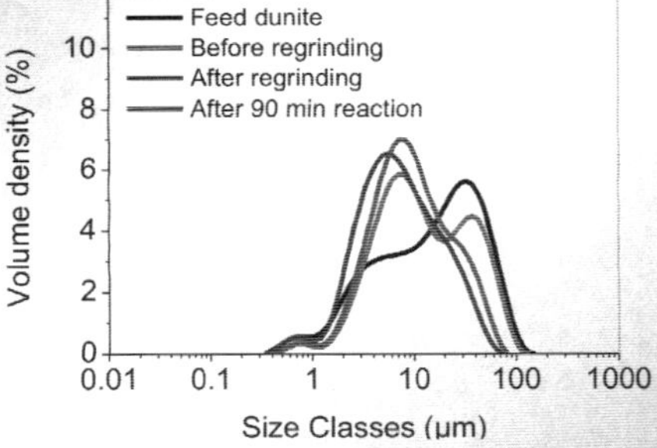

	d_{10} (µm)	d_{50} (µm)	d_{90} (µm)
Feed dunite	2.4	16	54
Before regrinding	3.3	11	49
After regrinding	2.3	6.7	24
90 min reaction	3.0	8.8	31

PSD increase after reground slurry carbonation

THE UNIVERSITY OF NEWCASTLE AUSTRALIA

Effectiveness of Regrinding

11

- Needle-like compounds (secondary material) observed after reground slurry carbonation
- They are observed in literature due to olivine preferential dissolution/degeneration of crystals[1]

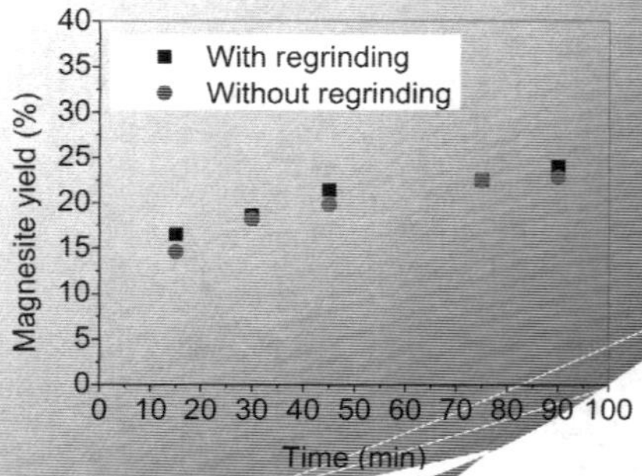

Regrinding ineffective in increasing magnesite yield

THE UNIVERSITY OF NEWCASTLE AUSTRALIA

[1] J. Olsson et.al Geochimica et Cosmochimica Acta, 2012.

Non heat-activated lizardite carbonation

12

- To confirm low reactivity of lizardite mineral in dunite rock, raw lizardite is carbonated

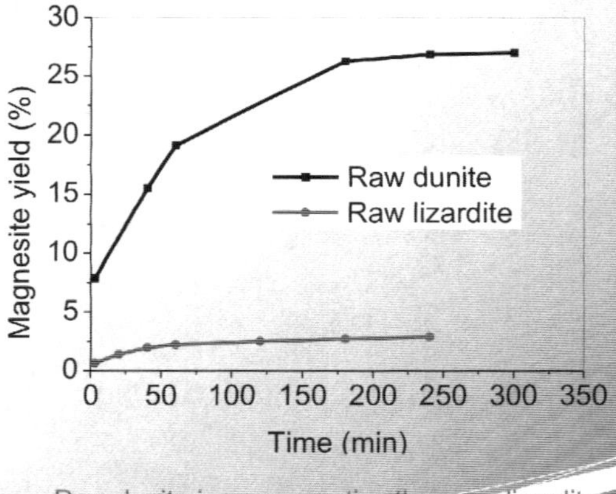

Raw dunite is more reactive than raw lizardite

THE UNIVERSITY OF NEWCASTLE AUSTRALIA

Dunite heat-activation

- Another approach for increasing magnesite yield
- Dunite was heat-activated at 630 °C for 4 h
- Heat-activation converted lizardite (present in dunite) into more reactive amorphous phase

Rotary kiln

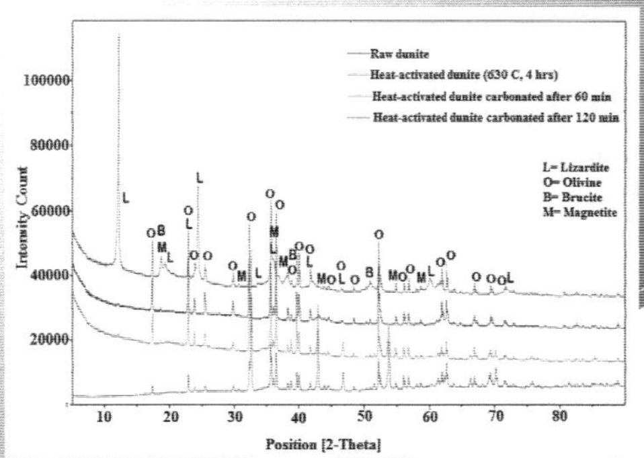

THE UNIVERSITY OF NEWCASTLE AUSTRALIA

Dunite heat-transformation to forsterite

- Dunite was heat-transformed in a rotary kiln at 800 °C for 3 h

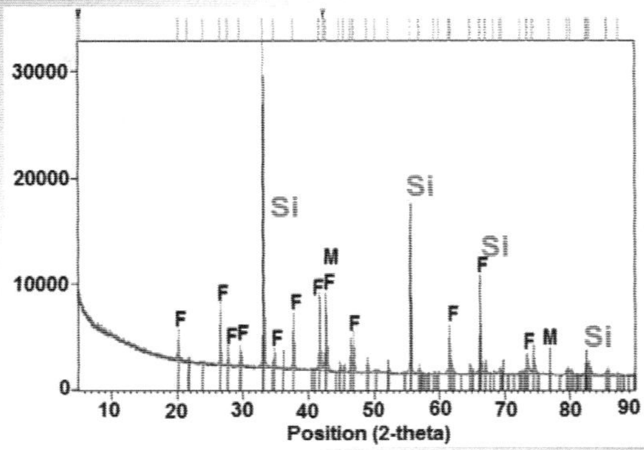

	Lizardite (%wt)	Forsterite (%wt)	Brucite (%wt)	Magnetite (%wt)
Raw Dunite	62	33	3.8	0.56
Heat-transformed Dunite	0	64	0	0.1

Aim is to study forsterite reactivity

THE UNIVERSITY OF NEWCASTLE AUSTRALIA

Heat-activated and heat-transformed dunite carbonation

15

- Heat-activated, heat-transformed and raw dunite was carbonated at 130 bar, 185 °C using 0.64 M $NaHCO_3$

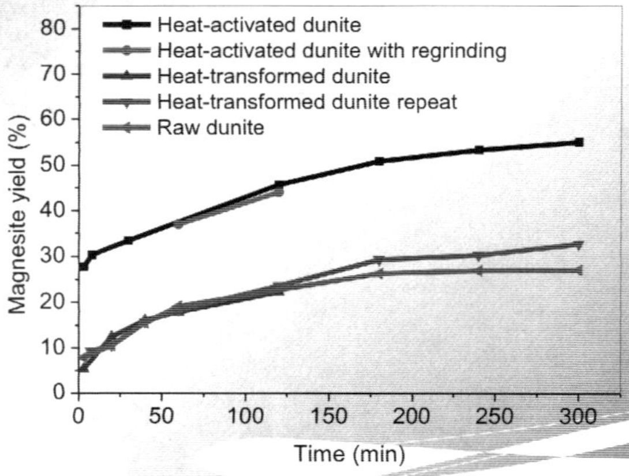

Heat-activated dunite provided 55% yield compared to 27% from raw dunite

Forsterite is less reactive

THE UNIVERSITY OF NEWCASTLE AUSTRALIA

Silanol nests formation during heat-activated dunite carbonation

16

- Sharp OH band present around 3690 cm^{-1} due to lizardite disappear during heat-activation but reappears following carbonation

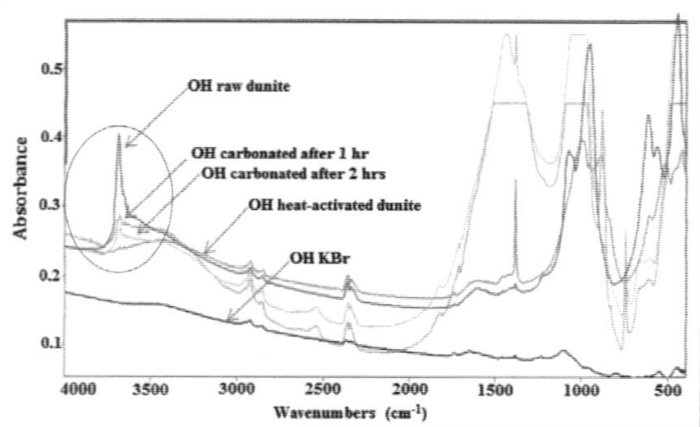

Nests develop at sites of silicon defects, present in the amorphous phase[1, 2]

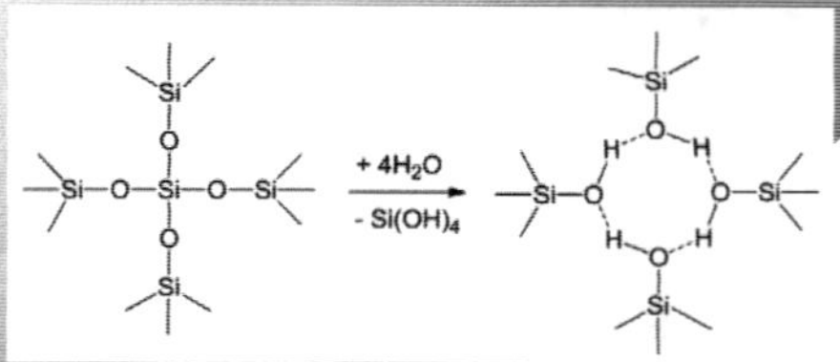

Removal of tetrahedral Si

[1] S. Bordiga et.al, Topics in Catalysis, 2001
[2] H. Bayahia et.al, Chemical Communications, 2013.

THE UNIVERSITY OF NEWCASTLE AUSTRALIA

Silanol nests formation (TGA-MS analysis)

17

- TGA-MS, 150-350 °C[1] and 550-850 °C[2] mass loss

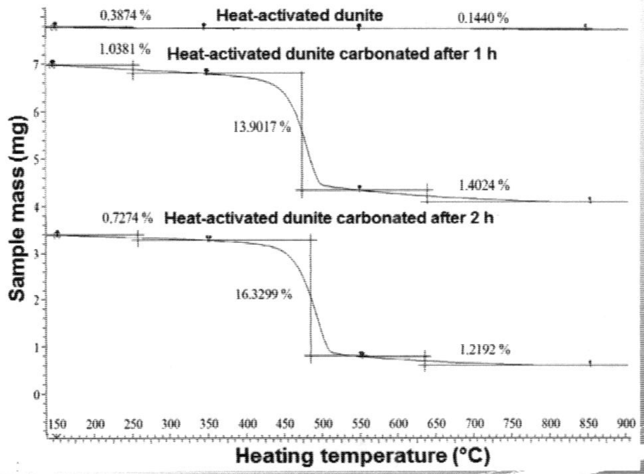

[1] S. Bordiga et.al, Topics in Catalysis, 2001.
[2] H. Bayahia et.al, Chemical Communications, 2013.

THE UNIVERSITY OF NEWCASTLE AUSTRALIA

Acid dissolution experiments

18

- Dissolution tests were performed in a buffer solution (pH=4.7) of acetic acid and sodium acetate

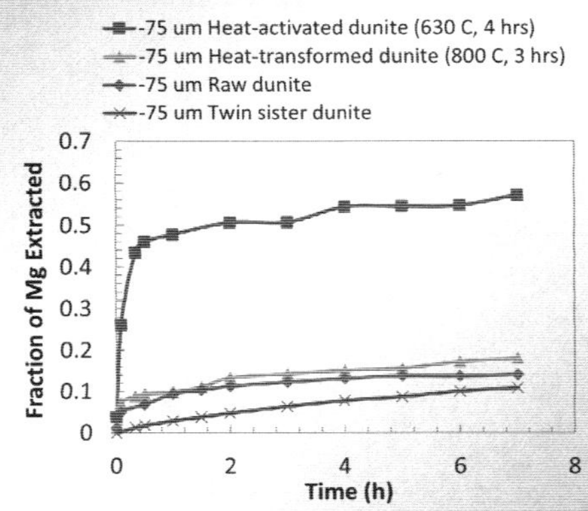

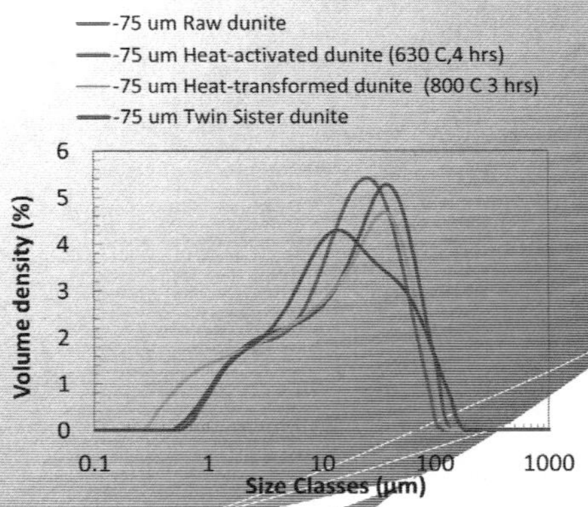

Mg extraction is 57% from heat-activated dunite compared to 14% from raw dunite

THE UNIVERSITY OF NEWCASTLE AUSTRALIA

Conclusions

19

- Raw dunite is more reactive than raw lizardite
- Heat-activated dunite provide 55% (after 5 h) magnesite yield compared to 27% (after 5 h) for raw dunite
- Silanol nests (undesired products) formation observed
- Heat-activated dunite showed highest Mg extraction followed by heat-transformed dunite and raw dunite
- Order of reactivity, heat-activated dunite > heat-transformed dunite > raw dunite > raw lizardite

THE UNIVERSITY OF NEWCASTLE AUSTRALIA

Acknowledgments

20

- The University of Newcastle for postgraduate research scholarship
- Orica Limited and Mineral Carbonation International for funding
- Staff at EM-X-Ray unit (University of Newcastle) for XRD and SEM/EDS

THE UNIVERSITY OF NEWCASTLE AUSTRALIA

21

Thank You

Any Questions?

THE UNIVERSITY OF NEWCASTLE AUSTRALIA

Calibration of Mass Spectrometer

22

- Sodium bicarbonate samples (different masses) were run in TGA coupled with MS from 0-300 °C
- Area under CO_2 peak was calculated. CO_2 mass was calculated from theoretical equation

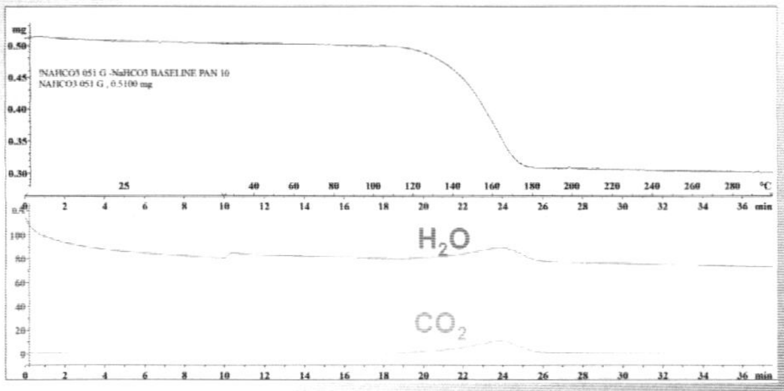

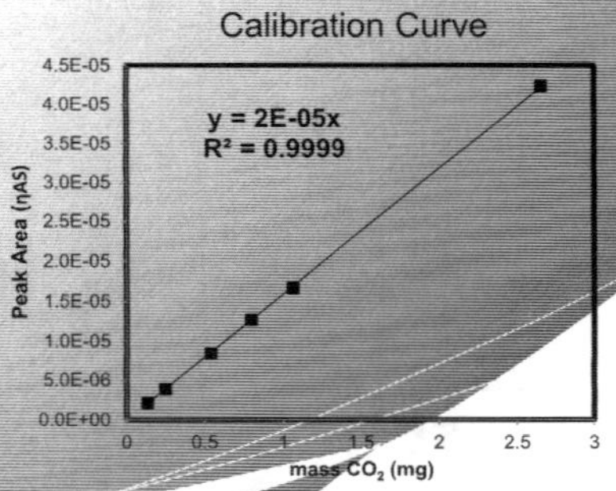

$2NaHCO_3 \rightarrow Na_2CO_3 + CO_2 + H_2O$

THE UNIVERSITY OF NEWCASTLE AUSTRALIA

Magnesite ($MgCO_3$) yield measurement

23

- A product sample (7-8 mg) was heated 25-1000 °C in TGA furnace
- Yield is amount of magnesium converted into magnesium carbonate
- Yield is calculated using calibration curve

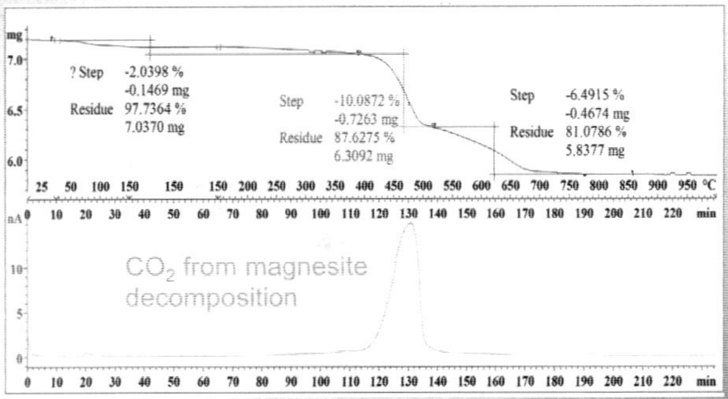

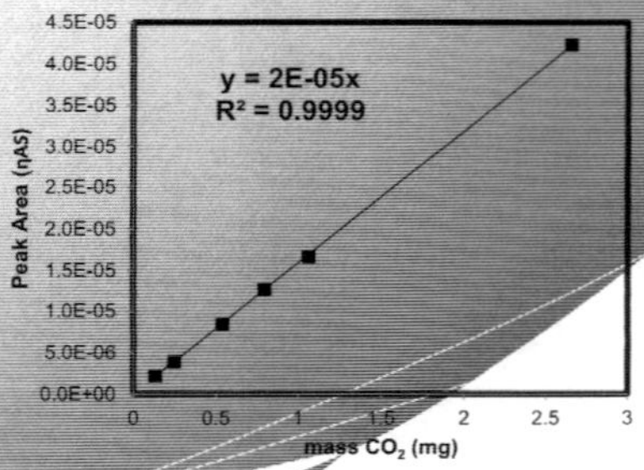

THE UNIVERSITY OF NEWCASTLE AUSTRALIA

Single-stage carbonation of dunite

24

- 5 h run to determine extent of reaction of brucite, olivine and lizardite

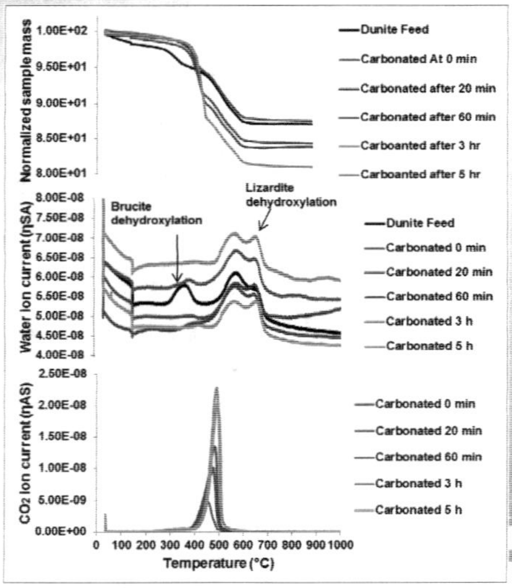

Brucite consumed completely during initial few min of reaction

Lizardite almost unreactive

THE UNIVERSITY OF NEWCASTLE AUSTRALIA

Maximum Theoretical Forsterite Formation

25

- $2Mg_3Si_2O_5(OH)_{4\,(s)} \rightarrow 3Mg_2SiO_{4\,(s)} + SiO_{2\,(s)} + 4H_2O_{(g)}$......1
- $Mg(OH)_{2\,(s)} \rightarrow MgO_{(s)} + H_2O_{(g)}$ 2

Input	Gram	Moles	Output	Moles	Gram
Forsterite	29	0.20	Forsterite	0.33	46.4
Lizardite	61	0.22	SiO_2	0.11	6.60
Brucite	8.3	0.14	Water (eq1)	0.44	7.93
Magnetite	1.3	0.0056	Magnetite	0.0056	1.30
			MgO	0.14	5.73
			Water (eq2)	0.14	2.56
			Forsterite from feed		29
Total	100		Total		99.6

Wt % theoretical forsterite formation = (Total Forsterite/(100-water (eq 1)-water (eq 2))*100)
= (75.4/(99.6-7.93-2.56))*100= 85%

THE UNIVERSITY OF NEWCASTLE AUSTRALIA

Acid dissolution

26

- Different fractions of heat-transformed dunite were dissolved in buffer solution (pH=4.7) of acetic acid and sodium acetate

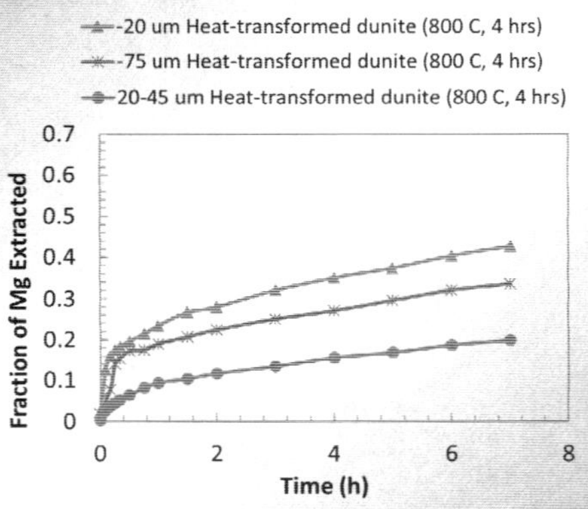

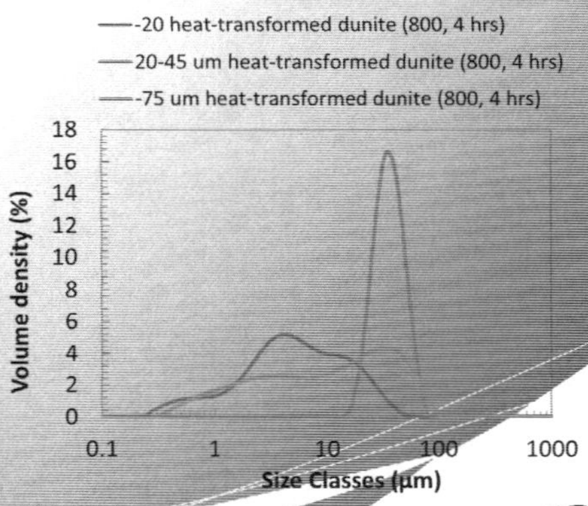

Higher extractions can be achieved by reducing particle size

THE UNIVERSITY OF NEWCASTLE AUSTRALIA

Incongruent dissolution

27

- Materials rich in forsterite ('twin sister' dunite, heat-transformed dunite, raw dunite) show preferential Mg release and exhibit incongruent dissolution similar to that of forsterite

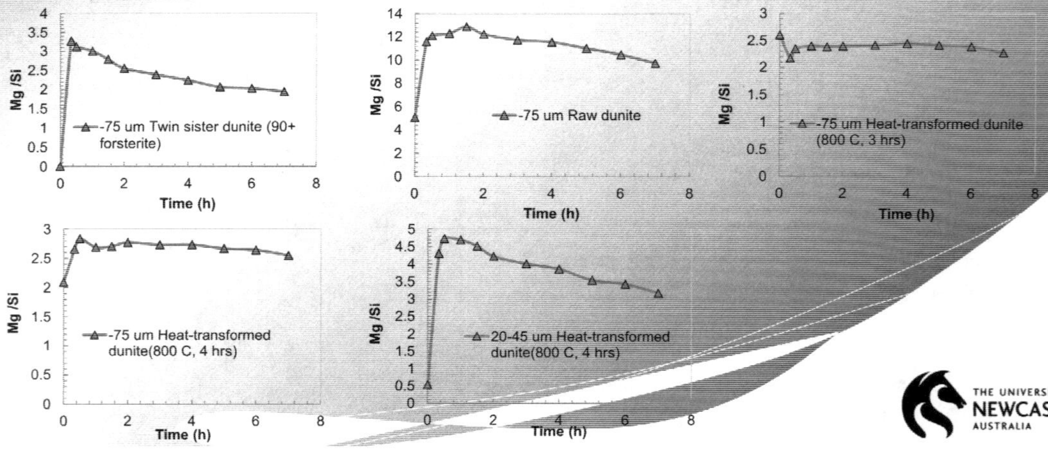

THE UNIVERSITY OF NEWCASTLE AUSTRALIA

Congruent dissolution

28

- Heat-activated dunite (amorphous magnesium silicate rich) behaves differently and shows congruent dissolution
- Heat-activated lizardite also show congruent dissolution[1]

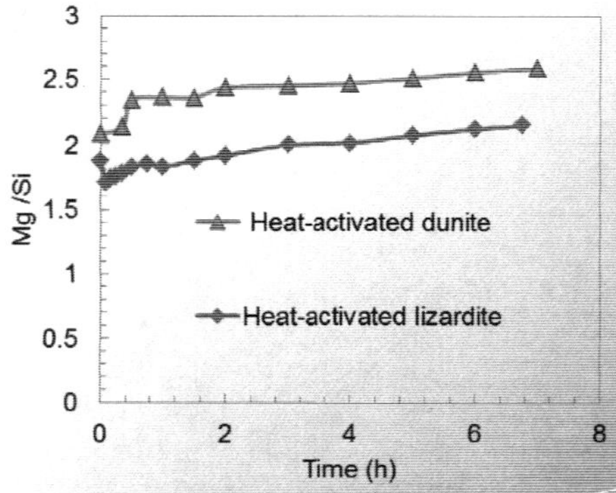

[1] Farhang et.al, Chemical Engineering Journal, 2017.

THE UNIVERSITY OF NEWCASTLE AUSTRALIA

Dynamic CO2 Sequestration in an Industrial-Scale Chrysotile Mining Waste Pile

Authors:

Beaudoin, G., Université Laval
Nowamooz, A., Université Laval
Larachi, F., LAVAL University
Dupuis, C., Université Lval
Molson, J. W. H., Université Laval
Lemieux, J. M., Université Laval
Horswill, M., Université Laval
Fortier, R., Université Laval
Maldague, X., Université Laval
Constantin, M., Université Laval
Duchesne, J., Université Laval
Therrien, R., Université Laval

Carbon mineralization in ultramafic mine wastes has been studied under controlled laboratory conditions and in small-scale experimental waste cells in the field. Mine waste mineralogy, water saturation, CO_2 availability, and temperature have been identified as key factors controlling the sequestration but can be optimized for more efficient long-term CO_2 storage.

To investigate CO_2 sequestration in an industrial-scale mining waste pile under natural environmental conditions, a 92-m deep borehole was drilled in a 110 Mt chrysotile mine waste pile at Thetford Mines, Québec, Canada. This borehole was instrumented with 20 thermistors for temperature measurements and 9 pneumatic ports for differential pressure measurements and for sampling CO_2 in the interstitial air. The carbon isotope composition of CO_2 in the samples was also measured periodically using a Wavelength Cavity Ring Down Spectroscopy instrument (Picaro G-1101-i). A weather station close to the borehole provided the local environmental conditions.

A nonlinear thermal regime with two thermal gradients was observed in the pile. The thermal gradient for the upper 50 m was 18 °C/km whereas the lower portion had a thermal gradient of 42 °C/km, 1.5 times higher than the expected regional thermal gradient (27 °C/km). These gradients suggest the presence of two thermo-hydro-chemical regimes for the transport and storage of CO_2 within the pile. During cold-weather periods, when the ambient air temperature is less than the temperatures prevailing in the pile (which typically varyfrom 10 to 14 °C), the pressure in the pile is higher than the atmospheric pressure due to thermal expansion of the interstitial air. This thermodynamic effect impedes the inflow of CO_2 from the atmosphere, while the carbonation reactions continue to consume CO_2 and reduce its concentration within the pile. During warm-weather periods, when the ambient air temperature is higher than the pile temperature, a general reversal of differential pressure enhances air inflow and migration of CO_2 into the pile, replenishing CO_2 concentrations. Nevertheless, during this period the carbonation mineralization reactions inside the pile reduce the CO_2 concentration of the inflowing air by 40%. The carbon isotope composition of CO_2 shows a transition from a kinetic, closed-system Rayleigh fractionation during the cold periods to an isotopic system at equilibrium between gas and dissolved CO_2 in interstitial water during the warm periods.

A thermo-hydro-aeraulic numerical model was developed to test the conceptual model and to assess the amount of CO_2 that can be sequestered in such industrial-scale waste piles. The numerical results suggest that about 100 tons of CO_2 per year are stored passively in the Thetford Mines waste pile which is in the same order of magnitude as the estimated storage capacities of other mining waste piles cited in the literature. Model results also suggest that a distributed heat source of at least 0.1 mW/m3 due to the heat released by the exothermic carbon mineralization reactions can explain the thermal regime observed in the pile.

International Conference on Accelerated Carbonation for Environmental and Material Engineering (ACEME)

Chemical and Crystallographic Controls on Wollastonite Carbonation

E. Ruiz-Agudo, F. di Lorenzo, **L. Monasterio** and C. Rodriguez-Navarro

Universidad de Granada

CEI Center for Energy Initiatives — An AIChE Technological Community

GOBIERNO DE ESPAÑA — MINISTERIO DE ECONOMÍA Y COMPETITIVIDAD

Introduction

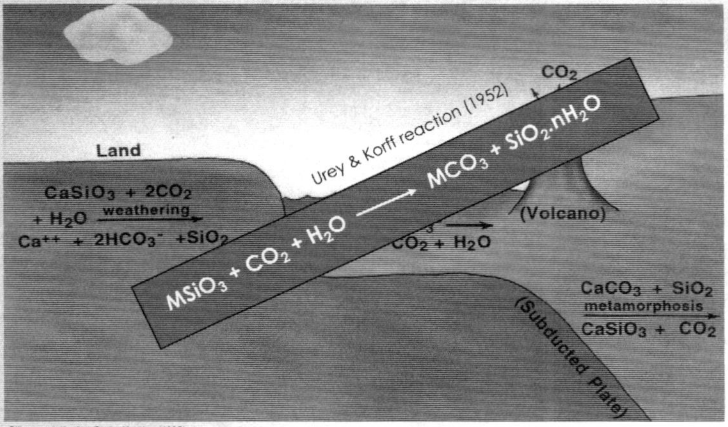

Silicate weathering Cycle. Kasting (1995)

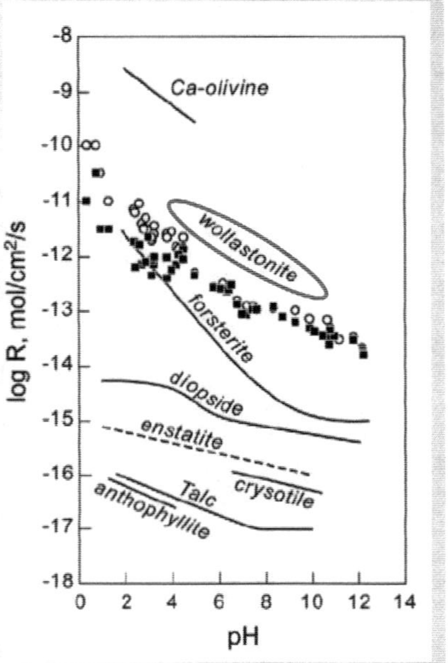

Comparison of dissolution rates of Ca/Mg silicates related to pH at 25°C. Schott et al. (2012)

Why Wollastonite?

- High dissolution rate
- Natural mineral
- Simple chemical composition ($CaSiO_3$)

Wollastonite dissolution / Interface-Coupled Dissolution-Precipitation

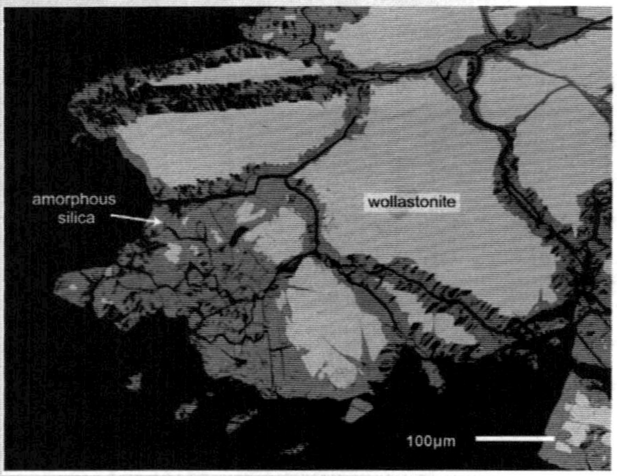

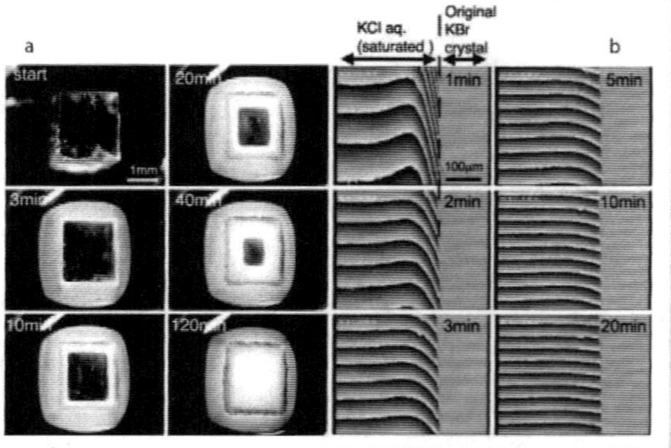

Mineral Replacement Process (MRP)

All the magic happens at the interface!

BIG PROBLEM!!

PASSIVATION LAYER

Objectives

Explore the role that the concentration gradients at the mineral-solution interface have on Mineral Replacement Processes (MRPs)

Understand under which conditions carbonation can be promoted by inhibiting the formation of passivation layers

Evaluate the role of the products' growth mechanism on porosity generation (key for the progress of the replacement reaction): epitaxial fit *vs* no-structural fit between reactant and product phases

Materials and methods

Natural Wollastonite from South Africa

Teflon reactors in hydrothermal autoclaves

90ºC / 2-7-15 days

pH 3, 6 (using HCl), 7.5 and 10 ($NaHCO_3$, Na_2CO_3)

XRD, FESEM, µ-RAMAN

PHREEQC for the modelling of the mineral solutions

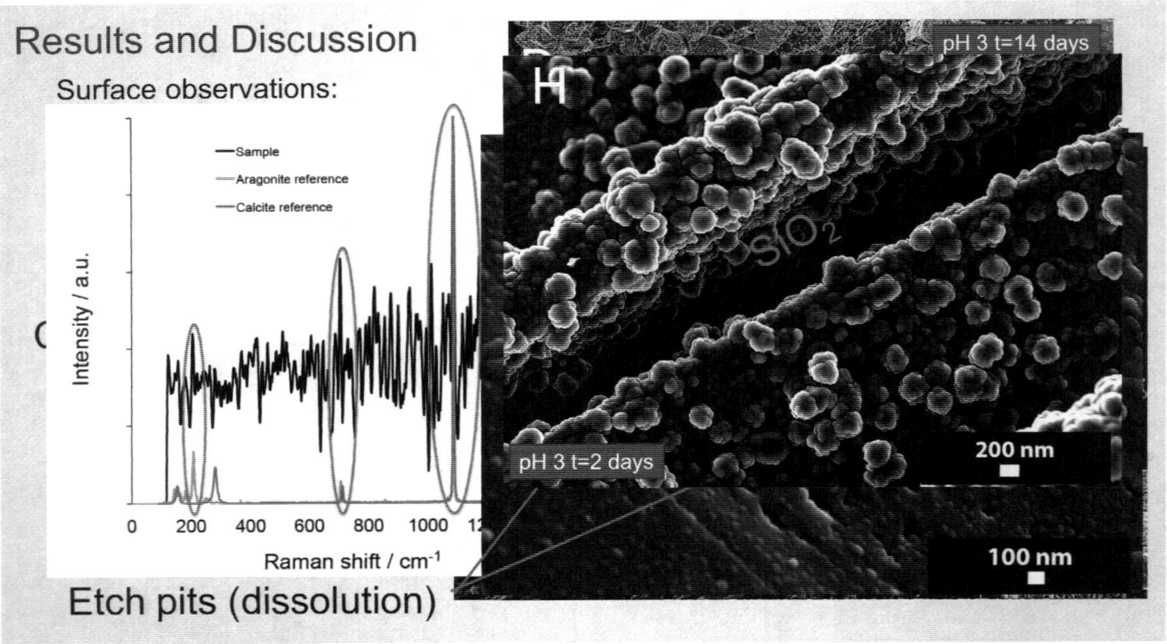
Results and Discussion
Surface observations:
Sample
Aragonite reference
Calcite reference
Intensity / a.u.
0
200
400
600
800
1000
Raman shift / cm-1
pH 3 t=14 days
SiO2
200 nm
pH 3 t=2 days
100 nm
Etch pits (dissolution)

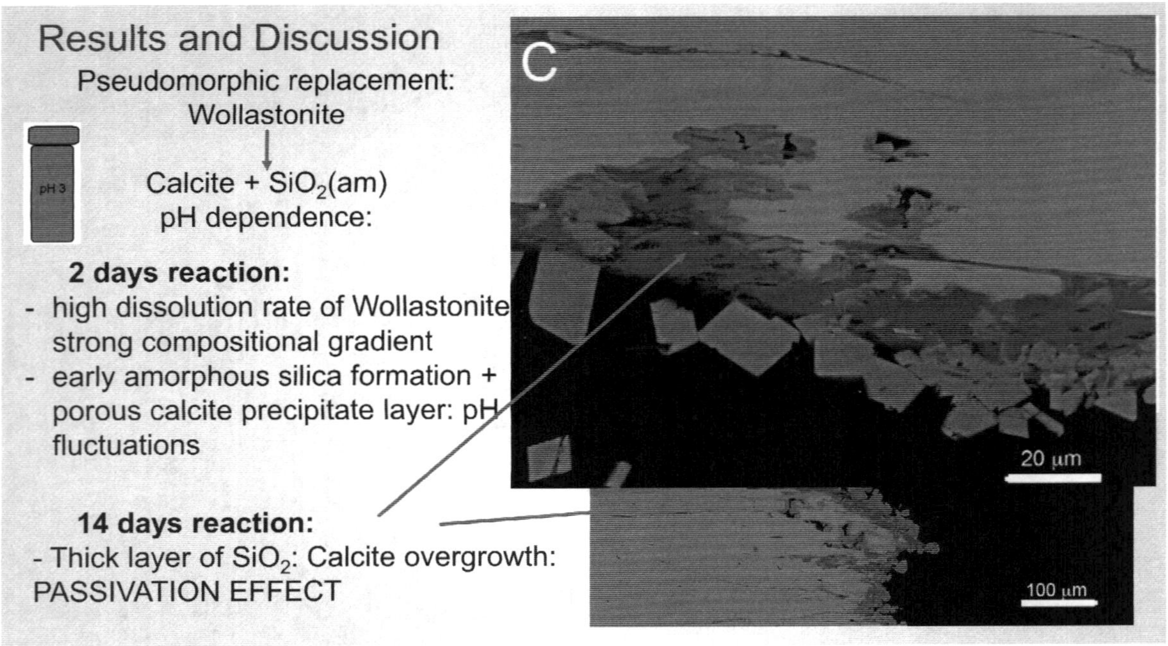
Results and Discussion
Pseudomorphic replacement:
Wollastonite
Calcite + SiO2(am)
pH dependence:
pH 3
2 days reaction:
- high dissolution rate of Wollastonite
strong compositional gradient
- early amorphous silica formation +
porous calcite precipitate layer: pH
fluctuations
14 days reaction:
- Thick layer of SiO2: Calcite overgrowth:
PASSIVATION EFFECT
C
20 μm
100 μm

Results and Discussion

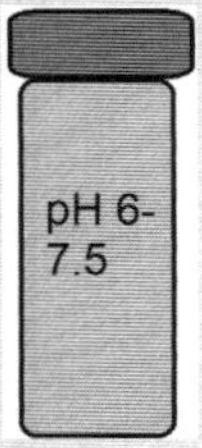

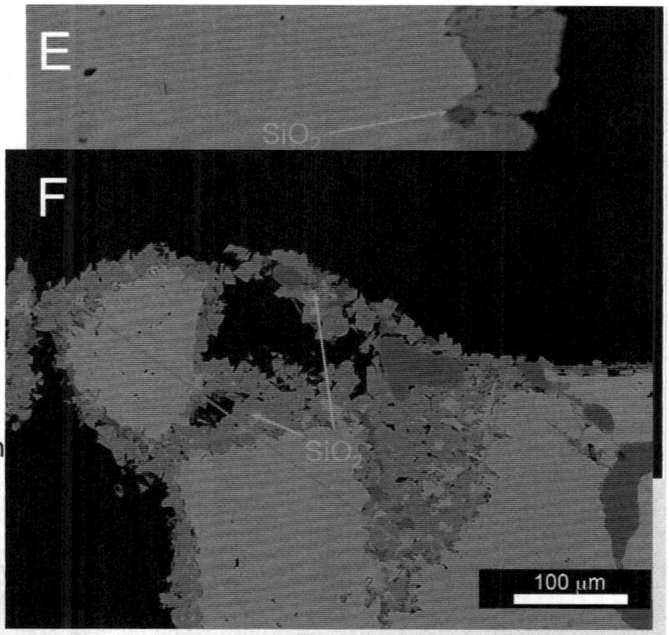

pH 6-7.5:

- Dominant calcite presence
- SiO_2 (am) as scarce patches intergrown with calcite and coverin calcite: simultaneous precipitation stop reaction = passivation. However.... we have cracks! → progress of the reaction

Results and Discussion

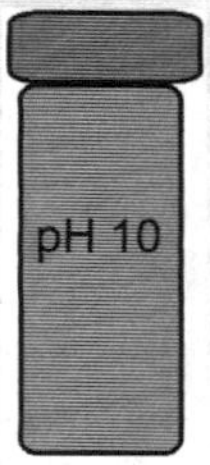

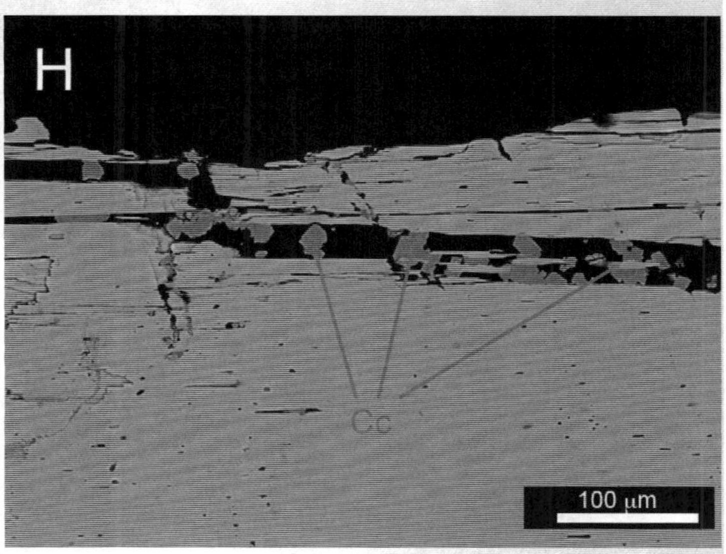

pH 10:

- Limited replacement
- Scarce calcite crystals in the cracks
- No silica detected: no passivation!

PROBLEM: KINETICS.
Alkaline pH = low wollastonite dissolution rate
= LIMITED CARBONATION

Results and Discussion

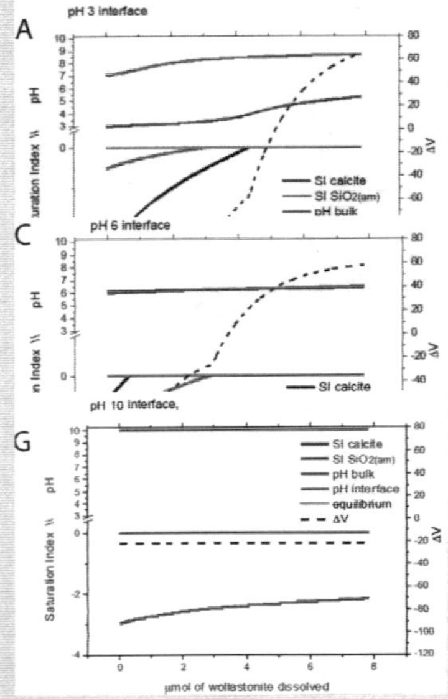

PHREEQC Simulations

$$\delta = \frac{D}{k}$$

δ= Boundary layer thickness
D= solute diffusion coefficient
k= mass transfer coefficient

Mass transfer process:
pH 3: - SiO_2 supersaturation
- Reaction procedes: Calcite supersaturation
pH 6-7.5: - Supersaturation Calcite>SiO_2
pH 10: Not supersaturation of SiO_2 : Not passivation

Results and Discussion

EBSD pH 7.5

1. Cracks formation along b-axis wollastonite
2. Calcite precipitation in cracks: Uniform crystallographic orientation {110}Cc with calcite twinning, parallel to {001}Wol
3. External calcite overgrowth: not preferencial orientation

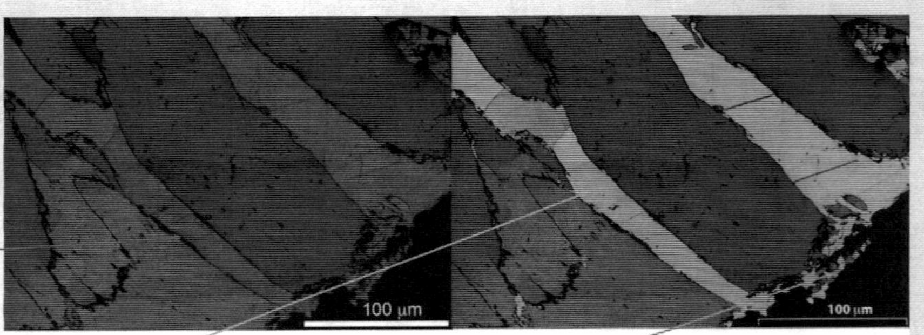

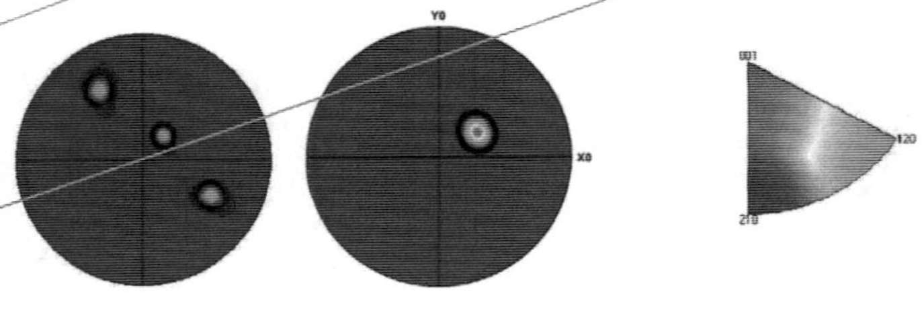

PF {110} Calcite   PF {001} Wollastonite

Results and Discussion

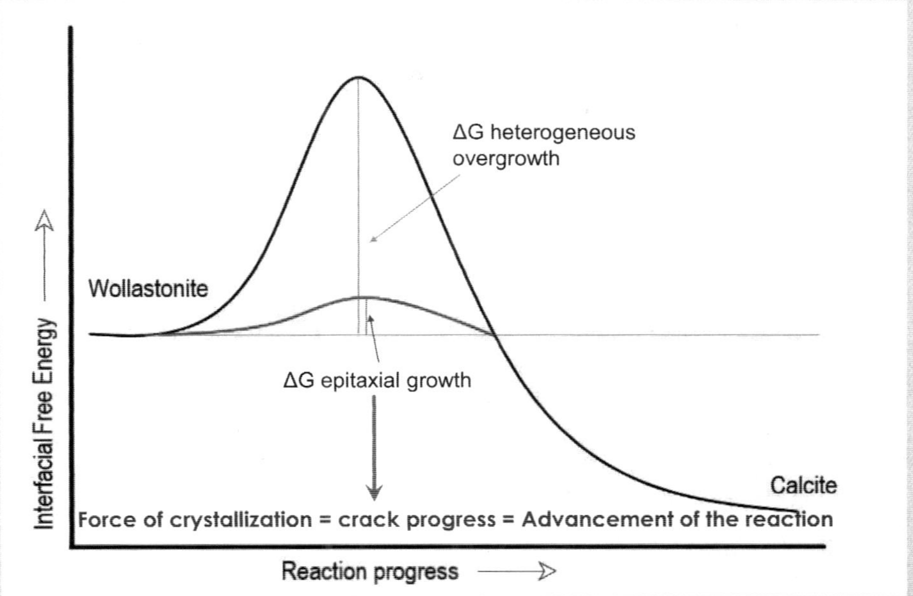

Conclusions

- Specific physico-chemical conditions during silicate carbonation control the development of passivated layer. A proper selection pf pH, close to neutral, maximize carbonation while preventing passivation
- Fine-tuning of porosity development could be achieved by controlling relative rates of mass transport / surface reaction during MRPs (e.g. hydrodynamic conditions, optimal pH selection)
- Knowledge of solution compositions and pH conditions allow prediction of the success of carbonation involving interface-coupled dissolution-precipitation (ICDP) reactions
- The presence of the silicate mineral surface during carbonation reactions influences calcium carbonate polymorph selection and epitaxial matching
- Passivation of wollastonite surface only occurs when simultaneous precipitation of amorphous silica and calcite takes place (extremely high dissolution rates at low pH)

TAKE AWAY

A balance between maximizing dissolution of wollastonite and minimizing passivation must be achieved to reach optimal carbonation

Acknowledgements

This research was funded by the Spanish Government (grants CGL2015-70642- R, CGL2015-73103-EXP), the European Commission (ERDF funds), the University of Granada ("Unidad Científica de Excelencia" UCE-PP2016-05) and the Junta de Andalucía (P11-RNM-7550).

Accelerated Mineral Carbonation Technology for Land Reclamation: A Bridge to Link Mitigation and Adaptation Actions

Authors:

Bu, J., Institute of Chemical and Engineering Sciences (ICES - A*Star)
Yeo, T. Y., Institute of chemical and engineering sciences
Sharratt, P., A*Star. Institute of Chemical and Engineering Sciences

With current greenhouse gas (GHG) emissions due to human activities already exceeding 40 Gt CO_2e per year, there is a large demand for CO_2 mitigation technologies. Unfortunately, despite the pledges, various recent studies agree that current efforts laid out in each countryâ€™s Nationally Determined Contributions (NDCs) are insufficient to meet the 2Â°C, much less the 1.5Â°C, target. It is estimated that the plans outlined in the current NDCs would still likely lead to a global warming scenario of up to 3.8Â°C in 2100, ultimately resulting in significant and catastrophic impacts on humanity. One of the most critical consequences of global warming is the rise in sea levels (SLR) due to melting ice from glaciers and the Earthâ€™s polar regions. The IPCC Fifth Assessment Report (AR5) projected a global SLR of up to 98 cm in 2100, but more recent studies suggest that the increase in SLR could be much higher than that due to collapsing ice sheets. This is a sobering warning, especially for coastal cities and low-lying areas that would be threatened by inundation and flooding from SLR. A higher SLR would certainly impact even more people and cause even greater losses.

This paper describes our effort over the past ten years to develop an integrated mineral carbonation (MC) technology, where the bulk carbonation process; producing carbonated materials could be used for land reclamation. In addition, the mineral upgrading technology also has been developed; this makes the whole process potentially to be self-funding through income generated from value-added products derived from minerals. Furthermore, instead of treating mitigation and adaptation as two mutually exclusive ideas, we propose a climate change policy based on a synergy between these two. This synergy is achieved through the development and implementation of MC technologies. Not only does MC have the capacity to sequester gigatonne-levels of CO_2, it can also produce materials such as mineral carbonates and silica that are useful for land reclamation purposes or the construction of dykes and raised river banks to protect inhabited areas against SLR and flooding. Thus, MC can contribute to both mitigation and adaptation measures simultaneously. MC Tech becomes more economically feasible and practical. In this paper, we present an in-depth discussion of the costs of MC and synergy between mitigation and adaption, as well as its benefits of protecting against the damage of SLR. Aspects that are crucial for the success of this synergy, such as process design and selection, material suitability, economics, public sector support and industry/commercial interests will also be discussed.

ACEME 2018

Newcastle (Australia), March 14, 2018

Scale up of the wet route carbonation of steel slag in a rotary kiln pilot-scale reactor

Renato Baciocchi, Giulia Costa, Paola Librandi

University of Rome Tor Vergata

Stefano Stendardo

ENEA, Research Center Casaccia

Università di Roma Tor Vergata

ENEA Agenzia nazionale per le nuove tecnologie, l'energia e lo sviluppo economico sostenibile

ACEME 2015: wet route carbonation of BOF slags

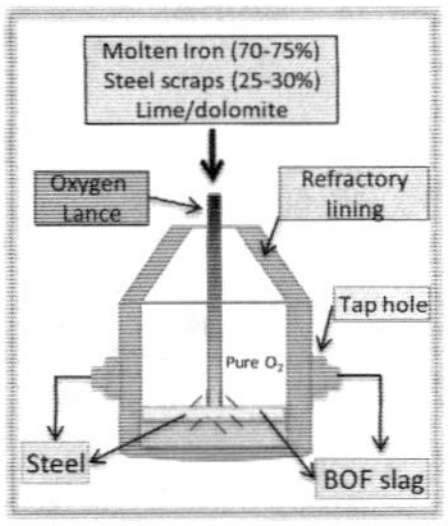

Basic Oxygen Furnace

Euroslag 2012

9.8 Mt BOFs/y

WET ROUTE (L/S<1 l/kg):

T=20-50 °C and p_{CO2}=1-10 bar

Lab scale batch tests

Energy requirements

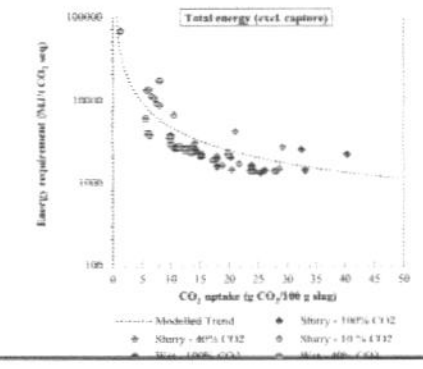

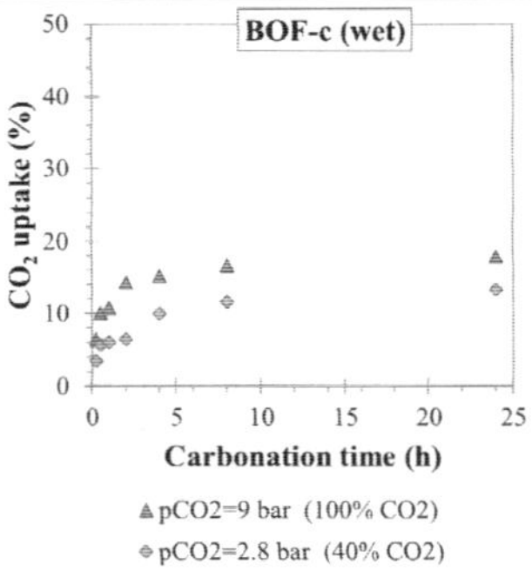

✓ Wet route carbonation feasible using diluted CO_2 sources (down to 40%)

✓ CO_2 uptake around 10% in 4h

ACEME 2015: Energy requirements of BOF slags carbonation

- Total energy requirement vs CO_2 uptake (excluding capture!!)

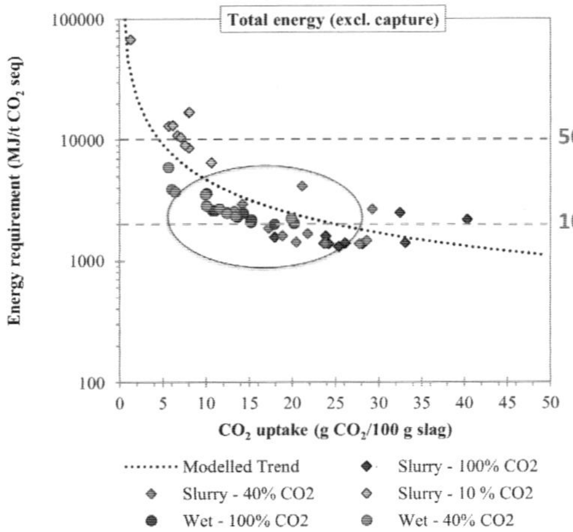

$$E_{tot} = a + \frac{b}{CO_{2_{upt}}^{1.5}}$$

a = 1524.2 MJ/t_{CO2}

b = 101.5 MJ $t_{CO2}^{0.5}/t_{slag}^{1.5}$

Scope of the work

WET ROUTE (L/S<1 l/kg):

T=20-50 °C and p_{CO2}=1-10 bar

Scale-up

Lab scale batch

Pilot scale plant

Scale-up of the wet route

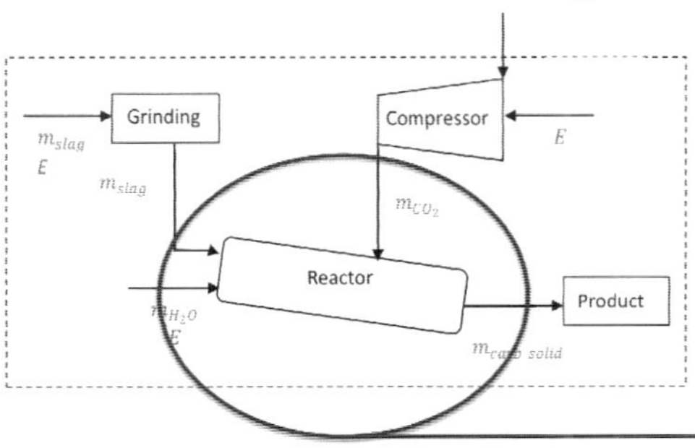

Pilot scale rotating drum

Pilot-scale facility: Valchiria

Features

- Rotating drum reactor
- Length= 150 cm
- Diameter = 25 cm
- Max Slope = 7°
- Max rotation = 8 rpm

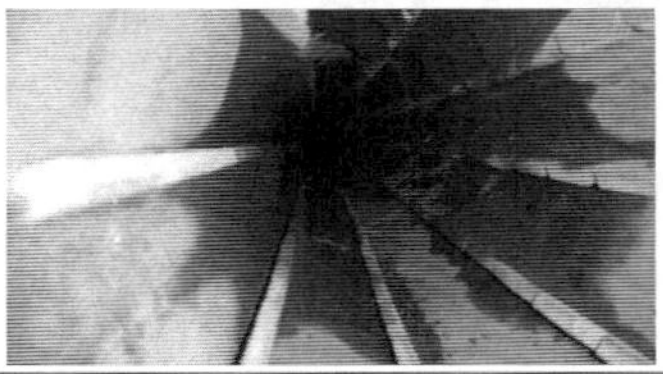

ENEA
Agenzia nazionale per le nuove tecnologie, l'energia e lo sviluppo economico sostenibile

Pilot-scale tests Valchiria

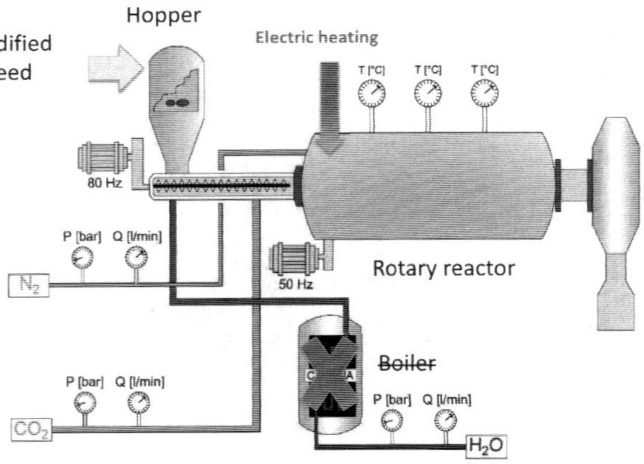

E tests

- electric heating (T controlled)
- slags humidified before feeding to the reactor

Pilot-scale tests Valchiria

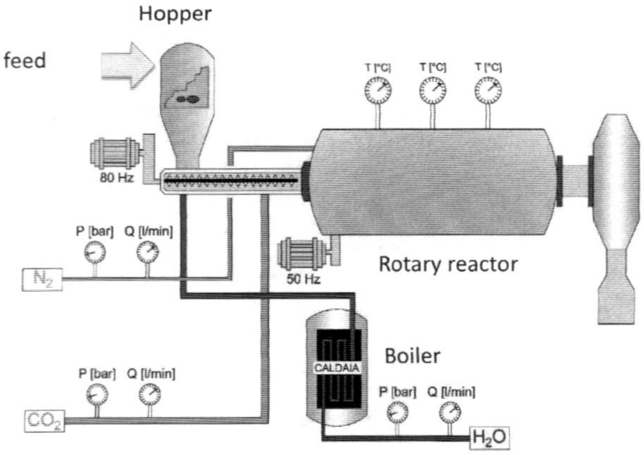

Z tests

- direct heating and slag humidification with **steam injection**
- on-line CO_2 outlet concentration measurement

Materials

BOF slag — By-product of a steelmaking plant employing the *Basic Oxygen Furnace* process

Element	Concentration [g/kg]		
	BOF1 slag	BOF2 slag	BOF3 slag
Ca	212.4	319.1	322.7
Fe	178.7	153.9	181.9
Mg	44.1	57.9	63.7
Si	39.1	59	65.1
Al	21.2	17.6	11.1
Mn	15.5	16.6	13.8
Total Carbon [%]	1.26	1.78	0.94
Inorganic Carbon [%]	0.64	1.36	0.84

BOF 1&2 used for E tests

BOF3 used for Z tests

- **Similar physical and chemical properties for the tested BOFs**
- **BOF2/3 Ca & Si content 50% higher than BOF1**
- **BOF2/3 inorganic carbon content higher than BOF1**

E tests: operating conditions

Experiment code	Target conditions	Material	Measured operating conditions	
			$CO_{2,IN}$ (%)	Temperature (° C)
E1 (LT, BOF1)	GC40%_LT_W	BOF1 slag	40	34
E2 (LT, BOF1)	GC40%_LT_W	BOF1 slag	46	37
E3 (LT, BOF2)	GC40%_LT_W	BOF2 slag	47	25
E4 (LT, BOF2)	GC40%_LT_W	BOF2 slag	47	27
E5 (HT, BOF1)	GC40%_HT_W	BOF1 slag	40	50
E6 (HT, BOF2)	GC40%_HT_W	BOF2 slag	42	45
E7 (HT, BOF2)	GC40%_HT_W	BOF2 slag	42	53

t = 30 min

L/S = 0.17 l/kg
Binder = deionized H_2O

CO_2 = 40%
p_{CO2} = 1 bar

Temperature
- low (LT)
- high (HT)

Treated material → *Curing in air* (*Room temperature*, $RH = 100\%$, $t = 28$ d) →

- **CO_2 uptake**
- **Particle size distribution**
- **Environmental behavior (EN 12457-2)**

E tests – Particle size distribution

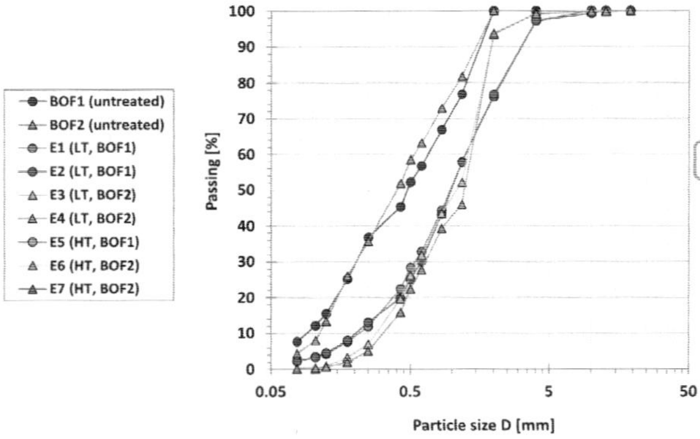

Pilot scale tests

D_{50} = 0.5 mm → D_{50} ≈ 1 mm
as received slag → treated slag (LT)

D_{50} = 0.5 mm → D_{50} < 0.8 mm
as received slag → treated slag (HT)

Lab scale tests

D_{50} = 0.5 mm → D_{50} > 3 mm
as received BOF1 slag → treated BOF1 slag

- **Particle size increase especially at lower operating temperature**

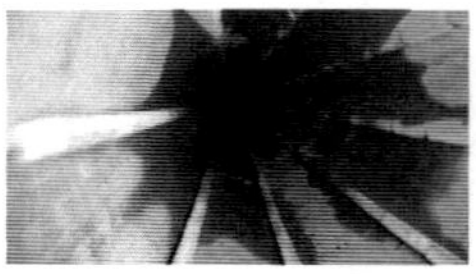

E tests – CO_2 uptake

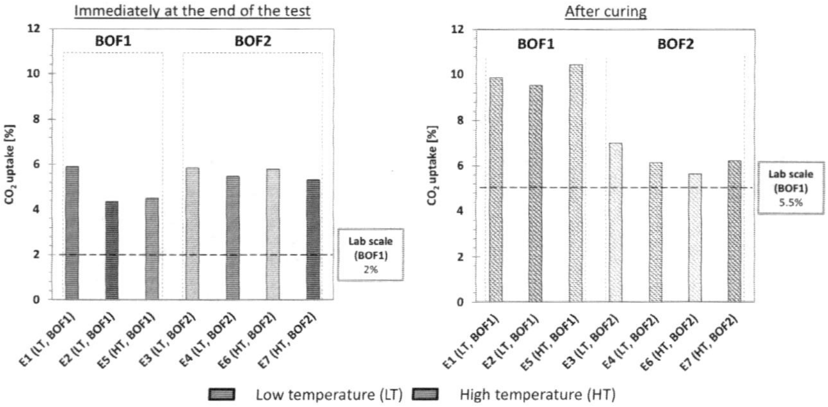

- **Temperature does not affect the CO_2 uptake**
- **CO_2 uptake values are similar immediately at the end of the test (4% - 6%)**
- **After curing, BOF1 CO_2 uptake values doubles (max value around 10%)**
- **CO_2 uptake values in the pilot scale tests twice the ones obtained at lab scale for BOF1**

E tests – Environmental behavior

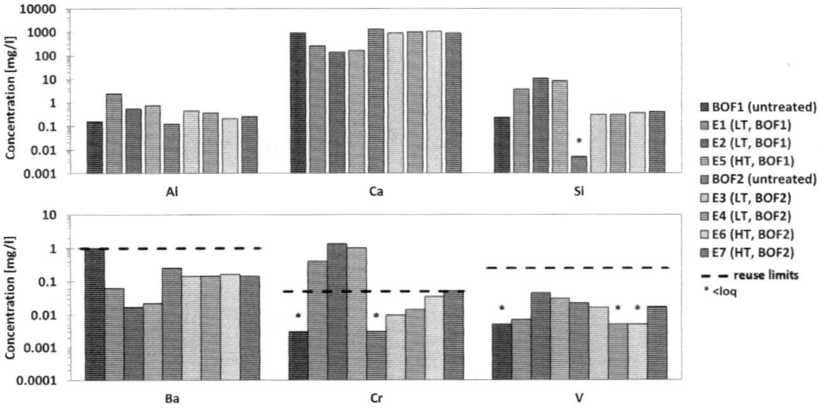

- ↑ **Al & Si, especially for BOF1 slag**
- ↓ **Ba, especially for BOF2 slag**
- ↑ **Cr, especially for BOF1 slag**

Z tests: operating conditions

Experiment code	Target conditions	Material	Measured Operating conditions	
			$CO_{2,IN}$ (%)	Temperature (° C)
Z1 (HT, BOF3)	GC40%_HT_W	2.4 kg BOF3 slag	39.9	45.4 max
Z2 (HT, BOF3)	GC40%_HT_W	2.4 kg BOF3 slag	46.5	49.9 max
Z3 (HT BOF3)	GC40%_HT_W	5.0 kg BOF3 slag	39.8	59.8 max

Test code Z: - steam injection (20 l/min)
- on-line CO_2 outlet concentration measurement

t = 60/100 min

Target L/S = 0.17 l/kg
Binder = deionized H_2O

CO_2 = 40%
p_{CO2} = 1 bar

Temperature
- high (HT)

Treated material → Curing in air (Room temperature, $RH = 100\%$, $t = 28$ d) →

- **CO_2 uptake**
- **Environmental behavior (EN 12457-2)**

Z tests – raw CO_2 and T data

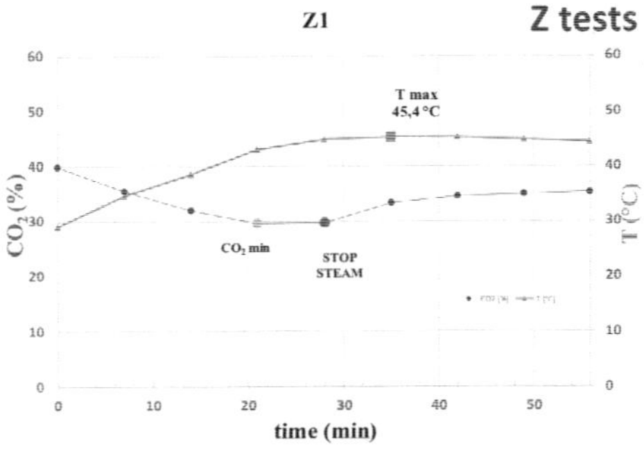

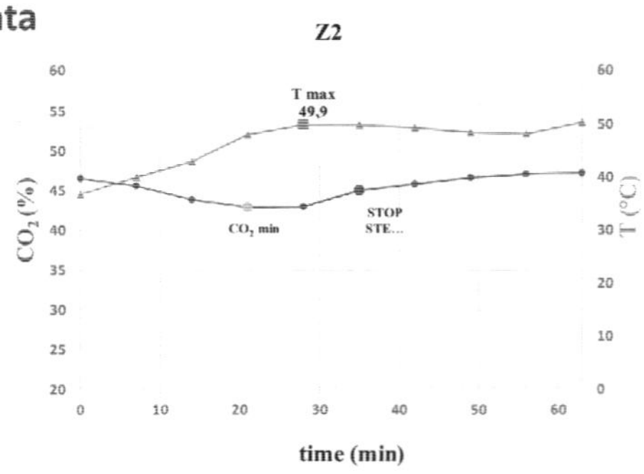

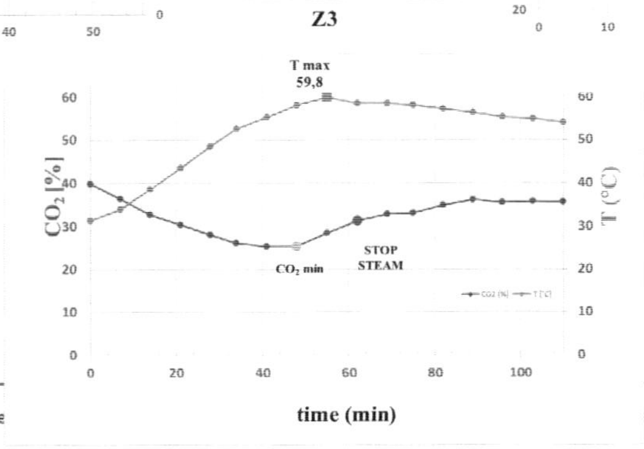

Z tests – CO_2 uptake

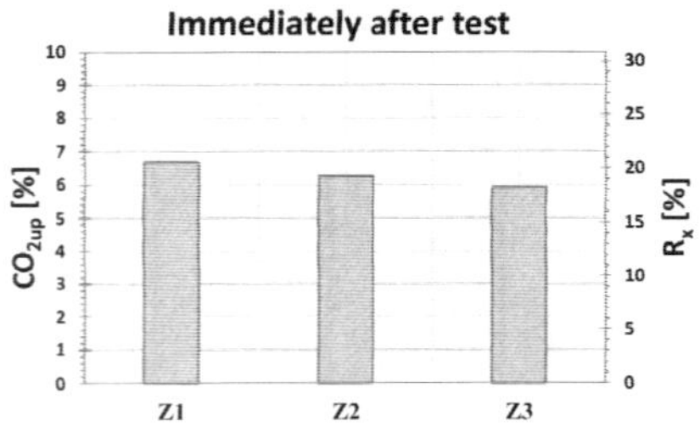

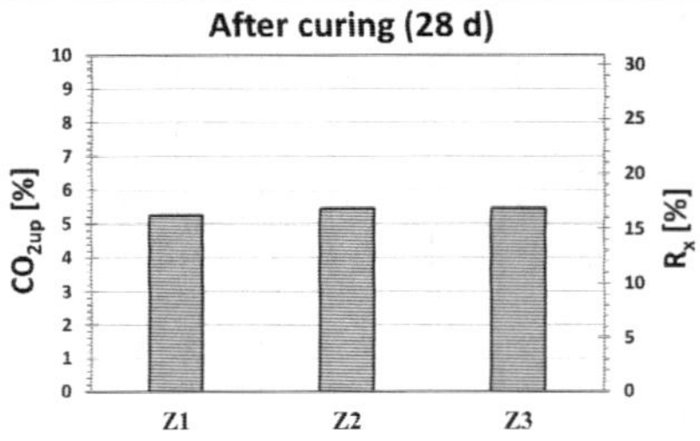

Slag humidity at the end of test

Test	U [%] t=0 d	L/S [l/kg]
Z1	10,01	0,11
Z2	16,39	0,20
Z3	6,06	0,06

Target L/S = 0.17 l/kg

CO_2 uptake / from gas phase mass balance

Test	CO_2 uptake [%] (t=0 d)	CO_2 mass balance [%]
Z1	6,80	7,46
Z2	6,20	2,73
Z3	5,90	8,22

Z tests – Environmental behavior

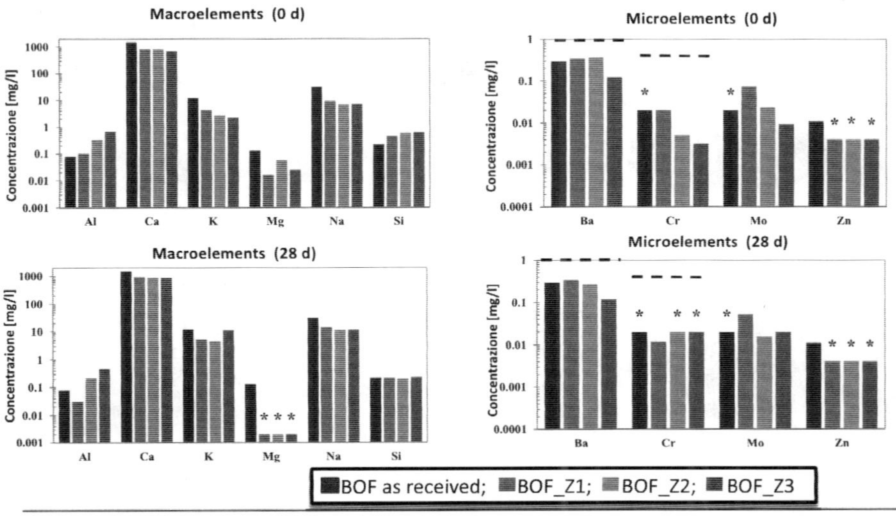

- ↑ Al & Si, especially at 0 d
- ↓ Ba
- ↓ Cr (or stable)

(*) non detect
---- reuse limit (Italy)

Remarks

- **The pilot scale results confirmed that the wet route scale up is feasible**
- **The CO_2 uptake was even higher than for laboratory tests**
- **Heating/Humidification using steam was also successfull**

Nevertheless....still some issues needs to be addressed

- **Extent of carbonation may negatively affect the environmental behavior** for oxyanions (especially chromium) although this depends on the slag batch produced;
- **Control of operating conditions (i.e. moisture, temperature) needs to be optimized**
- **Rotation speed needs to be increased to get larger particle size in view of slag valorization**
- **Need of increasing the performance of the CO_2 capture (improved gas/solid contact)**

School of Engineering and Information Technology, Murdoch University, Murdoch, Australia

Isoconversional Kinetic Modeling and In-Situ Synchrotron Powder Diffraction Analysis for Dehydroxylation of Antigorite

ACEME 2018

Presented by : Sana Zahid

Co-authors

Dr Hans C Oskierski

Assoc Prof Gamini Senanayake

Dr Mohammednoor Altarawneh

Dr Helen E A Brand

Dr Ibukun Oluwoye

Prof Bogdan Z Dlugogorski

14/03/2018 1

Synopsis of Presentation

1. **Introduction**
2. **Aims of Study**
3. **Methodology**
4. **Results and Discussion**
5. **Conclusions**
6. **Acknowledgement**

14/03/2018 2

Murdoch UNIVERSITY Perth | Singapore | Dubai

Mineral Carbonation

Australian Synchrotron | Australian Government | ansto

Introduction | Aims of study | Methodology | Results and discussions | Conclusions | Acknowledgement

Conversion of Mg, Ca, and Fe-rich silicate or oxide minerals to stable carbonates

Advantages

- Offers the safe and permanent disposal of CO_2
- Thermodynamically favourable
- Abundant and well-distributed raw materials

Disadvantages

- Slow reaction kinetics
- High energy cost

Peridotite and Serpentinite Ore Bodies

Magnesium resources far exceed world fossil fuel supplies

Source: Lackner et al.,1995

Murdoch UNIVERSITY Perth | Singapore | Dubai

Serpentine Minerals

Australian Synchrotron | Australian Government | ansto

Introduction | Aims of study | Methodology | Results and discussions | Conclusions | Acknowledgement

- Serpentine minerals are hydrated magnesium silicate minerals with composition of $Mg_3Si_2O_5(OH)_4$
- Consists of alternating tetrahedral (SiO_2) and octahedral sheets ($Mg(OH)_2$)
- Three polymorphs

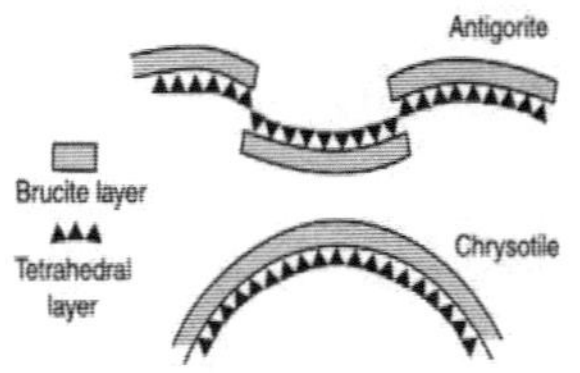

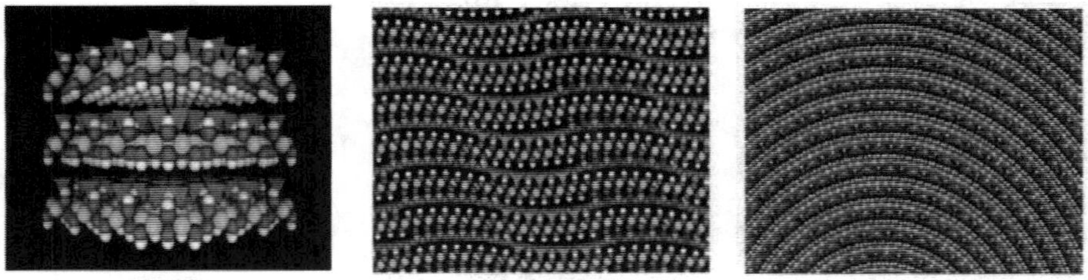

Lizardite | Antigorite | Chrysotile

The Mg, O and H atoms of the octahedral layer are represented by green, red and white spheres, respectively. The grey tetrahedra represents the silica layer.

Murdoch UNIVERSITY Perth | Singapore | Dubai

Dehydroxylation via Heat Treatment

Australian Synchrotron ansto

- Dehydroxylation of serpentine involves the removal of structurally bounded hydroxyl from the serpentine mineral at 700 °C.
- It results in higher reactivity through amorphicity in structure(Mckelvy et al., 2004)
- Dehydroxylation involves the following phases (Perrillat et al., 2005)

$$\text{serpentine} \xleftrightarrow{\text{Heat} < 800\ ^{\circ}\text{C}} \text{forsterite + intermediate phase} + H_2O$$

$$\text{forsterite + intermediate phase} \xrightarrow{\text{Heat} > 800\ ^{\circ}\text{C}} \text{enstatite} + H_2O$$

Murdoch UNIVERSITY Perth | Singapore | Dubai

Isoconversional vs Model Fitting approach for Kinetic Modeling

Australian Synchrotron ansto

Model Fitting Approach

- Involves the phenomenological understanding and best fit of experimental data.
- Assumes constant activation energy and single rate limiting step throughout the reaction progress.
- Suitable for homogeneous systems only.

Isoconversional Kinetic Modeling

- Calculates the activation energy independent of reaction model.
- Constant activation energy for a very small interval of extent of conversion.
- Suitable for homogeneous and heterogeneous systems.

Background of Kinetic Modeling of Dehdyroxylation of Serpentine

Introduction Aims of study Methodology Results and discussions Conclusions Acknowledgement

- Model fitting approach based activation energy values vary from 160- 577 kJ/mol for serpentine minerals (Alizadehhesari et al., 2012; Balucan et al., 2011; Gualtieri et al., 2012, Weber et al., 1965).

- Most phyllosilicates and other solid state reactions show the multistep reaction progress. So, isoconversional kinetics is favourable (Vyazovkin and Wight 1997)

- Trittschack and his colleagues studied the reaction kinetics for lizardite and chrysotile using Isoconversional kinetic modeling (Trittschack and Grobety 2012, Trittschack et al., 2014).

14/03/2018 7

Aims of Study

Introduction Aims of study Methodology Results and discussions Conclusions Acknowledgement

- To understand the structural and mineralogical changes during in-situ dehydroxylation of antigorite using Synchrotron PXRD.

- To estimate the activation energy based on the isoconversional (model free) kinetic modeling.

- To explore the potential of XRD data for the simultaneous study of structural changes and estimation of iso-conversional kinetic parameters.

- To provide the energy efficient solutions for the implementation of large scale mineral carbonation.

14/03/2018 8

Experimental Setup

Introduction | Aims of study | Methodology | Results and discussions | Conclusions | Acknowledgement

- Synchrotron radiations offers high signal to noise ratio
- High temporal and temperature resolution
- Rotating capillary

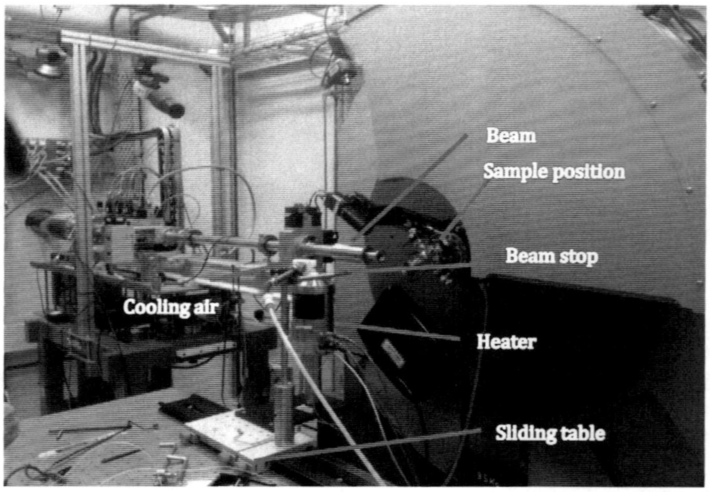

14/03/2018 9

Isoconversional Kinetic Modeling

Introduction | Aims of study | Methodology | Results and discussions | Conclusions | Acknowledgement

Basic kinetic equation is

$$\frac{d\alpha}{dt} = k(T)f(\alpha)$$

Where

$$k(T) = Ae^{-E/RT}$$

If temperature is increase by a constant heating rate $T = T_o + \beta t$

Then differentiating and putting the value of dt in equation (1) gives

$$\frac{d\alpha}{dT} = \frac{A}{\beta}e^{-Ea/RT}f(\alpha)$$

Integration gives

$$g(\alpha) = \frac{A}{\beta}\int_0^T e^{-Ea/RT}dT$$

14/03/2018 10

Murdoch UNIVERSITY Perth | Singapore | Dubai

Isoconversional Kinetic Modeling

Australian Synchrotron | Australian Government | ansto

Introduction | Aims of study | Methodology | Results and discussions | Conclusions | Acknowledgement

Starink isoconversional method

$$\ln\left(\frac{\beta_i}{T_{\alpha,i}^{B}}\right) = const - C\left(\frac{E_a}{RT_{\alpha,i}}\right)$$

B=1.92 and C=1.0008

Most advance method is Vyazovkin's isoconversional method which is based on the minimisation of the following function

$$n(n-1) - \sum_{i=1}^{n}\sum_{j\neq i}^{n} \frac{\beta_j I(E_\alpha, T_{\alpha,i})}{\beta_i I(E_\alpha, T_{\alpha,i})} = \sigma$$

where

$$I(E_\alpha, T_{\alpha,i}) = \int_{T_{\alpha-\Delta\alpha}}^{T_\alpha} e^{-E/RT_{\alpha,i}}\, dT$$

14/03/2018 11

Murdoch UNIVERSITY Perth | Singapore | Dubai

Isoconversional Kinetic Modeling

Australian Synchrotron | Australian Government | ansto

Introduction | Aims of study | Methodology | Results and discussions | Conclusions | Acknowledgement

1. Prograde heating run at 2, 4 and 6 °C/min from room temperature to 1000 °C under N_2 flow.

2. Peak area integration for non overlapping antigorite peak to estimate the extent of conversion using $\alpha = \frac{A_i - A_t}{A_i - A_f}$

3. Plot between T_α and extent of conversion (α) for different heating rates and estimation of temperatures for fixed conversion at different heating rates.

4. Linear graph is plotted between *1/T* and $\ln\left(\frac{\beta_i}{T_{\alpha,i}^{1.92}}\right)$

5. E_a is calculated from the slope of the straight line.

14/03/2018 12

Murdoch UNIVERSITY Perth | Singapore | Dubai

Results and Discussions

Australian Synchrotron

ansto

Introduction | Aims of study | Methodology | Results and discussions | Conclusions | Acknowledgement

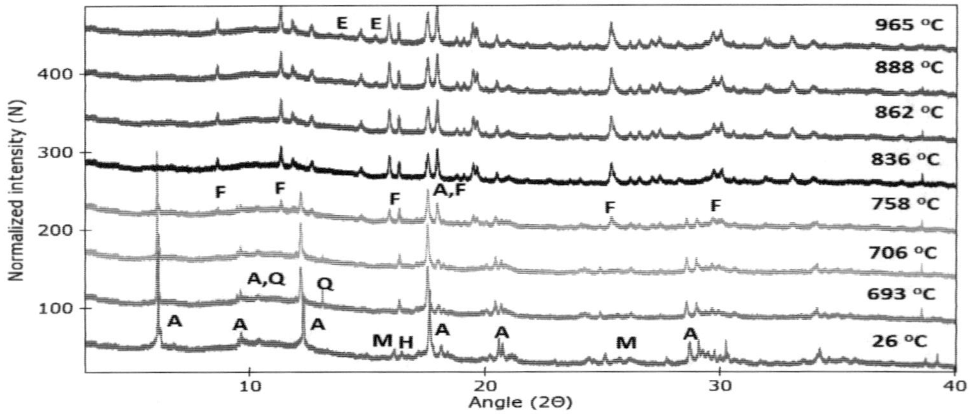

Selected in situ synchrotron PXRD patterns. Peaks are labelled as A=Antigorite, M=Magnesite, H=Hematite, F=Forsterite, E=Enstatite and Q=Quartz

Murdoch UNIVERSITY Perth | Singapore | Dubai

3 D view of S-PXRD data

Australian Synchrotron

ansto

Introduction | Aims of study | Methodology | Results and discussions | Conclusions | Acknowledgement

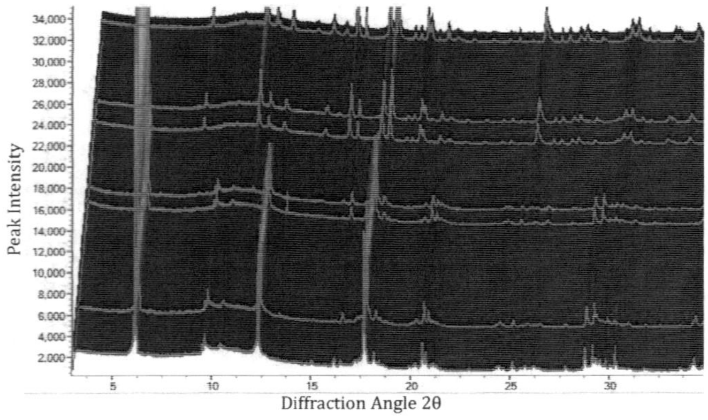

3 D S-PXRD

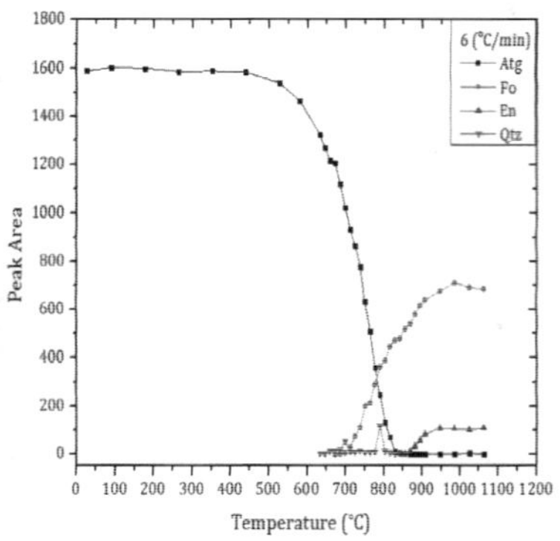

Integrated Peak Areas

Murdoch University
Perth | Singapore | Dubai

Isoconversional Kinetic Modeling by Starink Method

Australian Synchrotron | Australian Government | Ansto

Introduction | Aims of study | Methodology | Results and discussions | Conclusions | Acknowledgement

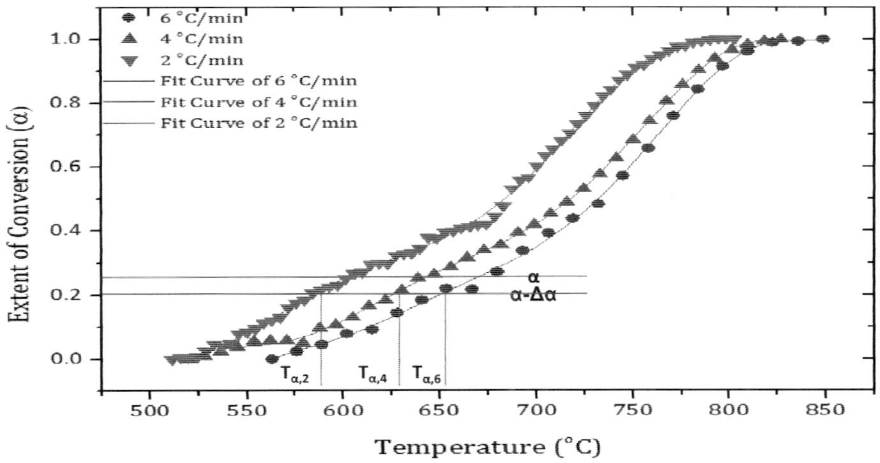

Murdoch University
Perth | Singapore | Dubai

Starink Method of Activation Energy Calculation

Australian Synchrotron | Australian Government | Ansto

Introduction | Aims of study | Methodology | Results and discussions | Conclusions | Acknowledgement

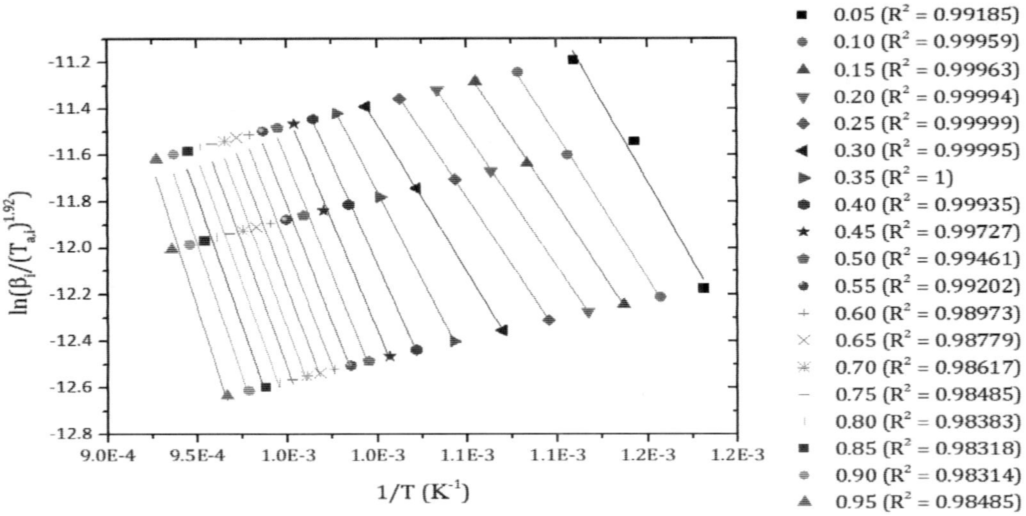

Murdoch UNIVERSITY Perth | Singapore | Dubai

Activation Energy vs Extent of Conversion (α)

Australian Synchrotron | Australian Government | ansto

Introduction | Aims of study | Methodology | **Results and discussions** | Conclusions | Acknowledgement

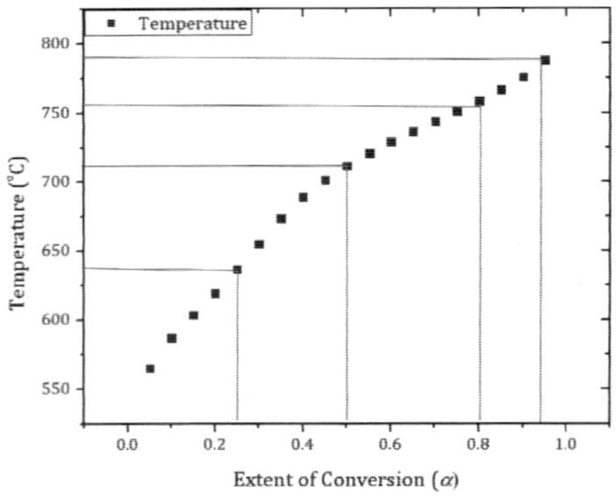

α vs Temperature

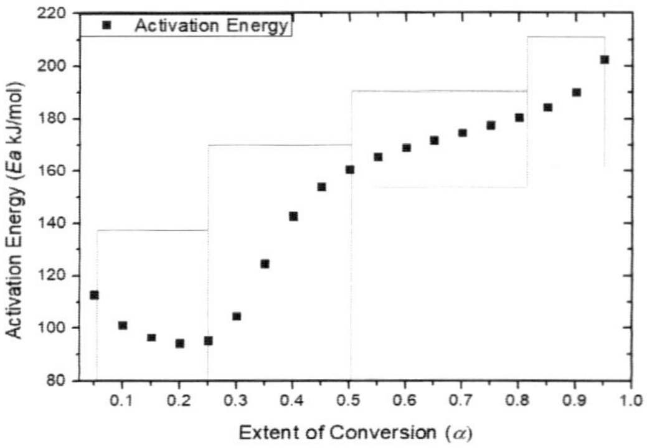

Variation in Activation energy with α

The temperature vs α is based on the average temperature for a given conversion at different heating rates. The standard deviation is ± 27 °C

14/03/2018 17

Murdoch UNIVERSITY Perth | Singapore | Dubai

Interpretation of Variation in Activation Energy

Australian Synchrotron | Australian Government | ansto

Introduction | Aims of study | Methodology | **Results and discussions** | Conclusions | Acknowledgement

Graph Section	Activation Energy	Possible Description	Reference
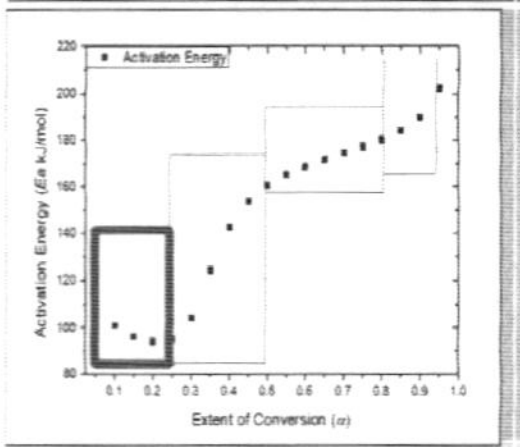	$\alpha \leq 0.25$ decrease in E_a (110-95 kJ/mol)	Condensation of hydroxyl group to form water or liberation of of H^+ and OH^- to diffuse along the inner layer.	Ortega et al., 2010, Triittschack et al., 2014, Evans, 2004
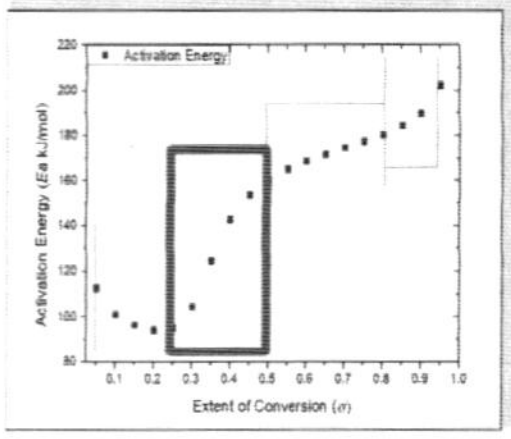	$0.25 \leq \alpha \leq 0.50$ fast increase in E_a (95-170 kJ/mol)	Formation of disordered antigorite or amorphous outer layer hinders the bulk diffusion	Ortega et al., 2010, Triittschack et al., 2014,

14/03/2018 18

Interpretation of Variation in Activation Energy

Murdoch UNIVERSITY Perth | Singapore | Dubai — Australian Synchrotron — ansto

Introduction | Aims of study | Methodology | Results and discussions | Conclusions | Acknowledgement

Graph Section	Activation Energy	Possible Description	Reference
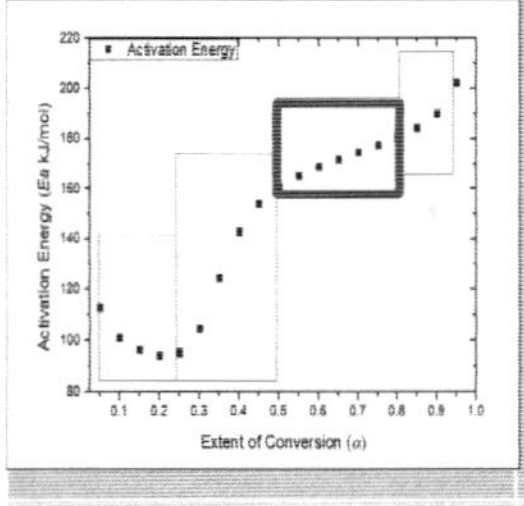	$0.50 \leq \alpha \leq 0.80$ nearly constant at 170 kJ/mol	Nucleation and growth of newly formed grain boundaries i.e. forsterite and talc like phases hence it facilitates the diffusion. supported by phase evolution at 700 °C in S-PXRD	Triittschack et al., 2014, Balucan et al., 2011
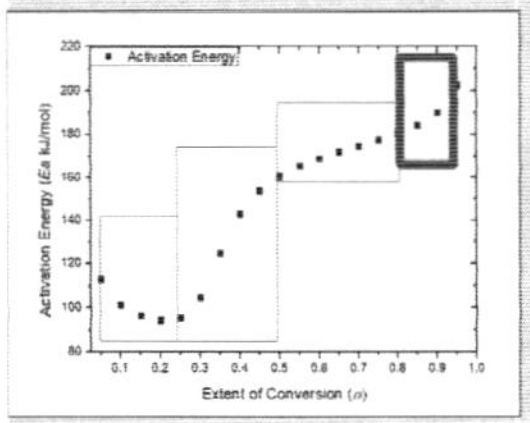	$\alpha > 0.80$ increase in E_a 80-210 kJ/mol	Simultaneous change of 2D diffusion to 3D diffusion due to the break down of final dehydroxylation of inner sheet and break down of talc like intermediate	Triittschack et al., 2014

14/03/2018 19

Conclusions

Murdoch UNIVERSITY Perth | Singapore | Dubai — Australian Synchrotron — ansto

Introduction | Aims of study | Methodology | Results and discussions | Conclusions | Acknowledgement

- Variation in activation energy shows the multistep nature of dehydroxylation of antigorite. Hence, isoconversional approach is preferable.
- E_a values changes in the range of 95-200 kJ/mol. Multiple factors such as particle size, impurities, mineralogy, vapor pressure effects the values of activation energy which hinders the comparison of actual values of activation energy.
- Ab initio calculation and thermodynamic studies should be included in order to support the in depth mechanistic understanding of the dehydroxylation of antigorite.

14/03/2018 20

Acknowledgement

Introduction | Aims of study | Methodology | Results and discussions | Conclusions | Acknowledgement

- Murdoch University for a Murdoch International Postgraduate Scholarship(MIPS) to Sana Zahid.
- Australian Institute of Nuclear Science (AINSE) Limited for providing Postgraduate Research Award (PGRA) to Sana Zahid
- The Australian Synchrotron (AS) for granting our research group Powder Diffraction (PD) beamtime under Proposal ID: p10844

14/03/2018 21

Murdoch
UNIVERSITY
Perth | Singapore | Dubai

References

Australian Synchrotron
ansto

1. Alizadehhesari, K., Golding, S.D., Bhatia, S.K. (2012) Kinetics of the Dehydroxylation of Serpentine. Energy & Fuels 26, 783-790.
2. Balucan, R.D., Dlugogorski, B.Z., Kennedy, E.M., Belova, I.V., Murch, G.E. (2013) Energy cost of heat activating serpentinites for CO2 storage by mineralisation. International Journal of Greenhouse Gas Control 17, 225-239.
3. Balucan, R.D., Kennedy, E.M., Mackie, J.F., Dlugogorski, B.Z. (2011) Optimization of antigorite heat pre-treatment via kinetic modeling of the dehydroxylation reaction for CO2 mineralization. Greenhouse Gases: Science and Technology 1, 294-304
4. Dlugogorski, B.Z., Balucan, R.D. (2014) Dehydroxylation of serpentine minerals: Implications for mineral carbonation. Renewable and Sustainable Energy Reviews 31, 353-367.
5. Gualtieri, A.F., Giacobbe, C., Viti, C. (2012) The dehydroxylation of serpentine group minerals. American Mineralogist 97, 666.
6. Lackner, K.S., Wendt, C.H., Butt, D.P., Joyce, E.L., Sharp, D.H. (1995) Carbon dioxide disposal in carbonate minerals. Energy 20, 1153-1170.
7. McKelvy, M.J., Chizmeshya, A.V.G., Diefenbacher, J., Béarat, H., Wolf, G. (2004) Exploration of the Role of Heat Activation in Enhancing Serpentine Carbon Sequestration Reactions. Environmental Science & Technology 38, 6897-6903.
8. Ortega, A., Macías, M., Gotor, F.J. (2010) The Multistep Nature of the Kaolinite Dehydroxylation: Kinetics and Mechanism. Journal of the American Ceramic Society 93, 197-203.
9. Perrillat, J.-P., Daniel, I., Koga, K.T., Reynard, B., Cardon, H., Crichton, W.A. (2005) Kinetics of antigorite dehydration: A real-time X-ray diffraction study. Earth and Planetary Science Letters 236, 899-913.
10. Power, I.M., Harrison, A.L., Dipple, G.M., Wilson, S.A., Kelemen, P.B., Hitch, M., Southam, G. (2013) Carbon Mineralization: From Natural Analogues to Engineered Systems. Reviews in Mineralogy and Geochemistry 77, 305.
11. Trittschack, R., Grobéty, B. (2012) Dehydroxylation kinetics of lizardite. European Journal of Mineralogy 24, 47-57.
12. Trittschack, R., Grobéty, B., Brodard, P. (2014) Kinetics of the chrysotile and brucite dehydroxylation reaction: a combined non-isothermal/isothermal thermogravimetric analysis and high-temperature X-ray powder diffraction study. Physics and Chemistry of Minerals 41, 197-214.
13. Vyazovkin, S., Burnham, A.K., Criado, J.M., Pérez-Maqueda, L.A., Popescu, C., Sbirrazzuoli, N. (2011) ICTAC Kinetics Committee recommendations for performing kinetic computations on thermal analysis data. Thermochimica Acta 520, 1-19.
14. Vyazovkin, S., Sbirrazzuoli, N. (2006) Isoconversional Kinetic Analysis of Thermally Stimulated Processes in Polymers. Macromolecular Rapid Communications 27, 1515-1532.
15. Vyazovkin, S., Wight, C.A. (1998) Isothermal and non-isothermal kinetics of thermally stimulated reactions of solids. International Reviews in Physical Chemistry 17, 407-433.
16. Weber, J.N., Greer, R.T. (1965) Dehydration of serpentine: heat of reaction and reaction kinetics at P_{H2O}=1 atm. American Minerology 50:450–64.

14/03/2018 22

14/03/2018 23

Economic Studies of Carbonation Processes, Turning Feasibility into Reality

Authors:

Pasquier, L. C., INRS-ETE, University of Quebec
Kemache, N., INRS-ETE, University of Quebec
Blais, J. F., INRS-ETE, University of Quebec
Kentish, S., University of Melbourne
Mercier, G., INRS-ETE, University of Quebec

Is carbonation a viable approach to climate change mitigation through CO_2 emissions reduction? Can carbonation overcome its well-known drawbacks, energy demand, a large amount of material handling etc? Most of these answers arise from the literature, and indicate that carbonation can be considered as viable CO_2 reduction approach from a technical point of view. Nevertheless, a technically demonstrated process needs to be economical (profitable) to be implemented.

Keys to economic success are becoming more and more evident. Carbonation cannot compete with underground storage and cannot only act as a sequestration technique. Processes should be efficient with both raw flue gas or pure CO_2 streams (sugar cane refineries or fertilizer sources). This way, CO_2 is removed from gaseous streams and is used/stored on site. This can be game changing in areas where underground storage is not an option. On the other hand, value added by-products need to be generated. Besides the potential income from such by-products, reuse of the waste material used as carbonation feedstock is needed to meet sustainability standards.

Such an approach has been triggering research made at INRS since 2012 to develop a direct flue gas carbonation process using waste mining residues. The process first developed under laboratory conditions showed the potential of using carbonic acid alone to leach Mg from heat activated serpentine. Combined with ambient temperature, moderate pressure, and successive recirculation, Mg was extracted in significant quantities with CO_2 removal around 65%. Furthermore, by separating the gas and pulp contacting steps from the carbonate precipitation, very pure hydrated magnesium carbonates are obtained. This high purity results in high potential sales value, and profitability. Later, the demonstration at a â €œpilot scaleâ€ using a flue gas within an operating cement plant confirmed laboratory scale results. Consequently, following the promising results, similar approaches were tested with steel slags and waste concrete.

A common point with all these research projects is the integration of economic evaluations at an early stage. As a result, process development is performed in a close loop with economic viability. The presentation will highlight the economic considerations and evaluation for the different approaches considered at INRS.

International Conference on Accelerated Carbonation for Environmental and Material Engineering

Newcastle, March 14th, 2018

Carbon Mineralization as one of the main CO_2 utilization strategies: Report on the Mission Innovation CCUS experts' workshop

Giulia Costa
Laboratory of Environmental Engineering,
DICII
University of Rome Tor Vergata

Mission Innovation

November 30th 2015 Paris 2015 UNCC conference: leaders from 22 Countries and the EU launched MI to dramatically accelerate innovation in clean energy

(missing leaders from Korea, Germany, Sweden and Australia)

Mission Innovation

ACEME 2018

Participating Countries:

G. Costa Mission innovation workshop – Carbon mineralization panel

MISSION INNOVATION
Accelerating the Clean Energy Revolution

Mission Innovation

ACEME 2018

*Accelerating widespread **clean energy** innovation is indispensable for responding to **climate challenge** and to provide **affordable** and **reliable** energy **for everyone** and promote economic growth.*

Progress has been made in cost reduction and deployment of clean energy technologies, but the pace of innovation and the scale of transformation and dissemination remains significantly short of what is needed.

Main points:

- Double Governmental Investment in R&D on Clean Energy Innovation (2016-2021)
- Private sector and business leadership to drive innovation (*Breakthrough energy coalition*)
- Implementation (joint research efforts through public-private partnerships as well as among participants)
- Information sharing (annual updates)

G. Costa Mission innovation workshop – Carbon mineralization panel

MISSION INNOVATION
Accelerating the Clean Energy Revolution

Mission Innovation

AEME 2018

Each country independently determines the best use of its R&D funding to reach the doubling goal

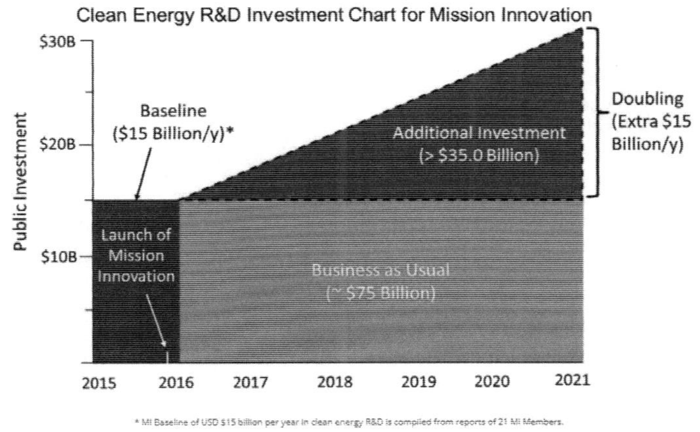

* MI Baseline of USD $15 billion per year in clean energy R&D is compiled from reports of 21 MI Members.

Mission Innovation

AEME 2018

And its own R&D priorities within **innovation challenges** that are developed through a collaborative process between MI members:

		Australia	Brazil	Canada	Chile	China	Denmark	EC	Finland	France	Germany	India	Indonesia	Italy	Japan	Mexico	Norway	Republic of Korea	Saudi Arabia	Sweden	The Netherlands	UAE	UK	USA
1	Smart Grids Innovation Challenge	Participant	Participant	Participant		Lead	Participant	Participant	Participant	Participant	Participant	Lead	Participant	Lead		Participant	Participant	Participant	Participant	Participant	Participant		Participant	Participant
2	Off Grid Access to Electricity Innovation Challenge	Participant	Participant	Participant		Participant		Participant	Participant	Lead		Lead	Participant	Participant		Participant	Participant	Participant	Participant	Participant	Participant		Participant	Participant
3	Carbon Capture Innovation Challenge	Participant		Participant		Participant	Participant	Participant	Participant	Participant	Participant	Participant	Participant	Participant	Participant	Participant	Participant	Participant	Lead	Participant	Participant	Participant	Participant	Lead
4	Sustainable Biofuels Innovation Challenge	Participant	Lead	Lead		Lead		Participant	Participant	Participant		Lead	Participant	Participant		Participant	Participant			Participant	Participant		Participant	Participant
5	Converting Sunlight Innovation Challenge	Participant	Participant	Participant	Participant	Participant	Participant	Lead	Participant	Participant	Lead	Participant		Participant	Participant	Participant	Participant		Participant	Participant	Participant	Participant	Participant	Participant
6	Clean Energy Materials Innovation Challenge	Participant		Lead			Participant	Participant	Participant	Participant	Participant	Participant		Participant		Lead	Participant	Participant	Participant	Participant	Participant	Participant	Participant	Lead
7	Affordable Heating and Cooling of Buildings Innovation Challenge	Participant	Participant	Participant		Participant	Participant	Lead	Participant	Participant	Participant	Participant		Participant		Participant	Participant		Participant	Participant	Participant	Lead	Lead	Participant

● Lead
● Participant

Mission Innovation

Carbon Capture Innovation Challenge: enable near-zero CO_2 emissions from power plants and carbon intensive industries. Focus: early stage breakthrough CCUS technologies.

Activities: **experts' workshop on R&D needs** (low TRL) for CCUS technologies held on September 25-29, 2017 in Houston (Texas) involving international experts from academia and industry to discuss breakthrough R&D opportunities and international synergies.

Next Steps: coordinate and align international R&D programs to best utilize and direct funding for CCUS.
Medium-term: explore ways to engage private sector interests.
Longer term: establish a standing platform for communications, information sharing, and building collaborations on R&D on breakthrough and innovative CCUS technologies.

G. Costa   Mission innovation workshop – Carbon mineralization panel   MISSION INNOVATION Accelerating the Clean Energy Revolution

Mission Innovation: CCUS experts' Workshop

The goal of the workshop was to identify breakthrough research directions (TRLs 1-3/4) needed to achieve long-term technologies for CCUS: **Priority Research Directions (PRDs)**

The workshop covered 3 major themes (focus areas): carbon capture, utilization and storage.

The product of the workshop is a **report** that will be issued shortly with two high level purposes:

- Inspire the global research community to develop breakthrough research concepts
- Convince the general readership that these research directions are critical to meeting future energy needs

Organized by US DOE and Saudi Arabia Ministry of Energy and Mineral Resources

Chair: Michelle Buchanan (ORNL)

G. Costa   Mission innovation workshop – Carbon mineralization panel   MISSION INNOVATION Accelerating the Clean Energy Revolution

Mission Innovation: CCUS experts' Workshop

240 experts from 22 countries attended the workshop and activities were carried out in panels:

CO_2 capture (co-chair Jorild Svalestuen, Gassnova)

- Solvents
- Sorbents and looping systems
- Membranes
- Combustion and other technologies

CO_2 utilization (co-chair Alissa Park, Columbia University)

- Thermochemical conversion and hydrogenation of CO_2
- Electrochemical and photochemical conversion of CO_2
- **CO_2 conversion to solid carbonates**
- Biological conversion of CO_2

CO_2 storage (co-chair Don De Paolo, Berkley University)

- Injectivity and capacity
- Monitoring, verification and performance metrics
- Forecasting and managing induced seismicity
- Well diagnostics

Crosscut (co-chair Joe Powell, Shell)

A total of 33 PRDs were identified and described by the panels

G. Costa **Mission innovation workshop – Carbon mineralization panel** MISSION INNOVATION Accelerating the Clean Energy Revolution

CO_2 conversion to solid carbonates panel

20 panelists from 10 Countries

Panel Co-Leads: Eric Kennedy & Giulia Costa

Bruce Adderley (Sheffield U.)	Youngjune Park (GIST Korea)
Georges Beaudoin (U. Laval)	Alissa Park (Columbia U.)
Florent Bourgeois (U. Toulouse III)	Louis Cesar Pasquier (U. Quebec)
Pen Chi-Chiang (National Taiwan U.)	Valentina Prigiobbe (Stevens Inst. Technol.)
Issam Dairaneih (CO_2 Sciences)	Richard Riman (Rutgers U.)
Gregory Dipple (U. British Columbia)	Rafael Santos (U. Guelph)
Geeshma Gadikota (U. Wisconsin)	Caijun Shi (Hunan U.)
Lars-Erik Gaertner (Linde Group)	Dirk Van Mechelen (ORBIX Belgium)
Sean Monkman (Carbon Cure)	Tao Wang (Zhejiang U.)
Tim Oliver (U. Newcastle)	Sean Zhou (GreenOre)

G. Costa **Mission innovation workshop – Carbon mineralization panel** MISSION INNOVATION Accelerating the Clean Energy Revolution

CO_2 conversion to solid carbonates panel

AEME2018

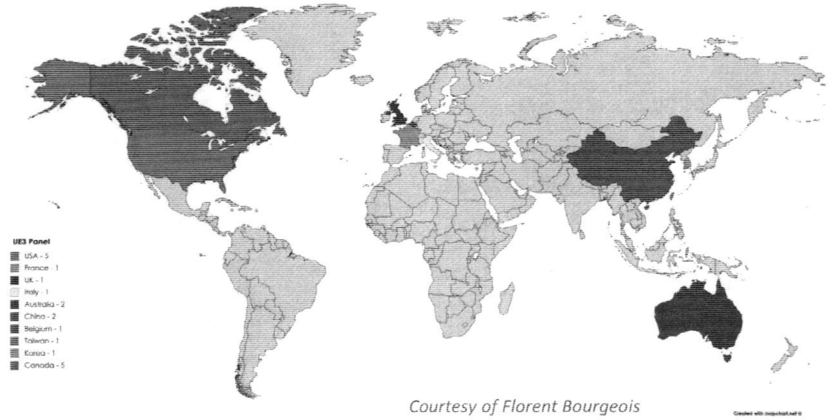

Courtesy of Florent Bourgeois

10+1 out of 20 +1 Countries that participate to MI's challenge on Carbon Capture (missing: DK, NL, FI, DE, IN, IND, JP, MX, NO, SA, SV, UAE)

G. Costa **Mission innovation workshop – Carbon mineralization panel** MISSION INNOVATION Accelerating the Clean Energy Revolution

CO_2 conversion to solid carbonates panel – main research fields

AEME2018

- Ex situ carbonation of minerals (mainly serpentinites) for CO_2 storage via direct/indirect routes
- Passive mineralization of atmospheric CO_2 in mine waste/tailings
- Accelerated carbonation of alkaline industrial residues (e.g. steel slag, red mud, waste incineration bottom ash) for:
  - CO_2 storage
  - Production of construction materials (aggregates or blocks) and recovery of metals
- CO_2 utilization for concrete and/or cement production
- Carbonation of alkaline brines for other applications (formation of metastable carbonate to provide bicarbonate ions to microalgae)

G. Costa **Mission innovation workshop – Carbon mineralization panel** MISSION INNOVATION Accelerating the Clean Energy Revolution

CO_2 conversion to solid carbonates panel – tasks

Starting from an analysis of the basic science challenges, opportunities, and research needs associated with carbonation of different materials as a CCU (CCS) strategy:

- identify **2 Priority Research Directions** for fundamental research in this field (all - 2.5 d)
- prepare 5-10 page **report** with relevant graphs, figures and sidebars (writers – 1.5 d +....)

CO_2 conversion to solid carbonates panel – PRD template

Title

Scientific challenges	Summary of research direction
Brief overview of the underlying science challenge	What fundamental research is needed to address the challenge? Why can this research be done now? (e.g. are there recently developed capabilities?)
Potential scientific impact	**Potential impact on CCUS technology**
What impact will this research have on the CCUS scientific field? What impact will it have on the general scientific community?	How will this impact CCUS-relevant technologies?

CO_2 conversion to solid carbonates panel – tasks

In order to reach these goals we decided to first examine what each one of us thinks the research needs/challenges in his/her own field of expertize are.

We also asked colleagues that were not able to join us in Houston ...

G. Costa Mission innovation workshop – Carbon mineralization panel

MISSION INNOVATION Accelerating the Clean Energy Revolution

Feedback from non participants: open issues

EU/ZEP: Explore options for combining CCS and CCU ⟶ Carbonation!

Ron Zevenhoven (Abo Akademi): a definition is needed for CCU vis á vis CCS, IPCC special report on CCU?

Marco Mazzotti (ETH): identify potential applications for carbonation, multi-phase reactors, advanced systems analysis techniques, need for CO_2 and use of additives. Feedstocks.

Mercedes Maroto Valer (Heriot-Watt U.): Regeneration of reagents, reactivity of mineral phases in rocks, utilisation of Mg carbonate products and industrial symbiosis

Jamie Knight (Coal Innovations NSW): use raw flue gas instead of captured CO_2

Jason Mann and Brian Haynes (U. Sydney): kinetic and thermodynamic limitations that govern Mg and Si extraction, operate carbonation below ARC conditions

Michael Hitch (U. NSW): development of particle preparation/pretreatment, and formal verification schemes to qualify mineral carbonation products for commercial Offset Credits/ Carbon Trading/ Cap and Trade. Develop Mineral Carbonation Product Utilisation

G. Costa Mission innovation workshop – Carbon mineralization panel

MISSION INNOVATION Accelerating the Clean Energy Revolution

Feedback from non participants: open issues

ACEME2018

Colin Hills (Greenwich and Carbon8 Systems): Low cost, widely available supplies of CO_2 suitable for the intended mineralisation process, low energy intensity - high yield processes with 'fit for purpose' products

Mai Uibu (Tallinn U.): Chemical-technological fundamentals of heterogeneous systems (oil shale ashe to PCC)

Mieke Quaghebeur, Ruben Snellings, Peter Nielsen (VITO, Belgium): relationship between mechanical properties of carbonated construction materials. Heat and moisture transfer during carbonation of porous compacts. Governing parameters on Si-speciation

Özlem Cizer (KU Leuven): binding process in building materials, enzyme aided carbonation (natural and accelerated). Carbonation and transport properties in cement-based materials

Yixin Shao (McGill University): precast and cast-in-place concretes to enhance utilization capacity. Carbonation of aggregates to promote carbon negative concrete products.

Carlos Rodriguez Navarro (U. Granada): Enhanced dissolution by Si-O bond-breaking with biomimetic catalysts (silicase and its biomimetics). Analysis at the molecular scale of reactant-solution-product interactions and the role of porosity generation for preventing passivation

G. Costa Mission innovation workshop – Carbon mineralization panel

MISSION INNOVATION
Accelerating the Clean Energy Revolution

Feedback from participants: open issues

ACEME2018

- Low energy activation & enhanced feedstock reactivity under lean CO_2 conditions
- Complete or near to complete utilisation of mineral feedstock
- Optimization of passive carbon mineralization process
- Molecular-scale simulation of carbonation reactions
- Thermodynamic properties of alkaline minerals for geochemical modeling
- Reusable carbonation additives/reagents
- Making use of carbonation exothermic heat
- Synthetic minerals for carbonation
- In-situ monitoring of carbonation reactions
- Material Reactivity: Flue gas conditions material behavior, kinetics enhancement, pre-treatment options
- Process integration: Energetic modeling, cost modeling
- Cost effective catalytic enhancement of dissolution, separation and carbonation of silicate materials (ligands, CA, seeding, etc.)
- Innovative system integration for recycle and reuse of energy (heat) and materials

G. Costa Mission innovation workshop – Carbon mineralization panel

MISSION INNOVATION
Accelerating the Clean Energy Revolution

Feedback from participants: open issues

- Combining various industrial waste materials for CO_2 carbonation technology (i.e. producing $NaHCO_3$ from desalination waste water)
- Role of the curing step on extent of carbonation and products' properties (e.g. in situ measurements, carbonation tests with model mineralogical phases)
- Scale up issues and data for LCA. Standardization of CO_2 life cycle analysis for carbonation processes
- Effects of operating conditions on the leaching behaviour of the product & solutions for reducing mobility of specific elements (metalloids) after carbonation
- New MC-based pathways for the transition to a circular economy of construction materials
- Innovative MC-based mining & mineral processing pathways for base metals production
- Breakthrough MC-based production of energy carriers
- Dry or low water consuming MC processes
- Effect of further water curing after CO_2 curing on the hydration products and microstructure of CO_2-cured concrete;
- Effect of carbonated recycled aggregate on the properties and microstructure of recycled aggregate concrete;

G. Costa Mission innovation workshop – Carbon mineralization panel

MISSION INNOVATION
Accelerating the Clean Energy Revolution

Feedback from participants: open issues

- Evolution of different carbonation product-layers (calcium silicates, lime, calcium hydroxide and C-S-H gel) and diffusion mechanism of gas and ions through the layer
- Restructure process of cementitious system during carbonation in the bulk solid phase
- Nucleation and growth mechanisms for carbonation/de-carbonation
- High CO_2 density precursors for *in situ* carbonation
- Ultra-fast carbonation
- Creating cheap CO_2
- Creating realistic time-lines for technology adoption and creating a climate model that evaluates its effectiveness

G. Costa Mission innovation workshop – Carbon mineralization panel

MISSION INNOVATION
Accelerating the Clean Energy Revolution

Key common themes

AC EME 2018

Feed

- Ca/Mg oxide stream - very diverse range of feeds
  - Ash, steel slag and other wastes
  - Serpentine, olivine feed
  - Extraction of valuable metal ores
  - Complications due to the presence of SiO_x
  - Concrete
  - Alkaline waste such as red mud
- CO_2 stream
  - Presence of impurities (NO_x, SO_x, Hg)
  - CO_2 content
- Aqueous media (in most processes)
  - pH buffer
  - Anions (chloride)
  - Cations (Na+, K+)
  - Development of additives, such as catalysts, chelates, microalgae

G. Costa Mission innovation workshop – Carbon mineralization panel

MISSION INNOVATION Accelerating the Clean Energy Revolution

Key common themes

AC EME 2018

Reaction

- Activation of Ca/Mg oxide stream
  - Crushing/grinding
  - Heat treatment for some input materials
- Passive carbonation
  - Mine waste; monitoring and modelling
  - CO_2 dissolution from air
- Active carbonation
  - Control of temperature, T
  - Single vs multiple reactors
  - Gas/liquid/solid reaction
- Reaction:
  - Mass transfer
  - Reaction kinetics and thermodynamic properties
  - Formation of passivation layer
  - Need for *in situ* analysis; offline and online measurements
  - Simulation using MM, elementary reaction mechanism development

G. Costa Mission innovation workshop – Carbon mineralization panel

MISSION INNOVATION Accelerating the Clean Energy Revolution

Key common themes

ACEME2018

Reaction

- Product separation
  - Recycling of unreacted CO_2
  - Recycling of aqueous solution, and accumulation of spectator species
  - Unreacted solid feed
- Product characterization
  - Chemical composition/purity and control
  - Product morphology
  - Leaching
  - Strength

Process

- LCA and process simulation and analysis (Aspen$^+$)
- Integration with other technologies and systems (industrial symbiosis)
- Pilot scale up

G. Costa **Mission innovation workshop – Carbon mineralization panel** MISSION INNOVATION Accelerating the Clean Energy Revolution

Research challenges for carbon mineralization

ACEME2018

Feedback from panelists + other experts on open fundamental research issues:

Specific key words

- Alkaline bearing materials feedstock
- Diluted CO_2 streams
- Product for high grade applications
- In situ monitoring
- Geochemical modelling: Si-speciation
- CO_2 binding mechanism
- Dissolution
- Nucleation and growth
- Diffusion mechanism
- Heat and moisture transfer
- Reactivity (activation)
- Molecular scale analysis
- Reactant-solution-product interactions
- Modeling
- Mineralogy
- Thermodynamic data
- Leaching behaviour

Cross-cutting issues

- System and process integration (industrial symbiosis)
- LCA
- Biomimetics (catalysts for dissolution)
- Online process monitoring in bio-chemical complex systems
- Bicarbonates for microalgae

G. Costa **Mission innovation workshop – Carbon mineralization panel** MISSION INNOVATION Accelerating the Clean Energy Revolution

Research challenges for carbon mineralization

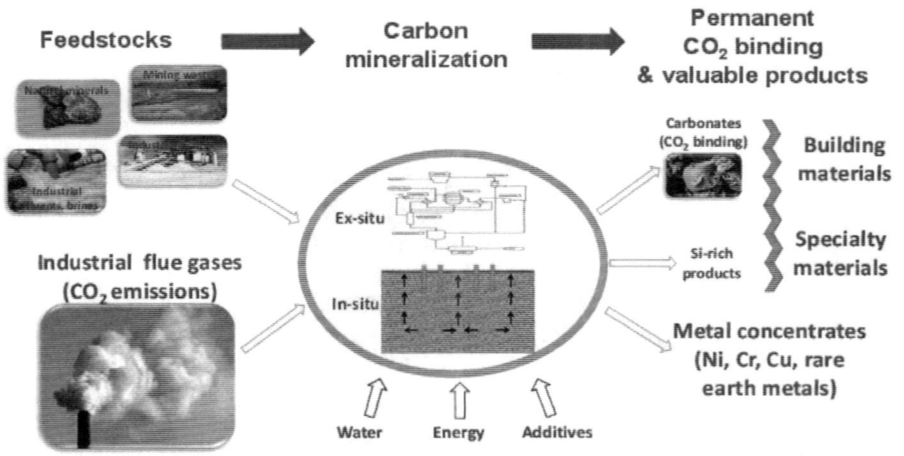

Florent Bourgeois

G. Costa Mission innovation workshop – Carbon mineralization panel

Hard work on PRDs...

G. Costa Mission innovation workshop – Carbon mineralization panel

Accelerating carbon mineralization by exploiting the complexity at solid-liquid-gas interfaces

ACEME2018

Scientific challenges

- Carbon mineralization, generally considered a slow phenomenon, has been observed to proceed at rates orders of magnitude faster than predicted by conventional models.
- Predicting the dynamic speciation and product properties of solid feedstock/fluid/CO_2 systems is the key to unlock carbon mineralization to control reaction pathways and rates.

Summary of research direction

- Broad-spectrum collection of thermodynamic and kinetic data associated with carbon mineralization of feedstocks, focusing on reactions and mass transfer at solid-fluid interfaces.
- Development of broadly available on-line high-resolution investigative tools for local mass transfer measurements and phase characterisation.
- Mechanistic understanding and predictive phenomenological modelling of carbon mineralization processes at the molecular level.

Potential scientific impact

- Provide internally-consistent and validated thermodynamic, transport and kinetic data for carbon mineralization systems.
- Novel high-resolution realtime instruments for characterizing dynamic solid-fluid systems.
- Improved understanding of solid-liquid-CO_2 reaction pathways under a broad range of conditions.

Potential impact on CCUS technology

- Fast-tracked novel market-ready net positive energy carbon mineralization processes for permanent CO_2 binding.
- Provide a ranked list of high potential carbon mineralization feedstocks.
- Provide input-output data for economic and environmental life cycle assessment of carbon mineralization processes

G. Costa Mission innovation workshop – Carbon mineralization panel

MISSION INNOVATION 27
Accelerating the Clean Energy Revolution

PRD 1: Accelerating carbon mineralization by exploiting the complexity at solid-liquid-gas interfaces

• Scientific Challenges

✓ Carbon mineralization, a slow phenomenon, has been observed to proceed at rates orders of magnitude faster than predicted by conventional models.

✓ Predicting the dynamic speciation and product properties of solid feedstock/fluid/CO_2 systems is the key to unlock carbon mineralization to control reaction pathways and rates.

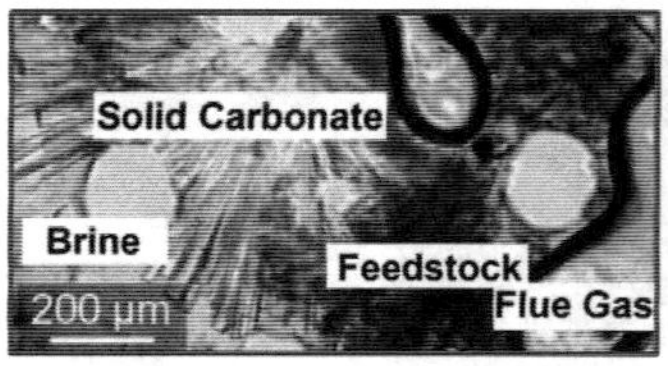

Microfluidics carbon mineralization experiment (after Harrison et al., 2017. Chem. Geol.)

• Research Directions

✓ Broad-spectrum collection of thermodynamic and kinetic data associated with carbon mineralization of feedstocks, focusing on reactions and mass transfer at solid-fluid interfaces.

✓ Development of broadly available on-line high-resolution investigative tools for local mass transfer measurements and phase characterization.

✓ Mechanistic understanding and predictive phenomenological modelling of carbon mineralization processes at the molecular level.

• Impacts

✓ Fast-tracked novel market ready net positive energy carbon mineralization processes for permanent CO_2 binding.

✓ Provide a ranked list of high potential carbon mineralization feedstocks.

✓ Provide internally-consistent and validated thermodynamic, transport and kinetic data for carbon mineralization systems.

✓ Improved understanding of solid-liquid-CO_2 reaction pathways under a broad range of conditions.

Tailoring material properties to enable carbon storage in products

AEME 2018

Scientific challenges

- To realize the potential of carbon mineralization for the built environment we must predict and control processes to consistently meet product performance criteria.
- Linking product performance to micro-structure, geochemistry and mineralogy is the key to achieve consistency.

Summary of research direction

- Develop and apply real time monitoring of carbon mineralization product properties during reaction.
- Exploit nano-characterization for deriving insight into solid and aqueous speciation and improve control on bulk physical and chemical properties.
- Understand and model structure-property-performance relationships.
- Develop pre/post processing treatments to improve product performance criteria.
- Predict and control long term durability for applications.

Potential scientific impact

- Provide a better understanding of process conditions on product properties (e.g. strength, leaching and durability).
- Extract and recover valuable by-products.
- Tailor products for new applications.

Potential impact on CCUS technology

- Reliable cost-effective processes for generating construction materials that meet performance criteria while permanently storing CO_2 on a Gigatonne scale.
- Insights applicable to the carbonation of industrial waste streams and sub-surface mineral trapping.

G. Costa Mission innovation workshop – Carbon mineralization panel

MISSION INNOVATION 29
Accelerating the Clean Energy Revolution

PRD 2: Tailoring material properties to enable carbon storage in products

- **Scientific Challenges**
  - ✓ To realize the potential of carbon mineralization for the built environment we must predict and control processes to consistently meet product performance criteria.
  - ✓ Linking product performance to micro-structure, geochemistry and mineralogy is the key to achieve consistency.
- **Research Directions**
  - ✓ Develop and apply real time monitoring of carbon mineralization product properties during reaction.
  - ✓ Exploit nano-characterization for deriving insight into solid and aqueous speciation and improve control on bulk physical and chemical properties.
  - ✓ Understand and model structure-property-performance relationships.
  - ✓ Develop pre/post processing treatments to improve product performance criteria.
  - ✓ Predict and control long term durability for applications.
- **Impacts**
  - ✓ Reliable cost-effective processes for generating construction materials that meet performance criteria while permanently storing CO_2 on a Gigatonne scale.
  - ✓ Provide a better understanding of process conditions on product properties (e.g. strength, leaching and durability).
  - ✓ Insights applicable to the carbonation of industrial waste streams and sub-surface mineral trapping.

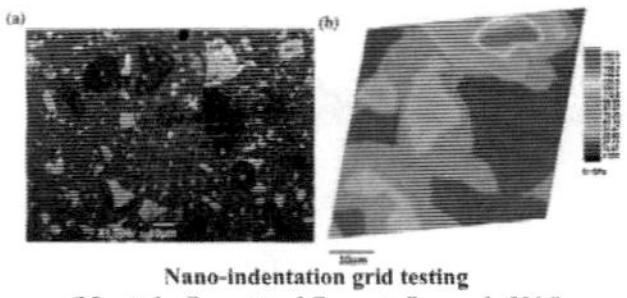

Nano-indentation grid testing
(Mo et al., *Cement and Concrete Research*, 2016)

Carbonated building block performance testing
(Columbia Dept. of Civil Engineering)

Concrete forest (The Who's Who in *Building & Construction*, Houston Spring 2016 Ed.)

Report and next steps

Report: *Permanent storage of CO_2 in valuable products through carbon mineralization*

- Introduction
- Status of Today's Field
- Addressing the different types of products and applications of carbon mineralization
- Identification of research challenges
- Why now
- Conclusions
- Sidebars on applications in construction and for enhanced metal recovery

+ 2 Reports on the PRDs

Currently being edited

Publication expected: end of March

Then....

Final messages

ACEME2018

- Watch out for funding opportunities in this area (**utilization**)
- Joint collaboration
- Networking
- ACEME!!!

(ACEME 2020)

		Australia	Brazil	Canada	Chile	China	Denmark	EC	Finland	France	Germany	India	Indonesia	Italy	Japan	Mexico	Norway	Republic of Korea	Saudi Arabia	Sweden	The Netherlands	UAE	UK	USA
1	Smart Grids Innovation Challenge	○	○	○		●	○	○	○	○	○	●	○	●		○	○	○	○	○	○		○	○
2	Off Grid Access to Electricity Innovation Challenge	○	○	○		○		○	○	●		●	○	○		○	○	○	○	○	○		○	○
3	Carbon Capture Innovation Challenge	○		○		○	○	○	○	○	○	○	○	○	○	○	○	○	●	○	○	○	○	●
4	Sustainable Biofuels Innovation Challenge	○	●	●		●		○	○	○		●	○	○		○	○			○	○		○	○
5	Converting Sunlight Innovation Challenge	○	○	○	○	○	○	●	○	○	●	○		○	○	○	○		○	○	○	○	○	○
6	Clean Energy Materials Innovation Challenge	○		●			○	○	○	○	○	○		○		●	○	○	○	○	○	○	○	●
7	Affordable Heating and Cooling of Buildings Innovation Challenge	○	○	○		○	○	●	○	○	○	○		○		○	○		○	○	○	●	●	○

● Lead

○ Participant

Improving Energy Efficiency of the Inrs Mineral Carbonation Process Using Pinch Analysis

Authors:

Tebbiche, I., INSTITUT NATIONAL DE LA RECHERCHE SCIENTIFIQUE
Pasquier, L. C., INRS-ETE, University of Quebec
Mercier, G., INRS-ETE, University of Quebec
Blais, J. F., INRS-ETE, University of Quebec
Kentish, S., University of Melbourne

Mineral carbonation is a key solution to tackle climate change issues. The process developed at INRS proposes to directly treat industrial CO_2 emissions without capture using serpentine mining waste. The process can be summarized into three separated steps, raw serpentine mineral processing, CO_2 and serpentine dissolution and carbonate precipitation. The efficiency observed at laboratory scale has been demonstrated during pilot plant tests where a real cement plant flue gas with CO_2 content of 18% was treated at 10 bar feed pressure and 22°C. While the process offers a viable and economic way to reduce CO_2 emissions, energy efficiency remains to be addressed.

In this work, pinch analysis was used with the aim to reducing the heat demand of the INRS process by energy integration. An Aspen Plus simulation flowsheet was first prepared for a base case plant with a capacity of 200 t of raw serpentine used per hour (350 000tCO_2/year) and the process conditions mentioned above. Cooling water and natural gas combustion were considered as external cooling and heating sources respectively. Secondly, serpentine and combustion air preheating temperatures were varied as optimisation parameters. For each case, the resulting mass and energy balances were used in a pinch analysis with the Aspen Energy Analyzer. Finally, results were compared in terms of required external heating, additional capital and operating costs.

Results show that preheating serpentine using the process waste heat before heat activation has the biggest impact on the external heat required. Raising the solid temperature by 100°C reduces the operating costs by 14%. On the other hand, a similar increase in the incoming gas temperature will only result in a 5% reduction. The best trade-off between energy integration and additional costs was obtained with solid and gas preheating temperatures of 400°C and 300°C respectively. This resulted in a reduction of global heat demand of more than 40 %. It is also worth mentioning that external cooling is also required in the INRS process, which is mainly used for cooling the recycled precipitation solution. This accounts for less than 15% of the total costs.

This work has demonstrated the ability to reduce mineral carbonation energy demand and proves the applicability of a viable and economical technology to address climate change.

Study of the Potential of Carbonated Solid Foams as Acoustic Insulation Using Sustainable Blowing Agents

EAN University, Bogotá, Colombia

- Duque, Oscar
- García, Leidy
- Bolívar, Nathalya
- García, Ángela
- Rodríguez, Soed
- Amado, Mónica

ACEME, 2018. Newcastle, Australia. March 14th 2018

1

EAN UNIVERSITY

- ✓Research group: ONTARE.
  - ✓Undergraduate Students.
  - ✓Research Hotbed
- ✓Research Areas:
  - ✓Information Technologies.
  - ✓Processes optimization.
  - ✓Automatization and electronic.

2

OUTLINE

- Colombian context
- Investigation aims
- Concrete Foams
- Chemical blowing agents
- Potential environmental impact
- Acoustic insulation

OUTLINE

- Colombian context
- Investigation aims
- Concrete Foams
- Chemical blowing agents
- Potential environmental impact
- Acoustic insulation

In Colombia, NTC 4945 says

Up to 70 dB → Negative psychologic effects, if concentration is required.
In between 80 and 90 dB → Mental problems and sleep alterations.
Up to 100 dB → Physical lesions.

A Bogota's average citizen is subject to more than 70 dB daily in urban areas. (Casas-Garcia, Betancur-Vargas & Montaño Erazo, 2015)

Zone	maximum level of sound pressure (dB)	
	Day Time	Night Time
Residential	65	45
Commercial	70	60
Industrial	75	75
Tranquility	45	45

Resolution 8321. (Health ministry, 1983)

In Colombia, specifically Bogotá D.C.

"Bogotá's 60% of acoustic contamination is derivate from cars, planes or mobile sources, the 40% remaining from commerce or another sources." (district secretariat of environment)

But, did you know that...

Continuous exposure to noise can cause:

- Stress
- Insomnia
- Anxiety
- Depression

OUTLINE

- Colombian context
- Investigation aims
- Concrete Foams
- Chemical blowing agents
- Potential environmental impact
- Acoustic insulation

INVESTIGATION AIMS

- ✓ To study environmental impact of two chemical blowing agents, which produce CO_2, in the concrete foams
- ✓ To study potential use of concrete foams as acoustic insulators.

OUTLINE

- Colombian context
- Investigation aims
- Concrete Foams
- Chemical blowing agents
- Potential environmental impact
- Acoustic insulation

9

Concrete Foams

Mohammed & Hamad (2014) define low-weight concrete as the one that meets the requirements of low density (between 300 to 2000 $\frac{kg}{m^3}$), compressive strength between 1 MPa and 60 MPa; and thermal conductivities less than 1 $\frac{W}{m*K}$.

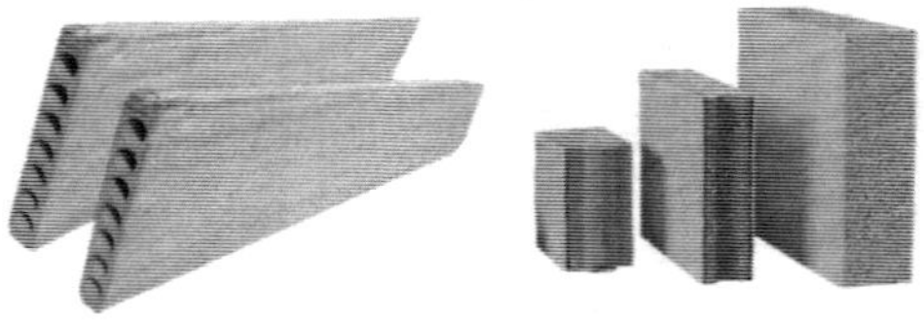

Retrieved from: http://www.wallpanel-supplier.com/es/products/Buy-foam-concrete-alternative-online-at-best-price.html

10

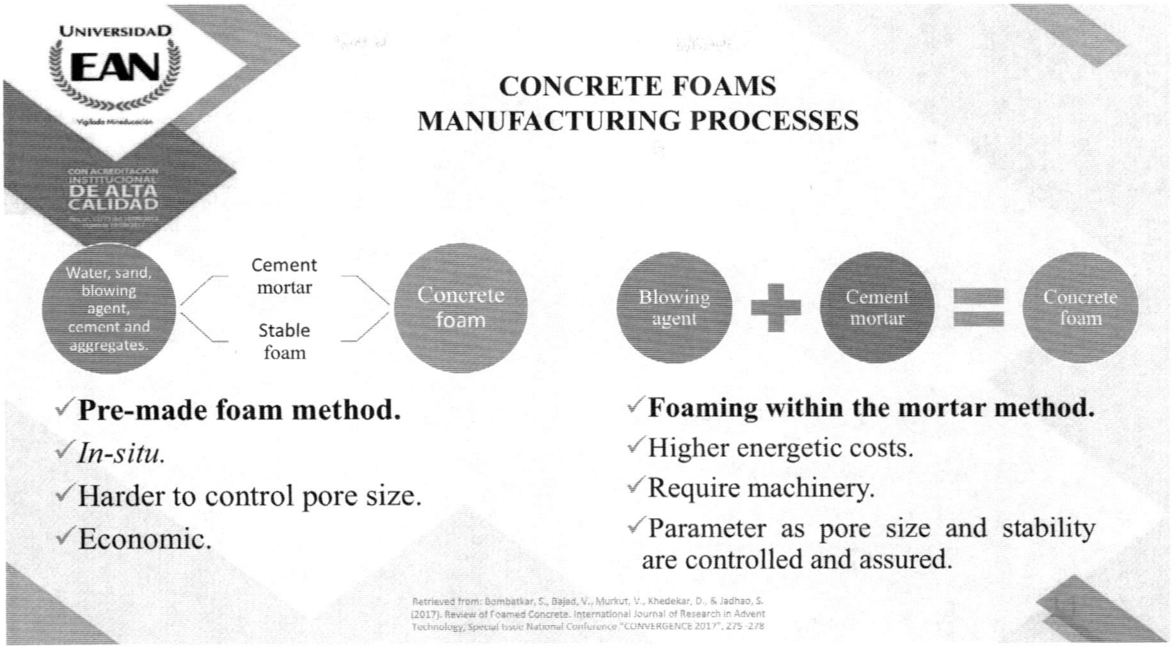
UNIVERSIDAD
EAN
Vigilada Mineducación
CON ACREDITACIÓN INSTITUCIONAL
DE ALTA CALIDAD
CONCRETE FOAMS
MANUFACTURING PROCESSES
Water, sand, blowing agent, cement and aggregates.
Cement mortar
Stable foam
Concrete foam
Blowing agent
Cement mortar
Concrete foam
✓Pre-made foam method.
✓In-situ.
✓Harder to control pore size.
✓Economic.
✓Foaming within the mortar method.
✓Higher energetic costs.
✓Require machinery.
✓Parameter as pore size and stability are controlled and assured.
Retrieved from: Bombatkar, S., Bajad, V., Murkut, V., Khedekar, D., & Jadhao, S. (2017). Review of Foamed Concrete. International Journal of Research in Advent Technology, Special Issue National Confurence "CONVERGENCE 2017", 275-278

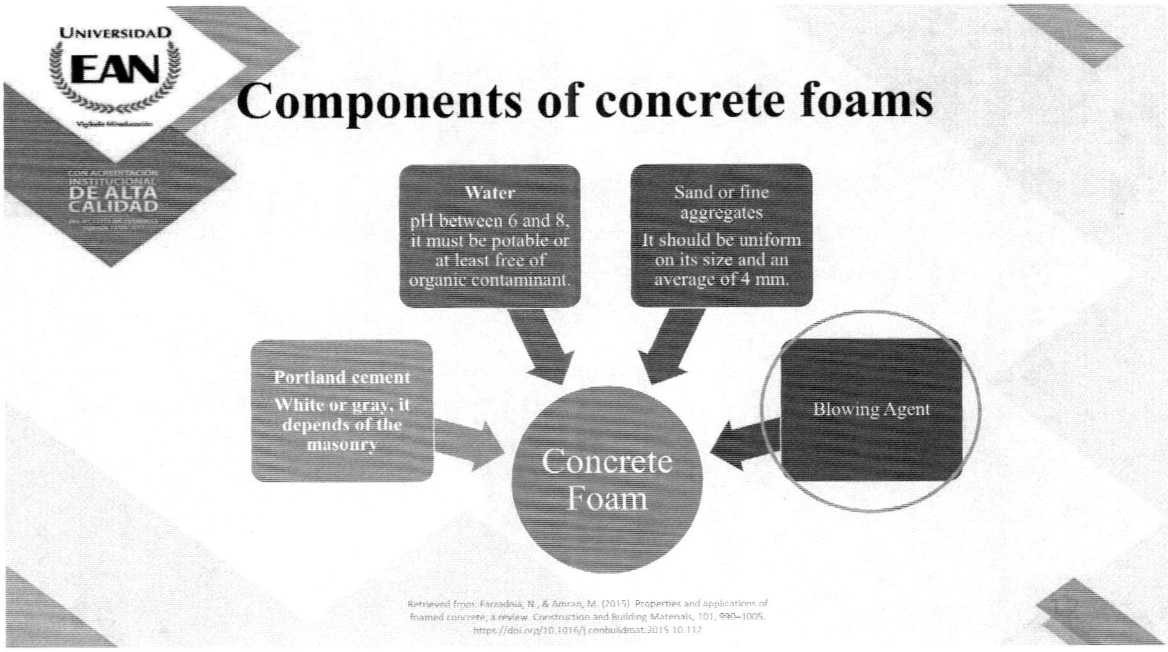
UNIVERSIDAD
EAN
Vigilado Mineducación
CON ACREDITACIÓN INSTITUCIONAL
DE ALTA CALIDAD
Components of concrete foams
Water
pH between 6 and 8, it must be potable or at least free of organic contaminant.
Sand or fine aggregates
It should be uniform on its size and an average of 4 mm.
Portland cement
White or gray, it depends of the masonry
Blowing Agent
Concrete Foam
Retrieved from: Farzadnia, N., & Amran, M. (2015). Properties and applications of foamed concrete; a review. Construction and Building Materials, 101, 990–1005. https://doi.org/10.1016/j.conbuildmat.2015.10.112

OUTLINE

- Colombian context
- Investigation aims
- Concrete Foams
- **Chemical blowing agents**
- Potential environmental impact
- Acoustic insulation

13

A Blowing agent is,

✓ Responsible for the gas phase inside the solid matrix.

✓ Non flammability and non corrosively as mandatory requirements.

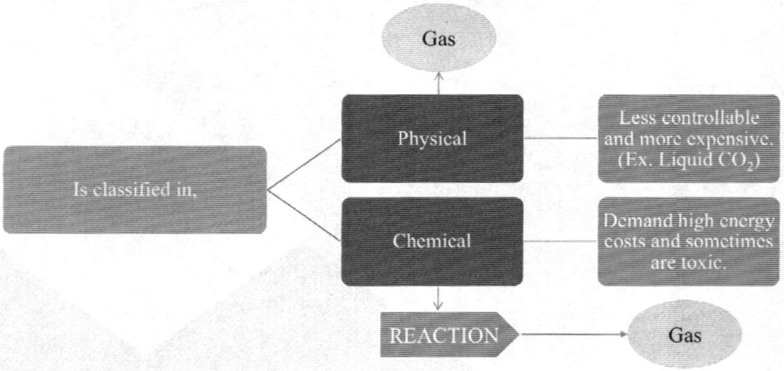

14

UNIVERSIDAD EAN

CON ACREDITACIÓN INSTITUCIONAL DE ALTA CALIDAD

A Chemical Blowing agent is,

- ✓Responsible for the gas phase inside the solid matrix <u>thanks to a chemical reaction, mostly thermal decomposition or deoxygenation</u>.
- ✓Temperature of reaction vs. stability.
- ✓Usually are powders or liquids.
- ✓Easy storage and direct relationship with final mechanical and physical properties.

Usual decomposition curve

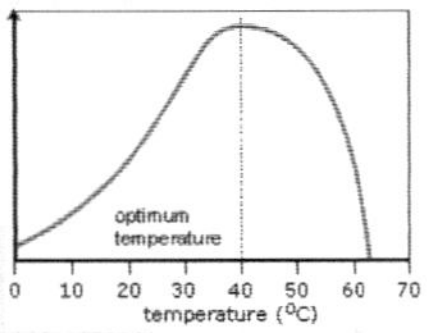

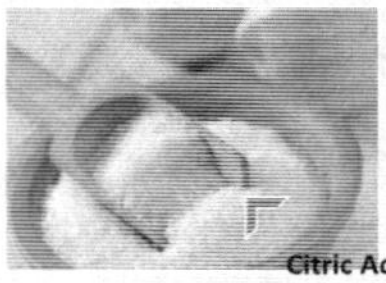

Retrieved from: https://www.amazon.co.uk/CITRIC-PRESERVATIVE-ACIDS-CITRATE-LEMON/dp/B00C7O4PPR
http://www.bbc.co.uk/schools/gcsebitesize/science/add_aqa_pre_2011/enzymes/enzymes3.shtml

UNIVERSIDAD EAN

CON ACREDITACIÓN INSTITUCIONAL DE ALTA CALIDAD

Selection criteria

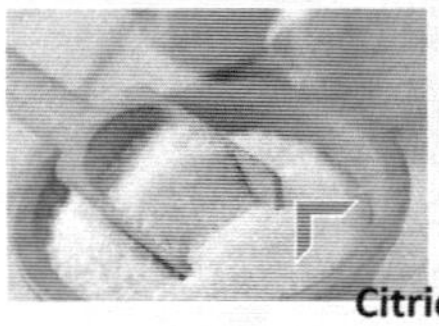

Retrieved from: https://www.amazon.co.uk/CITRIC-PRESERVATIVE-ACIDS-CITRATE-LEMON/dp/B00C7O4PPR

- ✓Non toxic.
- ✓Non corrosive.
- ✓Produce CO_2.
- ✓Alternative manufacturing process.
- ✓Low production of side gaseous products.

UNIVERSIDAD EAN

$NaHCO_3$ Thermal Decomposition

$$2NaHCO_{3(s)} \rightarrow Na_2CO_{3(s)} + H_2O_{(g)} + CO_{2(g)}$$

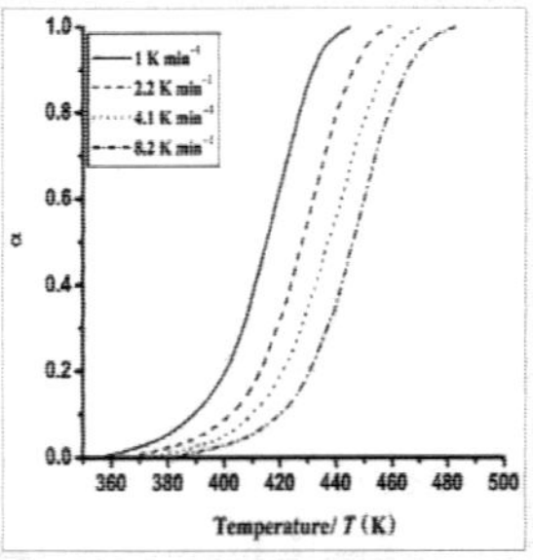

Extent of reaction [mol] vs. Temperature [°C].

- ✓Average optimum temperature: 158 °C (Wypych,2017).
- ✓An elevation in heating speed becomes in an elevation in optimum temperature (Janković & Adnađević, 2007).
- ✓Small particle sizes lower the optimum temperature (Heda et al., 1995).
- ✓And elevation in atmospheric water vapor makes the reaction go faster (Yamada & Koga, 2005).

Retrieved from: Janković, B., & Adnađević, B. (2007). The use of the IKP method for evaluating the kinetic parameters and the conversion function of the thermal decomposition of NaHCO3 from nonisothermal thermogravimetric data. International Journal of Chemical Kinetics, 39(8), 462–471. https://doi.org/10.1002/kin.20256

17

UNIVERSIDAD EAN

$C_6H_8O_7$ Thermal Decomposition

$$C_6H_8O_{7(s)} \rightarrow C_5H_4O_3 + CO_{2(g)} + 2H_2O_{(g)}$$

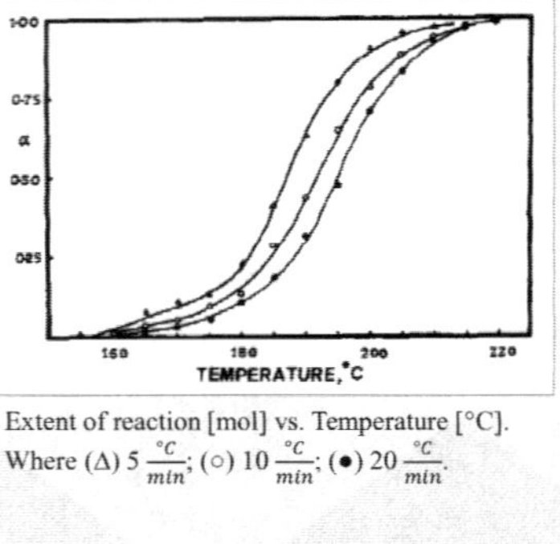

Extent of reaction [mol] vs. Temperature [°C]. Where (Δ) 5 $\frac{°C}{min}$; (○) 10 $\frac{°C}{min}$; (•) 20 $\frac{°C}{min}$.

- ✓Average optimum temperature: 234,8 °C (Stehr, 2016).
- ✓An elevation in heating speed becomes in an elevation in optimum temperature (Wyrzykowski et. al., 2011).
- ✓Small particle sizes lower the optimum temperature (Barbooti & Al-Sammerrai, 1986).

Retrieved from: Barbooti, M. M., & Al-Sammerrai, D. A. (1986). Thermal decomposition of citric acid. Thermochimica Acta, 98, 119–126. https://doi.org/10.1016/0040-6031(86)87081-2

18

UNIVERSIDAD EAN

OUTLINE

- Colombian context
- Investigation aims
- Concrete Foams
- Chemical blowing agents
- **Potential environmental impact**
- Acoustic insulation

19

UNIVERSIDAD EAN

Analysis boundaries
Concrete foam limited from cradle to fabrication

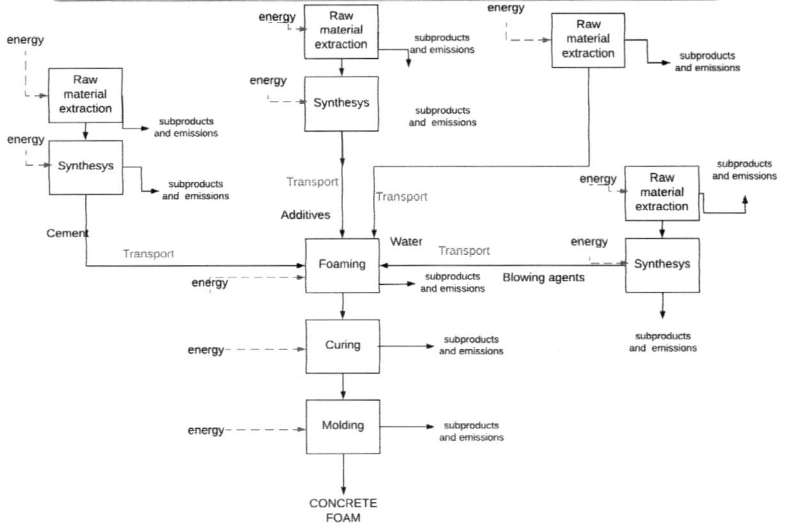

20

Analysis boundaries
Concrete foam limited from cradle to fabrication

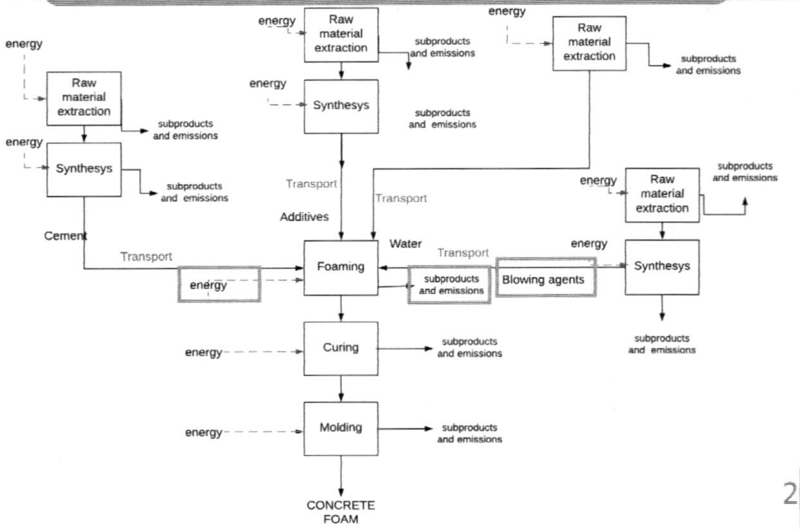

Potential environmental impact parameters

Energy consumption.

	Sodium bicarbonate	Citric acid
Working temperature [K]	431	507,8
Activation energy [kJ/mol]	90 – 120	96 – 103

Atmospheric emissions

Per every 1000g added	Sodium bicarbonate	Citric acid
CO_2 [g]	261,942	229,071
H_2O [g]	107,225	187,538

Process waste

Per every 1000g added	Sodium bicarbonate	Citric acid
Na_2CO_3 [g]	630,833	-
$C_5H_4O_3$ [g]	-	583,394

Potential environmental impact parameters

Toxicity

Compound	Risk parameter	Compound	Risk parameter
$NaHCO_{3(s)}$	$DL_{50}^{Rat} > 5000 \frac{mg}{kg}$	$C_6H_8O_{7(s)}$	$DL_{50}^{Rat} > 11700 \frac{mg}{kg}$
	Carcinogenic: No effects		Carcinogenic: No effects
$Na_2CO_{3(g)}$	$DL_{50}^{Rat} > 4090 \frac{mg}{kg}$	$C_5H_4O_{3(g)}$	$DL_{50}^{Rat} > 2600 \frac{mg}{kg}$
	Carcinogenic: No effects		Carcinogenic: No effects
$CO_{2(g)}$	$DL_{50}^{Rat} > 3000 \frac{mg}{kg}$	$CO_{2(g)}$	$DL_{50}^{Rat} > 3000 \frac{mg}{kg}$
	Carcinogenic: No effects		Carcinogenic: No effects
$H_2O_{(g)}$	$DL_{50}^{Rat} > 90000 \frac{mg}{kg}$	$H_2O_{(g)}$	$DL_{50}^{Rat} > 90000 \frac{mg}{kg}$
	Carcinogenic: No effects		Carcinogenic: No effects

Analysis boundaries
Concrete foam limited from cradle to fabrication

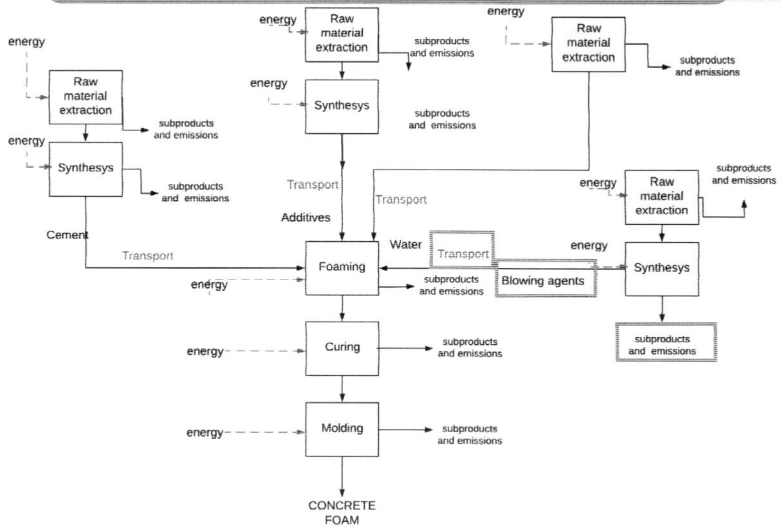

Raw materials analysis

Toxicity

Sodium Bicarbonate Production (Solvay Method)	
Compound	Risk parameter
HCl	$DL_{50}^{Rat} > 238\frac{mg}{kg}$
	Carcinogenic: No effects.
NH_4HCO_3	$DL_{50}^{Rat} > 1{,}576\frac{mg}{kg}$
	Carcinogenic: N/A.
NH_4Cl	$DL_{50}^{Rat} > 1300\frac{mg}{kg}$
	Carcinogenic: No effects.

Citric Acid Production (Fermentation using *Aspergillus Niger*)	
Compound	Risk parameter
Aspergillus Niger	Fungus
	Generally Recognized As Safe
	High exposure may cause Otomycosis

25

Raw materials analysis

Atmospheric emissions

Solvay Method	Fermentation using *Aspergillus Niger*
$CO_{2(g)}$	$CO_{2(g)}$
$NH_{3(g)}$	Spores

Process waste

Solvay Method	Fermentation using *Aspergillus Niger*
NH_4Cl	Biomass

Carbon footprint in transportation

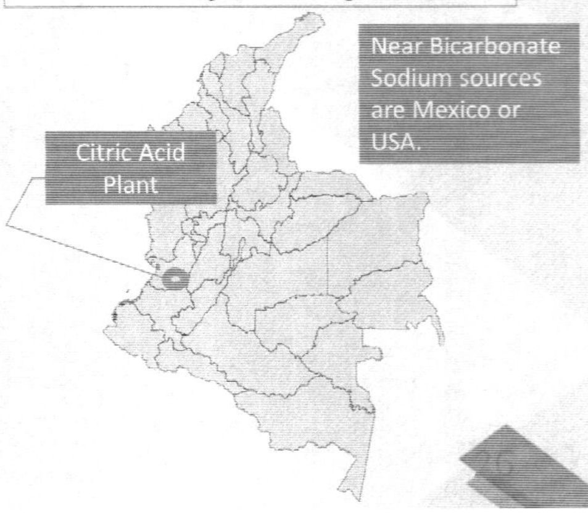

26

OUTLINE

- Colombian context
- Investigation aims
- Concrete Foams
- Chemical blowing agents
- Potential environmental impact
- Acoustic insulation

27

Acoustic Insulation

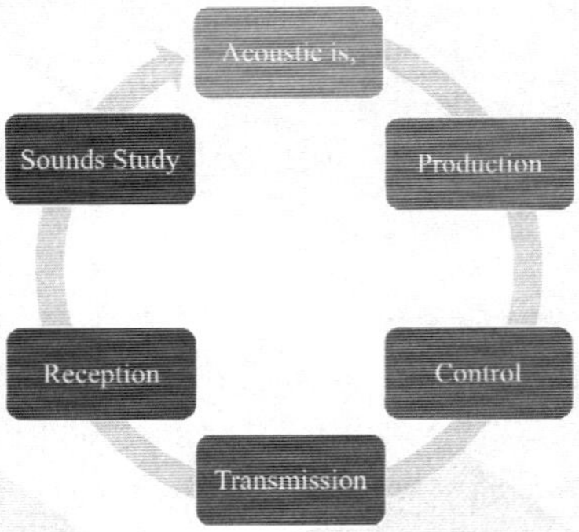

Two phenomena:

✓Absorption.

✓Reflection.

28

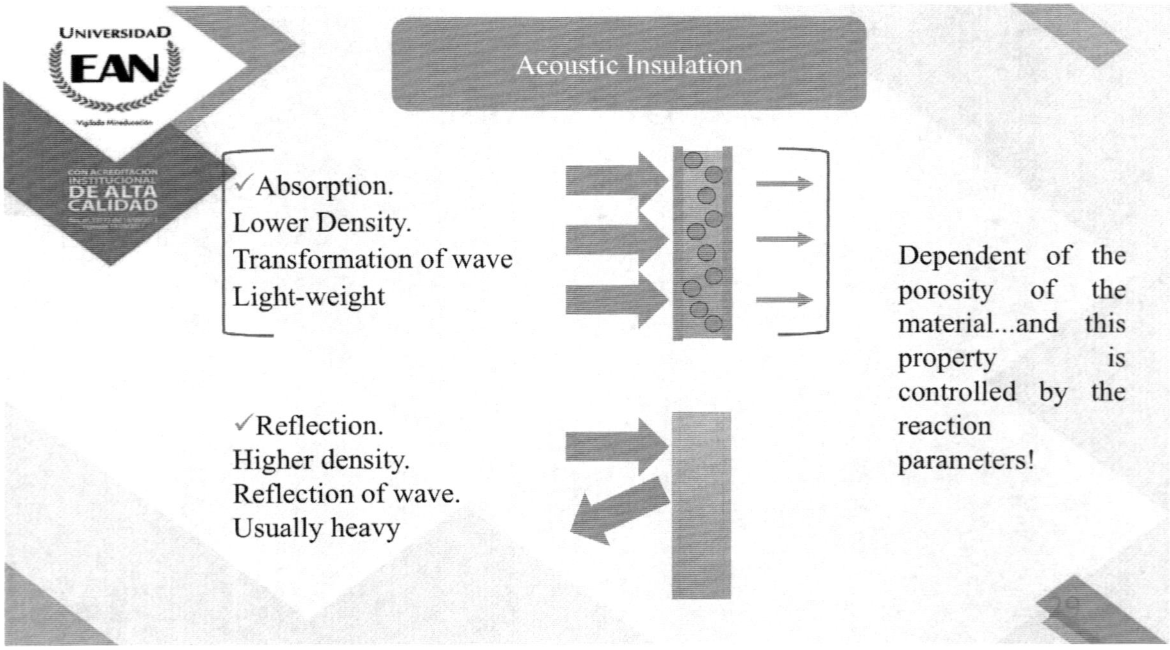
UNIVERSIDAD
EAN
CON ACREDITACIÓN INSTITUCIONAL
DE ALTA CALIDAD
Acoustic Insulation
✓Absorption.
Lower Density.
Transformation of wave
Light-weight
✓Reflection.
Higher density.
Reflection of wave.
Usually heavy
Dependent of the porosity of the material...and this property is controlled by the reaction parameters!
29

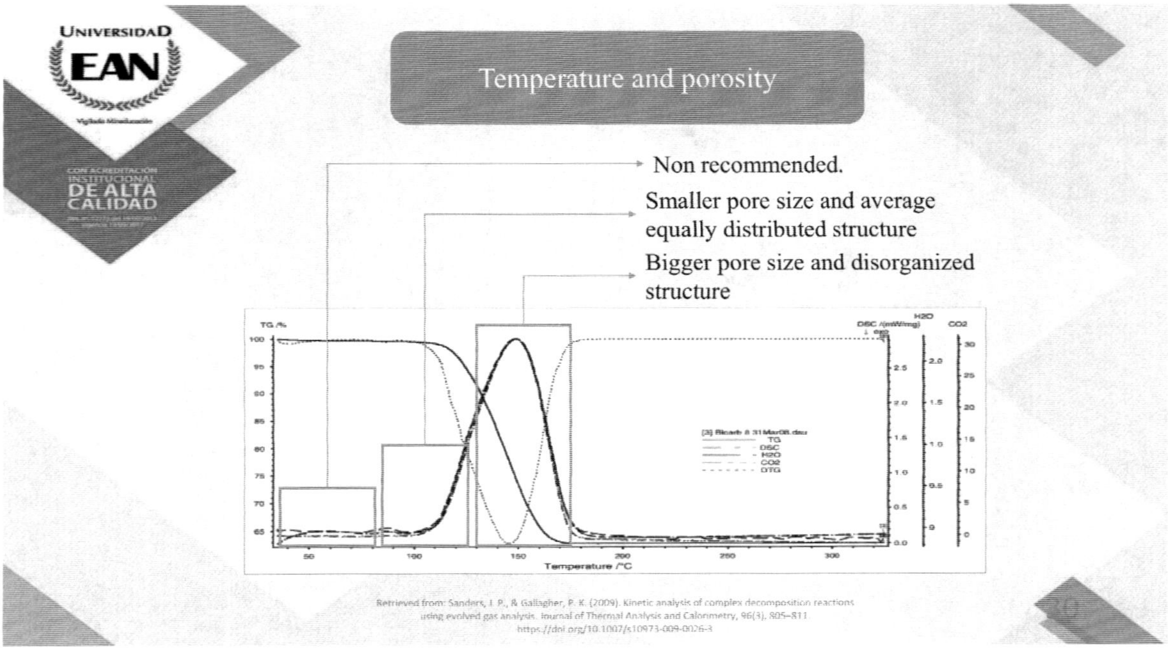
UNIVERSIDAD
EAN
CON ACREDITACIÓN INSTITUCIONAL
DE ALTA CALIDAD
Temperature and porosity
Non recommended.
Smaller pore size and average equally distributed structure
Bigger pore size and disorganized structure
TG /%
DSC /(mW/mg)
H2O
CO2
Temperature /°C
TG
DSC
H2O
CO2
DTG
Retrieved from: Sanders, J. P., & Gallagher, P. K. (2009). Kinetic analysis of complex decomposition reactions using evolved gas analysis. Journal of Thermal Analysis and Calorimetry, 96(3), 805–811. https://doi.org/10.1007/s10973-009-0026-3
30

UNIVERSIDAD EAN

Acoustic Insulators properties

Material	Density	Absorption coefficient (α)
Foam Concrete	330 Kg/m^3	0,20
Fiberglass (Thickness: 25,4 mm)	14,64 Kg/m^3	0,62
Wood agglomerate	230 Kg/m^3	0,44
Polyurethane foam (Thickness: 25 mm)	30 Kg/m^3	0,84

31

UNIVERSIDAD EAN

Conclusions

✓ Thermal insulation and low weight are the advantages of using foam concrete as acoustic insulator.

✓Sodium bicarbonate blowing agent presents higher levels of potential hazard by toxicity while citric acid presents higher energetic costs.

✓There is a direct relationship between acoustic insulation properties of concrete foam and foaming temperature.

✓To improve acoustic insulation foam concrete require reinforcement with another materials.

32

Future Research

✓ Comparative experiments using both chemical blowing agents to measure α-coefficient.
✓ Life cycle assessment.
✓ Use of CO2 produced in foaming stage as curing of foam concrete using mineral carbonation.
✓ Use of flue gas CO2 from industry emissions as a direct foaming agent and hence improve carbon footprint.

33

References

✓ Bombatkar, S., Bajad, V., Murkut, V., Khedekar, D., & Jadhao, S. (2017). Review of Foamed Concrete. International Journal of Research in Advent Technology, Special Issue National Conference "CONVERGENCE 2017", 275–278.
✓ Barbooti, M. M., & Al-Sammerrai, D. A. (1986). Thermal decomposition of citric acid. *Thermochimica Acta*, *98*, 119–126.
✓ Casas-García, O., Betancur-Vargas, C. M., & Montaño-Erazo, J. S. (2015). Revisión de la normatividad para el ruido acústico en Colombia y su aplicación. *Entramado*, *11*(1), 264–286.
✓ Everest, F. A., & Pohlmann, K. (2009). *Master Handbook of Acoustics* (5 edition). New York: McGraw-Hill/TAB Electronics.
✓ Farzadnia, N., & Amran, M. (2015). Properties and applications of foamed concrete; a review. Construction and Building Materials, 101, 990–1005.
✓ Heda, P. K., Dollimore, D., Alexander, K. S., Chen, D., Law, E., & Bicknell, P. (1995). A method of assessing solid state reactivity illustrated by thermal decomposition experiments on sodium bicarbonate. Thermochimica Acta, 255, 255–272.
✓ Janković, B., & Adnađević, B. (2007). The use of the IKP method for evaluating the kinetic parameters and the conversion function of the thermal decomposition of NaHCO3 from nonisothermal TM data. International Journal of Chemical Kinetics, 39(8), 462–471.
✓ Mohammed, J., & Hamad, A. J. (2014). Materials, properties and application review of lightweight concrete, 37, 10–15.
✓ Sanders, J. P., & Gallagher, P. K. (2009). Kinetic analysis of complex decomposition reactions using evolved gas analysis. Journal of Thermal Analysis and Calorimetry, 96(3), 805–811.
✓ Stehr, J. (2016). Chemical blowing agents in the rubber industry. Past - present - and future?, 43, 812–819.
✓ Wyrzykowski, D., Hebanowska, E., Nowak-Wiczk, G., Makowski, M., & Chmurzyński, L. (2011). Thermal behaviour of citric acid and isomeric aconitic acids. Journal of Thermal Analysis and Calorimetry, 104(2), 731–735.
✓ Yamada, S., & Koga, N. (2005). Kinetics of the thermal decomposition of sodium hydrogencarbonate evaluated by controlled rate evolved gas analysis coupled with thermogravimetry. Thermochimica Acta - THERMOCHIM ACTA, 431, 38–43

34

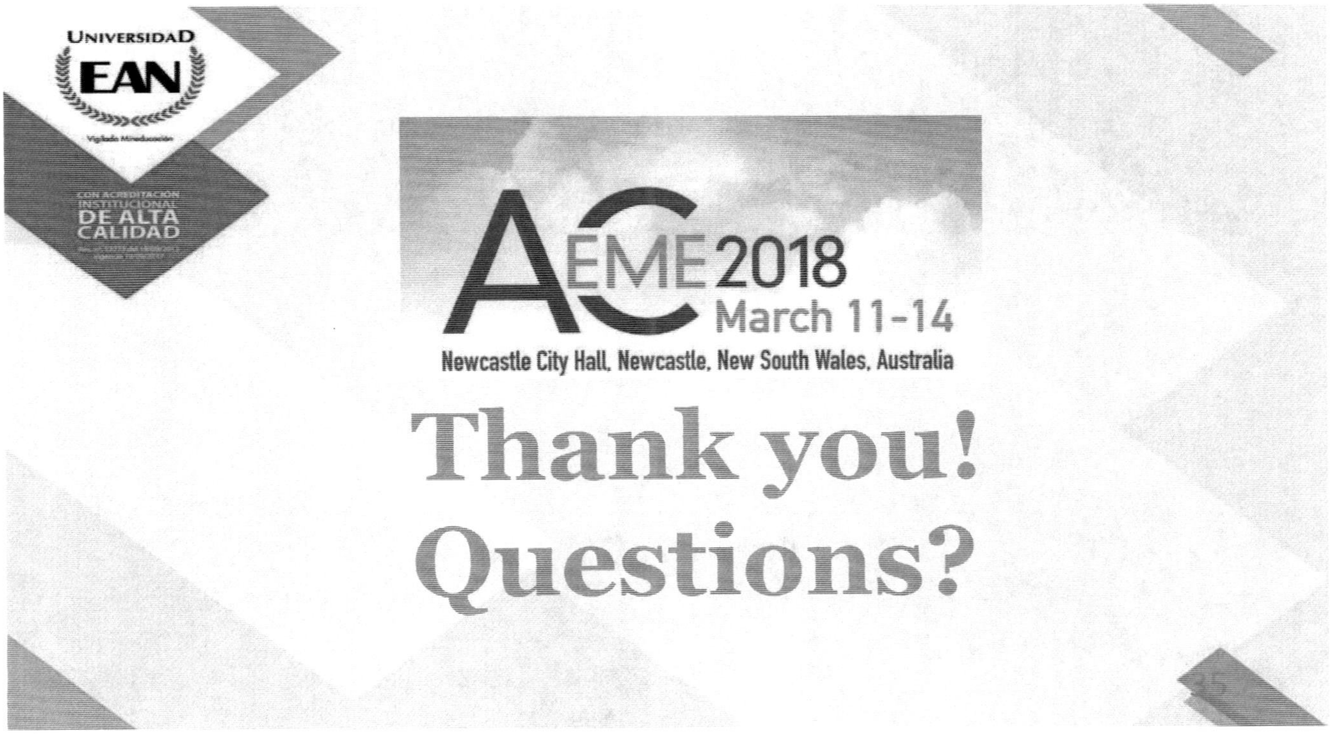
UNIVERSIDAD
EAN
CON ACREDITACIÓN INSTITUCIONAL
DE ALTA CALIDAD
AC EME 2018
March 11-14
Newcastle City Hall, Newcastle, New South Wales, Australia
Thank you!
Questions?

Synthesis and characterisation of reactive silica residue from MC process

Emad Benhelal[a], M.I.Rashid[a], M.Rayson[a,b], G.Brent[b], M.Stockenhuber[a], E.M.Kennedy[a]

[a] Department of Chemical Engineering, The University of Newcastle

[b] Orica Australia Pty Ltd

ACEME Conference, Newcastle, Australia March 2018

THE UNIVERSITY OF NEWCASTLE AUSTRALIA

Mineral Carbonation International

Presentation Overview

- Background
- Aims
- Materials preparation
- Results and discussion

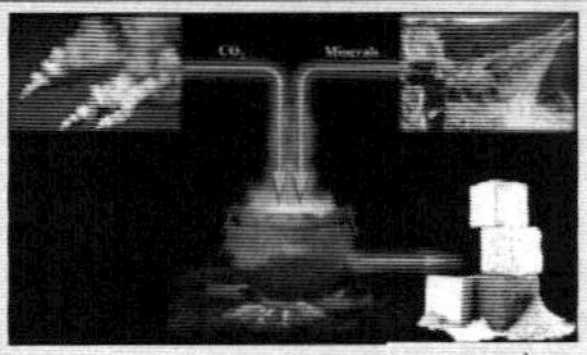

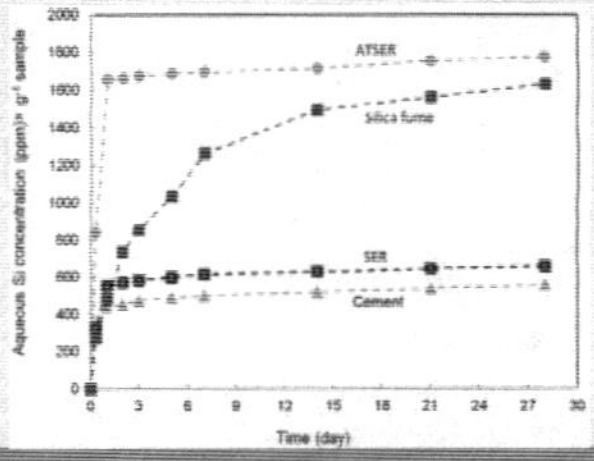

2

Background

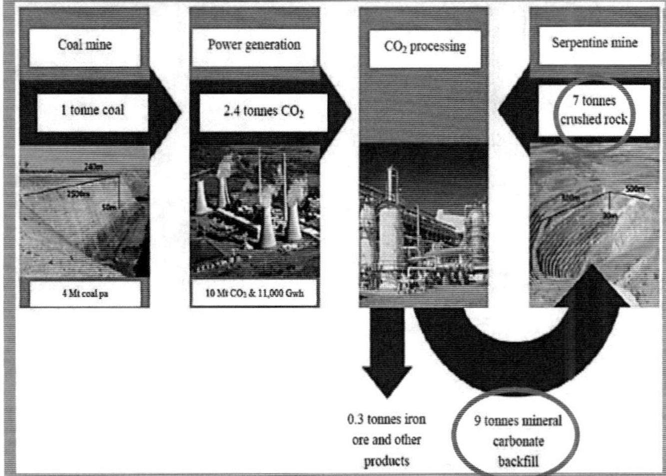

Relative material flows for serpentinite based mineral carbonation
(Brent et al., 2012)

The primary challenges MC faces:

1- High capital costs and energy demand;

2- Additional CO_2 emissions; and

3- Accumulation of reaction by-products.

Utilizing the by-products can offset these challenges.

Among by-products, silica enriched residue is more promising, can partially replace cement in concrete.

3

Single vs two stage aqueous mineral carbonation

Single stage (ARC process)

- High PCO_2 is required
- Both extraction of Mg and precipitation of $MgCO_3$ occurs in one vessel.
- By-product is a mixture of mainly $MgCO_3$, SiO_2 and unreacted serpentine.
- It is difficult to separate.

Two stage

- Much lower PCO_2 is needed.
- Extraction of Mg and precipitation of carbonate occur in two reactors.
- Solid residue of the first stage is mainly amorphous silica and unreacted lizardite.
- Products are relatively pure.

4

Amorphous silica

- Has potential to act as **pozzolanic material.**
- When added to cement and mixed with water, reacts with $Ca(OH)_2$ to produce **calcium silicate hydrate (C-S-H).**
- Accelerates cement hydration and **increases** compressive strength (Fu et al., 2002).
- The hydration reaction equation is shown as:

$$x Ca(OH)_2 + x\ SiO_2 + nH_2O = x\ CaO.SiO_2.(n+x)H_2O \quad \textbf{(CSH)}$$

5

Aims

- To **synthesise** and **characterise** amorphous silica enriched residue materials from 2 stage MC process;
- To investigate **pozzolanic activity** of synthesised materials in comparison with silica fume.

6

Materials preparation

Raw lizardite

Crushed lizardite from **Great Serpentine Belt, NSW** was wet milled in 2 L S.S ball mill and pass through **75 μm sieve**.

Heat activated lizardite

Dried milled lizardite was heat activated in a S.S rotary kiln at **630 °C, 4 h**.

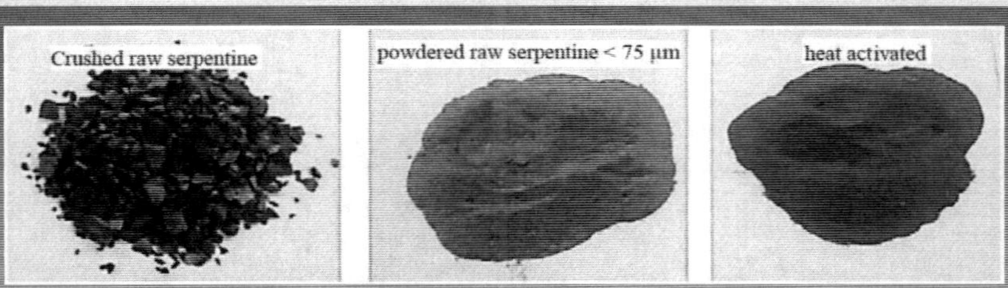

7

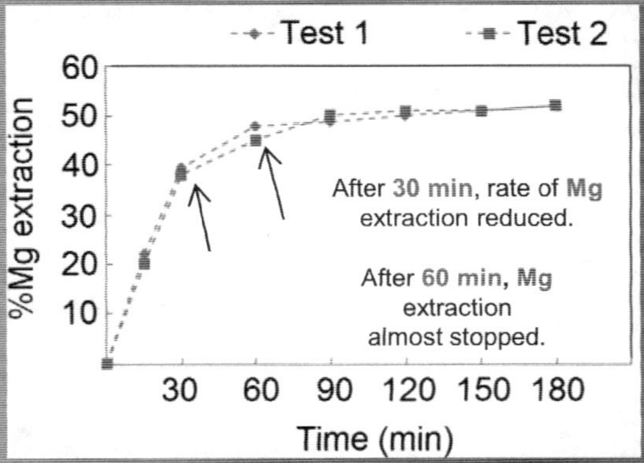

But why?

What's going on?

Dissolution residue

- Heat activated lizardite was dissolved in carbonic acid (**2 wt%** solid conc.);
- At PCO_2 = **6.5 bar** T = **45 °C**, t = **3 h**.

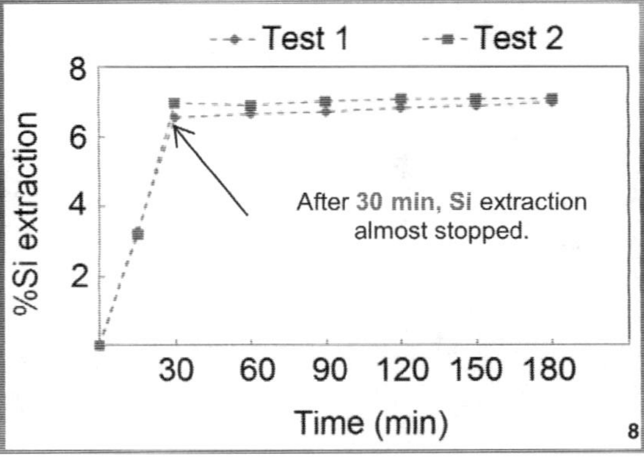

8

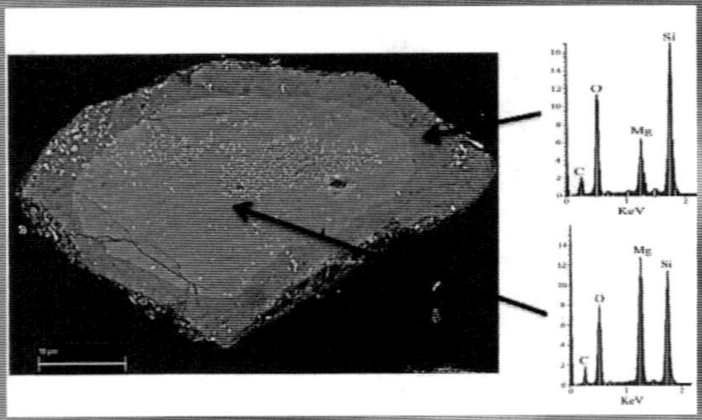

Cross section SEM image of residue particle after 3 h dissolution in carbonic acid

Silica-rich layer significantly **reduced** Mg and Si extraction

What is the effect of feed particle size on extraction?

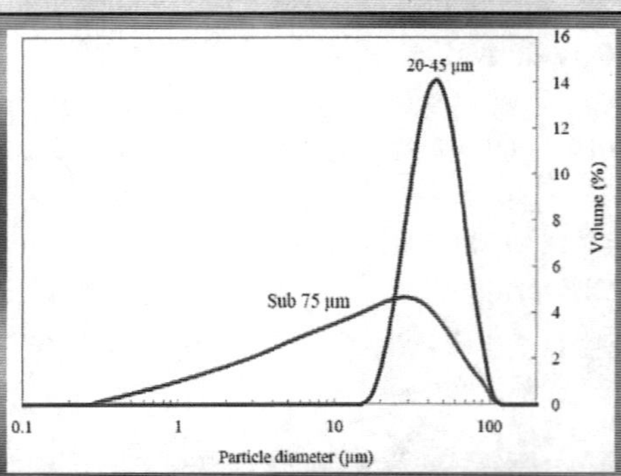

Sub 75 µm, d_{10} = 1.6 µm , d_{50} = 12.9 µm

20-45 µm, d_{10} =25.4 µm , d_{50} = 35.2 µm

9

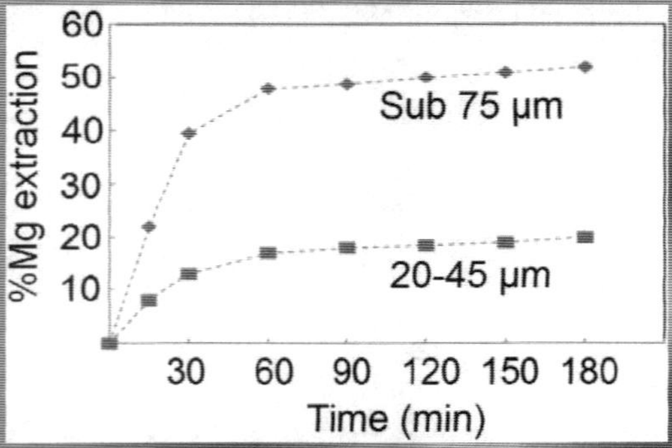

So they are two ways to **increase** the rate of extraction

1- Using finer particles as feed

2- Breaking silica rich layer

How to kill two birds with one stone?

Regrinding the dissolution residue

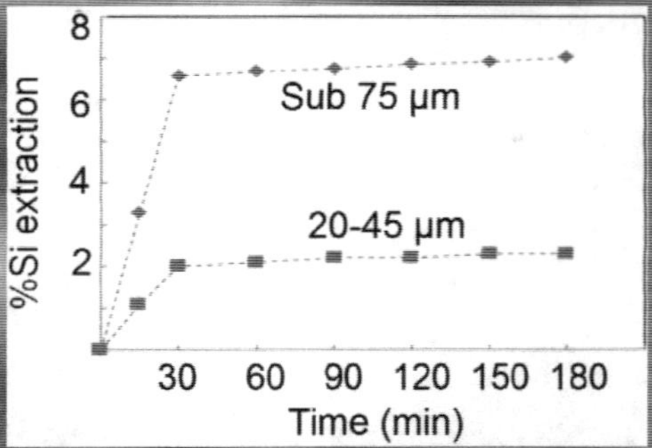

10

Multi-stage dissolution

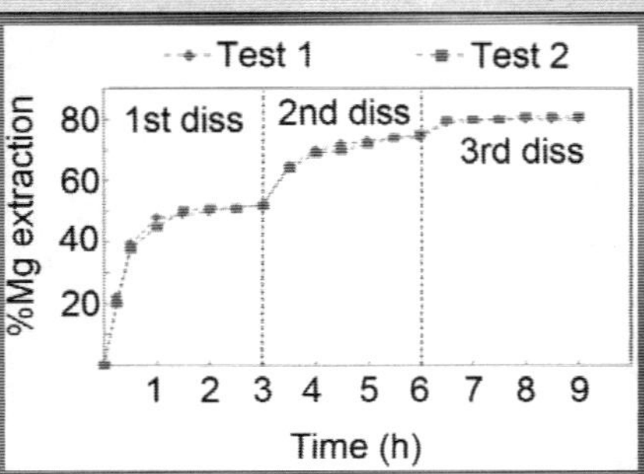

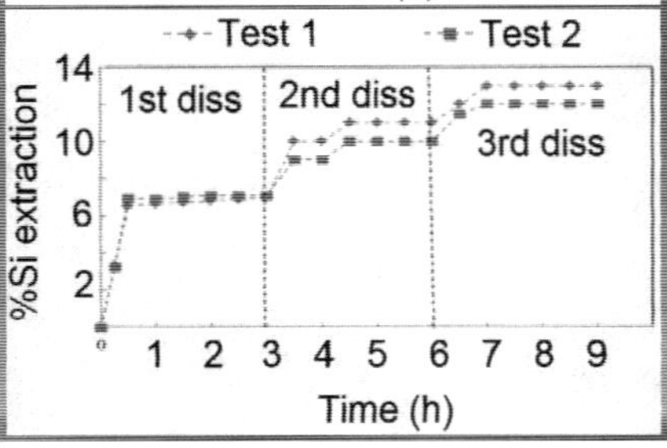

We developed a **multi-stage dissolution** process in carbonic acid;

80 wt% of Mg extracted. Solid residue **(SER)** was enriched with silica **(65 wt% amorphous silica)**

Even **2 stage of dissolution**, each for **1 h** is sufficient;

Can we enrich **further**?

1st regrinding
d_{10}=1.6 μm, d_{50}=12.9 μm to d_{10}=1.2 μm, d_{50}=4.4 μm

2nd regrinding
d_{10}=1.2 μm, d_{50}=4.4 μm to d_{10}=1.1 μm, d_{50}=3.4 μm

11

Acid treatment of SER

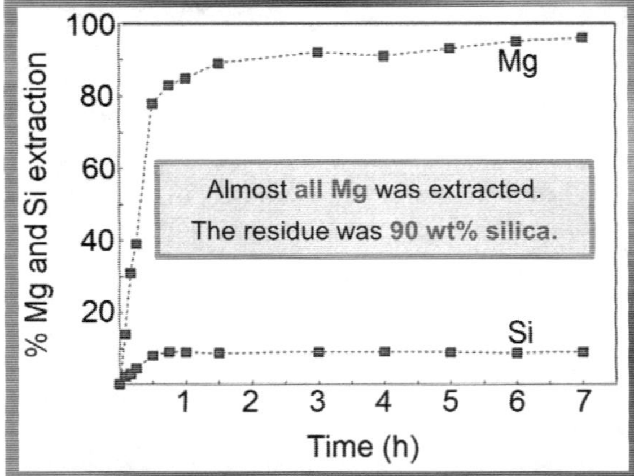

- SER was dissolved in **2 M nitric acid** (0.5 wt% solid conc.) for **7 h**.
- Solid residue was rinsed with distilled water, dried and named "acid treated silica enriched residue **(ATSER)**.

- For commercial application, different solid loading **(0.5 - 5 wt%)** and residence time **(30 min- 7 h)** tested.
- Even **5 wt%** solid loading and **30 min** acid treatment is sufficient.

12

Comparing ATSER and silica fume

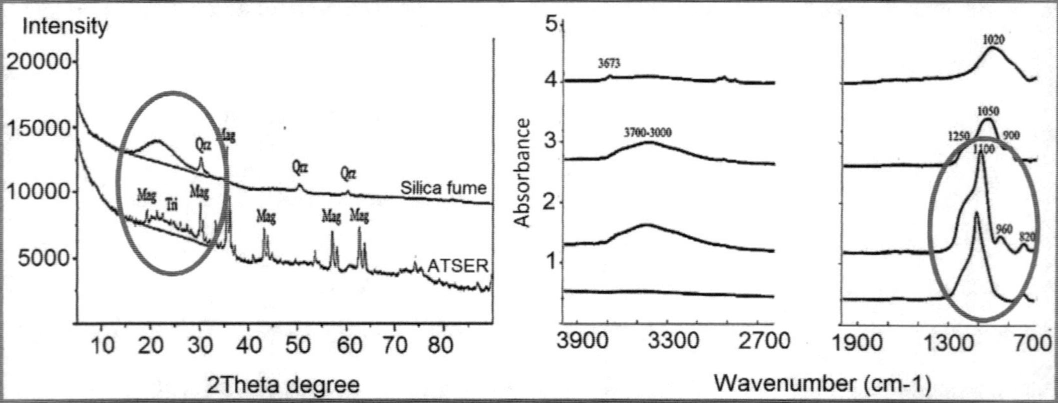

XRD patterns and **FTIR** spectrum are similar

We expect ATSER, **similar pozzolanic activity** to silica fume

13

Pozzolanic activity test

Pozzolanic activity is **capacity to release silicon** when water is added to cement mix and pH increases to 13-14 (Snellings et al., 2012);

Determining the **rate of silica dissolution** from pozzolanic materials is critical;

Due to **quick setting time** of the cement paste and formation of hydrated phases, periodic sampling to determine the dissolved silicon is difficult;

We developed an experiment to **mimic silicon extraction** from pozzolans in cement paste pore solution;

Materials were dissolved in **0.2 M NaOH (pH~13.5) for 28** days to determine the **rate of silicon extraction** in an environment similar to cement paste by ICP.

14

Comparing the rate of silica dissolution

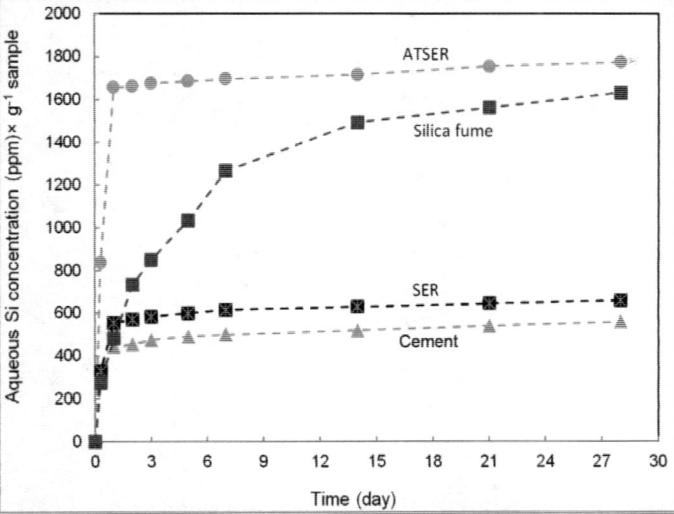

Rate of Si extraction from **ATSER** was the **fastest** among all materials studied. (92 wt% within 1 day, 98 wt% after 28 days).

Even faster than silica fume **(one of the best pozzolanic materials)**, 25 and 80 wt% after 1 and 28 days.

Surface area of **ATSER** and **silica fume** were **195** and **16** m^2/g respectively.

SER is slightly **faster than cement**, 40 and 50 wt% of Si content after 1 and 28 days.

We synthesised **highly reactive silica** (ATSER).

15

Conclusions

- Rate of Mg and Si extraction significantly **reduced** due to formation of silica rich layer.
- **Regrinding** increased Mg extraction due to **breaking silica rich layer** and **particle size reduction.**
- **80 wt%** of Mg was extracted under **very mild conditions.**
- Treating SER with **nitric acid** produced amorphous silica **(ATSER).**
- The rate of Si extraction from both **SER** and **ATSER** is **faster than cement** in alkaline solution.
- ATSER is a **highly reactive silica**, even more reactive than silica fume in alkaline solution.

16

Acknowledgment
• The University of Newcastle for Postgraduate research scholarships
• Mineral Carbonation international (MCi)
• EMX (The University of Newcastle) for XRD and SEM analyses
THE UNIVERSITY OF
NEWCASTLE
AUSTRALIA
MCi
Mineral Carbonation International
GreenMag Group
NSW Trade & Investment Resources & Energy
ORICA
17

THANKS FOR LISTENING
ANY QUESTIONS ?

International Conference on Accelerated Carbonation for Environmental and Material Engineering
Pilot and Full-Scale Applications, Utilization of the Carbonated Materials and Other II
Newcastle, March 14th, 2018

Influence of mineralogy and process parameters on microstructure development and mechanical properties of carbonated compacts

Paola Librandi, Giulia Costa, Renato Baciocchi
Laboratory of Environmental Engineering, DICII
University of Rome Tor Vergata

Mieke Quaghebeur, Peter Nielsen, Ruben Snellings
Sustainable Materials Management
VITO, Mol, Belgium

Introduction

Wet-route Accelerated Carbonation

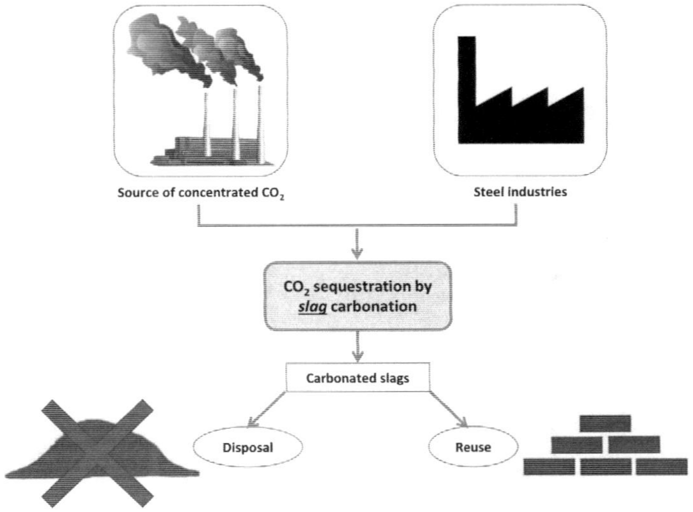

VITO Technique

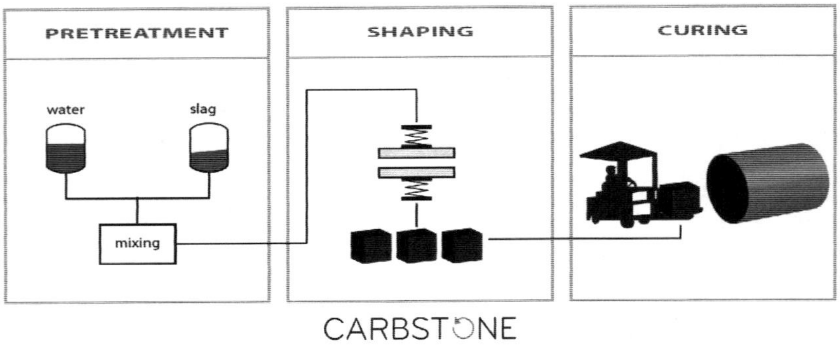

Aim of the work

- Fundamental study of the carbonation reaction during curing of compacts
- Identification of the main reactive phases (XRD, SEM)
- Correlation with the properties of the carbonated products:
  - CO_2 Uptake
  - Mechanical Strength
- Pure minerals carbonation (in progress)

Materials

Steel slags

- Basic Oxygen Furnace Slags (BOFs)
- Electric Arc Furnace Slags (EAFs)

Milled <125 μm by ball mill

Euroslag 2012

9.8 Mt BOFs/y → 45%

8.3 Mt EAFs/y → 38%

PSD

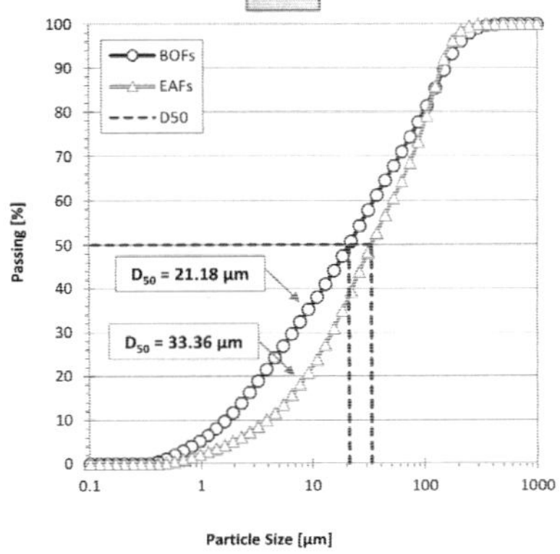

XRF

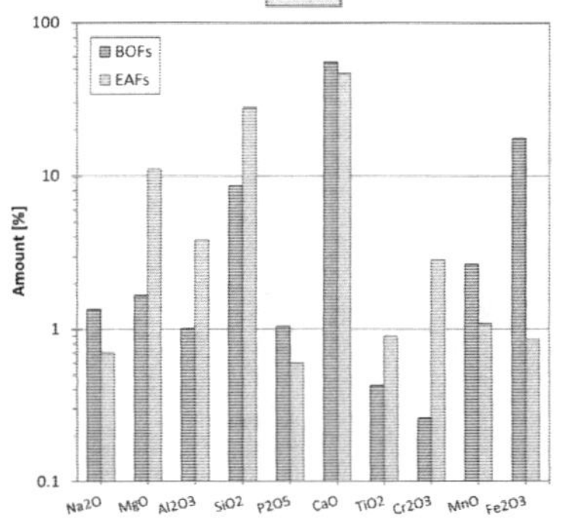

Total Inorganic Carbon	
Sample	TIC [%]
BOFs	0.6
EAFs	0.41

Materials

XRD

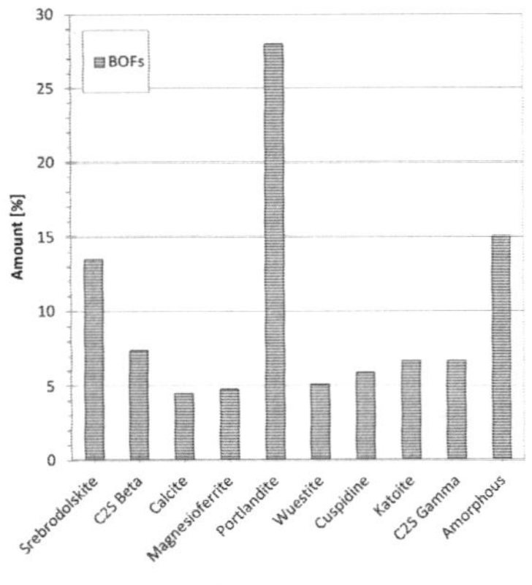

Main Phases:

Ca Si Fe

- $Ca(OH)_2$ Portlandite
- $Ca_2Fe_2O_5$ Srebrodolskite
- $Ca_3Al_2(SiO_4)_{1.5}(OH)_6$ Katoite
- Ca_2SiO_4 Dicalcium Silicate
- $Ca_4Si_2O_7F_{1.5}(OH)_{0.5}$ Cuspidine

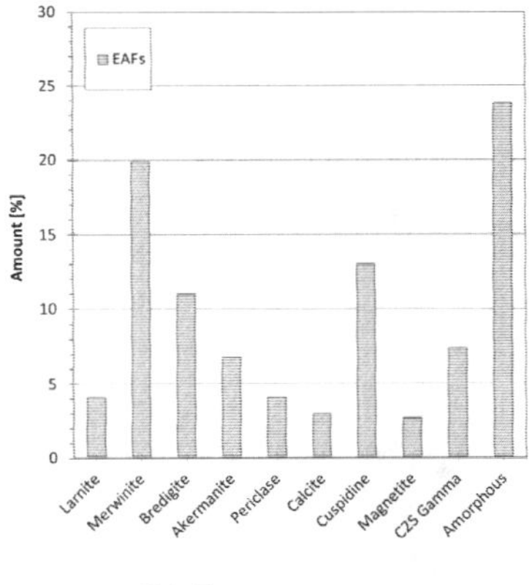

Main Phases:

Ca Si Mg

- $Ca_3Mg(SiO_4)_2$ Merwinite
- $Ca_7Mg(SiO_4)_4$ Bredigite
- $Ca_4Si_2O_7F_{1.5}(OH)_{0.5}$ Akermanite
- Ca_2SiO_4 Dicalcium Silicate
- $Ca_4Si_2O_7F_{1.5}(OH)_{0.5}$ Cuspidine

Methods: operating procedure and conditions

Steelmaking plant

Dried Slag

Moisture Content 10%

24 hr

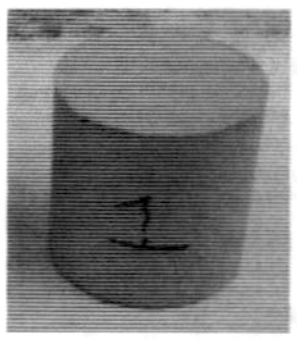

Compaction Pressure
150 kg/cm²

Carbonation Treatment

Reactor Volume 3.25 l

Carbonation conditions

Temperature [°C]	50
Gas Pressure [bar]	1.3 – 10
Gas Composition	100% CO_2
Reaction Time [min]	15 30 60 120 240 480

Methods: carbonated products characterization

TIC Analysis

Sample acidification at 200°C

$$CO_2[\%] = TIC\,[\%] \cdot \frac{MW_{CO_2}[g/mol]}{MW_C[g/mol]} \cdot 100$$

$$CO_{2uptake}[\%] = \frac{CO_{2fin}[\%] - CO_{2in}[\%]}{100 - CO_{2fin}[\%]} \cdot 100$$

$$R_X = \frac{CO_{2_{uptake}}(\%) \cdot \frac{MW_{Ca}}{MW_{CO_2}}}{Ca_{tot}(\%) - CaCO_{3_{in}}(\%) \cdot \frac{MW_{Ca}}{MWCaCO_3}} \cdot 100$$

XRD

SEM-EDX

Compressive Strength

BOF and EAF slags: CO_2 uptake

BOFs

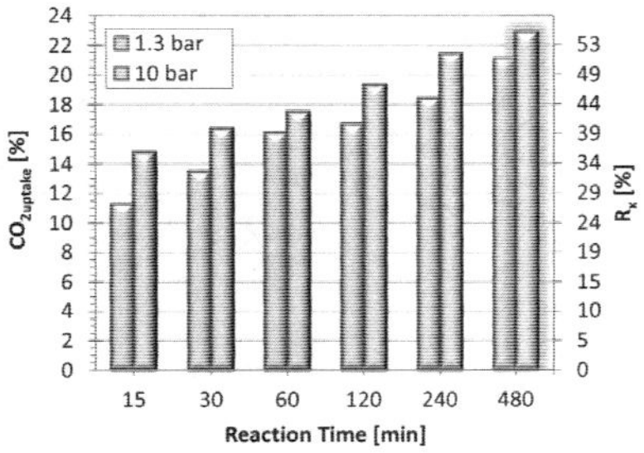

10bar 480 min

$CO_{2uptake}$ = 22.86%

R_x = 55.36%

EAFs

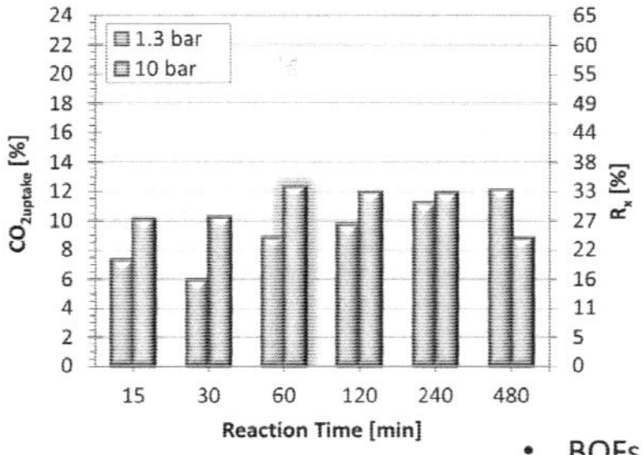

10bar 60 min

$CO_{2uptake}$ = 12.30%

R_x = 33.56%

- BOFs more reactive relative to EAFs
- BOFs carbonation kinetic very fast

BOF and EAF slags: compressive strength

BOF

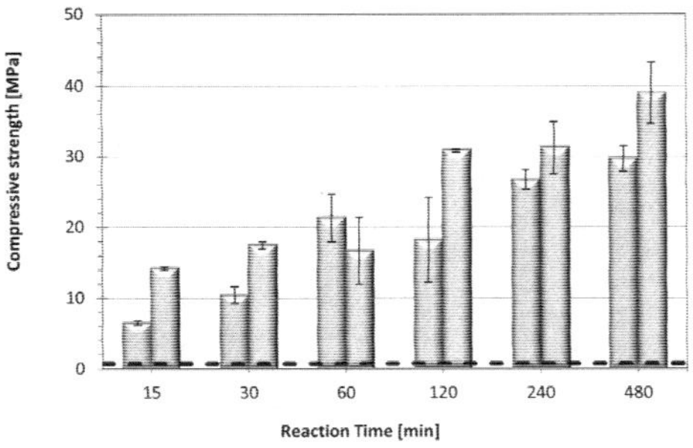

Comp. Strength [MPa] - 1.3 bar

Comp. Strength [MPa] - 10 bar

EAF

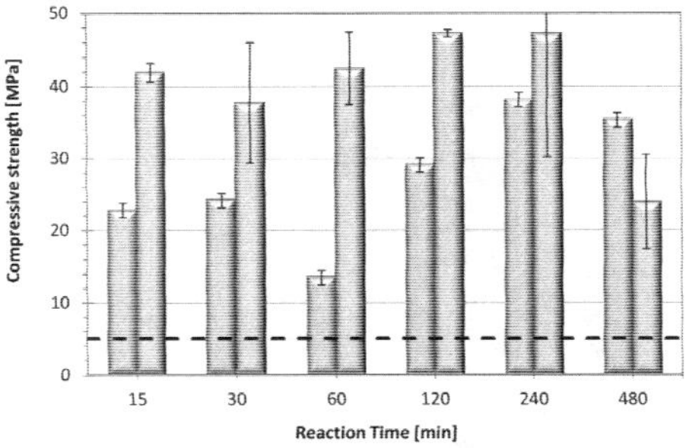

- Higher strength for EAF than BOF
- Starting strength also higher for EAF
- For BOF increasing strength with time
- For EAF strenght decrease at 480 min

BOF and EAF slags: compressive strength/CO_2 uptake correlation

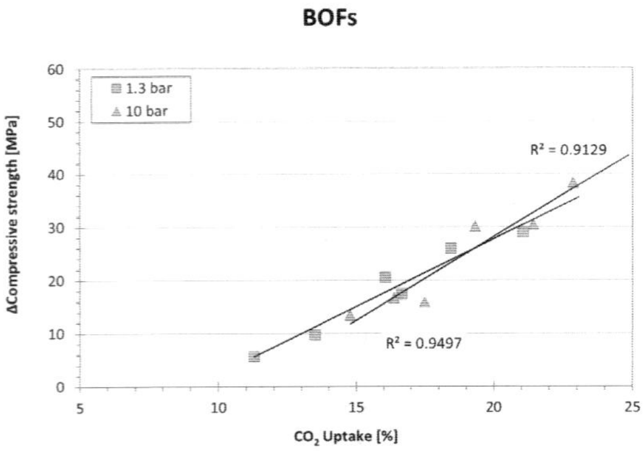

- Good correlation for BOF

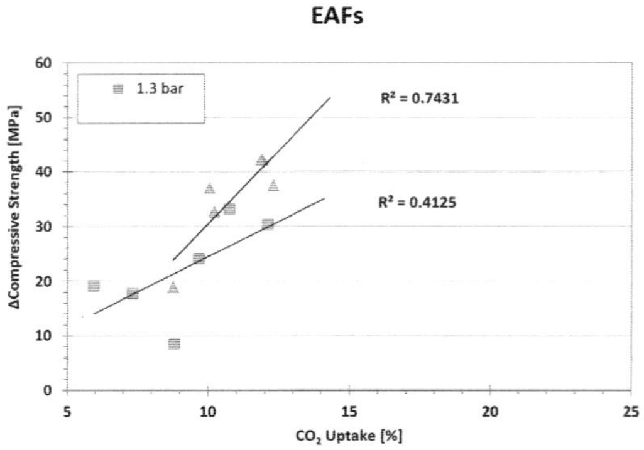

- Poor correlation for EAF

BOF slags: XRD with Rietveld

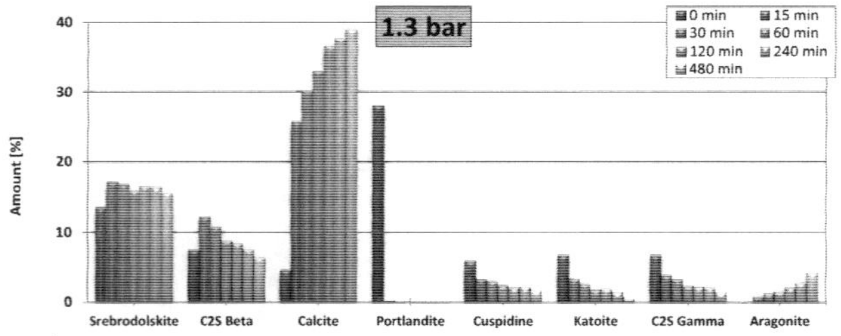

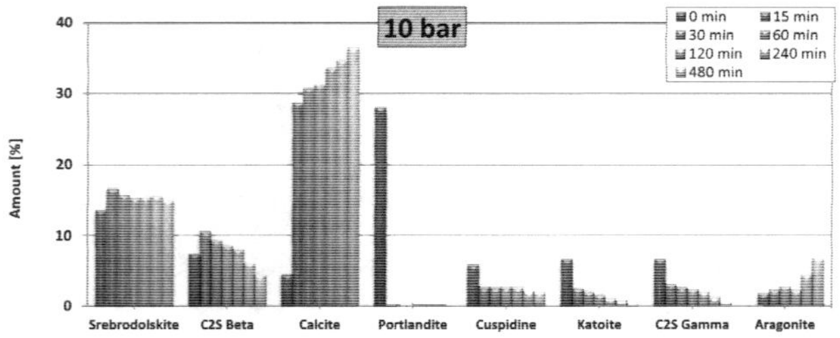

- ✓ Fast portlandite complete conversion
- ✓ Other reacting phases (slower rate)
  - ➢ C2S
  - ➢ Cuspidine
  - ➢ Katoite
- ✓ Calcite + aragonite increase

EAF slags: XRD with Rietveld

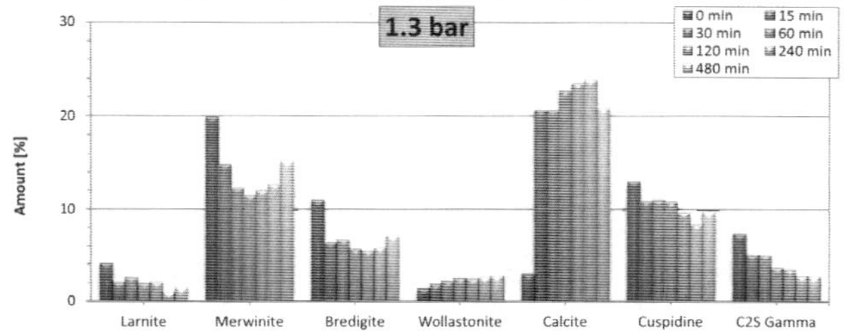

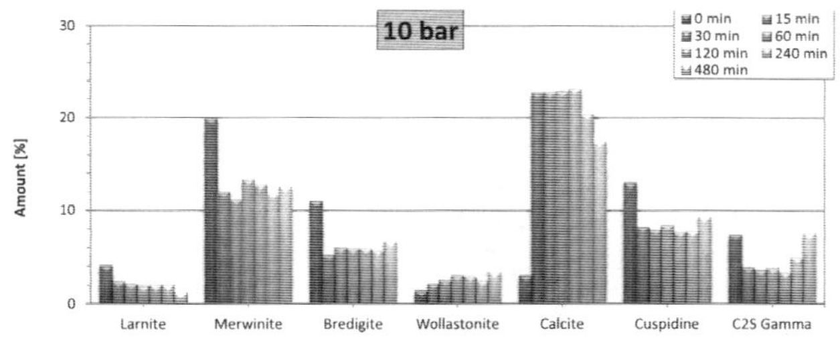

- ✓ Calcite increase mainly in first 15 min
- ✓ Reacting phases:
  - ➢ Merwinite
  - ➢ Bredigite
  - ➢ Cuspidine
  - ➢ C2S
- ✓ Wollastonite not reactive

BOF Slags: SEM

100% CO_2 50°C 1.3 bar 30min

BOF 100% CO_2 50°C 10 bar 480min

- ✓ Small voids (black) indicating a microporous matrix
- ✓ Matrix generally well carbonated (i.e. dense mass between the non-reactive slag grains).

BOF Slags: SEM

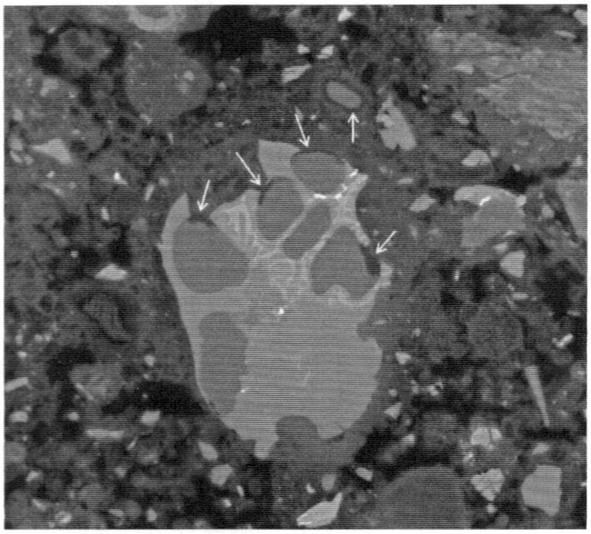

BOF 100% CO_2 50°C 10 bar 480min

Arrows show reactive grains, where Ca has been leached (dark areas) and has subsequently been precipitated as $CaCO_3$ around the slag grains.

BOF Slags: SEM 100% CO_2 50°C 10 bar 480min

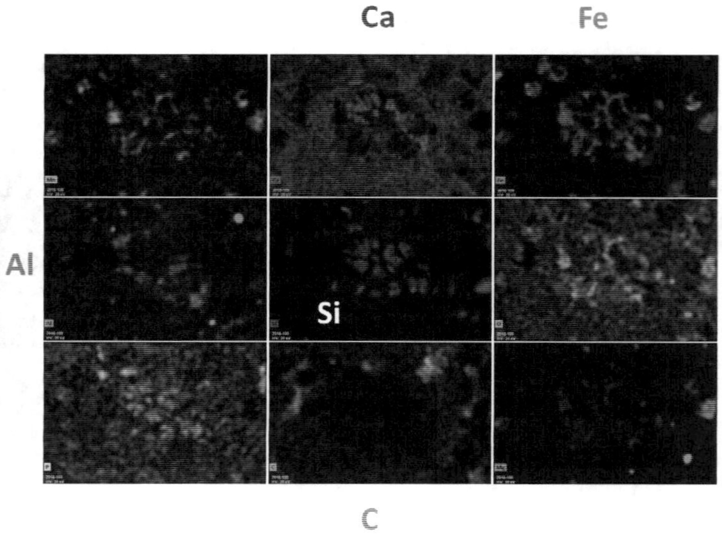

Pure minerals: C_2S and $Ca(OH)_2$

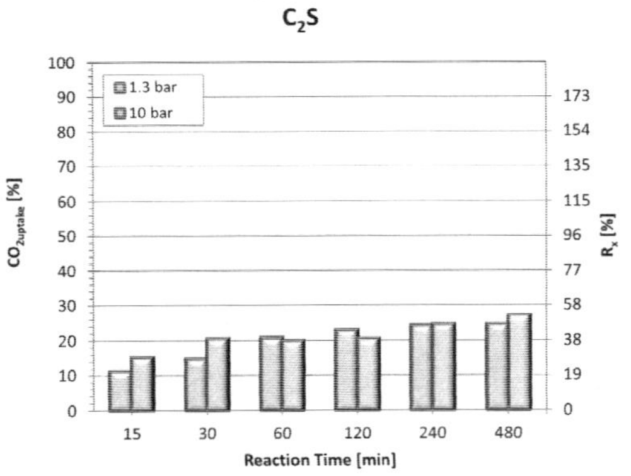

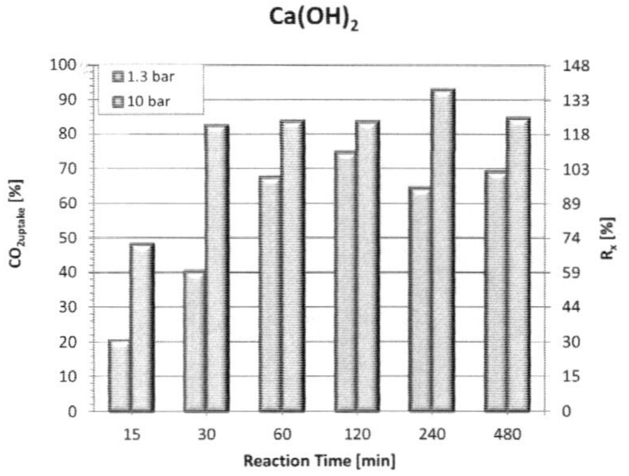

- Gradual slow CO_2 uptake increase
- Fast CO_2 uptake increase up to 30-60 min

Pure minerals: C_2S and $Ca(OH)_2$

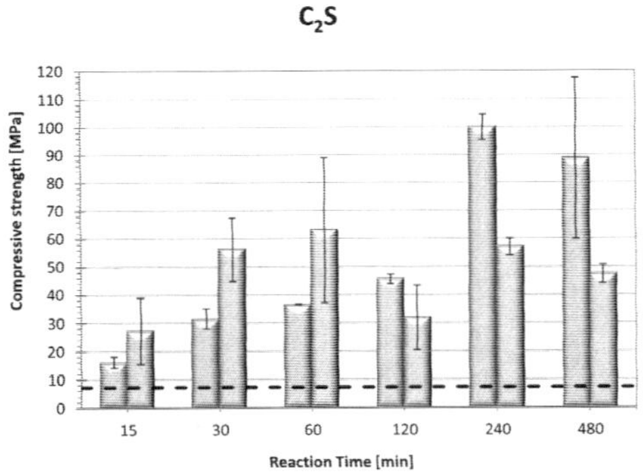

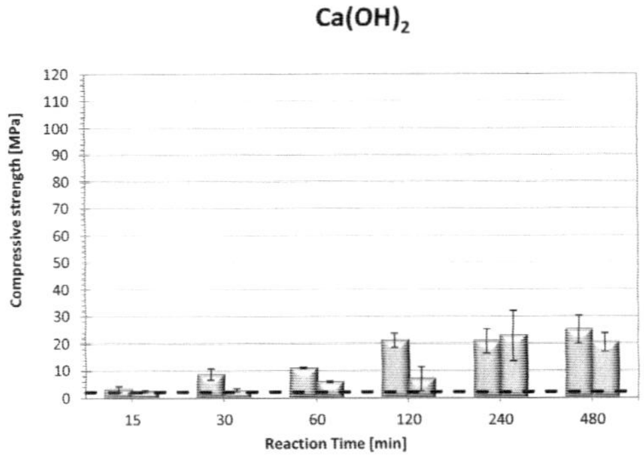

- Compressive strength in line with those of BOF slags at same reaction time

Comp. Strength [MPa] - 1.3 bar
Comp. Strength [MPa] - 10 bar
- - Comp.Strength [MPa] - Uncarbonated

- Compressive strenght increasing gradually;
- Relatively low values even for long reaction time

Conclusions

BOF

- ✓ CO_2 uptake and strength increase with reaction time and well correlated
- ✓ Behaviour in line with pure C_2S carbonation
- ✓ Compressive strenght values in line with those of carbonated pure C_2S
- ✓ Contribution to strength development of $Ca(OH)_2$ conversion looks negligible

EAF

- ✓ CO_2 uptake and strength basically steady after 15 minutes
- ✓ Decrease of calcite concentration coupled with strenght decrease at 480 min
- ✓ No indications from available pure mineral phases carbonation

Accelerated Carbon Mineralization in MINE Tailings

Authors:

Dipple, G. M., The University of British Columbia
Power, I. M., The University of British Columbia
Carrol, K., The University of British Columbia
De Baere, B., The University of British Columbia
Vanderzee, S.

Alkaline waste generated from mining of magnesium silicate rocks reacts spontaneously with atmospheric carbon dioxide (CO_2) to precipitate carbon in solid mineral form. The total capacity of these mine tailings to sequester carbon is about ten times greater than greenhouse gas emissions of associated mining and mineral processing. Waste from mining activity globally has capacity to sequester 100-200 Mt of CO_2 per year. However passive, or unintentional, CO_2 mineralization at individual mine sites is modest (1-50 kt/yr), and typically limited by CO_2 supply. Acceleration of these reactions represents an opportunity to generate considerable greenhouse gas offsets for the industry, and to develop expertise in carbon mineralization that is relevant to accelerated weathering at Earthâ€™s surface and mineral trapping in low temperature aquifers and reservoirs.

Experimental acceleration of carbon mineralization is readily achieved through enhanced delivery of CO_2, wherein reaction rates are limited by rates of cation (e.g., Mg^{2+}) supply from mineral dissolution. Further acceleration requires optimization of mineral dissolution processes. Continuous-flow dissolution experiments on minerals and mine tailings exhibit rapid, transient cation release rates that decay to slower rates indicative of conventional steady-state bulk mineral dissolution processes. The transient initial phase of the experiments can release a significant amount (5-10%) of the total cation content of the material. It reflects the dissolution of highly soluble trace minerals, and surface processes in sheet silicate minerals which together we take to represent the labile cation capacity of the material. Longer-term steady-state cation release is much slower and represents recalcitrant cation capacity indicative of bulk mineral dissolution. The labile cation content represents the carbon mineralization capacity of alkaline mine wastes accessible with existing low-cost technologies while recalcitrant cation content is unlikely to be tapped at existing carbon prices (Fig. 1B). Measured labile cation content of mine tailings varies substantially between and within deposits, with implications for how carbon mineralization capacity should be characterized and how carbonation intervention would be incorporated into mine operations. Specific mines and specific alteration types with high labile cation content, which for some mines is sufficient to offset total mine greenhouse gas emissions, should be the focus of pilot scale carbon mineralization projects.

Effect of Accelerated Carbonation on the Performance of MgO-Based Binder

Authors:

Ahn, J. Y., Pusan National University
Kim, T. Y., Pusan National University
Hwang, I., Pusan National University
Yun, D. Y., Busan Institute of S&T Evaluation and Planning

Magnesium oxide (MgO) and industrial by-products recently have been regarded as the alternative material for replacement of cement in terms of CO_2 reduction, industrial waste recycling and stabilization of pollutants. The aim of this study is to investigate the effects of accelerated carbonation on the strength development and CO_2 storage of MgO-based binder which is binary mixtures of MgO with portland cement (PC) or ground granulated blast furnace slag (GGBS) or fly ash (FA). The carbonation mechanism was also assessed by analyses of porosity and morphological/mineralogical characteristics using X-ray diffraction (XRD), scanning electron microscopy (SEM) and nuclear magnetic resonance spectroscopy (NMR). The compressive strengths of all binders were higher in the 20% CO_2 condition and for longer curing time. The strength was generally higher as in the order of MgO/PC > MgO/GGBS > MgO/FA system. The binder, mixed of 20% MgO and 80% PC, showed the highest compressive strength (38.0 MPa) and it was higher than that of PC. As a result of correlation analysis between porosity and compressive strength, compressive strength was higher when porosity was lower. The binder, composed of 50% MgO and 50% PC, was found to be capable of storing 0.34 kg of CO_2 at 20% CO_2 condition through thermogravimetric analysis. The hydration and carbonation products of MgO including brucite, magnesite and nesquehonite presumably filled the pores and contributed to strength development.

Making Biogas a CO2 Sink Using Carbon Mineralization

Authors:

Sakowski, B. A., TU Dresden

In order to reduce carbon emissions, lots of research has been conducted in the field of carbon capture and storage technologies.

Thereby the focus is mainly on technologies in which carbon dioxide is sequestrated and pressed into depleted oil fields, unmineable coal seams or saline aquifers. What all these technologies have in common is that they are energy-intensive and reduce the efficiency of fossil fired power plants. Also these technologies include long-term monitoring, verification, permanence and other unknown side effects and potential risks.

But there is another possibility of carbon capture, so far not so much in focus: Carbon mineralization processes use alkaline earth metals to capture and store carbon dioxide in the form of ecofriendly carbonates. The advantage of a solid reaction product makes liquefaction as well as the search for secure storage location unnecessary. Carbon dioxide is stored environmentally compatible and is stable long term. These are only some advantages of this alternative process.

The herewith introduced research project has the goal to use alkaline earth metals from combustion ashes like calcium and magnesium to upgrade biogas to natural gas level.

At the same time, calcium carbonate shall be separated as a recyclable material and process water pollution shall be reduced. Therefore a test field concept was developed and realized.

The carbon mineralization process is conducted as described and accelerated in a packed column. In this test field separating and storing carbon dioxide from continuous simulated biogas volume flows up to $5Nm^3/h$ are possible.

The research includes an investigation of balance of test parameters in order to find the optimum of the upgrading process.

Additionally research in the field of heavy metal reduction from the process water was done using hydrogen sulfide from simulated biogas. The idea is to find the best environment for lead sulfide reaction. Thereby heavy metal pollution in ash and process water but also the hydrogen sulfide content in biogas is reduced.

The research project is part of the interdisciplinary Boysen-TU Dresden-research training group. In twelve projects from engineering and social science junior scientists are working and collaborating on sustainable energy systems. Future research collaboration in the field of economics, political and communication studies are planed.

AUTHOR INDEX

AUTHOR INDEX

AIChE
120 Wall Street, FL 23
New York, NY 10005-4020

ISBN 978-1-5108-7281-3

CURRAN ASSOCIATES INC.
proceedings.com

5th Process Safety Management Mentoring (PSM2) Forum 2016

Topical Conference at the 2016 AIChE Spring Meeting and 12th Global Congress on Process Safety

Houston, Texas, USA
10-14 April 2016

AIChE
The Global Home of Chemical Engineers